Crystal Radio
Design and Construction

Crystal Radio Design and Construction

David I. Barker

ISBN: 978-0-6457346-1-4 (Paperback Black and White)

First edition 2023

Published by David I. Barker

Preface

During the late 1960s, commercially produced Japanese-made crystal radios became readily available. They typically resembled pocket-sized transistor radios sold at the time, but could be purchased for a fraction of the cost of their powered counterpart. You could also buy crystal radios in the shape of a rocket, as the Space-race was in full swing. I was fascinated by the idea of listening to music plucked out of the air without the need for a power source such as batteries. A few years ago, this fascination resurfaced, and I decided to build my own crystal radio.

Searching the internet for crystal radio circuits resulted in a plethora of designs. Although simple in design and construction, I wanted to know why some designs used single coils and some used two coils. Others had capacitors attached to the antenna input, and others didn't. Coils were wound on different-sized formers with different numbers of loops. The explanations given for the circuits were both simplistic and generic in nature and did not offer much insight into the reasons for the various designs. At this time, I decided to determine how crystal radios worked at a fundamental level.

Although I have a number of books on electronics and circuit analysis, they needed to go deeper for a complete understanding, and this became more of a problem in physics than in circuit analysis. The internet was both a blessing and a curse; seeking the answer to a single question often resulted in ten different answers. I then decided to create a computer model based on fundamental established principles and equations. As the model developed, I would wind coils and test out the predictions made by the program against actual measurements. The results of some of these measurements are included in this book in the form of graphs and tables. In this way, my confidence that the model reflected reality grew.

This was a time-consuming and sometimes very frustrating process. To give an example of just one of the many problems I faced was determining the mutual inductance between two separate coils for different coil dimensions and separation distances. I calculated the mutual inductance using a popularly given equation. I then measured the mutual inductance between various coils at different distances. I found the calculated values to be so far from the measured values that the equation was quite useless. Trying to find a better equation was an exercise in futility, so in the end, I developed my own method. I performed the calculations and found the results to be in very good agreement with the measurements.

In a step-by-step fashion, this book develops a model that can be used to design crystal radios. The various phenomena and the necessary equations are explained. Starting with a discussion on basic physics concepts, circuit analysis, transmission line theory and antennas, diodes, coil design and more, culminating in a complete model and hopefully a complete understanding of how crystal radios work. You will have an understanding of each of the components used, their use and effect on the performance of the radio. The book then moves on to designing and constructing various crystal radios with step-by-step instructions.

The equations and procedures developed in this book have been incorporated into a computer program that is available for you to use freely. As the internet is dynamic in nature, I will not provide a direct link to the program. Instead, search the internet for a program called "CrystalRadioDesign.exe". I do not expect the reader to manually perform calculations using the equations and procedures given in this book. They are provided as an aid to understanding. A description is given explaining what the equations are trying to accomplish, so even if the reader can not follow the mathematics, then hopefully, the concept will be understood. The concept is the important thing. Some fully worked-out examples will be given in some Chapters as an introduction or a refresher on the topic covered. I will not provide worked-out examples for all topics and procedures, as the book would be 90% worked examples and not particularly interesting.

You will also come across the phrases DC current and AC current. Even though DC stands for direct current and AC for alternating current, they are used to describe both current and voltage as in DC voltage or AC voltage, and DC and AC have really become terms that mean direct or alternating. Therefore in this book, we will use the phrases DC current and AC current even though, by definition, they contain the word current and may appear redundant.

CONTENTS

1

Introduction

1.1 A BRIEF HISTORY OF LIGHT AND VISION

The development of radio results from a long line of incremental improvements in our understanding of the natural world, coupled with technological advancements. The journey spans thousands of years and started in antiquity with the investigation into the nature of light and vision. Ancient Greek philosophers pondered this question, and philosophers such as Pythagoras (570 - 495 BC), Plato (428 - 348 BC) and Ptolemy (100 - 170 AD) believed that light emanated from the eye, and vision was, in a sense caused by the eye reaching out to touch an object. Democritus (460 – 370 BC) believed light consisted of indivisible particles he called atoms, and vision was the result of the interaction between light coming from the object and light coming from the eye, meeting and forming an impression in the air (like a stamp). Epicurus (341 - 270 BC) believed that vision was the result of atoms of light emanating from an object and entering the eye. A Persian philosopher, Alhazen (965 - 1040 AD), believed any hypothesis needed to be supported by experimentation. Combining logic and experimentation, he showed vision was due to light entering the eye. His argument held sway, and this became the prevailing view.

Although geometric and algebraic models of light were developed to describe various phenomena, such as reflection (light bouncing off a surface), refraction (bending of light as it enters or leaves a different medium), and diffraction (light bending around the corners of objects or through a hole), the nature of light itself was still not well understood. Isaac Newton (1642–1727) showed that light from the sun consisted of various colours. Using a prism, he split the incoming white light into a rainbow of colours and then, using another prism, recombined the coloured light back into white light. Like the atomists, Isaac Newton believed light to be corpuscular in nature, with light corpuscles being smaller than matter corpuscles. Christian Huygens (1629 - 1695) was a proponent of the wave nature of light, as the observed properties of light could readily be explained by treating light as waves. Diffraction was difficult to explain by treating light as little balls. Little balls going through a hole would move in straight lines forming a cone, and have no reason to bend. Newton believed diffraction was just a special case of refraction and was due to some unobservable atmosphere near the surface of a body. Due to

Newton's status, the corpuscular nature of light held sway and was widely accepted.

Eventually, the corpuscular theory of light was overturned in favour of the wave theory, mainly due to the work of Thomas Young (1773 – 1829). Using a shallow tank of water with a paddle at one end (known as a ripple tank), Young generated two separate waves that travelled outward in a concentric circular pattern from their source of origin. When these two separate wave fronts met, they interfered with one another to produce a distinctive interference pattern; peaks meeting with peaks, and troughs meeting with troughs, created larger peaks and larger troughs (constructive interference). Peaks meeting with troughs generated much smaller peaks (or no peaks, destructive interference).

Young then devised an experiment to examine whether light would produce the same interference pattern. Shining a single light source onto a screen, Young placed an opaque barrier, with two closely spaced slits, between the light source and the screen. A series of light and dark bands were observed, similar to the interference patterns produced by water waves. This experiment is called the "double-slit experiment" and is one of the most famous and widely reproduced experiments of all time.

As the light from a single source passes through each of the slits, the light waves spread out concentrically from each of the slits due to diffraction. There are now two separate wave sources of the same light (as was done in the ripple tank). As these light waves spread out, they constructively and destructively interfere with one another to produce the observed interference pattern.

1.2 ELECTROMAGNETIC WAVES

In 1820 Hans Christian Ørsted (1777 - 1851) discovered that if a compass were placed near a current-carrying wire, the compass needle would deviate from magnetic north. Mapping these deviations, he showed that the magnetic field flowed in a circular pattern around a straight current-carrying wire. Ørsted's experiment demonstrated a relationship between electricity and magnetism and sparked a new line of study.

While investigating the relationship between electricity and magnetism, in 1831, Michael Faraday (1791 – 1867) wound two insulated wire coils around an iron ring. He found that at the instant when one coil was connected to a battery, a momentary current flowed through the other coil. And when the battery was disconnected, a momentary current also flowed through the other coil. This current, Faraday postulated, was due to the change in the magnetic field. Expanding on his experiments, he discovered that if a magnet is passed through a stationary coil, or a coil is moved over a stationary magnet, an electric current is also induced in the coil. This led to the concept of magnetic induction that a current is induced in a wire or coil due to a changing magnetic field. Faraday also introduced the concept of electric and magnetic lines of flux to understand electric and magnetic fields and their relationship.

James Clerk Maxwell (1831 – 1879) used Faraday's results (and others) to produce a set of equations describing the electric and magnetic forces and their relationships with each other. In particular, a time-varying electric field will have an associated time-varying magnetic field, and a time-varying magnetic field will have an associated time-varying electric field. Maxwell's equations predicted the existence of coupled oscillating magnetic and electric fields in the absence of any charge or current, which travelled outward from the source. As these electric and magnetic waves travelled together, they were called ***electromagnetic waves***. Maxwell calculated the speed of propagation of these coupled waves, and it came out to the speed at which light travels through a vacuum. He concluded the substance that the electromagnetic waves propagate through must be the same as the substance that light waves propagate through, so light and electromagnetic waves were disturbances in the same medium. The implications were obvious, that light itself was nothing more than an electromagnetic wave. His equations not only suggested that light was an electromagnetic wave, but that these waves could be produced by time-varying currents in a wire.

Maxwell produced a set of 20 equations that Oliver Heaviside (1850 - 1925) reduced to a set of four partial differential equations (there is also a set of four integral versions, the same equations expressed in a different form) that we use today.

These waves, according to Maxell and others, required a medium through which to travel; in the same way, sound waves travel through air and ocean waves through water. This medium was known as the ***luminiferous aether***. Using the Earth's motion around the sun, experimenters reasoned that a difference in the speed of light should be measurable for light travelling in the same direction as the Earth as opposed to light travelling at right angles to the Earth's motion. However, experiments could not find the existence of the luminiferous aether, as no difference in speed was detected. When Einstein showed in his Special Theory of Relativity that the measured speed of light would always come out to the same value regardless of your frame of reference, the aether was dispensed with, and electromagnetic waves were considered self-propagating waves through space. Strangely though, in Quantum Mechanics, the Standard Model of particle physics requires each fundamental particle to have its own field, and a particle is nothing more than a disturbance in this field. So photons (electromagnetic waves) are just a disturbance in the Quantum electromagnetic field. They should have called this field the luminiferous aether; it sounds better.

1.3 THE BIRTH OF RADIO

In 1886, in an attempt to detect Maxwell's electromagnetic waves Heinrich Rudolf Hertz (1857 - 1894) created the first crude radio transmitter and receiver. His transmitter consisted of a step-up transformer with the secondary coil attached to two lengths of wire bent at 180 degrees (a dipole antenna). At the outer ends of the antenna, he attached two large spheres for capacitance and placed two smaller metal spheres on each wire at the bend that formed a gap between the antenna wires (a spark gap). The receiver consisted of a single loop of wire with a spark gap separating the ends.

When Hertz applied power to the step-up transformer, the momentary surge in the current created a large voltage in the secondary coil producing a spark that jumped across the transmitter antenna's spark gap. Whenever a spark was produced by the transmitter, a similar spark jumped across the receiving antenna.

Between 1886 and 1889, Hertz conducted a series of experiments to exclude simple magnetic induction as the cause and thus conclusively proved the existence of Maxwell's electromagnetic waves. Hertz also measured the velocity of these waves and found the speed of propagation to be equal to the speed of light.

Hertz did not see any practical use for his apparatus and was simply a device to prove the existence of Maxwell's electromagnetic waves. Unfortunately, Hertz died at the young age of 36 after complications from surgery and did not live to see the revolution that would eventually occur.

Hertz's creation and detection of these waves, now referred to as "Hertzian waves" (later called radio waves around 1910), resulted in feverish experimentation and development. Building upon Hertz's apparatus, Guglielmo Marconi (1874 - 1937) developed the first practical radio transmitter and receiver system. Marconi used his system to transmit and receive Morse code signals, known as Wireless telegraphy.

Marconi made several improvements, particularly in the antenna used for transmitting and receiving. Hertz used a short horizontal dipole (two horizontal wires), whereas Marconi used a long vertical monopole and the Earth to form the antenna. Using long vertical antennas, Marconi reduced the frequency of the electromagnetic waves and found he could transmit and receive signals at much greater distances.

Unfortunately, Marconi found little support for his ideas in his native Italy. Eventually, Marconi moved to Britain, where support was more forthcoming. In July 1896, Marconi demonstrated his wireless telegraphy system to the British government. On May 13 1897, Marconi transmitted a message over the Bristol Channel from Flat Holm Island to Lavernock Point near Cardiff, a distance of 6 kilometres (3.7 mi). Improvements over the years saw these distances expand dramatically, and Marconi's system eventually became widely used on ocean-going vessels and land stations alike. And in a bizarre twist of fate Marconi was offered free passage on the ill-fated Titanic. He had work to do, and the Lusitania had a public stenographer on board. He chose to sail with the Lusitania. The Titanic sank on her maiden voyage. The resulting death toll would have been much higher if not for Marconi's wireless system.

1.4 AM BROADCAST RADIO

Wireless telegraphy systems, using spark gap transmitters, produce a burst during transmission and then stop. A short burst is treated as a dot, and a long burst is treated as a dash. A receiver just needed to detect these bursts. A device called a coherer was initially used to detect these signals. The coherer was made using a glass tube filled with metal filings with electrodes attached at both ends. In the presence of a radio wave, the metal filings would cling (cohere) together, decreasing the resistance between the electrodes and allowing current to pass easily. When the filings were not sticking together, the resistance between the electrodes was very high. The coherer was attached across the antenna input and across a DC circuit. When a radio signal was detected, the filings would cohere, allowing the DC current to pass. The DC current would then drive a device such as a bell, produce a click in headphones or drive a device to place marks on paper tape. However, the filings would continue to cohere after the signal stopped, so a decoherer device was created to automatically tap the tube to un-stick the filings before the next signal was received. These devices were broadband in nature and could be triggered by a wide range of frequencies, so they were susceptible to interference. Also, due to needing a decoherer, they were quite slow. Finally, they could not be used to demodulate amplitude-modulated (AM) signals.

In 1874 Karl Ferdinand Braun discovered that a thin wire in contact with certain crystals, such as galena, could rectify a signal, allowing current to pass in one direction and not another. This device was known as a cat's whisker and was first used by Jagadish Chandra Bose in 1894 as an experimental device to detect microwaves. By 1902 it was found that many materials could be used with a cat's whisker to detect signals. These crystal detectors relied on the cat's whisker to be in contact with a sensitive spot on the crystal. If the assembly was bumped, the cat's whisker could reposition itself so that the signal would no longer be detected. This form of detection was not ideal for moving ships but was often used for land-based receivers.

For voice and audio communications, it was realised that a continuous wave transmitter would be required rather than the burst produced by Marconi's system. A continuous wave was needed to modulate an audio signal (amplitude modulation or AM). The first voice transmission was claimed by Reginald Fessenden (1866 - 1932) using a high-frequency spark gap transmitter. As the time interval between sparks decreased, the greater it approximated a continuous wave (CW), and in 1900 he reported transmitting distorted speech over a distance of 1.6 kilometres (1 mile).

On December 21 1906, using a high-speed alternator-transmitter that produced a continuous sine wave, Reginald Fessenden broadcasted speech from Brant Rock to a listening site at Plymouth, some 18 kilometres (11 miles) distance, the first truly AM signal. On Christmas Eve 1906, Fessenden claimed he successfully broadcast an audio program from Brant Rock, which was heard by ships out at sea. The program included Fessenden playing "O Holy Night" on the violin and reading a passage from the Bible.

In 1904 John Ambrose Fleming (1849 - 1945) invented the diode vacuum tube. The diode allowed current to flow in one direction only and was largely used as a replacement in receivers using coherers. Crystal detectors continued to be used, however, due to their simplicity and low cost. Both the diode and the crystal could be used to demodulate AM signals.

Following the invention of the vacuum tube diode, the triode vacuum tube was invented by Lee de Forest (1873 - 1961) in 1907. De Forest invented the triode to amplify detected signals in receivers, making them more sensitive. Initially, the triode would provide only a modest gain, but over the years, improvements were made to increase the gain. De Forest experimented with the triode and found that if a portion of the output was fed back into the input of the triode, a circuit could be made to oscillate. No longer were transmission frequencies dictated by the mechanical actions of the power supply or the physical geometry of the system. Transmitters could easily be designed to oscillate at any frequency desired. Triodes could also be used in mixing (modulating) audio signals with the radio carrier wave signal to produce AM audio transmission. The triode eventually became the game changer that gave birth to broadcast radio.

From 1907 to 1920, many experimental and amateur radio stations popped into and out of existence worldwide. Some experimental stations morphed into licensed commercial radio stations. Commercial in this context means privately owned rather than public (or state-owned). However, the station commonly referred to as the first "commercial radio station" started broadcasting on November 2 1920, with the call sign KDKA in Pittsburgh, Pennsylvania. This station was created by the Westinghouse Company, a leading supplier of radio receivers, to boost sales. The early 1920s saw an explosion of AM broadcast radio stations opening in many countries throughout the world. Listeners could buy commercially available radios or simply build their own crystal radio, as radio receivers using a Cat's whisker detector were called.

Crystal radios remained popular for many decades, and small pocket-sized commercially produced crystal radios could be purchased well into the early 1970s. As Frequency Modulated (FM) broadcasting became more popular and with the advent of very cheap transistor radios, these pocket-sized crystal radios eventually disappeared. From the 1980s onward, the number of AM radio stations started to decrease in many countries. And now, in the 21st century, there is competition from the internet and digital radio. However, AM refuses to die. In countries like Australia and the U.S.A., AM is still alive and well and clinging to life.

2

Prerequisites in Mathematics

2.1 SYSTEM OF UNITS

This book will use the SI (international standard) system of units. The SI system consists of a small set of fundamental base units from which all other units are derived. Unit names are treated as common nouns, and all start in lowercase, even if named after a person. For example, the unit name for temperature was named after Lord Kelvin, but when writing the unit name, it's written as kelvin. Unit ***symbols*** are also written in lowercase, except when the unit name is named after a person. So the unit symbol for kelvin is in upper case K. Quantities of units are written with the value of the quantity followed by a space and then the unit symbol. For example, twenty-one kilograms would be written as 21.0 kg. Table 2.1 describes these base units. Some derived units of interest are given below in Table 2.2.

Quantity	Unit Name	Unit Symbol
Length	metre	m
Mass	kilogram	kg
Temperature	kelvin	K
Time	seconds	s
Electric current	ampere	A
Amount of substance	mole	mol
Luminous intensity	candela	cd

Table 2.1
Fundamental units.

Prefixes are used with the base or derived symbols to denote some power of 10 multiple. For example, one million hertz would be written as 1 MHz (megahertz). The kilogram is an exception to the rule. Although the kilogram is a base unit, prefixes are added as though the gram was the base unit. For instance, a kilogram is written as kg, and a milligram is written as mg. These prefixes are given in Table 2.3.

Quite often, symbols are used from the Greek alphabet. The Greek alphabet, along with its lower and uppercase symbols, is given in Table 2.4.

Quantity	Unit Name	Unit Symbol	Base Units	Other Units
Frequency	hertz	Hz	$1/s$	
Force	newton	N	$kg \cdot m/s^2$	
Pressure	pascal	Pa	$kg/m \cdot s^2$	N/m^2
Energy/work/heat	joule	J	$kg \cdot m^2/s^2$	$N \cdot m$
Power	watt	W	$kg \cdot m^2/s^3$	J/s
Electric charge	coulomb	C	$A \cdot s$	
Electric potential difference	volt	V	$kg \cdot m^2/s^3 A$	W/A
Capacitance	farad	F	$s^4 A^2/kg \cdot m^2$	C/V
Resistance	ohm	Ω	$kg \cdot m^2/s^3 A^2$	V/A
Magnetic Flux	weber	Wb	$kg \cdot m^2/s^2 A$	$V \cdot s$
Magnetic Flux Density	tesla	T	$kg/s^2 A$	Wb/m^2
Inductance	henry	H	$kg \cdot m^2/s^2 A^2$	Wb/A
Temperature relative to 273.15K	celsius	°C	K	

Table 2.2
Derived units.

Base 10 value	Word	Symbol
10^{-12}	pico	p
10^{-9}	nano	n
10^{-6}	micro	μ
10^{-3}	milli	m
10^{3}	kilo	k
10^{6}	mega	M
10^{9}	giga	G
10^{12}	tera	T
10^{15}	peta	P

Table 2.3
Unit prefixes.

Letter	Lowercase	Uppercase
Alpha	α	A
Beta	β	B
Gamma	γ	Γ
Delta	δ	Δ
Epsilon	ε	E
Zeta	ζ	Z
Eta	η	H
Theta	θ	Θ
Iota	ι	I
Kappa	κ	K
Lambda	λ	Λ
Mu	μ	M
Nu	ν	N
Xi	ξ	Ξ
Omicron	ο	O
Pi	π	Π
Rho	ρ	P
Sigma	σ	Σ
Tau	τ	T
Upsilon	υ	Y
Phi	φ	Φ
Chi	χ	X
Psi	ψ	Ψ
Omega	ω	Ω

Table 2.4
The Greek alphabet.

2.2 TRIGONOMETRIC FUNCTIONS

Basic trigonometric functions are based on the right-angle triangle. They define relationships between angles and the length of the sides of the triangle, as shown in Figure 2.1. You will find a button on your calculator to give the values of these trigonometric functions for any given angle.

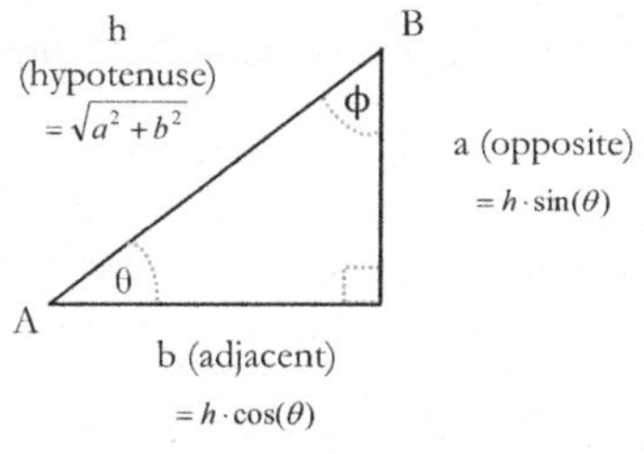

Figure 2.1
Defining trigonometric functions using the right triangle.

Sine

The sine (shortened to sin) of angle θ is defined as the ratio of the length of the opposite side to the length of the hypotenuse.

$$\sin(\theta) = \frac{\text{opposite}}{\text{hypotenuse}} = \frac{a}{h} \quad 2.1$$

And the inverse is the cosecant and is given by

$$\csc(\theta) = \frac{\text{hypotenuse}}{\text{opposite}} = \frac{h}{a} = \frac{1}{\sin(\theta)} \quad 2.2$$

Rearranging Equation 2.1, we can calculate the length of side a.

$$a = h\sin(\theta) \quad 2.3$$

If we know the value of the sine function x, we can determine the angle θ between the sides by taking the ***inverse sine function*** (written as arcsine or $\sin^{-1}$). There will be a button on your calculator for the inverse functions as well.

$$\theta = \sin^{-1}(x) = \sin^{-1}\left(\frac{\text{opposite}}{\text{hypotenuse}}\right) \quad 2.4$$

Cosine

The cosine (cos) of angle θ is defined as the ratio of the length of the adjacent side to the length of the hypotenuse.

$$\cos(\theta) = \frac{\text{adjacent}}{\text{hypotenuse}} = \frac{b}{h} \quad 2.5$$

And the inverse is the secant, given by

$$\sec(\theta) = \frac{\text{hypotenuse}}{\text{adjacent}} = \frac{h}{b} = \frac{1}{\cos(\theta)} \quad 2.6$$

Rearranging Equation 2.5, we can calculate the length of side b.

$$b = h\cos(\theta) \quad 2.7$$

If we know the value of the cosine function, we can determine the angle θ between the sides by taking the inverse cosine function (written as arccos or $\cos^{-1}$).

$$\theta = \cos^{-1}(x) = \cos^{-1}\left(\frac{\text{adjacent}}{\text{hypotenuse}}\right) \quad 2.8$$

Tangent

The tangent (tan) of angle θ is defined as the ratio of the length of the opposite side over the adjacent side.

$$\tan(\theta) = \frac{\text{opposite}}{\text{adjacent}} = \frac{a}{b} = \frac{\sin(\theta)}{\cos(\theta)} \quad 2.9$$

And the inverse cotangent is

$$\cot(\theta) = \frac{\text{adjacent}}{\text{opposite}} = \frac{b}{a} = \frac{1}{\tan(\theta)} \quad 2.10$$

The length of the hypotenuse is given by the Pythagoras equation

$$h = \sqrt{a^2 + b^2} \quad 2.11$$

The inverse tan function (written as arctan or $\tan^{-1}$) is

$$\theta = \tan^{-1}(x) = \tan^{-1}\left(\frac{\text{opposite}}{\text{adjacent}}\right) \quad 2.12$$

2.3 HYPERBOLIC FUNCTIONS

Hyperbolic functions are similar to trigonometric functions but are based on the hyperbola rather than the right triangle. They are related to the natural exponential function $\mathbf{e^x}$. The natural exponential function is used to describe the natural growth or decay of many systems, and its value is approximately 2.71828. The hyperbolic functions are defined as follows.

$$\sinh(x) = \frac{e^x - e^{-x}}{2} \quad 2.13$$

$$\cosh(x) = \frac{e^x + e^{-x}}{2} \quad 2.14$$

$$\tanh(x) = \frac{e^x - e^{-x}}{e^x + e^{-x}} = \frac{\sinh(x)}{\cosh(x)} \quad 2.15$$

2.4 VECTORS

In our daily lives, we constantly use numerical values to describe the amount of something we need to know. For instance, the amount of petrol needed to fill up a car's petrol tank and the amount of money needed to pay for the petrol. They all have one thing in common, the use of a single bit of information, the magnitude of a numerical value to completely tell us what we need to know. Something that has a magnitude only is called a ***scalar***.

What if you were in an unfamiliar neighbourhood and needed petrol? So you stopped and asked someone for directions to the nearest petrol station. And they responded about 1.5 kilometres away. That tells you how far away it is but not how to get there. You need more information. Not only do you need the distance, but also the direction(s) to travel. Something that has both a magnitude (e.g. 1.5 kilometres) and direction (e.g. eastward) is called a ***vector***.

Graphically, vectors are depicted as arrows, with the length of the arrow indicating the magnitude, as shown in Figure 2.2.

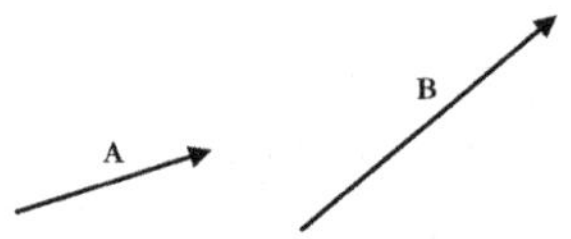

Figure 2.2
Vector representation.

Analytically, vectors are typically represented by a letter with an arrow on top such as ($\vec{A}$). In books and printed works, a bold font is commonly used to indicate a vector, such as **A**. In this book, bold letters will be used to represent vectors. The magnitude of the vector will either be represented by the non-bold letter (e.g. A) or the vector within straight brackets $|\mathbf{A}|$. The representation chosen will be based on clarity.

A vector can be split into vector components, as shown in Figure 2.3. The two-dimensional vector **A** can be split into a vector in the x-direction ($\mathbf{A_X}$) and a component vector in the y-direction ($\mathbf{A_Y}$). These components are at right angles to each other and are called ***component vectors***.

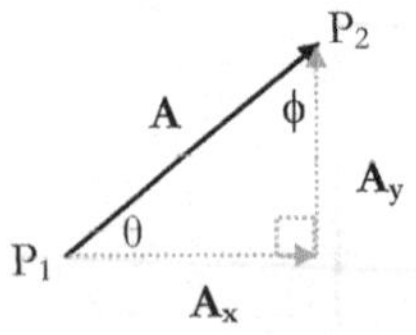

Figure 2.3
Vector components.

To determine the component vectors, we can use Equations 2.3 and 2.7.

$$\boxed{\mathbf{A_X} = \mathbf{A}\cos(\theta)} \qquad 2.16$$

And

$$\boxed{\mathbf{A_Y} = \mathbf{A}\sin(\theta)} \qquad 2.17$$

The magnitudes of the component vectors A_X and A_Y are just called the components, and the magnitude of the vector **A** can be found using Pythagoras' theorem

$$A = |\mathbf{A}| = \sqrt{A_x^2 + A_y^2}$$

Vectors can be represented in a notational format as a combination of their vector components as follows:

$$\mathbf{A} = < A_x, A_y > \qquad \text{Vector notational form}$$

Note the use of the angled bracket and not the round bracket. Values enclosed in round brackets represent points. If vector **A** were a three-dimensional vector, then its representation would be

$$\mathbf{A} = < A_x, A_y, A_z >$$

and the magnitude would be given by Pythagoras' theorem in three dimensions.

$$\boxed{A = |\mathbf{A}| = \sqrt{A_x^2 + A_y^2 + A_z^2}} \qquad \text{Calculating the magnitude of a vector} \qquad 2.18$$

Consider Figure 2.4a. We have drawn a vector on a two-dimensional Cartesian coordinate system. Vector **A** has an initial point at $P_1(1,1)$ and a terminal point at $P_2(4,2)$. The components of

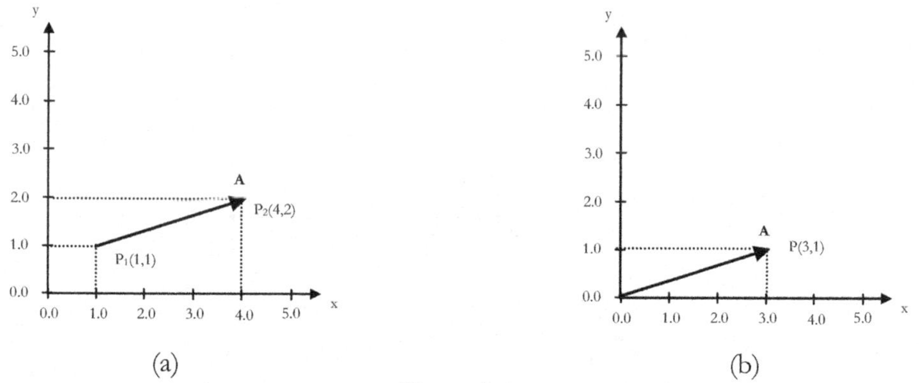

Figure 2.4
Position vectors.

vector **A** are the displacements along the x and y axis. We can calculate the x and y components by subtracting the x and y values of the starting and ending points. For this example,

$$\begin{aligned} A_x &= P_{2_x} - P_{1_x} \\ &= 4 - 1 \\ &= 3 \end{aligned}$$

And

$$\begin{aligned} A_y &= P_{2_y} - P_{1_y} \\ &= 2 - 1 \\ &= 1 \end{aligned}$$

So we can express vector **A** as

$$\mathbf{A} = <3,1>$$

There is, however, an infinite number of positions where **A** could be placed to give the same component values. To define the vector as given, we need to specify the initial point. If we now say vector **A** = <3,1> with respect to point P_1 we have defined a unique vector. A vector defined with respect to an initial point is called a ***position vector***. Unless otherwise stated, a position vector is defined with respect to the origin, point (0,0), as in Figure 2.4b. A vector that has not been specified with respect to an initial point is called a ***displacement vector***.

Vector Addition

Vectors can be algebraically manipulated using addition, subtraction and multiplication. Given the two vectors in Figure 2.5, we can add vectors **A** and **B** to find the resultant vector **C**. In vector addition, the resultant vector is drawn by placing the initial point of **B** on the terminal point of **A** and drawing a line from the initial point of **A** to the terminal point of **B** to form vector **C**.

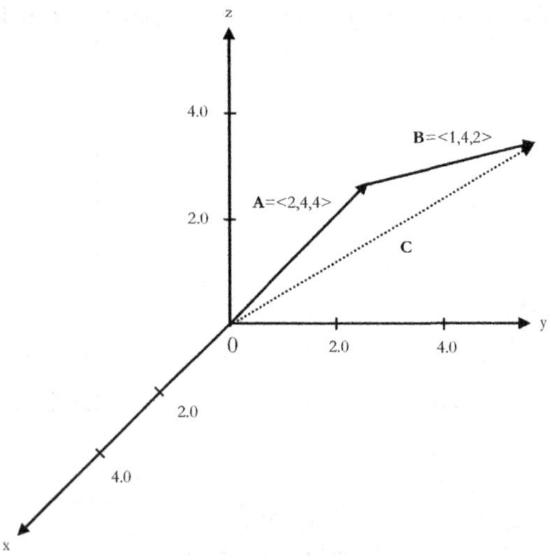

Figure 2.5
Vector addition of three-dimensional vectors.

Algebraically we add the x components together and the y components together, and if three-dimensional, the z components to determine the resultant vector.

$$\boxed{\mathbf{C} = \mathbf{A} + \mathbf{B} = < A_x + B_x, A_y + B_y, A_z + B_z >} \qquad \text{Vector addition} \qquad 2.19$$

Adding vectors **A** and **B** given above would be

$$\mathbf{C} = \mathbf{A} + \mathbf{B}$$

$$= < 2+1, 4+4, 4+2 > \qquad \text{Adding vectors } \mathbf{A} \text{ and } \mathbf{B} \text{ to get vector } \mathbf{C}$$

$$= < 3,8,6 >$$

Vector Subtraction

Subtraction is carried out in a similar fashion to addition, except now we subtract the component values.

$$\boxed{\mathbf{C} = \mathbf{A} - \mathbf{B} = < A_x - B_x, A_y - B_y, A_z - B_z >} \qquad \text{Vector subtraction} \qquad 2.20$$

Subtracting vector **B** from vector **C,** we should get vector **A**

$$\mathbf{A} = \mathbf{C} - \mathbf{B}$$

$$= < 3-1, 8-4, 6-2 > \qquad \text{Subtracting vector } \mathbf{B} \text{ from } \mathbf{C} \text{ to get vector } \mathbf{A}$$

$$= < 2,4,4 >$$

Vector Scalar Multiplication

The simplest form of vector multiplication is multiplying a vector by a scalar.

$$\boxed{a\mathbf{A} = a < A_x, A_y, A_z > = < aA_x, aA_y, aA_z >} \qquad \text{Vector scalar multiplication} \qquad 2.21$$

Say we multiply vector **A** by 3.

$$3\mathbf{A} = 3 < 2,4,4 >$$

$$= < 6,12,12 > \qquad \text{Multiplying vector } \mathbf{A} \text{ by a scalar}$$

Multiplication of vectors with vectors will be covered in the dot and cross product sections to follow.

Algebraic Vector Notation

In Figure 2.6a, we have drawn a three-dimensional rectangular coordinate system with three vectors along each of the three axes. These vectors each have a magnitude (length) of 1.0. These are special vectors known as the rectangular unit vectors, standard unit vectors or standard basis vectors. The symbol (^) above the vector is called a hat, so the unit vectors are referred to as "i hat", "j hat", and "k hat". These vectors represent the direction of each component of a vector in the standard rectangular coordinate system. In vector notation, these unit vectors are defined as

$$\hat{\mathbf{i}} = < 1,0,0 > \hat{\mathbf{j}} = < 0,1,0 > \hat{\mathbf{k}} = < 0,0,1 > \qquad \text{Standard unit vectors} \qquad 2.22$$

Since a vector can be found by the addition of its component vectors and using vector scalar multiplication, we can use these unit vectors to define a vector in algebraic format as a vector sum of its component vectors.

We can express vector **A** in algebraic vector notation as

$$\mathbf{A} = A_x\hat{\mathbf{i}} + A_y\hat{\mathbf{j}} + A_z\hat{\mathbf{k}}$$ Algebraic vector notation 2.23

This notation resembles normal algebraic equations and makes vector algebraic manipulation easier. From Figure 2.6b, we can read off the component values from the graph and express vector **A** as the sum of its component vectors.

$$\mathbf{A} = 2\hat{\mathbf{i}} + 4\hat{\mathbf{j}} + 4\hat{\mathbf{k}}$$

Applying this vector notation to the addition of vectors **A** and **B** in Figure 2.5, we would write

$$\begin{aligned}\mathbf{C} &= \mathbf{A} + \mathbf{B}\\ &= (2\hat{\mathbf{i}} + 4\hat{\mathbf{j}} + 4\hat{\mathbf{k}}) + (1\hat{\mathbf{i}} + 4\hat{\mathbf{j}} + 2\hat{\mathbf{k}})\\ &= (2+1)\hat{\mathbf{i}} + (4+4)\hat{\mathbf{j}} + (4+2)\hat{\mathbf{k}}\\ &= 3\hat{\mathbf{i}} + 8\hat{\mathbf{j}} + 6\hat{\mathbf{k}}\end{aligned}$$

This form of vector notation is very important and will be used when analysing AC circuits later in this book.

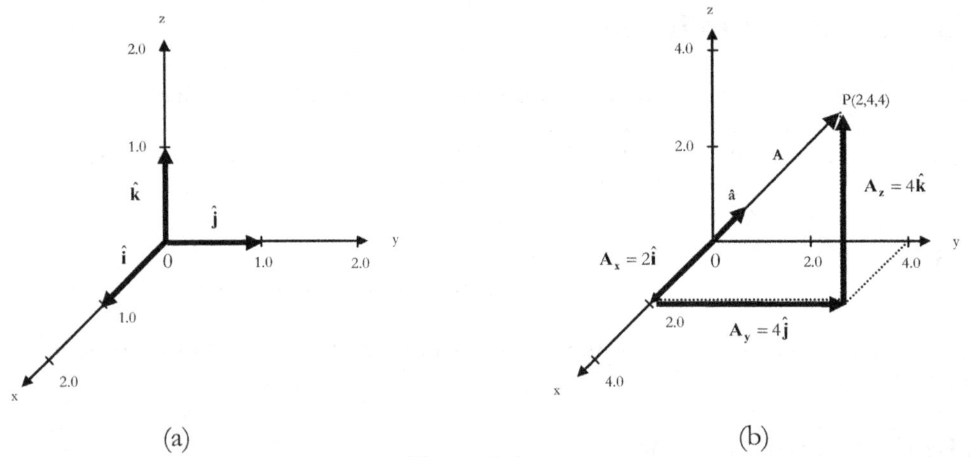

Figure 2.6
Standard unit vectors.

The Unit Vector

Any vector can have its own unit vector that points along its direction, as shown in Figure 2.6b. This unit vector points in the same direction as vector **A**. Just the same way the standard unit vector $\hat{\mathbf{i}}$ points along the x-axis and standard unit vector $\hat{\mathbf{j}}$ points along the y-axis. We can calculate a unit vector by dividing a vector by its magnitude.

$$\hat{\mathbf{a}} = \frac{\mathbf{A}}{|\mathbf{A}|}$$ The unit vector 2.24

So for the vector **A** in Figure 2.6b, the unit vector is

$$\hat{\mathbf{a}} = \frac{2\hat{i} + 4\hat{j} + 4\hat{k}}{\sqrt{2^2 + 4^2 + 4^2}}$$
$$= \frac{2\hat{i} + 4\hat{j} + 4\hat{k}}{6}$$
$$= 0.333\hat{i} + 0.667\hat{j} + 0.667\hat{k}$$

The components of the unit vector are the components in the x, y and z directions as for normal vectors. For this to be a unit vector, the magnitude should be 1. Checking the magnitude of a-hat using Equation 2.18.

$$a = |\hat{\mathbf{a}}| = \sqrt{a_x^2 + a_y^2 + a_z^2}$$
$$= \sqrt{0.333^2 + 0.667^2 + 0.667^2}$$
$$= 1.0$$

We can now express the vector **A** by its magnitude and direction using the unit vector. This resembles the definition of a vector having both a magnitude and direction.

$$\boxed{\mathbf{A} = |\mathbf{A}|\hat{\mathbf{a}}}$$ Defining a vector using unit vectors 2.25

In this example, we can express vector **A** as

$$\mathbf{A} = \sqrt{2^2 + 4^2 + 4^2}(0.333\hat{i} + 0.667\hat{j} + 0.667\hat{k})$$
$$= 6(0.333\hat{i} + 0.667\hat{j} + 0.667\hat{k})$$
$$= 2\hat{i} + 4\hat{j} + 4\hat{k}$$

The Unit Normal Vector

Consider Figure 2.7a; the vector **A** is a function of position. That is, the magnitude and direction change depending on the x, y and z values. It is not a straight line like the other vectors we have considered. We can represent this as ***vector function*** **A**(x,y,z). If we draw a tangent at any location along the curve, we get a vector tangential to the curve, $\mathbf{A}^{\mathbf{T}}$. If we now draw a vector at right angles to the tangent, we get another vector, n-hat ($\hat{\mathrm{n}}$). This vector, n-hat, is called the ***unit normal vector*** and has a magnitude (length) of one. The unit normal vector is at a right angle (or orthogonal) to a given vector. Surfaces can also have unit normal vectors, as shown in Figure 2.7b. Just as unit vectors describe the direction of a vector, the unit normal vector represents the direction perpendicular or orthogonal to some vector or surface.

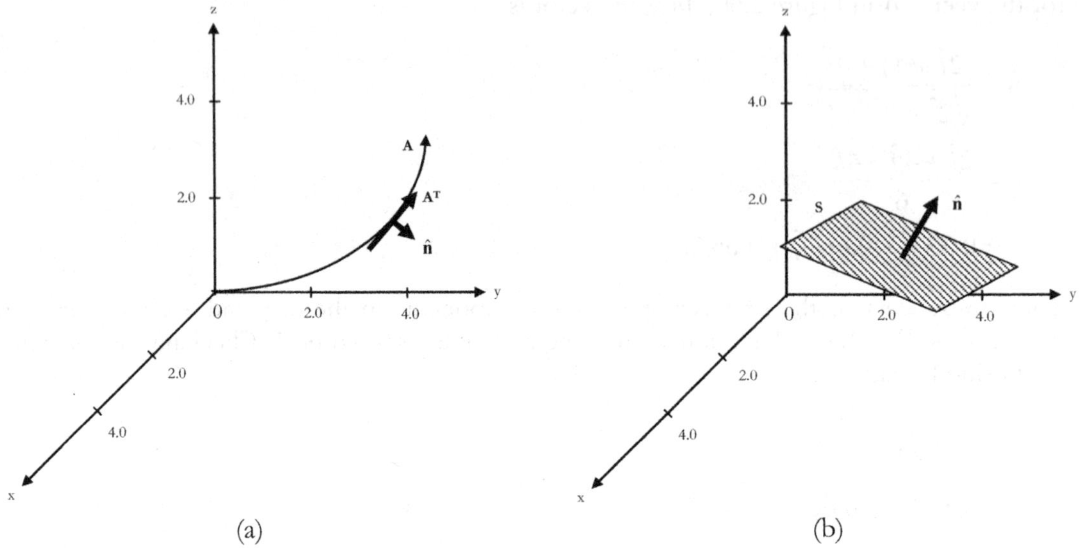

Figure 2.7
Unit normal vectors.

The Dot Product

Consider Figure 2.8. A block is resting on a table, and a force represented by the vector **A** is applied to the block. The force applied is at an angle to the block, where a portion of the force is in the downward direction **Ay** and a portion of the force in the horizontal direction **Ax**. As the block is resting on a table, the block will only be displaced horizontally in the x-direction. Therefore the component of the force causing the horizontal displacement is also in the x-direction, that is, the component vector **Ax**. We can determine the x-component vector using Equation 2.16. The vector **B** is a vector representing the direction and magnitude of the displacement of the block due to the force applied. Multiplying the magnitude of the displacement vector **B** by the magnitude of the force that caused the displacement **Ax** is the definition of ***work***. We can write this as

$$\mathbf{A} \bullet \mathbf{B} = |\mathbf{A_X}||\mathbf{B}| = (|\mathbf{A}|\cos\theta)|\mathbf{B}|$$ The dot product of two vectors 2.26

where θ is the angle between the two vectors. This operation is called the ***dot product*** and is represented by a large dot. Note the result of the dot product is a scalar value and not a vector.

The dot product can also be calculated by multiplying the components of the vectors as follows

$$\mathbf{A} \bullet \mathbf{B} = (A_x B_x) + (A_y B_y) + (A_z B_z)$$ The dot product of two vectors 2.27

If we apply the dot product of the unit vector **î** with each of the other unit vectors, we arrive at the following relationships.

$$\hat{\mathbf{i}} \bullet \hat{\mathbf{i}} = |\hat{\mathbf{i}}||\hat{\mathbf{i}}|\cos(0) = (1)(1) = 1$$

$$\hat{\mathbf{i}} \bullet \hat{\mathbf{j}} = |\hat{\mathbf{i}}||\hat{\mathbf{j}}|\cos(90) = (1)(0) = 0$$

$$\hat{\mathbf{i}} \bullet \hat{\mathbf{k}} = |\hat{\mathbf{i}}||\hat{\mathbf{k}}|\cos(90) = (1)(0) = 0$$

The dot product is commutative, so we arrive at the following useful relationships

$$\hat{\mathbf{i}} \bullet \hat{\mathbf{i}} = \hat{\mathbf{j}} \bullet \hat{\mathbf{j}} = \hat{\mathbf{k}} \bullet \hat{\mathbf{k}} = 1$$

Standard unit vector dot products 2.28

$$\hat{\mathbf{i}} \bullet \hat{\mathbf{j}} = \hat{\mathbf{i}} \bullet \hat{\mathbf{k}} = \hat{\mathbf{j}} \bullet \hat{\mathbf{k}} = 0$$

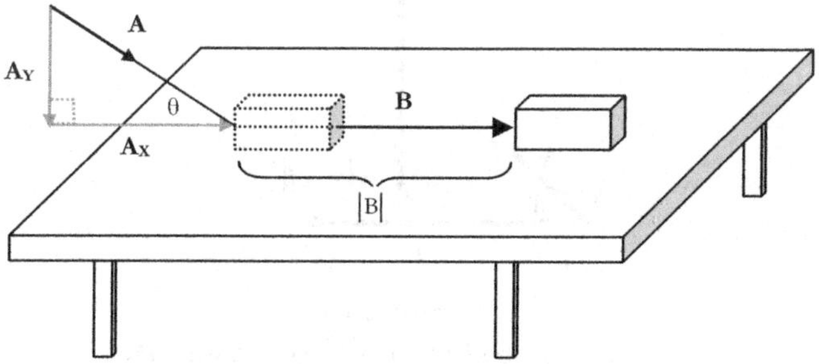

Figure 2.8
Using the dot product to find the work done on an object.

The Cross Product

While the dot product can be thought of as an indication of how well a vector **A** is in line with vector **B**, there is another form of vector multiplication known as the ***cross product*** that indicates how orthogonal two vectors are. As the angle between the vectors approaches 90°, the greater the cross product will be.

The cross product of two vectors gives a third vector perpendicular to the plane of **A** and **B,** as shown in Figure 2.10 and is in the unit normal vector (n-hat) direction. Also, note the area of the plane is the magnitude of the cross product. The cross product is defined as

$$\boxed{\mathbf{A} \times \mathbf{B} = |\mathbf{A}||\mathbf{B}|\sin(\theta)\hat{\mathbf{n}}}$$

The cross product of two vectors 2.29

where $\hat{\mathbf{n}}$ is the unit normal vector. The cross product also has an algebraic component form, however somewhat complicated. Use the matrix diagram in Figure 2.9 to select the components to multiply.

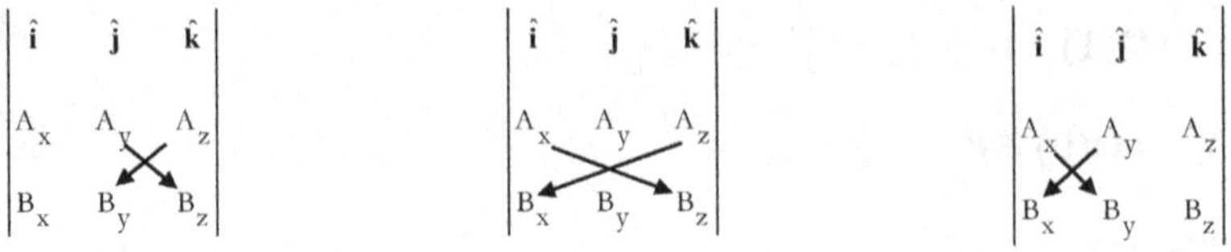

Figure 2.9
Components multiplied by the cross product.

$$\mathbf{A}\times\mathbf{B} = (A_yB_z - A_zB_y)\hat{\mathbf{i}} - (A_xB_z - A_zB_x)\hat{\mathbf{j}} + (A_xB_y - A_yB_x)\hat{\mathbf{k}}$$

The cross product of two vectors 2.30

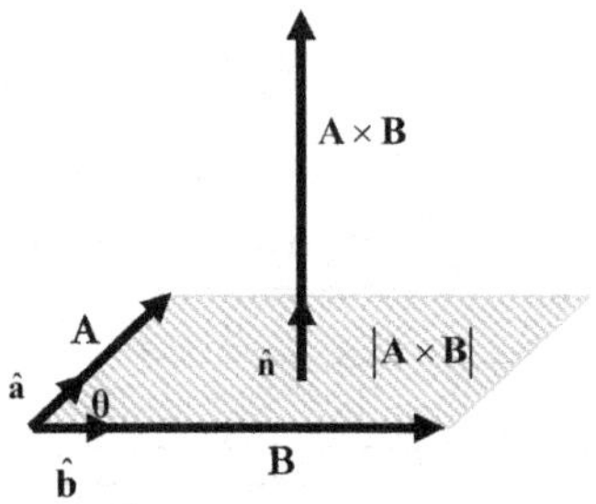

Figure 2.10
The cross product of vectors **A** and **B.**

The cross product of the standard unit vectors gives the following values

$$\hat{\mathbf{i}}\times\hat{\mathbf{i}} = \hat{\mathbf{j}}\times\hat{\mathbf{j}} = \hat{\mathbf{k}}\times\hat{\mathbf{k}} = 0$$
$$\hat{\mathbf{i}}\times\hat{\mathbf{j}} = \hat{\mathbf{k}}$$
$$\hat{\mathbf{j}}\times\hat{\mathbf{i}} = -\hat{\mathbf{k}}$$
$$\hat{\mathbf{i}}\times\hat{\mathbf{k}} = -\hat{\mathbf{j}}$$
$$\hat{\mathbf{k}}\times\hat{\mathbf{i}} = \hat{\mathbf{j}}$$
$$\hat{\mathbf{j}}\times\hat{\mathbf{k}} = \hat{\mathbf{i}}$$
$$\hat{\mathbf{k}}\times\hat{\mathbf{j}} = -\hat{\mathbf{i}}$$

The cross product of the standard unit vectors 2.31

2.5 CALCULUS

Differential Calculus

Say the graph shown in Figure 2.11a represents the distance an object travels as a function of time, given by the equation

$$f(t) = 1.875t^2$$

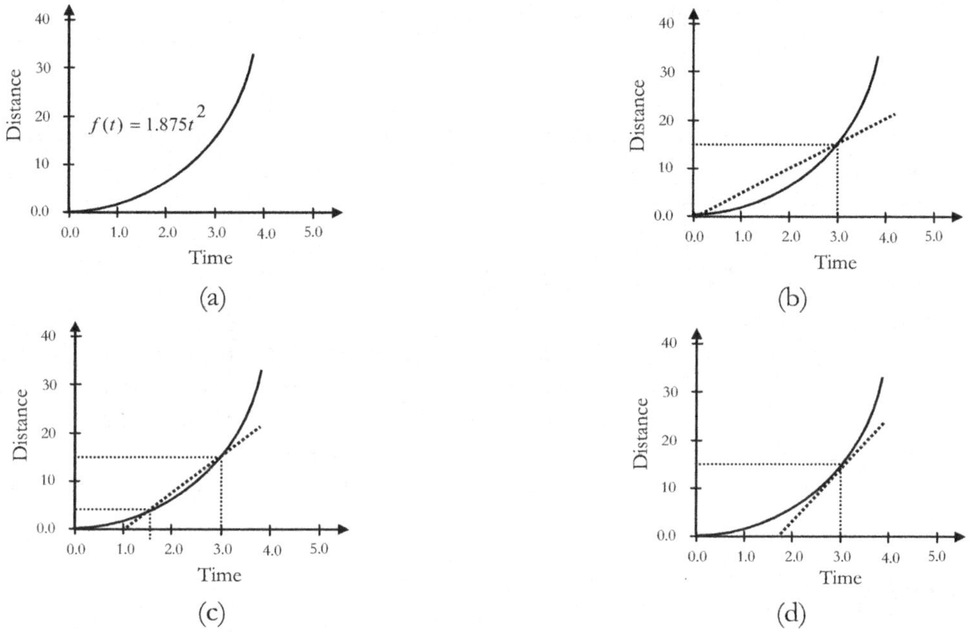

Figure 2.11
A tangent to the curve as the time difference tends to zero.

Say we want to know the speed at which the object travels in three seconds. We could determine the speed by dividing the distance travelled by the time taken to travel that distance; that is, the change in distance with respect to the change in time. We could write

$$s(t) = \frac{\Delta f(t)}{\Delta t}$$

$$= \frac{f(t_2) - f(t_1)}{t_2 - t_1}$$

where Δ (delta) means a change. This represents an average speed and not the speed precisely at three seconds (called instantaneous speed). If we take the difference in time starting from zero seconds and ending in three seconds and draw a straight line through the values of the function f(t) at these times, we get the dotted line shown in Figure 2.11b. If we decrease the time interval by setting the starting time at 1.5 seconds and draw a line through the distance values, we get the dotted line in Figure 2.11c. We can see that as we decrease the time interval, the straight line between the two points of the function starts to approximate the tangent of the curve at three seconds. We can set this time interval as small as we like, and as the time interval approaches zero, we get the tangent to the curve at three seconds, as shown in Figure 2.11d. This concept of making an interval approach some number is called a ***limit***, and we can write the time interval approaching zero as

$$\lim_{\Delta t \to 0}$$

Using the concept of the limit in the speed equation above, we can write

$$s(t) = \lim_{\Delta t \to 0} \frac{f(t+\Delta t) - f(t)}{\Delta t}$$

This function is called a ***derivative***. The derivative of a function f is another function f' whose value at any point x in the domain of f is

$$f'(x) = \lim_{\Delta x \to 0} \frac{f(x+\Delta x) - f(x)}{\Delta x} \qquad \text{The derivative} \qquad 2.32$$

provided that the limit exists.

In the above example, we selected a value for the time, which is 3 seconds. Instead of specifying a value, we could use the above definition of the derivative to arrive at another function. We could use this derivative function to find the speed at any given time. Also, note there is nothing special about using the variable x; we could use t or any other variable to denote time, distance etc. Using the given function for speed and setting the independent variable as t, and substituting in 2.32 we get

$$\begin{aligned} f'(t) &= \lim_{\Delta t \to 0} \frac{1.875(t+\Delta t)^2 - 1.875t^2}{\Delta t} \\ &= \lim_{\Delta t \to 0} \frac{1.875(t^2 + 2t\Delta t + \Delta t^2) - 1.875t^2}{\Delta t} \\ &= \lim_{\Delta t \to 0} \frac{1.875\Delta t(2t + \Delta t)}{\Delta t} \\ &= \lim_{\Delta t \to 0}[1.875(2t+\Delta t)] \\ &= 1.875(2t) \\ &= 3.75t \end{aligned}$$

We now have a function that will give us the object's speed at any time, not just at 3 seconds. For t=3 seconds, the speed is

$$\begin{aligned} f'(3) &= 3.75(3) \\ &= 11.25 \end{aligned}$$

There are many rules for determining derivatives that provide shortcuts for writing derivatives without having to go through the steps in Equation 2.32. We will not state these shortcuts and rules as it is beyond the scope of this book. No matter what rules are used, a derivative will always tell us how a function changes with respect to a change in the independent variable. In this example, the derivative tells us how the distance changes with a change in time; that is, it gives us the instantaneous speed at any time.

A common way to express a derivative is by using the Leibnitz notation. The Leibnitz notation can either be thought of as representing a derivative or as a ratio of two differentials, whichever is most convenient for a given situation

$$\frac{dy}{dx} = f'(x)$$

The concept of the derivative remains the same; just a different way of writing it. We also take the derivative of a derivative. That is, we take the derivative twice. This is called a second-order derivative and is written as

$$\frac{d^2y}{dx^2}$$

Derivatives can apply to functions dependent on more than one variable. Typically, however, we often want to know how the function changes with respect to a change in a single variable. In this case, we treat all other variables as constants and take the derivative with respect to a single variable. This is called a ***partial derivative***, and it is written as

$$\frac{\partial y}{\partial x}$$

And the second order form is

$$\frac{\partial^2 y}{\partial x^2}$$

Integral Calculus

Consider Figure 2.12. If we want to know the area under the curve for the time spanning 0 to 4 seconds, we could split the curve into sections, calculate the area for each section and sum the areas to get the total area under the curve. If we split the interval, in this case, time, into four equal intervals, the total area would be

$$A = f(t_1)\Delta t + f(t_2)\Delta t + f(t_3)\Delta t + f(t_4)\Delta t$$

In Figure 2.12a, we can see a lot of area under the curve not included in the sum. If we make the interval smaller, the sum of the rectangular areas becomes a better approximation of the area under the curve, as shown in Figure 2.12b. We could make this interval as small as we like. Splitting the number of areas into equal widths of n sections gives us

$$A = f(t_1)\Delta t + f(t_2)\Delta t + f(t_3)\Delta t + f(t_4)\Delta t ... f(t_n)\Delta t$$

In summation notation, this is written as

$$A = \sum_{i=1}^{n} f(t_i)\Delta t$$

We can decrease the interval so that the interval width approaches zero and the number of intervals (n) approaches infinity, and we have

$$A = \lim_{n\to\infty} \sum_{i=1}^{n} f(t_i)\Delta t$$

If we replace the summation to infinity with the integral $\int$ sign and use "dt" to indicate an infinitesimally small length of time, we would write for the area under the curve between t=0 and t=4 seconds

$$A = \int_0^4 f(t)dt$$

This is called a ***definite integral*** over the interval 0 to 4. For any definite integral, we would write

$$\boxed{\int_a^b f(x)dx = F(b) - F(a)}$$ The definite integral 2.33

where F(b) and F(a) are values of the function we get after integration at value b and at value a, respectively. If the limits of integration are not specified, then the integral is called an ***indefinite integral*** or simply the ***integral*** and has the form

$$\boxed{\int f(x)dx = F(x) + c}$$ The indefinite integral 2.34

where c is called the constant of integration. Integration is nothing more than summing up a bunch of very tiny bits to find the whole or the portion we are interested in. These tiny bits could be areas, tiny bits of line segments, tiny bits of volume or even tiny bits of an electric field.

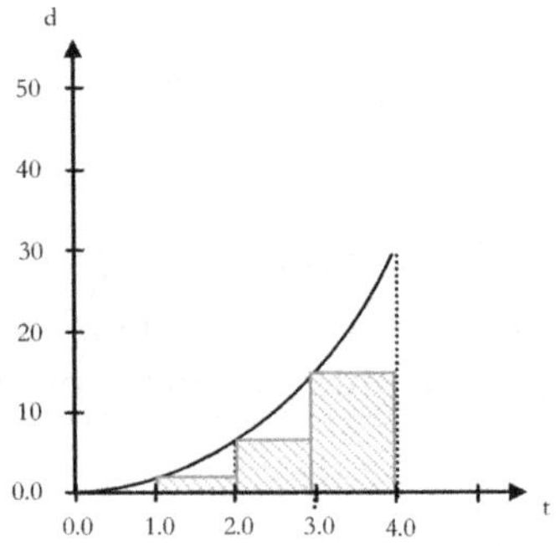

(a)

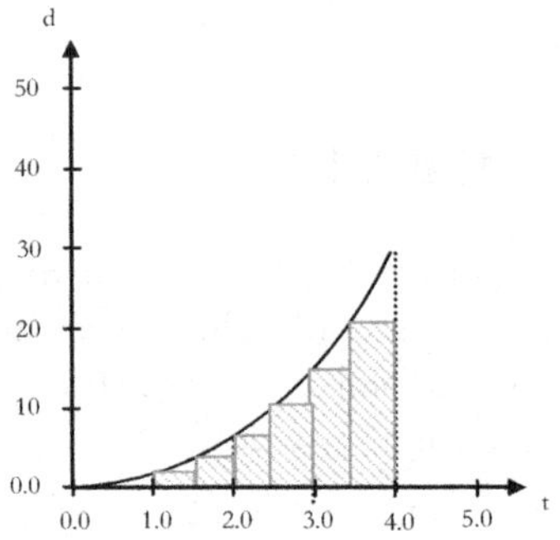

(b)

Figure 2.12
As the number of rectangular areas increases, the closer we get to the true area under the curve.

As with derivatives, there are many rules and shortcuts used to calculate integrals. We will state one of them; the integration power rule.

$$\int f(x^n)dx = \frac{x^{n+1}}{n+1} + c$$

In the previous example, we found a function for speed by differentiating the distance equation. Given the equation for speed, what if now we wanted to know the total distance travelled over a period from 0 to 4 seconds? We can calculate the distance by multiplying speed by time. By multiplying the instantaneous speed at some time by an infinitely small amount of time, we get the distance travelled for this small amount of time. We then add these little bits of distance over the period 0 to 4 seconds to get the total distance travelled in this interval of time. Integrating this equation and applying the power rule, we get

$$d = \int_0^4 (3.75t)dt$$

$$= \frac{1}{1+1}(3.75t^{1+1})\Big|_0^4$$

$$= \frac{1}{2}(3.75t^2)\Big|_0^4$$

$$= 1.875t^2\Big|_0^4$$

$$= 1.875(4)^2 - 1.875(0)^2$$

$$= 30$$

Notice that the function found after integrating was the original distance function; which is $1.875t^2$. This highlights an important relationship between integration and differentiation. An integral is sometimes referred to as the ***anti-derivative*** and is the inverse of differentiation. If we integrate a derivative, we get back the original function. If a function F(x) has a derivative F´(x) such that F´(x) = f(x), then the integral is an anti-derivative.

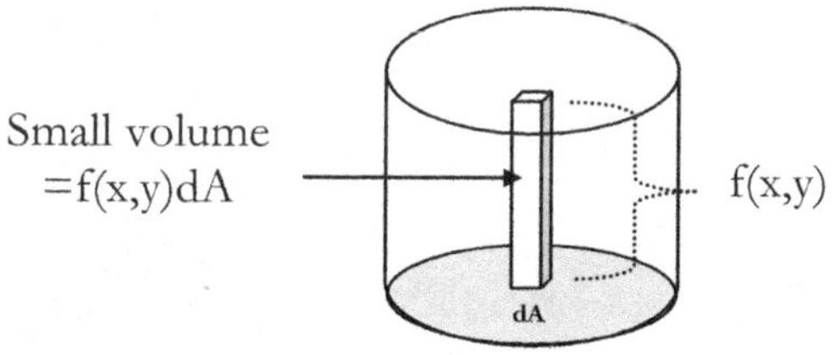

Figure 2.13
Integral of a volume.

Finding the indefinite integral of the speed equation

$$F(t) = \int (3.75t)dt$$

$$= \frac{1}{2}(3.75t^2) + c$$

$$= 1.875t^2 + c$$

Notice the constant of integration c. At time t=0, we have not travelled any distance, so we can set c to 0 (in this instance). So we end up with

$$f(t) = 1.875t^2$$

which is the original distance equation.

Integration also applies to functions with multiple variables. Consider Figure 2.13. In the X-Y plane, we have an area (region) signified by dA. This is just dx times dy. The height is given by the function f(x,y), so f(x,y)dA is just a small volume. Add up all the small volumes, and we have the total volume of the object. A two-dimensional integral (two independent variables) has the form, with R indicating a region

$$V = \iint_R f(x,y)dA = \int_{x_1}^{x_2}\int_{y_1}^{y_2} f(x,y)dxdy$$ Two-dimensional integral

Where the inner integral is performed first, and then the outer integral is performed on the result of the inner integral.

There are other forms of integrals. In Figure 2.14a, we have a small vector d**r**. If we add up all the tiny line segments, we get the length of the line. This is called a line integral, and it is designated by a C, for curve, at the bottom of the integral sign.

$$\int_C d\mathbf{r} \qquad \text{(A line integral)}$$

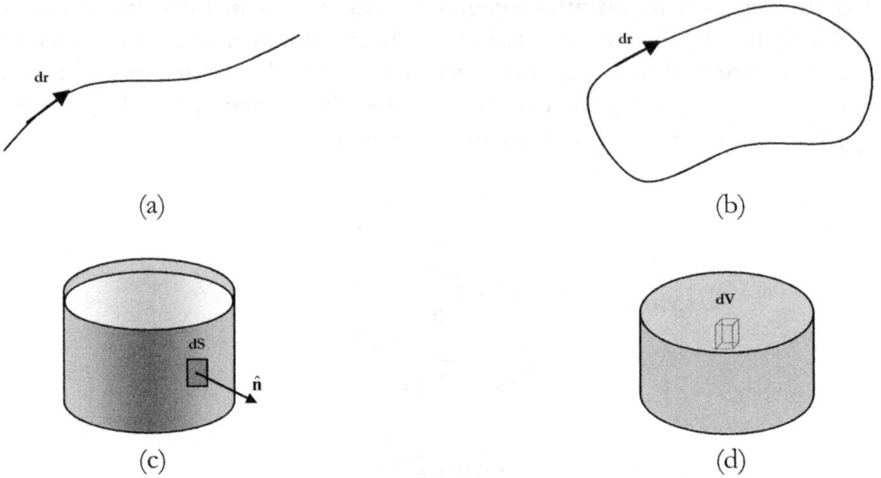

Figure 2.14
Different forms of integration.

If the curve is closed, as in Figure 2.14b, then the integral symbol includes a circle to indicate a closed path

$$\oint_C d\mathbf{r} \qquad \text{(A closed line integral)}$$

In Figure 2.14c, we are adding up the little bits of surface area of a three-dimensional object, where the direction of d**S** is given by the unit normal vector **n̂**. This type of integral is called a surface integral. Sometimes the notation uses two integral signs, and sometimes only one is used; they mean the same thing.

$$\int_S d\mathbf{S} = \iint_S d\mathbf{S} = \iint_S \hat{\mathbf{n}} \bullet d|\mathbf{S}| \qquad \text{A surface integral}$$

If it is a closed surface, which this example is, then the integral notation becomes

$$\oiint_S d\mathbf{S} \qquad \text{A closed surface integral}$$

This extends out to volumes, as shown in Figure 2.14d, and the notation is

$$\int_V d\mathbf{V} = \iiint_V d\mathbf{V} \qquad \text{A volume integral}$$

Note the volume integral adds up tiny bits of volume (dx.dy.dz) to give the total volume, whereas the two-dimensional integral multiplies tiny areas (dxdy) with the function giving small volumes, which are then summed to get the total volume.

2.6 GRADIENT

Say we measured the temperature in a room at many locations to build up a three-dimensional model showing the temperature variations within the room depending on the x, y and z locations. We derive a function T(x,y,z) describing the temperature value as a function of x, y and z. Temperature is a scalar value as it has no direction, so the function T(x,y,z) is said to describe a ***scalar field***. We now wish to know how the temperature changes with respect to a change in position. From our study of differential calculus, we know that a derivative gives us the change in the dependent variable with respect to a change in the independent variable; which is a slope or gradient. If we wish to know the contributions of each of the x, y and z values separately, we can use partial differentiation. If we consider the partial derivative with respect to x as the magnitude in the x direction, and the partial with respect to y as the magnitude in the y direction and the same for z, then we can define a vector giving not only the rate of change in temperature but the direction at which the maximum change occurs. We can do this using a special differential operator known as the ***del*** operator. Its symbol is the upside-down Greek delta, that is ∇. The del operator is defined as

$$\nabla = \frac{\partial}{\partial_x}\hat{\mathbf{i}} + \frac{\partial}{\partial_y}\hat{\mathbf{j}} + \frac{\partial}{\partial_z}\hat{\mathbf{k}}$$ The del operator 2.35

A mathematical operator is something that is done to or performed on an equation. So if we apply the del operator to our function T, we would write

$$\nabla \mathbf{T} = \frac{\partial T}{\partial_x}\hat{\mathbf{i}} + \frac{\partial T}{\partial_y}\hat{\mathbf{j}} + \frac{\partial T}{\partial_z}\hat{\mathbf{k}}$$ The gradient 2.36

The del operator applied to a scalar field (function) is called the ***grad*** or ***gradient***. So we would describe the above equation by saying "the grad of T". Note this takes a scalar function and returns a vector function. The magnitude of the rate of change is given by

$$|\nabla \mathbf{T}(x, y)|$$ The magnitude of the rate of change 2.37

If we want to know the rate of change of temperature in a particular direction, then dot the gradient with the unit vector of the vector of interest. That is

$$\nabla \mathbf{T} \bullet \hat{\mathbf{u}} = |\nabla \mathbf{T}| \cos \theta$$ The directional derivative 2.38

This is called a directional derivative. If the angle theta is 0, then the directional derivative is maximal; that is, we are going in the same direction as the vector $\nabla \mathbf{T}$. This is why grad returns the maximum rate of change. In this example, we have used temperature, but the scalar field may describe any phenomena such as height, pressure, electrostatic charge etc.

2.7 DIVERGENCE

Consider a sink with a plug hole in the middle. We start to fill the sink with water by placing a hose under the plug hole. Water spreads out in all directions from the hole. We can represent the motion of water about this point (the plug hole) as a collection of vectors radiating outwards, as shown in Figure 2.15a. We could say that the water diverges from this point.

We now lift the tank so that some of the water flows across the hole, but water is still coming in from the hose. Some water flows across the hole, but more is coming out in one direction than the other, vectorially shown in Figure 2.15b. As more water is flowing away from the hole than approaching the hole, overall, the region around this point still has divergence, just a little less than before.

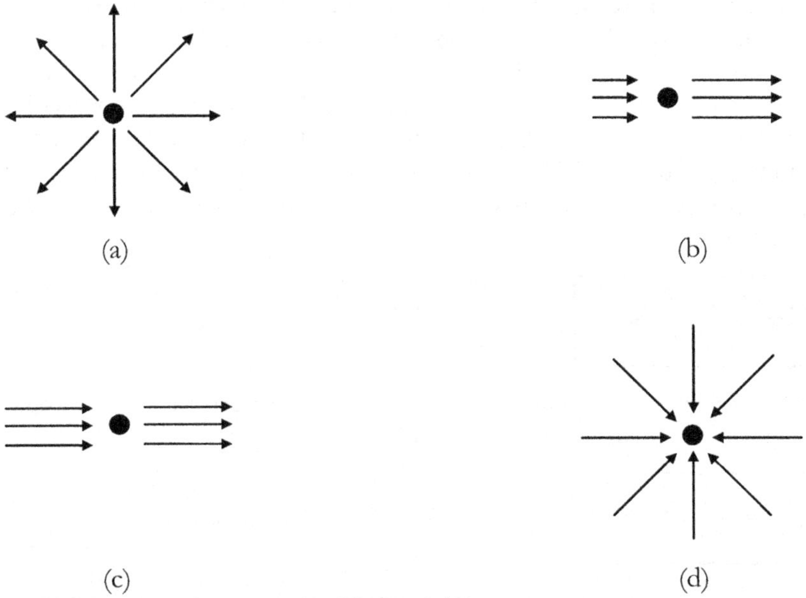

Figure 2.15
Divergence.

If we now turn off the hose and plug the hole while tilting the sink. In this case, there is no overall divergence around this point as just as much water is moving toward the plug as is moving away, Figure 2.15c.

Finally, we lay the sink flat, and the plug is removed. Water starts to pour out of the sink in all directions. All the water is now moving toward the hole; that is, it is converging towards this point. Expressed another way we could call this negative divergence. ***Divergence*** can be thought of as how much a vector field gravitates or moves away from a point.

Say the vector function **V**(x,y,z) describes the flow of water in our tank and each of the components of the vector is described by functions $\mathbf{V_X}$,(x,y,z), $\mathbf{V_Y}$,(x,y,z) and $\mathbf{V_Z}$,(x,y,z). We can express this vector in terms of standard unit vectors as

$$\mathbf{V}(x,y,z) = \mathrm{V_X}(x,y,z)\hat{\mathbf{i}} + \mathrm{V_Y}(x,y,z)\hat{\mathbf{j}} + \mathrm{V_Z}(x,y,z)\hat{\mathbf{k}}$$

The divergence of **V** or the ***div*** of **V** is defined as

$$div\mathbf{V} = \nabla \bullet \mathbf{V} = \left(\frac{\partial}{\partial_x}\hat{\mathbf{i}} + \frac{\partial}{\partial_y}\hat{\mathbf{j}} + \frac{\partial}{\partial_z}\hat{\mathbf{k}} \right) \bullet \left(V_x\hat{\mathbf{i}} + V_y\hat{\mathbf{j}} + V_z\hat{\mathbf{k}} \right)$$

and

$$\boxed{div\mathbf{V} = \frac{\partial V_x}{\partial_x} + \frac{\partial V_y}{\partial_y} + \frac{\partial V_z}{\partial_z}} \qquad \text{Divergence} \qquad 2.39$$

Notice the resemblance to the dot product. The result of the div**V** (the div of V) is a scalar function. The value is greater than zero when the field in the neighbourhood of a point has a tendency to move away from that point, zero when there is no divergence at all, and negative when the tendency is to move toward or gravitate to the point.

2.8 CURL

Using the water tank example described above, when we talk of divergence, we are talking about the specific motion of the water as it moves away or toward the plug. The water will also have another type of motion in that it will swirl or circulate around the plug hole as it moves towards or away from the hole.

In Figure 2.16a, we have drawn the motion of water swirling around the plug hole in an anti-clockwise direction. If we consider the water to be flowing into the tank in Figure 2.16a, then we can think of the hole as a source of water, and the direction of the source is toward you, as indicated by the black dot (representing the tip of an arrow). If we point our thumb in the direction of the source and then curl our fingers, the tips of the fingers point in the direction of the curl (the right-hand rule). In this case, we say the curl is positive. In Figure 2.16b, the water is flowing out of the tank as indicated by the cross (representing the end of an arrow), and the curl is negative.

Figure 2.16
Curl around a point.

Now consider Figure 2.17a. It would appear that there is no curl, as the vectors are pointing across the point in a straight line. However, if we place an object across the hole, say a Popsicle stick, the water flow will cause the Popsicle stick to rotate as the magnitude of the water flowing across one side of the stick is greater than the other. In this case, curl exists around this point. In Figure 2.17b, there is no difference in magnitude between the top and bottom vectors, and a

Popsicle stick placed in the flow will not rotate. In this case, there is no curl around this point. Therefore ***curl*** can be thought of as something that will cause a rotation around a point. The vector field may represent a magnetic field or an electric field; it does not matter, just imagine it to be the flow of water, and if dropping a stick across it causes the stick to rotate, then the curl is non-zero.

Curl is similar to the cross product of a vector. The cross product of two vectors gives us another vector pointing perpendicular to the plane of the crossed vectors. The result of the curl is also a vector orthogonal to the plane of the circulating vector field, see Figure 2.18.

(a) (b)

Figure 2.17
Curl around a point.

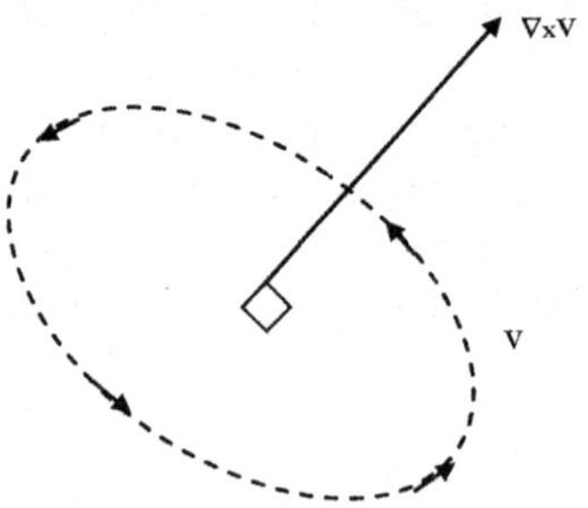

Figure 2.18
Taking the curl of a vector results in another vector pointing orthogonally.

The curl is the cross-product of the del operator with the vector. Given a vector field function **V**, then

$$curl\mathbf{V} = \nabla \times \mathbf{V} = \left(\frac{\partial}{\partial_x}\hat{\mathbf{i}} + \frac{\partial}{\partial_y}\hat{\mathbf{j}} + \frac{\partial}{\partial_z}\hat{\mathbf{k}}\right) \times \left(V_x\hat{\mathbf{i}} + V_y\hat{\mathbf{j}} + V_z\hat{\mathbf{k}}\right)$$

Solving gives

$$curl\mathbf{V} = \nabla \times \mathbf{V} = \left(\frac{\partial V_z}{\partial_y} - \frac{\partial V_y}{\partial_z}\right)\hat{\mathbf{i}} - \left(\frac{\partial V_z}{\partial_x} - \frac{\partial V_x}{\partial_z}\right)\hat{\mathbf{j}} + \left(\frac{\partial V_y}{\partial_x} - \frac{\partial V_x}{\partial_y}\right)\hat{\mathbf{k}} \quad \text{Curl} \quad 2.40$$

3

Prerequisites in Physics

3.1 SPEED AND VELOCITY

Speed is a concept we are all familiar with. It is simply the distance travelled per period of time. That is

$$S = \frac{x}{t}$$

where S is the speed in metres per second, x is the distance travelled in metres, and t is the time taken to travel the distance in seconds. If you walk to a shop a hundred metres away in twenty seconds, then your speed is

$$S = \frac{100}{20}$$
$$= 5$$

That is 5 metres per second. This is, in fact, the average speed, as your walking speed may not be constant throughout the twenty seconds. If you record the distance travelled every one second, then you will have some detail about the variation in speed during the walk. If you record the distance travelled every 1/10th of a second, then you will have a greater amount of detail about your variation in speed. Decreasing the time period between distance measurements will enable you to determine your speed at any given moment with greater accuracy. We can write this as

$$s = \frac{\Delta x}{\Delta t}$$

where Δd represents some small change (a differential) in distance, and Δt represents some small change in time. When the change in time Δt approaches zero, we write the equation as

$$\boxed{s = \frac{dx}{dt}} \qquad \text{Instantaneous speed} \qquad 3.1$$

The ratio of these two differentials is called a differential equation, and we can use differential calculus to solve the equation giving us an equation that describes instantaneous speed for any given time value. If we have information on both the speed and direction, we have a vector and speed or direction, or both may change at any given time. This combination of speed and direction is called ***velocity*** and is a vector, whereas speed is a scalar. The average velocity is

$$\mathbf{V} = \frac{\mathbf{x}}{t}$$

where **x** is now a vector as it implies both distance and direction. The instantaneous velocity is given by

$$\boxed{\mathbf{V}(t) = \frac{d\mathbf{x}}{dt}}$$ Instantaneous velocity 3.2

3.2 ACCELERATION

According to Isaac Newton's first law of motion, a body remains at rest or in uniform motion unless acted upon by a net external force. A body at rest has a velocity of zero, and a body in uniform motion has some velocity greater than zero. A net force applied to a body that results in a change of velocity, either by a change of speed, direction or both, results in the object undergoing ***acceleration***. And we can write this as

$$\boxed{\mathbf{a}(t) = \frac{d\mathbf{v}}{dt}}$$ Instantaneous acceleration 3.3

Acceleration is a vector and so will have a numerical value and direction. The units are in metres per second squared (m/s^2). For instance, an object will accelerate near the surface of the Earth at 9.8 metres per second squared (9.8 m/s^2); that is, the velocity will increase by 9.8 m/s for every second it is falling.

3.3 FORCE

Newton's second law of motion defines a force as something that results in a body undergoing acceleration. Expressed as

$$\boxed{\mathbf{F} = m\mathbf{a}}$$ Force 3.4

where **F** is the net force in newtons applied to the body, m is the mass of the body in kilograms, and **a** is the acceleration in m/s^2. Note the use of the term net. If two forces act on the body with the same value but in opposite directions, then the net force is zero, and no acceleration is experienced. This is probably one of the most often used equations in physics and is certainly one of the most important physical laws ever expressed.

3.4 MOMENTUM AND INERTIA

Before Newton, it was generally believed that a moving body would eventually come to rest if no external force was applied. After all, a cart set in motion will eventually stop if no one pushes it, and a sailing boat will eventually stop when the wind ceases to blow. That is, the natural state of a body is to be at rest. After Newton developed his three laws of motion, it was clear that objects stop due to external forces such as friction, wind resistance and gravitational forces. If no external force were applied to a cart, it would maintain its current velocity, speed and direction forever. If we multiply the mass of a body by its velocity, we get a property called ***momentum***. You can think of momentum as *the time required to bring a body to rest for a given force.* The greater the momentum of an object, the longer it takes to come to rest for a given constant opposing force. The momentum of an object is given by

$$\boxed{\boldsymbol{\rho} = m\mathbf{v}}$$ Momentum 3.5

where $\boldsymbol{\rho}$ is the momentum (a vector) in kg•m/s, m is the mass in kilograms, and **v** is the velocity in metres per second.

Momentum is an important concept and is fundamental to understanding many physical phenomena. It is one of the few properties in the universe that is conserved. For example, if no external force is applied and two bodies collide with one another and smash into tiny pieces, the addition of all the momenta of the little bits will be the same as the initial momentum of both of the objects before the collision. The total momentum before the collision is equal to the total momentum after the collision, meaning the total momentum of a system is conserved. We can express this as

$$\boldsymbol{\rho}_T = m_1\mathbf{v}_1 + m_2\mathbf{v}_2 + \ldots m_n\mathbf{v}_\mathbf{n}$$ Conservation of momentum 3.6

This is known as the conservation of momentum. Furthermore, if we substitute the acceleration Equation 3.3 into the force Equation 3.4, we get

$$\mathbf{F} = m\frac{d\mathbf{v}}{dt}$$

Using some algebraic manipulation, we can rewrite this as

$$\mathbf{F} = \frac{d(m\mathbf{v})}{dt}$$ Force is the time-rate change in momentum 3.7

This tells us that force can be defined as the time-rate change in momentum. Normally mass does not change, and the force is typically given by Equation 3.4. Rewriting Equation 3.4 for acceleration, we get

$$\boxed{\mathbf{a} = \frac{\mathbf{F}}{m}}$$ acceleration 3.8

We can see that for a given constant force, the acceleration of a body depends on its mass. The greater the mass, the lower the acceleration. And for the same force, a lighter mass will acquire a larger acceleration. Therefore the greater the mass of a body, the more it resists a change in its velocity. *The tendency of a body to resist a change in its motion is called* ***inertia.*** And the amount of inertia a body has is directly proportional to the amount of mass a body has, sometimes referred to as ***inertial mass***.

3.5 WORK

If a force acts on a body causing the object to be displaced from its original position, then work has been performed. In Chapter 2, we discussed how to use the dot product to find the work done. ***Work*** is defined as the product of the displacement of an object multiplied by the magnitude of the force that caused the displacement.

$$W = \mathbf{F} \bullet \mathbf{d} = (|\mathbf{F}|\cos\theta)|\mathbf{d}| \qquad \text{Work (joules)} \qquad 3.9$$

where **F** is the applied force in newtons, **d** is the displacement in metres that occurs while the force is applied, and θ is the angle between the direction of the applied force and the direction of the displacement, see Figure 2.8. Note that both force and displacement are vectors; however, work is a scalar value in joules.

If the applied force is not constant and varies over time, then we can use integral calculus to get the total work. We would add up all the tiny bits of work done, **F•dS**, along the path.

$$W = \int_C \mathbf{F} \bullet d\mathbf{S} \qquad 3.10$$

Note that this is a line integral, and we have indicated the infinitesimal displacement as d**S.**

3.6 ENERGY

Energy is defined as the ability of an object to do work or to transfer heat.

$$E = W + Q \qquad \text{Energy} \qquad 3.11$$

where E is the energy transferred, W is the work done, and Q is the thermal energy (heat) transferred. The units for each are in joules. A joule is a unit of energy defined as the energy transferred (work done) on an object when a force of one newton is applied for a distance of one metre. If no thermal energy (heat) is transferred, then the energy transferred is just equal to the amount of work done.

$$E = W = \mathbf{F} \bullet \mathbf{d} = \int_C \mathbf{F} \bullet d\mathbf{S} \qquad \text{Energy without heat transfer} \qquad 3.12$$

Like momentum, energy is a conserved quantity in a closed system. A closed system is any arbitrary set of objects you desire where you assume no energy, matter or force is coming in from the outside or leaving the system. That is, you consider it to be isolated from the rest of the universe. No system is truly isolated, but for the period of time under consideration, if no significant external energy, force or matter is entering or leaving the system, then it can be considered isolated or closed for analysis.

Energy can come in several forms:

- Nuclear energy
- Electromagnetic energy
- Chemical energy
- Kinetic energy
- Potential energy

Here we will only be considering Kinetic and Potential energy in our discussion. The only other forms of energy of interest to us are the energy contained in the Electric and Magnetic fields and the energy contained in an Electromagnetic wave.

Kinetic Energy

Kinetic energy is defined as the energy a body acquires due to its motion, or the work done to change the velocity of a body. Consider a system of two isolated particles. One particle is moving with some velocity **v**, and the other particle is at rest. The particles collide, and say all of the kinetic energy of the moving particle is transferred to the stationary particle causing it to be displaced, so work has been done. We can not assume that the force is constant, so if we divide the work done into infinitesimally small sections F•dS. We can use calculus to add these small increments of work to give us the total work done, which is the total amount of energy transferred. We could write this as

$$W = E_K = \int_C \mathbf{F} \bullet d\mathbf{S}$$

By solving this equation, we can develop a formula for kinetic energy.

$$\begin{aligned} E_K &= \int_C \mathbf{F} \bullet d\mathbf{S} \\ &= \int_C m\mathbf{a} \bullet d\mathbf{S} \\ &= \int_C m\frac{d\mathbf{v}}{dt} \bullet d\mathbf{S} \\ &= \int_C m\frac{d\mathbf{S}}{dt} \bullet d\mathbf{v} \\ &= \int m\mathbf{v} \bullet d\mathbf{v} \\ &= \frac{1}{2}mv^2 + c \end{aligned}$$

The constant of integration, c, is the initial transfer of energy before the collision, which is zero. So we can write the kinetic energy of a body of mass m, moving with velocity V as

$$\boxed{E_K = \frac{1}{2}mv^2} \qquad \text{Kinetic energy (joules)} \qquad 3.13$$

The units of energy are in joules.

Potential Energy

Consider a bowling ball sitting on the ground. The bowling ball is then lifted and placed on a table. A force was applied to overcome gravity and displace the bowling ball; by definition, work has been performed. However, after performing this work, the bowling ball has zero velocity. It is just sitting on the table. From our previous discussion, we know that energy must be conserved. The energy was transferred to the ball when it was placed on the table, but the bowling ball has zero kinetic energy. So where did the energy go?

Say the ball is bumped, and it falls off the table. It starts to accelerate downwards, and as it gains speed, it gains kinetic energy, increasing to a maximum just before it strikes the ground. In this instance, gravity acted on the ball and performed work. The kinetic energy the ball acquired just before striking the ground equals the work done on the ball required to place it on the table. The work done on the ball by gravity is given by

$$W = F \bullet d = (mg)h$$

where m is the mass, g is the acceleration due to gravity, and h is the distance displaced (its height above ground). The gravitational force applied to the ball is mg.

While the ball was resting on the table, we say that it had the potential to gain kinetic energy due to its height above the ground. The potential of an object to gain kinetic energy due to its position within a force field is called ***Potential Energy***. For energy to be conserved, the total energy of a system must be the sum of the kinetic energy and the potential energy. That is

$$\boxed{E_{total} = E_K + E_P} \qquad \text{Total energy of a system} \qquad 3.14$$

This relationship holds for all objects and particles. It is just as true for electrons as well as for bowling balls. In the above example, the work done and so the energy to lift the object to a certain height is given by mgh. The kinetic energy of the bowling ball at any given moment is $\frac{1}{2}mv^2$. So the total energy is given by

$$E_{total} = \frac{1}{2}mv^2 + mgh$$

While the bowling ball sits on the table, it is at rest, so v is zero, and all the energy is in the form of potential energy. When the ball is knocked off the table, it accelerates downward, losing some potential energy and gaining that amount in kinetic energy. The total energy will always remain the same; as v increases, h decreases. The moment just before the ball strikes the ground, all of the potential energy is converted to kinetic energy, h is zero, and v is at a maximum.

4

The Electric Field

4.1 DEFINING THE ELECTRIC FIELD

The standard atomic model describes an atom consisting of a nucleus, made up of positively charged protons and neutral neutrons, surrounded by orbiting negatively charged electrons. As the neutrons have no charge, the charge on the nucleus will be positive, see Figure 4.1. The electrons orbit the nucleus in one or more shells. This is a simplified model; a model closer to reality would show the electrons as a collection of standing waves surrounding the nucleus, but this model will do just fine for our purposes.

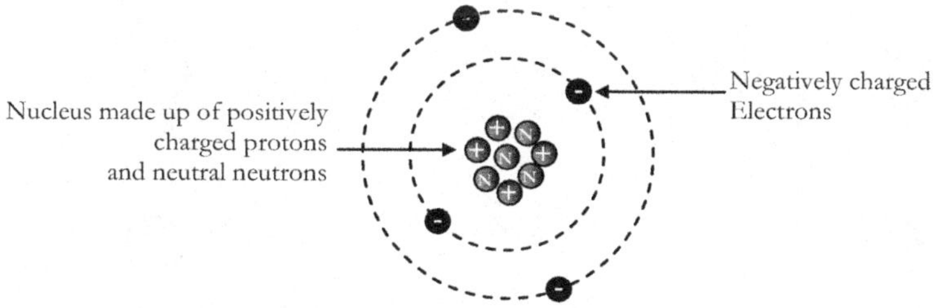

Figure 4.1
The standard atomic model.

In a neutral atom, an atom without any overall (net) charge, there will be as many electrons as there are protons. When an atom has more electrons than protons, it will have an overall negative charge, and when an atom has more protons than electrons, it will have a net positive charge. An atom with a net charge is called an ***ion***.

Each charged particle has an associated field that attracts oppositely charged particles and repels like-charged particles. Protons are repelled by other protons, and electrons are repelled by other electrons, but a proton and an electron will be attracted to one another. This field is called the ***electric field***, and this attraction and repulsion is felt as a force between particles that will cause them to accelerate toward or away from one another.

Static electric fields do not vary with time and are called electrostatic fields. In this discussion, we will look at static electric fields due to charged particles. Electric fields, by convention, are represented as vectors that radiate spherically from the charged particle. For positively charged particles, vectors point away from the particle, and for negatively charged particles, the vector points toward the particle. See Figure 4.2. These vectors are referred to as lines of flux, a concept introduced by Faraday. Note that there are no gaps in the electric field as suggested by the lines of flux; it is simply an easy way to picture fields.

(a) (b)

Figure 4.2
The direction of the lines of electric flux.

If a positively charged particle is placed near a negatively charged particle, the lines of force flow from the positive to the negative direction, representing an attractive force, Figure 4.3a. If two like-charged particles are placed next to each other, the lines of force bunch up between the particles representing a repulsive force. Figure 4.3b shows the repulsive field lines for two electrons. The shape of the flux lines would be the same if two protons were shown, except the direction of the arrows would be pointing away from the particle instead of towards it.

(a) (b)

Figure 4.3
Lines of flux representing attraction and repulsion.

The force felt between two individual particles, either attractive or repulsive, is given by Coulomb's law. The vector form of Coulomb's law is

$$\boxed{\mathbf{F} = \frac{q_1 q_2 \hat{\mathbf{r}}}{4\pi\varepsilon r^2}} \qquad \text{Force between two charged particles} \qquad 4.1$$

where F is the force in newtons, ε is the permittivity of the medium, q_1 and q_2 are the magnitudes of the charges in coulombs, r is the distance between q_1 and q_2 in metres, and $\hat{\mathbf{r}}$ is the unit vector giving the direction between charges q_1 and q_2. Permittivity will be discussed later, but for now, consider it to be a numeric value indicating the ability of a medium to store energy in the electric field. When you are only interested in the magnitude of the force between charges, use the scalar form of the equation

$$|\mathbf{F}| = \frac{q_1 q_2}{4\pi\varepsilon r^2} \qquad \text{Scalar form} \qquad 4.2$$

A coulomb of charge is equal to 6.2415×10^{18} elementary charges. The electron and proton are elementary charges. The charge of a proton is 1.602×10^{-19} coulombs, and for an electron, it is -1.602×10^{-19} coulombs. Both have the same magnitude but different signs indicating positive and negative charges. Note that the force is proportional to $1/r^2$, so if we double the distance, the force reduces by a quarter.

For objects consisting of multiple charges, each individual charge acts upon one another, and the individual forces are summed to get the total force. If we have two such bodies and wish to determine the force between them, so long as the separation distance is large with respect to the size of the bodies, we can treat them as single-point charges and use Equation 4.1.

Equation 4.1 determines the force experienced between two charged particles in an electric field but does not describe the field shown in Figure 4.2. The field exists, but no forces are experienced until another charged object enters this field. How do we then describe this field? We have as yet not defined what the electric field is.

Say we have a charge particle Q, and we place another particle q_{test} (a test particle) in this field. Say the charge on this theoretical test particle is so small that the acceleration of Q due to the force between Q and q_{test} is negligible. That is q_{test} does not disturb the system in any measurable fashion. The force between the particles will be

$$\mathbf{F} = \frac{Q q_{test} \hat{\mathbf{r}}}{4\pi\varepsilon r^2}$$

If we divide the force by the charge of q_{test}, we get a force per unit charge that a particle will feel at some distance r from Q.

$$\frac{\mathbf{F}}{q_{test}} = \frac{Q\hat{\mathbf{r}}}{4\pi\varepsilon r^2}$$

The electric force per unit charge is called the *electric field strength*, the *electric field density* or simply the *electric field*. Therefore the definition of the ***electric field*** is the electric force per unit charge.

$$\boxed{\mathbf{E} = \frac{Q\hat{\mathbf{r}}}{4\pi\varepsilon r^2}} \qquad \text{Electric field of a point charge (Volts/Metre)} \qquad 4.3$$

and the SI units are volts per metre V/m (alternatively, newtons per coulomb). For a system of more than one charge, the electric field at a given location is the vector addition of the components of the individual electric fields for each charge at that location. That is, for a system of N charges

$$\mathbf{E} = \frac{1}{4\pi\varepsilon}\sum_{i=1}^{N}\frac{Q_i\hat{\mathbf{r}}_i}{r_i^2}$$

If we know the electric field strength for a given distance, we can calculate the force felt by some charged object Q_0 placed in the electric field at that distance.

$$\boxed{\mathbf{F} = Q_0\mathbf{E}}$$ Electric force felt by a charged particle 4.4

Electric Flux

The electric field represents the magnitude of the electric force per unit charge. Diagrammatically the electric field strength is represented by the number of lines per area. These lines are referred to as lines of flux; the higher the field strength, the greater the number of lines of flux per unit area. Figure 4.4a represents a lower electric field strength compared to the field strength in Figure 4.4b, as the number of lines entering the same area is greater in Figure 4.4b. In this way, the electric field (**E**) represents a density and is sometimes called the ***electric field density***. There seem to be a lot of names describing the same phenomenon.

E = electric field = electric field strength =electric field density

If we multiply the field density by the surface area in which the flux is passing, we get the total magnitude of the **electric flux** (Φ_E) passing through that surface. If the electric field is constant, then the electric flux is given by the dot product of the electric field density and the area.

$$\boxed{\Phi_E = \mathbf{E}\bullet\mathbf{A} = (E\cos\theta)A}$$ Constant electric flux (Volt-metre) 4.5

The units of electric flux are Volt-metres (Vm). The electric flux is a scalar value, where A is the surface area in square metres and $\mathbf{A} = A\hat{\mathbf{n}}$. For non-uniform electric fields, the electric flux is simply the sum of the electric field densities passing through some infinitesimally small area and is given by

$$\boxed{\Phi_E = \iint_S \mathbf{E}\bullet d\mathbf{S}}$$ Varying electric flux 4.6

Figure 4.4
Electric flux.

Just to add to the confusion of terms, Electric flux can also be defined using ***electric flux density***.

The electric flux density is almost the same as the electric field density. Given a charge Q enclosed by a sphere, the total electric lines of flux in terms of coulombs is just the charge Q. The flux density is then Q divided by the surface area of the sphere. We will denote the flux density by the letter D. So

$$\mathbf{D} = \frac{Q\hat{\mathbf{r}}}{4\pi r^2}$$

The units are in coulombs per square metre (C/m²). Notice that the flux density is almost the same as the equation for the electric field. If differs only by 1/ε. Multiplying both sides by 1/ε, we get

$$\frac{\mathbf{D}}{\varepsilon} = \frac{Q\hat{\mathbf{r}}}{\varepsilon 4\pi r^2}$$
$$= \mathbf{E}$$

Therefore the electric flux density D is the permittivity times the electric field strength.

$$\boxed{\mathbf{D} = \varepsilon\mathbf{E}}$$ Electric flux density (Coulombs /m²) 4.7

The electric flux defined using D is the dot product of D with the area. We will use the Greek letter ψ (psi) to refer to this definition of electric flux.

$$\psi = \mathbf{D} \bullet \mathbf{A} = \varepsilon\mathbf{E} \bullet \mathbf{A}$$ Electric flux defined using flux density (Coulombs) 4.8

The flux ψ is expressed in terms of charge, whereas Φ_E is expressed in terms of volts. In this book, we will use the electric field **E** and the electric flux Φ_E. These correspond closely to the magnetic field and magnetic field flux.

4.2 ELECTRIC POTENTIAL (VOLTAGE)

Consider two metal plates separated by a distance d, the top plate having a net positive charge and the bottom plate having a net negative charge, as shown in Figure 4.5. There will be an electric field **E** between the plates. If we place a positive charge, q, near the top plate. The positive charge q will feel the electric force and move toward the negative plate. The charge q will accelerate,

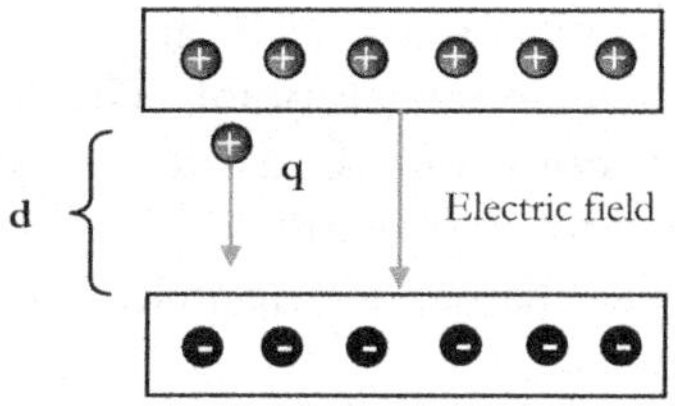

Figure 4.5
The movement of a positive charge within an electric field.

gaining velocity and therefore increasing kinetic energy. This is directly analogous to the ball falling from the table, as described in our discussion on potential energy. Work is done on the charge by the electric field in moving q from the positive side to the negative side. If no heat is produced, then the work done is equal to energy, and we can write

$$W = U = \int_C \mathbf{F} \bullet d\mathbf{S}$$

where W is work, and U is energy. From Equation 4.4, we know the force felt by the charge q in the electric field is q**E**. Also, when the charge was at its initial position and had not started to accelerate, the charge had potential energy E_P. We can write the potential energy as

$$U_P = \int_C q\mathbf{E} \bullet d\mathbf{S}$$

If we divide the potential energy by the magnitude of the charge, we get the potential energy per unit charge.

$$\frac{U_P}{q} = \frac{q\int_C \mathbf{E} \bullet d\mathbf{S}}{q}$$

$$= \int_C \mathbf{E} \bullet d\mathbf{S}$$

Electric potential energy per unit charge is called ***voltage*** *or* ***electric potential***, and the units are in volts.

$$\boxed{V = \int_C \mathbf{E} \bullet d\mathbf{S}} \quad \text{Voltage or electric potential difference} \quad 4.9$$

Voltage is a scalar value. If the electric field is constant, then

$$\boxed{V = \mathbf{E} \bullet \mathbf{d}} \quad \text{Voltage for constant electric field} \quad 4.10$$

where **d** is the displacement vector. If the electric field is constant and the direction of the displacement is along the direction of the electric field, then the voltage is the magnitude of the electric field times the distance.

$$V = |\mathbf{E}||\mathbf{d}| \quad 4.11$$

Note that potential energy is always measured with reference to some location within the field. By convention, the positive side is considered to have a higher potential than the negative side. In the case of gravitational potential energy, we chose the reference location as the ground; that is ground was considered to be at zero potential. The same is done for electric potential and voltage. For instance, if we measure the voltage of a battery, the negative terminal is considered to be at zero potential (zero volts), and the positive terminal is at a higher potential (above zero). In electronics, we are free to select our own reference point. The reference point which all circuit voltages are measured from is called ***common***. Usually, common is connected to an earth (ground), and common is simply referred to as the earth.

Substituting the equation for a single charge electric field 4.3 and using the variable r as the distance from the charge, we can rewrite Equation 4.11 as

$$V = \frac{Q\hat{\mathbf{r}}}{4\pi\varepsilon r^2} r\hat{\mathbf{r}}$$

Simplifying gives

$$\boxed{V = Q/4\pi\varepsilon r} \quad \text{Voltage at a distance from a single charge} \quad 4.12$$

This gives the electric potential (voltage) at a distance r from the charge. For a system of more than one charge, the voltage is the resultant of the sum of the potentials for each charge, and

$$V = \frac{1}{4\pi\varepsilon}\sum_{i=1}^{N}\frac{Q_i}{r_i}$$

If we place a charge of Q_0 at this location, then the electrical potential energy the charge has is given by

$$U_P = VQ_0$$ Potential energy of a charge

Consider a battery where the charges are separated due to chemical reactions. This separation of charges acts as a voltage source. A battery can perform work on a charged particle (create current). The source of the force used to perform work on a charged particle is called the ***electromotive force*** (emf), and the units are in volts. Emf refers to the *work done on an electric charge by a non-electrical device.* A battery uses chemical energy to separate charges to produce an emf; a generator produces emf by mechanical work. Emf only refers to voltage sources that are created by other forms of non-electrical energy.

4.3 CONDUCTORS IN ELECTROSTATIC EQUILIBRIUM

Consider a conductor situated inside an external static electric field (E_{ext}), as shown in Figure 4.6. The electric field runs across the gap and through the conductor. In a conductor, the electrons are free to move and distribute themselves toward the positive side around the ends and surface. The protons bound up in the nucleus are not free to move in the crystalline lattice, so there will be a separation of charge. This charge separation will create an electric field inside the conductor (E_{INT}) independent and opposing the external field. As more charges separate, the opposing electric field increases in magnitude until the internally generated electric field is equal in strength and opposite to the external field. At this stage, the movement of charges stops, and the net electric field inside the conductor is zero. This is called ***electrostatic equilibrium*** and is mathematically described as

$$E_{net} = E_{ext} - E_{\text{int}} = 0$$

This process occurs extremely quickly in a small fraction of a second.

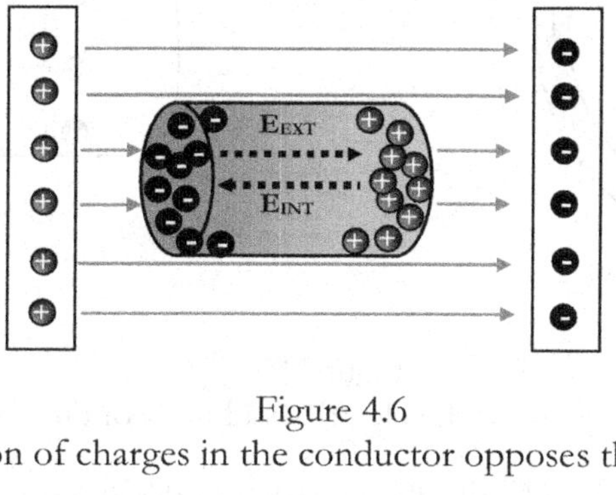

Figure 4.6
The separation of charges in the conductor opposes the external field.

There seems to be a commonly held belief that electric fields do not penetrate a conductor and that they only exist on the surface. This statement is entirely false! Electric fields penetrate all the way to the core of a conductor, as described above. For conductors in electrostatic equilibrium, you can say that "there is "no NET electric field inside the conductor". Notice the use of the word

"NET". This is not just being pedantic; there is a big distinction between "no electric field" and "no net electric field". The first statement leaves you with the impression that there is some mysterious phenomenon at work that repels the electric field and that it is forbidden for electric fields to enter the bulk of a conductor. Having such a belief would preclude you from explaining such simple phenomenon as current, where conductors are not in a state of electrostatic equilibrium. However, the idea that current only flows on the surface of a conductor is another commonly held belief that is also completely false. It is an easy matter to show that both of these statements are false. Simply ask the person making such statements to explain the skin effect (we shall discuss this very important effect later). If current only flows on the surface of a conductor, there is no need for the (very real) concept of the skin effect and the change in the density distribution of current. And if they then accept the existence of the skin effect, then what is the cause of the current in the bulk of the conductor? Well, the electric field!

4.4 CONDUCTORS CONNECTED TO A VOLTAGE SOURCE

When a conductor (a length of wire) is connected across a voltage source such as a battery, the electrons touching the positive terminal feel the positive electric field and migrate to the positive terminal. The electrons near the negative terminal are repelled. However, the electric field emanating from the battery spreads out spherically from the terminals and does not follow along the path of the wire. The initial movement of electrons causes an uneven charge distribution in the wire near the terminals setting up another electric field inside the wire that is felt by other electrons that migrate along the wire. Eventually, in a very small period of time, a charge gradient is created on the surface of the wire, as shown in Figure 4.7a. The gradient is highly positive near the positive side of the voltage source, decreasing to neutral and then becoming more negative, with the greatest negative charge near the negative side of the battery.

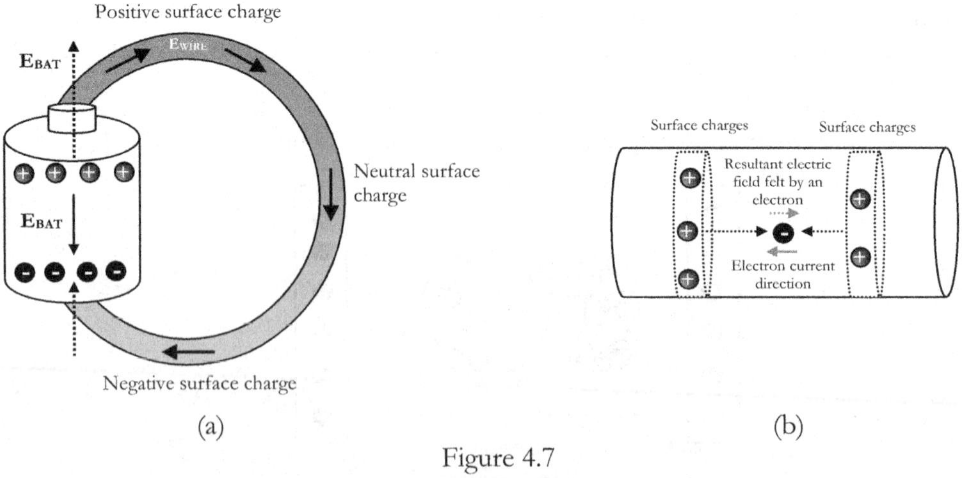

Figure 4.7
The cause of the electric field in a conductor.

The inside of the conductor is electrically neutral, and the electric field inside the conductor is due to the surface charge distribution gradient. This surface charge gradient is felt by the electrons in the wire, which causes them to flow through the wire to produce a current, as shown in Figure 4.7b. Therefore the electric field inside the conductor is not the electric field from the voltage source, but it is a field created due to the surface distribution of charges along the conductor. In this way, the electric field inside the wire follows the contours of the wire, and the wire can be twisted in any fashion, and current will still flow. For a DC voltage source, the charge distribution creates a constant electric field inside the conductor, as would be expected. For AC voltage sources, the charge distribution varies with time.

Say we connect a half metre of wire across a 1.5-volt battery. What is the electric field strength inside the wire? The voltage of the battery is dropped across the wire so

$$V_{bat} = V_{wire}$$
$$1.5 = E_{wire} L$$

where L is the length of the wire. Rewriting in terms of the electric field in the wire, E_{wire} becomes

$$\begin{aligned} E_{wire} &= \frac{1.5}{L} \\ &= \frac{1.5}{0.5} \\ &= 3.0 \end{aligned}$$

So the magnitude of the electric field strength in the wire is 3 volts per metre.

4.5 DRIFT VELOCITY AND CURRENT

In the atomic model, electrons orbit the nucleus at different distances with different energy levels. These energy levels determine how tightly they are bound to the atom. In metals, there is an electron energy level called the conduction band. Electrons occupying this band are free to move around the metal's crystalline structure while the atoms are held firmly in place. In a wire, at room temperature, these free electrons are zipping around in random directions with an average speed of about 10^6 metres per second; that's fast.

When a voltage source is connected across a wire, an electric field is created inside the wire, causing the electrons to move toward the positive side of the voltage source. It is not long, however, before an electron collides with another electron or an atom inside the wire. The average time between collisions is in the order of 2.5×10^{-14} seconds; a very small time period indeed. The collision causes the electrons to slow down or to be deflected in different directions. They then start to accelerate once again toward the positive voltage side. This has the effect of slowing the progression of the electrons through the wire. This gradual progression of the electrons through a wire is called ***drift velocity***.

To determine this drift velocity start with the force felt by an electron due to the electric field,

$$\mathbf{F} = q_e \mathbf{E}$$

where q_e is the charge of one electron 1.602177×10^{-19} coulombs. The average acceleration between collisions is given by Newton's second law

$$\mathbf{a} = \frac{\mathbf{F}}{m_e}$$

where m_e is the mass of a single electron, $9.1093836 \times 10^{-31}$ kg. The average velocity obtained before a collision is the drift velocity S_D.

And S_D is

$$\mathbf{S_D} = \mathbf{a}\tau$$
$$= \frac{\mathbf{F}}{m_e}\tau$$

where τ is the average time between collations. Substituting Equation 4.4 for the force, the average drift velocity becomes

$$\mathbf{S_D} = \frac{q_e\mathbf{E}}{m_e}\tau$$ The average drift velocity of an electron in a wire 4.13

If a voltage source of nine volts is applied across a wire, half a metre in length, the electric field strength is

$$E = \frac{V}{d}$$
$$= \frac{9}{0.5}$$
$$= 18 volts/metre$$

Substituting this value for E into 4.13, the drift velocity becomes

$$S_D = \frac{q_e E}{m_e}\tau$$
$$= \frac{(1.602177\text{x}10^{-19})(18)}{9.1093836\text{x}10^{-31}}(2.5\text{x}10^{-14})$$
$$= 0.07915$$

That is 79.15mm per second; a snail's pace. So how is it possible that a signal of say 1 MHz can be transmitted through a long length of wire? The signal passing along the wire is the electric field or voltage, and a changing voltage can propagate, depending on the cable configuration, from close to the speed of light down to about 2/3rd the speed of light. We will discuss signal propagation speeds in a later Chapter. So as long as there are enough free electrons in the volume of wire given by the drift velocity and the diameter of the wire to pass any given cross-section of a wire per second, then the current will follow the signal frequency.

Current is defined as *the number of coulombs of charge passing a given cross-section along a wire every second.*

$$\boxed{I = \frac{Q}{t}}$$ Current (amps) 4.14

The units of current are amperes or amps. There is approximately one free electron for every atom in a copper wire. Given there are about 8.4 x 10^{28} atoms per cubic metre of copper, we can get an estimate for the number of free electrons in a volume of wire. Say we have a wire diameter of 1 mm, so the number of free electrons of interest in our example will be given by the cross-sectional area of the wire times the length covered by the electron's drift velocity in one second. This will give a volume in cubic metres. Multiply this volume by the number of atoms per cubic metre, and we shall have an estimate of the number of free electrons (n_e) in this volume of copper wire. We can write this as

$$n_e = nAS_Dt \qquad \text{Number of free electrons in a wire} \qquad 4.15$$

Using the calculated values

$$\begin{aligned} n_e &= (8.4x10^{28})\pi r^2 S_D t \\ &= (8.4x10^{28})(\pi(0.0005)^2(0.07915)) \\ &= (8.4x10^{28})(6.2164x10^{-8}) \\ &= 5.2x10^{21} \end{aligned}$$

where n_e is the number of electrons available per second, and for this example, is $5.2x10^{21}$. That's a total charge, in coulombs, of

$$\begin{aligned} Q &= n_e q_e \\ &= (5.2x10^{21})(1.602177\text{x}10^{-19}) \\ &= 833 \end{aligned}$$

Using the definition of current

$$\begin{aligned} I &= Q/t \\ &= n_e q_e/t \\ &= 833/1 \\ &= 833 \end{aligned}$$

Therefore, a wire with a 1 mm diameter has enough free electrons to draw a current of 833 amps. Of course, the wire would melt well before a current of this magnitude was drawn.

4.6 CONDUCTIVITY AND RESISTIVITY

Using Equations 4.14 and 4.15, the current can be written as

$$\begin{aligned} I &= Q/t \\ &= n_e q_e/t \\ &= (nAS_Dt)q_e/t \\ &= nAq_e S_D \end{aligned}$$

Substituting Equation 4.13 for the drift velocity S_D

$$\begin{aligned} I &= nAq_e S_D \\ &= nAq_e\left[q_e E/m_e\right]\tau \\ &= [q_e^2 n\tau/m_e]AE \end{aligned} \qquad 4.16$$

The electron charge q_e and the electron mass m_e are constant. The number of free electrons per cubic metre (n) and the average time between collisions (τ) is a property of the material, and at a given temperature, will also remain constant. Therefore the term inside the square brackets is intrinsic to the material regardless of the cross-sectional area, the length, or the voltage applied. It is directly proportional to the current and therefore indicates how well a material will conduct current and is called ***conductivity***. The symbol σ (sigma) is used to indicate conductivity and is given by

$$\sigma = \frac{q_e^2 n\tau}{m_e}$$ Conductivity (siemens per metre S/m) 4.17

If we take the inverse of conductivity, we get an indicator of how a material resists the flow of current or ***resistivity***. The symbol ρ (rho) is used to indicate resistivity.

$$\rho = \frac{m_e}{q_e^2 n\tau}$$ Resistivity (ohms-metre) 4.18

If we multiply resistivity by the length of a wire divided by the cross-sectional surface area, we get the total resistivity or ***resistance*** of a wire.

$$\boxed{R = \rho\frac{\ell}{A}}$$ Resistance (ohms) 4.19

where R is the resistance of a wire, ℓ is the length of the wire, A is the cross-sectional surface area of the wire, and ρ is the resistivity. The units for resistivity are in ohm-metres, and resistance is in ohms. Tables can be found giving the resistivity of many materials. The resistivity for copper is 0.0168 micro ohms-metre.

Resistance is due to the collisions of electrons in the wire (as discussed above), and for a given material, the longer the wire, the greater the resistance. Also, wires with smaller diameters will have greater resistance than wires with larger diameters.

4.7 OHM'S LAW

Expressed using conductivity, Equation 4.16 can be written as

$$I = \sigma\frac{A}{\ell}V$$

Rearranging for voltage

$$V = I\frac{1}{\sigma}\frac{\ell}{A}$$

As resistivity is the inverse of conductivity, we can write

$$V = I[\rho\frac{\ell}{A}]$$

The term in square brackets is resistance, so now the equation can be written as

$$\boxed{V = IR}$$ Ohm's Law 4.20

This is the most important relationship for circuit analysis and is called ***Ohm's Law***. It will be used extensively throughout the remainder of the book, so memorize it! It states that the voltage drop across some material is equal to the current flowing through the material times the resistance across the material. It applies to entire circuits and equally well to components in a circuit.

This relationship can be expressed in terms of current or resistance. That is

$$I = \frac{V}{R} \quad \text{Ohm's Law for current} \quad 4.21$$

If we know the voltage drop across a component and its resistance, we can calculate the current through the component. And if we know the voltage drop and current, we can calculate the resistance of a component.

$$R = \frac{V}{I} \quad \text{Ohm's Law for resistance} \quad 4.22$$

4.8 CAPACITANCE

Consider two electrically neutral metal plates separated by some small distance compared to the dimensions of the plates. Now connect both ends of the plates to a voltage source, such as a battery, as in Figure 4.8.

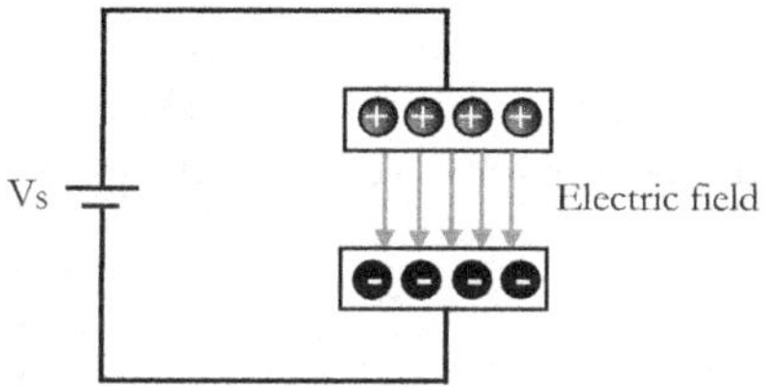

Figure 4.8
Separated metal plates connected to a voltage source.

Electrons will move out from the plate attached to the positive side of the battery. The reduction in the number of electrons on this plate will result in an overall positive charge on the plate. At the same time, electrons will move from the battery's negative side to the other plate making this plate negatively charged. The charge separation on the two plates will result in an electric field existing between the plates. The electrons will continue to move until the voltage across the plates equals the voltage source, the voltage on the battery. The capacity or ability of the plates to hold an electric charge is called ***capacitance*** and is proportional to the charge on the plates and the voltage potential between the plates and is given by

$$C = \frac{Q}{V} \quad \text{capacitance (farads)} \quad 4.23$$

where C is the capacitance in farads, Q is the charge in coulombs, and V is the voltage across the plates. Capacitance can also be thought of as the ability of a medium to store energy in an electric field.

4.9 PERMITTIVITY

In section 4.1, the concept of permittivity was introduced and was defined as *the ability of a medium to store energy in the electric field.* This is one of several definitions of permittivity. Trying to find a single consistent definition of permittivity was a fruitless task. Many books define permittivity simply as the constant of proportionality between the electric field **E** and the displacement field **D** (oh yes, another name for **D** that is also called the electric flux density). If we rearrange Equation 4.7, we get

$$\varepsilon = \frac{\mathbf{D}}{\mathbf{E}} \qquad \text{Permittivity (farads/metre)}$$

This is not particularly insightful. Sometimes permittivity is defined as *how easily a medium permits the lines of flux to form.* Others define permittivity as the *opposition to the formation of an electric field.* And some authors give different definitions for "permittivity" and the "permittivity of a vacuum" that are logically inconsistent, where permittivity is the result of the formation of dipoles (separated particles with opposite charges). Although true for dielectrics, the vacuum has no matter and does not form dipoles. There should only be one definition that applies to all, regardless of the medium. There are only two definitions I consider to be descriptive and accurate and only one that is useful. We will now discuss these two definitions.

Opposition to the Formation of an Electric Field

In Figure 4.9a, two groups of separate charges on metal plates form an electric field between them. There is no material between the charges, and they are separated by a vacuum. In Figure 4.9.b, a dielectric was introduced between the charges. A dielectric is an insulator that has the ability to

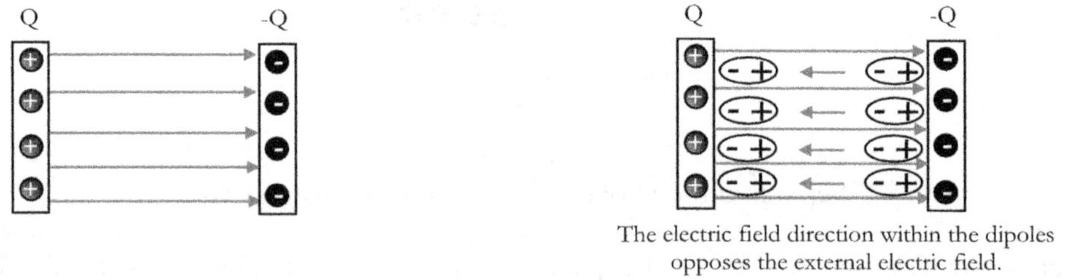

(a) (b)

Figure 4.9
The change in the electric field due to a dielectric.

form electric dipoles (oppositely charged regions). The negative parts of the dipoles align themselves with the positive charge Q, and the positive parts of the dipoles align themselves with the negative charge –Q. These dipoles set up their own electric field that is opposite to the existing electric field and reduces the overall net electric field. In this sense, the dielectric has opposed the existing electric field, and the net **E** has decreased between the plates. The capacitance is given by

$$C = \frac{Q}{V} = \frac{Q}{E \bullet d}$$

In this situation, as the charges have nowhere to go, Q remains constant, and E decreases. From this equation, if the electric field decreases, then the capacitance increases for a given charge.

Now if the capacitor in 4.9a were connected across a 9-volt battery, the voltage across the capacitor would be 9 volts. If the dielectric is now slipped between the plates, the voltage across the capacitor will still remain the same (9 volts) and for the voltage to remain the same, the electric field strength must remain the same. However, the dielectric has now reduced the net electric field between the plates. In order to bring the electric field up again so that 9 volts exist between the plates, more charge is required to be placed on the plates. So Q increases, and V remains the same; therefore, C (the capacitance) increases.

The capacitance between two plates is given by

$$\boxed{C = \varepsilon \frac{A}{d}} \qquad \text{Capacitance between plates (farads)} \qquad 4.24$$

where A is the surface area of the plates and d is the distance between the places. For a given geometry of a capacitor, for the capacitance to increase, the permittivity must increase. Therefore the introduction of the dielectric in the capacitor decreased the net electric field but has increased the permittivity, as would be expected from Equation 4.3. Therefore the description of permittivity as the opposition to the formation of the electric field is valid and good.

Stored Energy

Once again, consider the capacitor as shown in Figure 4.9. If a wire were connected across the terminals of the capacitor, the charges would move off the plates. The electrons would migrate across the wire to the positive side, reducing the number of charges on both plates. Eventually, all the charges would be removed, and the voltage across the capacitor would be zero. The energy to perform this work is stored in the electric field (**E**) between the plates. The work done to move a tiny bit of charge dQ is

$$dW = (E \bullet dQ)\ell$$

where ℓ is the length of the wire. Now $E\ell$ is voltage from Equation 4.10, so we can write

$$\begin{aligned} dW &= E\ell \bullet dQ \\ &= V \bullet dQ \end{aligned}$$

As the charge decreases on the plates, the voltage across the capacitor will also decrease. Therefore the voltage across the capacitor is a function of charge. From Equation 4.23, we can express the voltage in terms of charge, V=Q/C. Note that the capacitance will remain constant as the reduction in charge will be followed by an equal reduction in voltage.

Integrating from no charge on the plates (0) to all the charge on the plates (Q), we get the total work done in moving these charges.

$$\begin{aligned} W &= \int_0^Q V \bullet dQ \\ &= \int_0^Q \frac{Q}{C} \bullet dQ \\ &= \frac{Q^2}{2C} \end{aligned}$$

Equating work done to energy and expressing Equation 4.23 in terms of charge Q=VC, we can rewrite this as

$$\boxed{E = \frac{1}{2}V^2C} \qquad \text{Energy stored in a capacitor (joules)} \qquad 4.25$$

As capacitance is directly proportional to permittivity, we can also say that permittivity represents the ability to store energy in the electric field. The notion of stored energy in the electric field of a capacitor is essential when performing circuit analysis, particularly when we develop the model of a crystal radio. This is the definition that is more applicable to many electronic problems and is the definition I consider to be the most useful. In particular, when the permittivity of the vacuum is considered, the vacuum must store energy in the electric field. Otherwise, there would be no propagation of the electromagnetic wave (photons) and the universe as we know it would not exist. Electromagnetic waves will be discussed in a later Chapter, and you will see the importance of the permittivity of the vacuum.

Relative Permittivity

The permittivity of a medium is generally expressed as a ratio with respect to the permittivity of a vacuum. The permittivity of a vacuum is constant and is denoted by ε_0.

$$\varepsilon_R = \frac{\varepsilon}{\varepsilon_0} \qquad \text{Relative permittivity}$$

This ratio is called ***relative permittivity***. The relative permittivity for many dielectrics is known, and to find the permittivity (sometimes called the absolute permittivity), just multiply the relative permittivity by the constant ε_0.

$$\boxed{\varepsilon = \varepsilon_R \varepsilon_0} \qquad \text{Permittivity (Farads per metre)} \qquad 4.26$$

The permittivity of the vacuum is

$$\varepsilon_0 = 8.8541878128x10^{-12} \qquad \text{Permittivity of a vacuum (Farads per metre)}$$

4.10 GAUSS' LAW FOR THE ELECTRIC FIELD

Consider a charged particle Q inside an enclosing sphere, as shown in Figure 4.10. The charged particle has an associated electric field emanating spherically out from the charge.
If we examine some small surface area dA on the surface of the sphere, the electric flux for this small section of area is the dot product of the electric field with the surface area, E•dA. Adding up all these tiny flux values gives us the total flux through the surface of the sphere.

$$\oiint_S \mathbf{E} \bullet d\mathbf{A}$$

Gauss's law states that *the total electric flux through the surface of an object is equal to the charge inside the object divided by the permittivity.*

$$\oiint_S \mathbf{E} \bullet d\mathbf{A} = \frac{Q_{inside}}{\varepsilon} \qquad 4.27$$

This means we can treat any charged object as if it contains all its charge in the centre. Also, the enclosing surface does not have to be a sphere but can be any shape. Any enclosing surface through which a vector flux is calculated is called a ***Gaussian surface***. This is analogous to how gravity is treated. When calculating the effects of gravity between two planets, the entire mass of the planet is treated as though the source of gravity is concentrated at its centre, and the distance between planets is calculated as the distance from the centre of one planet to the centre of the other planet. The differential form of Gauss' law is

$$\nabla \bullet \mathbf{E} = \frac{\rho}{\varepsilon} \qquad 4.28$$

where $\nabla \bullet \mathbf{E}$ is the divergence of the electric field, ρ (rho) is the charge density (Q/volume). The divergence is a measure of the net flow of the electric field going in or coming out of the charged surface.

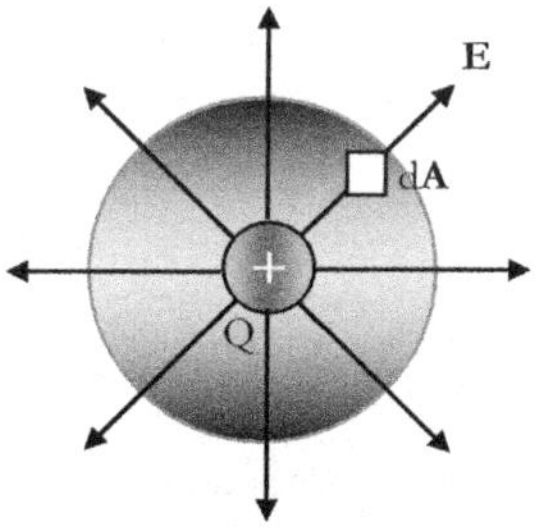

Figure 4.10
Gauss' law for the electric field.

5

The Magnetic Field

5.1 MAGNETS

The electric field associated with charged particles is inherent to the particle and always exists. There is another field associated with charged particles that arise due to a particle's motion. This field is called the ***magnetic field*** and is orthogonal (at right angles) to the direction of motion of the charged particle. It does not matter if the particle has a constant velocity or is accelerating; if the particle is in motion, the magnetic field exits.

As with the electric, field lines of flux describe the magnetic field. Magnets have a north and south pole, and the direction of the lines of flux point outward from the north pole and inward to the south pole, see Figure 5.1.

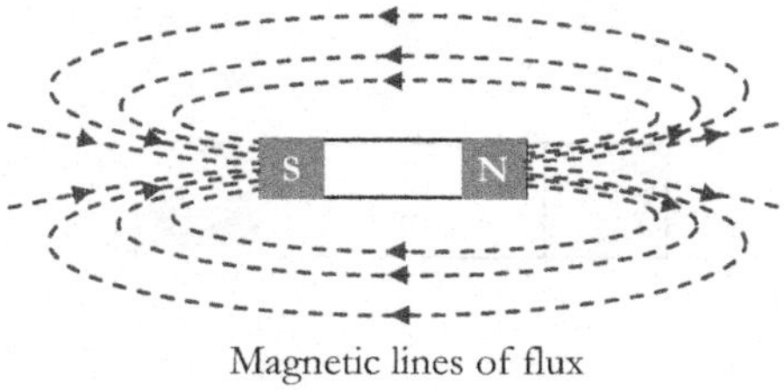

Figure 5.1
The magnetic field of a bar magnet.

In the classical model of an atom, the atoms can jiggle back and forth and up and down due to thermal energy. The electrons also move in orbits around the atom. Inside the atom, the fundamental particles such as electrons, protons and neutrons have another type of motion called spin. In a classical interpretation, this is analogous to a ball spinning around on its axis, Figure 5.2a. Particles with spin produce a magnetic field called a spin magnetic moment. The electron has both a spin magnetic moment and a magnetic field due to its orbit. The spin magnetic moment for the electron is much higher than for the other particles as the magnetic moment is inversely proportional to mass, and the electron's mass is very much smaller than protons and neutrons. Electron spin is, therefore, the major contributor to the net magnetic moment of an atom.

In ferromagnetic materials, such as iron, there are regions where the spin direction of the electrons is the same, creating a strong magnetic field in a particular direction. These regions are called domains and typically contain around 10^{12} atoms. If the directions of the domains are random, the magnetic fields tend to cancel one another, and the object's net magnetic field is very small or non-existent, Figure 5.2b.

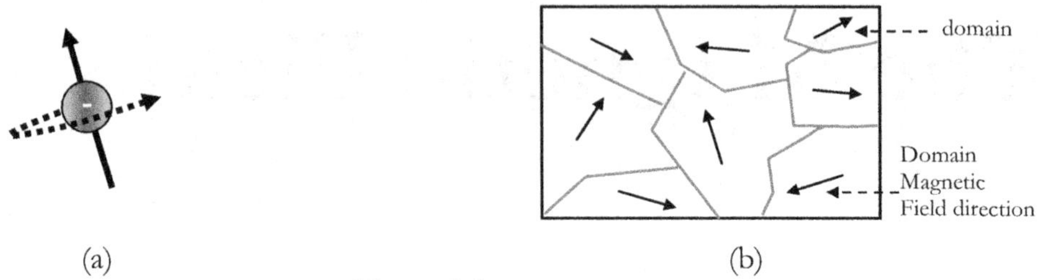

Figure 5.2
The spin of electrons causes the creation of magnetic domains.

If a ferromagnetic object is brought close to a permanent magnet, the external magnetic field will cause the domains to line up, causing the object to be magnetised, Figure 5.3. The magnetic fields of the domains will add up, creating a strong magnetic field and causing the object to be attracted to the permanent magnet. The north pole of the object will be attracted to the south pole of the magnet and visa versa. If the domains remained aligned in the object after being removed from the external magnetic field, the object remains magnetised and is now a permanent magnet itself.

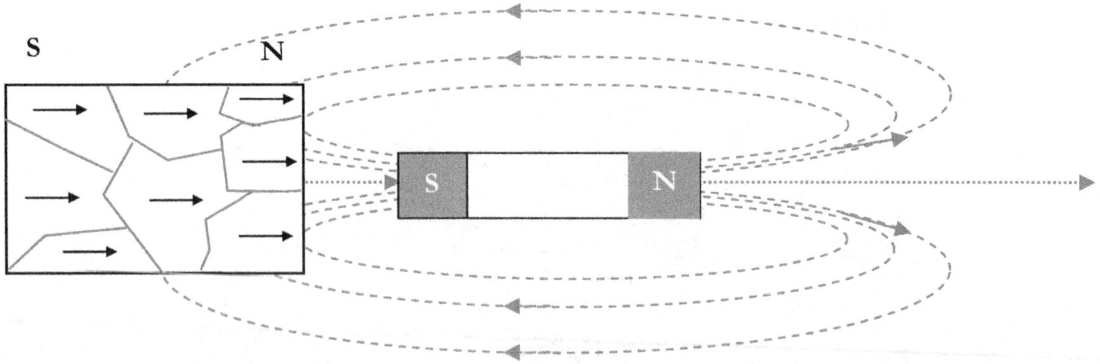

Figure 5.3
The directions of the magnetic domains are aligned, and the ferromagnetic object is magnetised.

5.2 THE MAGNETIC FORCE

Charged particles in an electric field feel a force exerted on them by the electric field. Similarly, a charged particle feels a force exerted by the magnetic field. The strength of the force is proportional to the strength of the magnetic field, the magnitude of the charge and the relative velocity of the charged particle with respect to the magnetic field. Note that it is the relative motion that counts. The magnetic field could be static while the charged particle moves through it, or the particle could be static while the magnetic field moves. The magnetic force felt is orthogonal to the relative velocity of the charged particle and the magnetic field and so is given by the cross product of the relative velocity **v** and the magnetic field strength **B**, times the value of the charge

$$\boxed{\mathbf{F}_M = Q(\mathbf{V}\times\mathbf{B}) = Q|\mathbf{v}||\mathbf{B}|(\sin\theta)\hat{\mathbf{n}}}$$ Magnetic force felt by a charged particle 5.1

where θ is the angle between the velocity vector **v** and the magnetic field **B,** and $\hat{\mathbf{n}}$ is the unit normal vector pointing in the direction of the force felt by the particle. To determine the direction of the magnetic field force, use the right-hand rule described below.

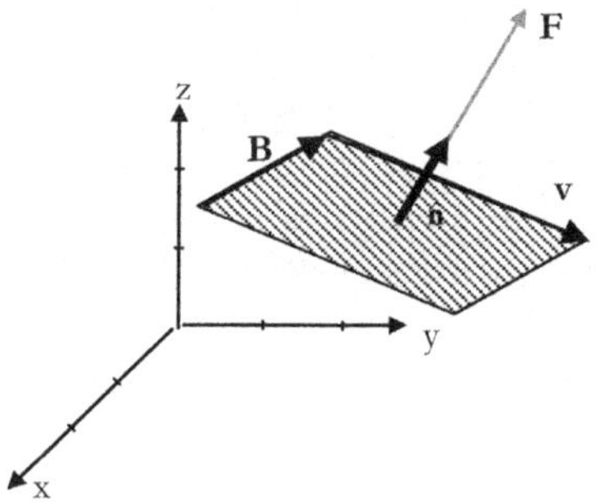

Figure 5.4
The force felt by a moving charged particle in a magnetic field.

The total force felt by a charged particle is given by the sum of the forces due to the electric and the external magnetic field.

$$\mathbf{F} = \mathbf{F_E} + \mathbf{F_M}$$

The total force **F** is known as the Lorenz Force and is given by

$$\boxed{\mathbf{F} = Q\mathbf{E} + Q(\mathbf{v}\times\mathbf{B})}$$ The total force felt by a charged particle (Newtons) 5.2

The Right-hand Rule – Force on a Charge

The right-hand rule determines the direction of the force felt by a positively charged particle moving in a magnetic field. For a negatively charged particle, you can use your left hand in the same fashion or use the right-hand rule treating the negatively charged particle as if it had a positive charge and then flip your hand around at the end of the procedure.

To use the right-hand rule, point your pointing finger in the direction of the velocity vector (relative velocity) of the charged particle. Now extend your middle finger in the direction of the magnetic field (pointing away from the north pole as given by the direction of the lines of magnetic flux) and extend your thumb. Your extended thumb should now point in the direction of the magnetic force felt by the particle. See Figure 5.5.

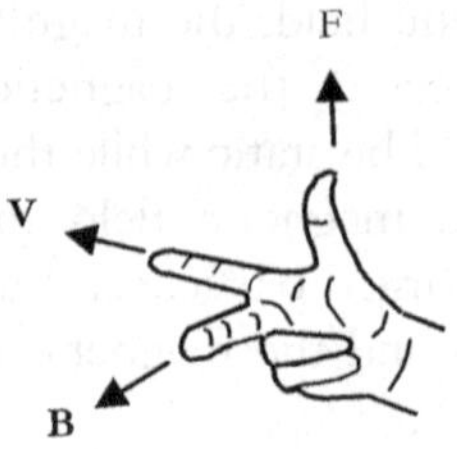

Figure 5.5
The right-hand rule for the magnetic force felt by a charge.

5.3 THE MAGNETIC FIELD STRENGTH

Section 4.1 introduced the concept of electric flux Φ_E and the electric field strength E. Similar concepts are used when dealing with magnetic fields, the magnetic flux Φ_M and the magnetic field strength (or magnetic flux density) **B**. The strength of the magnetic field is proportional to the number of lines of flux passing through a given area. In Figure 5.6a, fewer lines of flux are passing through the area, so the magnetic flux density is smaller than that shown in Figure 5.6b, where more lines of flux are passing through the same area. Rearranging Equation 5.1 to determine the magnetic field strength, we get

$$\boxed{\mathbf{B} = \frac{|\mathbf{F}_M|}{Q|\mathbf{v}|(\sin\theta)}\hat{\mathbf{b}}}$$ Magnetic field strength (tesla)

where $\hat{\mathbf{b}}$ is the magnetic field unit vector giving the direction of the magnetic field. Performing dimensional analysis, we get

$$\frac{N}{C}\frac{1}{\frac{m}{s}} = \frac{N}{m}\frac{1}{\frac{C}{s}}$$

$$= \frac{N}{m}\frac{1}{A}$$

And the units for magnetic field strength are newtons per metre per amp, called a **tesla,** and the symbol for a tesla is **T**.

Just like the electric field, there are several confusing names for the magnetic field **B.** It can be called the ***magnetic field***, the ***magnetic field strength*** or the ***magnetic flux density***. The ***magnetic flux*** (Φ_M) is the dot product of the magnetic field strength by the surface area. That is

$$\boxed{\Phi_M = \mathbf{B} \bullet \mathbf{A}}$$ Magnetic flux (weba) 5.3

For non-uniform magnetic fields, the magnetic flux is given by

$$\boxed{\Phi_M = \iint_S \mathbf{B} \bullet d\mathbf{S}}$$ Magnetic flux (weba) 5.4

The unit of magnetic flux is the ***weba***, and the symbol is **Wb**. Note that this is a dot product of the magnetic field with the surface area, so the angle formed between the magnetic field and the surface matters. The component of the magnetic field that is perpendicular to the plane of the surface only contributes to the flux.

Figure 5.6
Magnetic flux.

5.4 THE MAGNETIC FIELD DUE TO CURRENT

A current through a wire is the movement of charged particles (electrons), and moving charged particles generate a magnetic field. Therefore a current-carrying wire will have an electric field surrounding the wire. For a straight length of wire, this magnetic field flows concentrically around the wire, as shown in Figure 5.7. Generally, when analysing circuits, the current is treated as the flow of positive charges (even though it is the electrons that are moving). This is a through-back that is still with us, and it is the result of a wrong assumption made very early on in the study of electricity. This type of current is referred to as ***conventional current***.

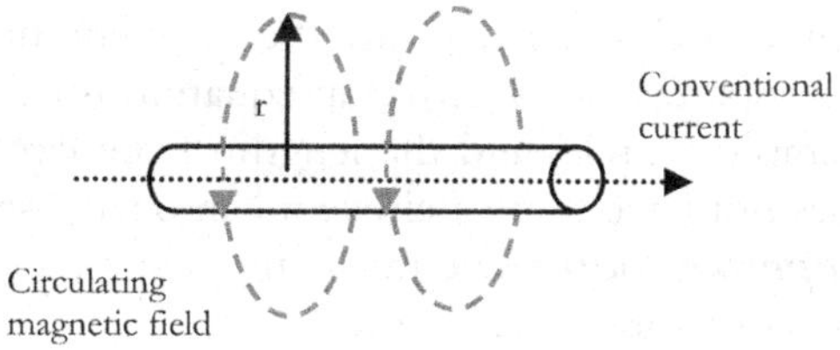

Figure 5.7
The magnetic field around a straight wire.

The Right-hand Rule for a Straight Wire

To determine the direction of the magnetic field circulating around a straight current-carrying wire, point your thumb in the direction of the conventional current flow and then curl your fingers around the wire. Your fingertips point in the direction of the magnetic field flow.

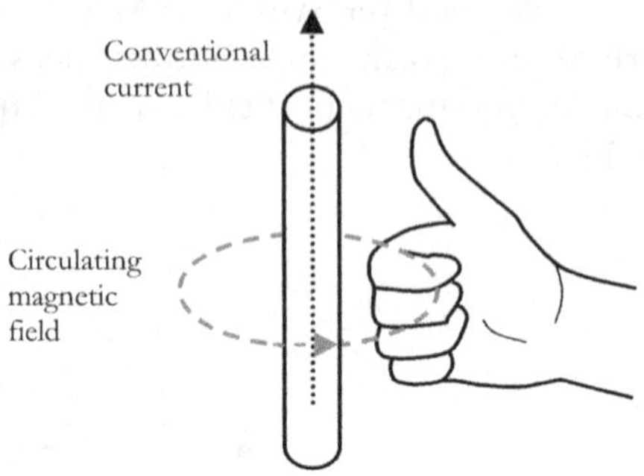

Figure 5.8
The right-hand rule for a straight wire.

Ampere's Law

Ampere's law describes the relationship between the magnetic field strength and the current flowing through a wire as follows.

$$\oint_C \mathbf{B} \bullet d\mathbf{l} = \mu I_{enclosed}$$ Ampere's law 5.5

It states the sum of magnetic field strength along a closed curve is equal to the current in the wire passing through this enclosed curve times the permeability of the medium. Permeability is a similar concept to permittivity; it is the ability of a material to store energy in a magnetic field.

Consider the current flowing through a wire, as shown in Figure 5.7. The magnetic field radiates outward in concentric circles. To calculate the magnetic field strength at some distance r from the wire, we can use Amperes law to derive an equation for the magnetic field strength. The curve around the wire is simply a circle, and the length of the circumference of the circle is 2πr. Note the enclosed path does not have to be a circle; it can be any shape required. These imaginary closed paths are called ***Amperian loops***. The magnetic field strength is the same value along the circumference of the circle. Therefore we can write

$$\oint_C \mathbf{B} \bullet d\mathbf{l} = \mu I_{enclosed}$$

$$|\mathbf{B}|(2\pi r) = \mu I_{enclosed}$$

The magnetic field strength around a straight wire at some distance r from the wire, is given by

$$|\mathbf{B}| = \frac{\mu I}{2\pi r}$$ Magnetic field strength at a distance r around a straight wire (tesla) 5.6

where μ is the permeability of the surrounding material (generally just free space), I is the current in amps, and r is the distance from the centre of the wire in metres.

Biot-Savart's Law

The Biot-Savart law is an empirically derived formula that relates the magnetic field strength contribution, at a distance r, due to a small length of current carrying wire, as shown in Figure 5.9. The relationship is given by,

$$dB = \frac{\mu I \sin(\theta)}{4\pi r^2} dl$$ The Biot-Savart law 5.7

If we add up all the contributions along the length of the wire for all angles from 0 to π, we can recover Equation 5.6 given by Ampere's law (see Appendix 1).

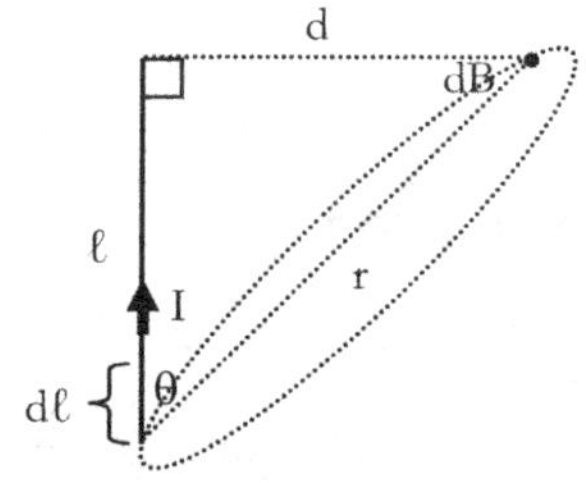

Figure 5.9
The magnetic flux density due to a small segment of a current-carrying wire as a function of distance r.

The Magnetic Field in a Coil

If a straight wire is bent into a loop, the magnetic field circulates around the wire, with the field converging at the centre, Figure 5.10a. If we add more loops of wire, the magnetic field tends to run axially through the loops and takes on the shape shown in Figure 5.10b

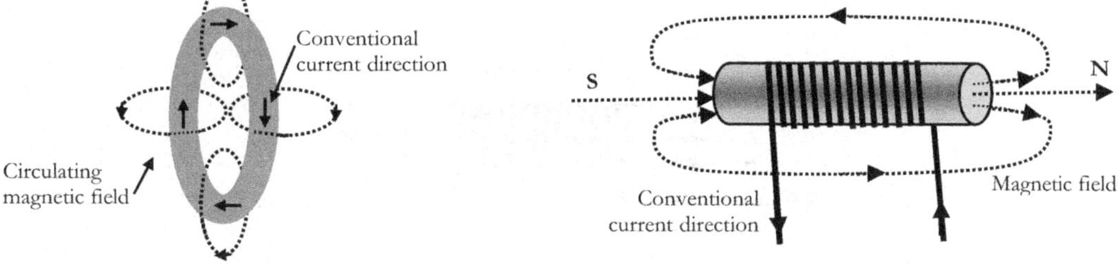

(a) (b)

Figure 5.10
The magnetic field around a coil.

The following is a commonly used derivation of the magnetic field inside a coil. It makes some assumptions that are not strictly true, namely that the magnetic lines of flux are all inside the coil pointing axially and that there is no magnetic field outside the coil. In reality, the result of this derivation is an approximation and the longer the coil is to its diameter, the better the approximation is.

Consider the rectangular region enclosing the top wires of a coil (an Amperian loop), shown in Figure 5.11. Using Ampere's equation, we can calculate the magnetic field inside the coil as follows. From segment c to d, the magnetic field is zero as it is outside the coil. The dot product of the magnetic field between a and d and between b and c is zero as they are at right angles to the magnetic field. That only leaves a magnetic field value for the segment a to b, and the length of ab is ℓ. We can write

$$\oint_C \mathbf{B} \bullet d\mathbf{l} = \mu I_{enclosed}$$

And

$$\oint_C \mathbf{B} \bullet d\mathbf{l} = \int_a^b \mathbf{B} \bullet d\mathbf{l} + \int_b^c \mathbf{B} \bullet d\mathbf{l} + \int_c^d \mathbf{B} \bullet d\mathbf{l} + \int_d^a \mathbf{B} \bullet d\mathbf{l}$$

$$= B\ell + 0 + 0 + 0$$

where ℓ is the length of the solenoid. $I_{enclosed}$ is the sum of the current that passes through the Amperian surface. The current through each loop is I, and there are N loops passing through the surface, so we can write

$$\mu I_{enclosed} = \mu NI$$

Now we can write

$$B\ell = \mu NI$$

And the magnetic field strength becomes

$$\boxed{|\mathbf{B}| = \frac{\mu NI}{\ell}}$$ Magnetic field strength in a coil (tesla) 5.8

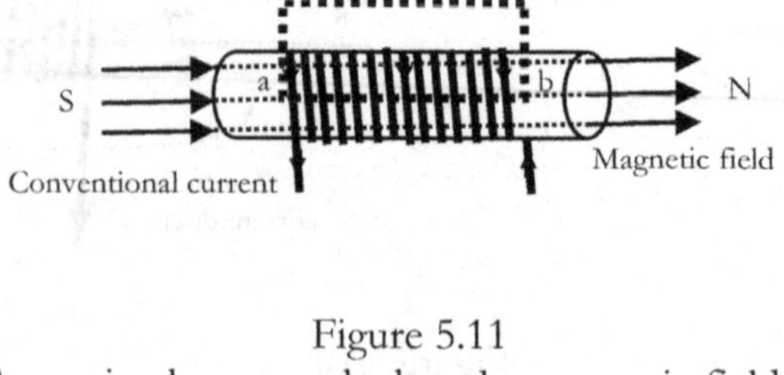

Figure 5.11
Using an Amperian loop to calculate the magnetic field inside a coil.

The Right-hand Rule for a Coil

To determine the direction of the magnetic field created by a current flowing through a coil, curl your fingers around the coil with your fingertips pointing in the direction of the conventional current flow. Your thumb will indicate the north magnetic pole.

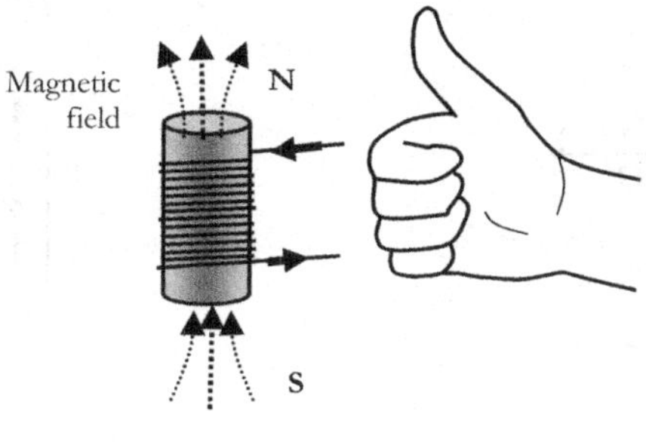

Figure 5.12
The right-hand rule for a coil.

5.5 INDUCTION DUE TO RELATIVE MOTION

Consider a straight length of wire (or metal bar) placed across a rail, sitting in a static magnetic field of constant strength **B**, as shown in Figure 5.13a. The clear circles indicate that the magnetic field direction is coming out of the page. If the bar is moved across the magnetic field in the direction given by the velocity vector *v*, the electrons in the wire will feel a magnetic force $\mathbf{F_M}$, given by Equation 5.1. Note the use of lower-case **v** to indicate velocity to distinguish it from voltage.

$$\begin{aligned}\mathbf{F}_M &= Q(\mathbf{v}\times\mathbf{B})\\ &= Q|\mathbf{v}||\mathbf{B}|(\sin 90)\hat{\mathbf{n}}\\ &= Q|\mathbf{v}||\mathbf{B}|\hat{\mathbf{n}}\end{aligned}$$

The direction of the force felt by the electrons will be parallel to the wire, thus causing a current (I) to flow. Note that the directions of the force and current shown in the diagrams were obtained using the right-hand rule for a magnetic force on a charge.

As the bar is not connected to a circuit, the charges have nowhere to go, and the movement of electrons will cause a separation of charge to occur, as shown in Figure 5.13b. One end of the wire will be positive, and the other negative. This separation of charge will produce an electrostatic potential (voltage) across the wire and an associated electric field. The force felt by electrons in the electric field created will be opposite to the force due to the magnetic field. When the magnetic force equals the electric force, current flow ceases. Expressed as

$$\begin{aligned}F_E &= F_M\\ QE &= QvB\end{aligned}$$

And

$$E = vB$$

The voltage induced (emf) in this fashion due to the relative movement between the magnetic field and the charges in the wire is called ***motional emf.***

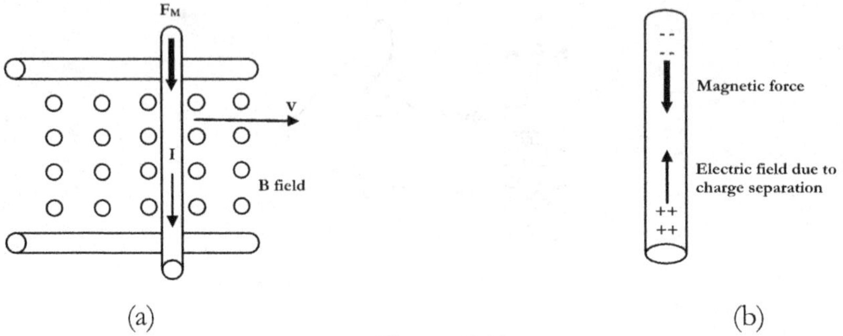

Figure 5.13
Current and charge separation due to magnetic force.

Say another metallic bar is fixed across the two rails making a complete circuit, as shown in Figure 5.14. Now if the original bar is moved, current will flow around the circuit for as long as the bar is moving in the magnetic field. The current flowing through the circuit is called the ***induced current***, and the process is called ***motional induction***. The emf (voltage) induced across the wire is

$$Ed = vBd = emf \qquad \text{Motional emf} \qquad 5.9$$

where d is the length of the wire in the magnetic field. The magnetic flux through the loop (closed circuit) is given by the magnetic field strength times the area of the loop.

$$\Phi_M = B(Ld)$$

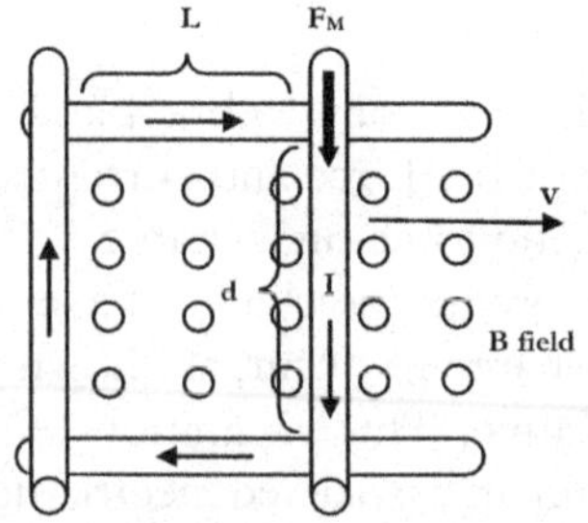

Figure 5.14

As the wire is moved across the rails, the area changes, and so the magnetic flux changes.

$$\frac{\Delta\Phi_M}{\Delta t} = Bd\frac{\Delta L}{\Delta t}$$

The change in length with respect to time is just the velocity of the wire, **v**. So we can write

$$\frac{\Delta\Phi_M}{\Delta t} = vBd$$

From Equation 5.9, we see that the term "vBd" is just the induced emf (voltage). If we take the limit as Δt→0, we get

$$\boxed{emf = \frac{d\Phi_M}{dt}} \qquad \text{emf due to motion (volts)} \qquad 5.10$$

5.6 INDUCTION DUE TO A CHANGING MAGNETIC FIELD

Consider the case where a loop of wire is stationary, and a magnet is pushed through the loop, as shown in Figure 5.15. The magnetic force felt by a charged particle is given by

$$\mathbf{F}_M = Q|\mathbf{v}||\mathbf{B}|(\sin\theta)\hat{\mathbf{n}}$$

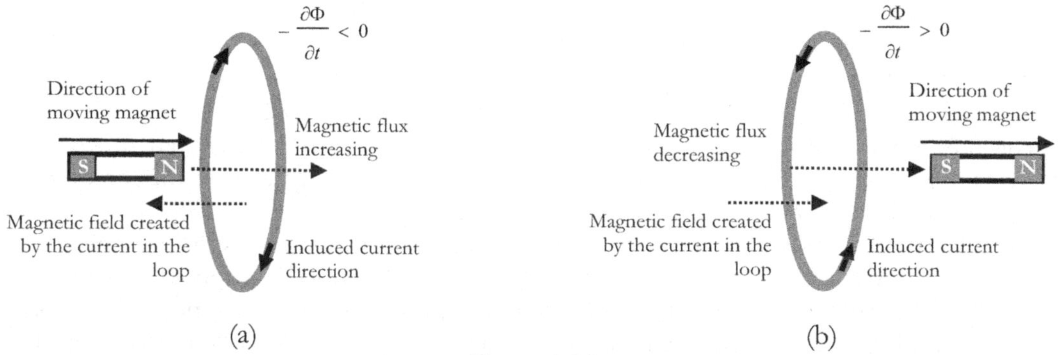

Figure 5.15
Electromagnetic induction.

The direction of the velocity vector **v** is in the same direction as the magnetic field. Therefore the magnetic force felt by a charged particle Q in the wire is

$$\begin{aligned}\mathbf{F}_M &= Q|\mathbf{v}||\mathbf{B}|(\sin 0)\hat{\mathbf{n}}\\ &= 0\end{aligned}$$

However, current will flow through the wire. Given the total force felt by a charged particle is the addition of the magnetic force and the electric force, the force felt by the free electrons in the wire must be due to an electric field. That is

$$\mathbf{F} = (\mathbf{F_E} + \mathbf{F_M}) = (\mathbf{F_E} + 0) = \mathbf{F_E}$$

According to Faraday, a time-varying magnetic field always has an associated electric field. And for any given arbitrarily closed loop, the induced emf can be calculated using Faraday's Equation

5.10, the same equation used for motional induction. Note that the electric field exists regardless if a wire is present or not. This emf is called ***induced emf***, and the process is called ***electromagnetic induction*** or just ***induction***.

5.7 FARADAY'S LAW OF INDUCTION

Faraday's law of induction states that *the emf around a closed path equals the negative of the time rate change of the magnetic flux enclosed by that path*. This can be written in integral form as follows.

$$\boxed{\oint_C \mathbf{E} \bullet dl = -\frac{\partial \Phi}{\partial t}} \quad \text{Faraday's law} \tag{5.11}$$

This is just another way of writing Equation 5.10, where the negative sign on the partial derivative indicates that the induced emf is set up in such a way as to oppose the time-rate change in magnetic flux that created the emf in the first place. The current due to this emf (called a back emf) will create its own magnetic field in an attempt to stop any change of the magnetic flux from occurring; the loop wants to keep the magnetic flux the same. This is known as ***Lentz's law*** and can be succinctly expressed as; *nature abhors a change in magnetic flux*.

Applying Lentz's law to Equation 5.10, the equation becomes

$$\boxed{emf = -\frac{\partial \Phi}{\partial t}} \tag{5.12}$$

This is sometimes called Lentz's equation, but mostly it is just called Faraday's equation. If the magnetic flux is increasing, a magnetic field in the opposite direction is created to oppose the increase. If the magnetic flux is decreasing, a magnetic field will be created in the same direction that adds to the magnetic field to stop it from decreasing.

Consider a magnet moving towards a loop of wire, as shown in Figure 5.15a. The magnetic flux inside the loop increases as the magnet gets closer to the loop. To oppose this increase in magnetic flux, the current must travel in a direction that produces a magnetic field in the opposite direction. To find the direction of the induced current, use the right-hand rule for loops. In Figure 5.15b, the magnet is moving away from the loop, and the magnetic flux is decreasing. An emf will be induced such that the current will create a magnetic field that will aid the external magnetic field to keep the field from decreasing. As the induced current is proportional to the rate of change of the magnetic flux, the faster the magnet or loop is moved in relation to one another, the greater the magnitude of the induced emf and current.

Another variation of Faraday's equation relates the electric field intensity to the magnetic field intensity.

$$\boxed{\nabla \times \mathbf{E} = -\frac{\partial \mathbf{B}}{\partial t}} \tag{5.13}$$

This states that a varying magnetic field will have an associated varying electric field that circulates orthogonally (the curl) around the magnetic field, see Figure 5.16. Note, once again, the existence of this electric field does not require a wire or any object to be present; it just exists. If we place a loop of wire where the electric field exists, then an emf (voltage) will be induced due to work done on the charged particles in the wire, as discussed above.

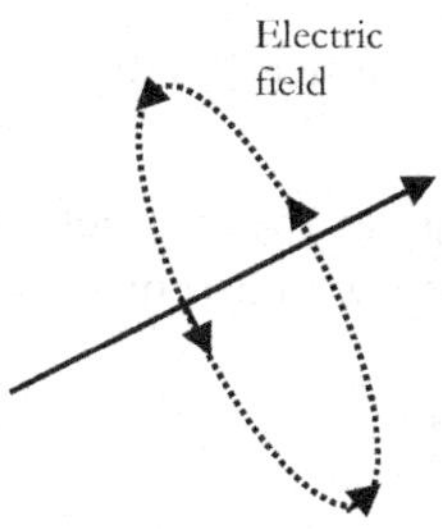

Figure 5.16
A time-varying magnetic field has an associated time-varying electric field.

5.8 INDUCTANCE

A time-varying (AC) current will create a time-varying magnetic field around the wire. And a time-varying magnetic field will induce an emf in the wire to oppose the change in the magnetic field and thereby oppose the change in current as well. This is called the ***back emf***. The tendency of a material, including the vacuum, to *oppose a change in magnetic flux* (or current) is called ***inductance*** (not to be confused with induction). The symbol L is used to signify inductance, and the SI unit of inductance is the henry (H). In a vacuum, no current flows; however, as we shall see when discussing transmission lines, that free space has an opposition to the creation of electric and magnetic fields known as the characteristic impedance.

The time rate change in the induced current is directly proportional to the induced emf. That is

$$emf = -\frac{\partial \Phi}{\partial t} \propto \frac{\partial i}{\partial t}$$

The time rate change in magnetic flux can be equated to the time rate change in current by introducing a constant of proportionality. As the constant of proportionality is defined as an opposition to the time-rate change in magnetic flux, the negative sign in Lentz's equation is dropped and,

$$\frac{\partial \Phi}{\partial t} = L\frac{\partial i}{\partial t} \qquad 5.14$$

where L is the constant of proportionality and is called inductance, and is ***a measure of the opposition to the change of current***. The time rate change in magnetic flux is equal to the emf, so

$$emf = L\frac{\partial i}{\partial t} \qquad \text{emf due to changing current (volts)} \qquad 5.15$$

Solving the partial differential Equation 5.14, gives

$$\int d\Phi = \int L di$$

$$\Phi = Li$$

Using the above relationship and Equation 5.14, we can write inductance as

$$L = \frac{\Phi}{i} = \frac{d\Phi}{di}$$ Inductance (Henrys) 5.16

A device that has inductance and stores energy in the magnetic field (a coil) is called an ***inductor***. The inductance of a material in response to its own current flow is called ***self-inductance***.

Self-inductance of a Straight Wire

The self-inductance of a current-carrying straight wire can be subdivided into two components; the self-inductance of the wire due to the magnetic field inside the wire and the portion of the self-inductance due to the magnetic field outside the wire.

The internal self-inductance is calculated by adding up each contribution of the changing magnetic flux inside the wire through an infinitesimally small section of wire for the entire length of the wire. Using Ampere's law, it can be shown that the internal self-inductance of a current-carrying wire at low frequencies approaching DC, is approximately

$$L_{INT} = \frac{\mu_0 \ell}{8\pi}$$ DC Internal self-inductance of a wire (Henrys) 5.17

where ℓ is the length of the wire. Notice that the internal inductance is independent of the wire diameter. The internal inductance of a wire is frequency-dependent due to the skin effect. And is given by

$$L_{INT} = \frac{\mu_0 \ell}{2\pi} \frac{\delta}{W_D}$$ AC Internal self-inductance of a wire (Henrys) 5.18

where δ sigma is the skin depth (to be described next), and W_D is the diameter of the wire. At near DC, the internal self-inductance is about 50 nH per metre, and at 50 Hz, about 4 nH per metre.

The external inductance is the summation of the contribution of the external magnetic field along the length of the wire. If we assume that the length of the wire is very much greater than the wire diameter, the external inductance becomes

$$L_{EXT} = \frac{\mu_0 \ell}{2\pi}[\ln(\frac{2\ell}{W_r}) - 1)$$ AC External self-inductance of a wire (Henrys) 5.19

where W_r is the wire radius in metres. The total self-inductance of a wire is the sum of the internal self-inductance and the external self-inductance.

$$L = L_{INT} + L_{EXT}$$

And

$$L_{DC} = \mu_0 \ell / 2\pi [\ln(2\ell / W_r) - 0.75)$$ DC self-inductance of a wire (Henrys) 5.20

This formula is for frequencies below 50Hz and is commonly called the DC inductance. At higher frequencies, the internal self-inductance approaches zero due to the skin effect and is dropped, and the total self-inductance just becomes the external self-inductance; that is

$$L_{AC} = \frac{\mu_0 \ell}{2\pi}[\ln(\frac{2\ell}{W_r}) - 1)$$ AC self-inductance of a wire (Henrys) 5.21

The geometry of the metal affects the inductance. For example, the inductance of a thin metal plate is

$$L_{AC} = \frac{\mu_0 \ell}{2\pi}[\ln(\frac{2\ell}{W_W + W_T}) + 0.5 + 0.2235(\frac{W_W + W_T}{\ell}))$$ AC self-inductance of a thin metal plate 5.22

where W_W is the width of the plate in metres and W_T is the thickness of the plate in metres.

5.9 THE SKIN EFFECT

The electric field produced by self-inductance (the back emf) is largely due to the time-varying external magnetic field, with a vanishingly small contribution from the internal magnetic field. The electric field lines of flux produced by the external time-varying magnetic flux have the greatest magnitude at the centre of the conductor, with the back emf in the wire decreasing as we move toward the surface of the wire. This produces a back emf gradient along the cross-sectional surface of the wire. This opposition to the change in current forms a cross-sectional current density distribution across the wire, with the highest density at the edges and the lowest at the centre, see Figure 5.17.

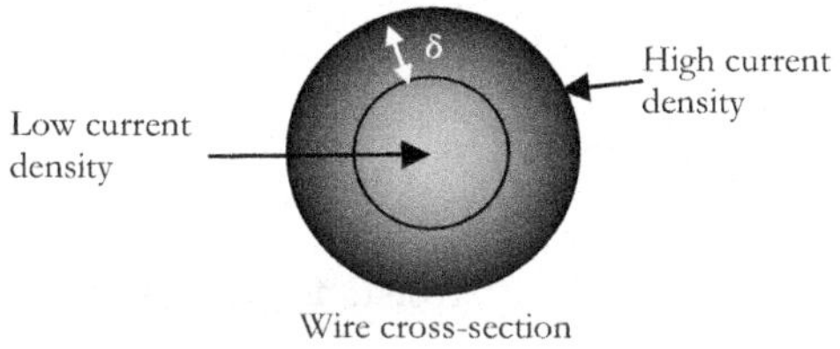

Figure 5.17
The skin effect and the distribution of current due to the back emf.

This distribution of the current density is called the ***skin effect***, as the effect moves the bulk of the current flow out towards the skin of the conductor. The same phenomenon responsible for self-inductance is also responsible for the skin effect. Furthermore, this skin effect is frequency-dependent. For a DC current, there will be no skin effect, and the charge distribution will be the same across the wire. As the frequency increases, the back emf increases, forcing more of the current to the outside.

The distance from the surface of the conductor to where the current density falls to 1/e=0.368 (37%) of its value at the surface is called the ***skin depth***, and the higher the frequency, the smaller the skin depth. The skin depth is calculated using the following formula.

$$\delta = \sqrt{\frac{2\rho}{\omega\mu}}$$ The skin depth(metres) 5.23

where δ (sigma) is the skin depth in metres, ρ rho is the resistivity of the conductor, ω omega is the angular velocity (2πf), and μ mu is the absolute permeability ($\mu_0\mu_R$).

AC Wire Resistance

The current distribution in a wire, due to the skin effect, reduces the effective cross-sectional surface area of the wire in which the current flows. The resistance of a length of wire is inversely proportional to the cross-sectional surface area of a wire. So as the AC frequency increases, the resistance of the wire increases. The DC resistance of a length of wire is given by Equation 4.19. This equation needs to be modified to consider the reduction of the cross-sectional area for AC currents. Although the current doesn't abruptly stop at the skin depth, we can get a very good value for the AC resistance when the wire diameter is much greater than the skin depth, $W_D >> \delta$. For the copper wire diameters and RF frequencies of interest, this will be the case, therefore

$$R_{AC} = \frac{\rho \ell}{A_{effective}}$$

The effective cross-sectional area is the cross-sectional area of a wire minus the area of wire no longer used for current flow, see Figure 5.17. That is

$$\begin{aligned} A_{effective} &= \pi r^2 - \pi (r-\delta)^2 \\ &= \pi (r^2 - r^2 + 2\delta r - \delta^2) \\ &= \pi (2\delta r - \delta^2) \\ &= \pi (2r - \delta)\delta \end{aligned}$$

The AC resistance of a wire can now be written as

$$\boxed{R_{AC} = \frac{\rho \ell}{\pi (2r - \delta)\delta}} \qquad \text{AC resistance of a wire} \qquad 5.24$$

For a flat conductor, this would be

$$\boxed{R_{AC} = \frac{\rho \ell}{(W + D - \delta)\delta}} \qquad \text{AC resistance of a flat conductor} \qquad 5.25$$

where W is the width of the conductor, and D is the depth of the conductor.

5.10 PERMEABILITY

Permeability is an indicator of *the ability of a medium to store energy in the magnetic field*; the greater the permeability, the greater the magnetic flux density. The units of permeability can be derived using dimensional analysis. Starting with Equation 5.6, the magnetic field around a wire

$$B = \frac{\mu I}{2\pi r}$$

Rearranging for μ

$$\mu = \frac{2\pi r B}{I}$$

Performing dimensional analysis

$$\mu_0 = \frac{2\pi(metre)(teslsa)}{amp}$$

$$= 2\pi \frac{metre}{amp} \frac{weber}{metre^2}$$

$$= 2\pi \frac{weber}{amp} \frac{1}{metre}$$

We get weber per amp per metre. Weber per amp is inductance in henrys, so the units of permeability are ***henrys per metre*** (H/m). Notice the similarity with permittivity, which is in capacitance per metre. Another definition of permeability is *a measure of the ease with which a magnetic field can be established in a material.*

Stored Energy

From Equation 5.15, the voltage across an inductor is given by the inductance of the coil multiplied by the time rate change in current.

$$V = L\frac{\partial I}{\partial t}$$

The power across a device is given by the voltage times the current flowing through the device. Multiplying both sides by the current, we get

$$P = IL\frac{\partial I}{\partial t}$$

Solve the differential equation by integrating both sides

$$\int Pdt = \int ILdI$$

$$Pt = \frac{1}{2}LI^2$$

Power is energy per second, so multiplying power by time gives energy. The energy stored in an inductor becomes

$$\boxed{E_B = \frac{1}{2}LI^2}$$ Energy stored in an inductor (joules) 5.26

Relative Permeability

The permeability of a medium is generally expressed as a ratio with respect to the permeability of a vacuum. The permeability of a vacuum is constant and is denoted by μ_0.

$$\mu_R = \frac{\mu}{\mu_0}$$ Relative permeability

This ratio is called ***relative permeability***. To find the permeability multiply the relative permeability by the constant μ_0.

$$\boxed{\mu = \mu_R \mu_0}$$ 5.27

The modern value of the permeability of the vacuum is

$$\mu_0 = 1.25663706212x10^{-6}$$ Permeability of a vacuum (Henrys per metre)

The old value of permeability of a vacuum was defined to be

$$\mu_0 = 4\pi x10^{-7}$$ The old definition of the permeability of a vacuum (Henrys per metre)

and is still occasionally used in equations.

6

The Electromagnet Field

6.1 MAXWELL'S EQUATIONS

Maxwell's equations are a set of equations describing aspects of the electric and magnetic fields and how they are related to each other. Six physical principles underline these equations. They are:

1. Electric fields surround charged particles.
2. Moving charged particles create magnetic fields.
3. Accelerating charged particles create electromagnetic waves.
4. Electric fields impart a force on charged particles.
5. Magnetic fields impart a force on moving (relative movement) on charged particles.
6. Time-varying magnetic fields have an associated electric field.

Maxwell's equations consist of a set of four integral equations. These equations are also expressed in differential form. Which form to use depends on the problem to be solved, and each gives a slightly different way of viewing the same physical process. Some of Maxwell's equations have already been introduced, but they shall be restated here for clarity and completeness.

Gauss' Law for the Electric Field

Gauss's law states that the total electric flux through the surface of an object is equal to the charge inside the object divided by the permittivity. The integral form is

$$\boxed{\oiint_S \mathbf{E} \bullet d\mathbf{A} = \frac{Q_{inside}}{\varepsilon}} \qquad \text{Gauss' law for electric field}$$

And the differential form is

$$\nabla \bullet \mathbf{E} = \frac{\rho}{\varepsilon}$$

Gauss' Law for the Magnetic Field

Consider a magnetic enclosed by some imaginary sphere, Figure 6.1. Magnetic lines of flux flow out from the north pole of the magnet through the surface of the sphere and then flow back through the surface to connect to the south pole of the magnet. This means the magnetic field has no divergence, unlike the electric field, which has divergence. There is just as much magnetic flux going out of the surface as coming in, so the net flow through the surface is zero. In integral form, this is

$$\oiint_S \mathbf{B} \bullet d\mathbf{A} = 0$$ Gauss' law for magnetic field 6.1

In differential form

$$\nabla \bullet \mathbf{B} = 0$$ 6.2

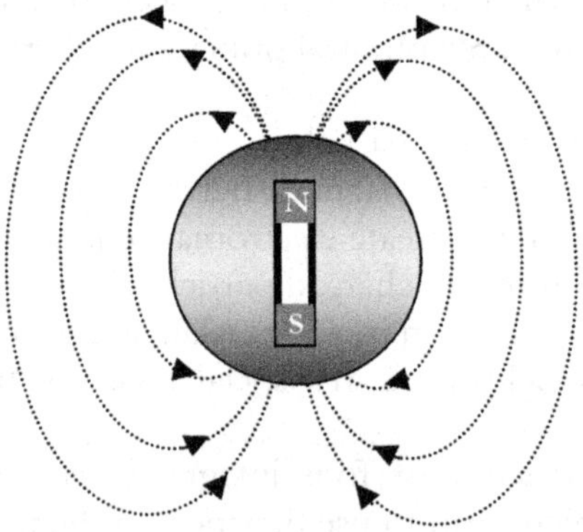

Figure 6.1
Magnetic flux lines through an enclosing sphere.

Faraday's Law of Induction

Faraday's law of induction states that *the emf around a closed path equals the negative of the time rate change of the magnetic flux enclosed by that path.* This can be written in integral form as follows.

$$\oint_C \mathbf{E} \bullet dl = -\frac{\partial \Phi}{\partial t}$$ Faraday's law

In differential form

$$\boxed{\nabla \times \mathbf{E} = -\frac{\partial \mathbf{B}}{\partial t}}$$

Ampere-Maxwell Law

Ampere's law states *the magnetic field around a closed path is equal to the enclosed current times the permeability of the medium.* This describes the magnetic field around a current-carrying wire.

$$\oint_C \mathbf{B} \bullet d\mathbf{l} = \mu I_{enclosed}$$

Maxwell modified this equation to describe the behaviour of capacitors in a circuit. Maxwell believed the interactions between electrical currents and magnetism were through the interaction of rotating cells and the stress on a substance known as the luminiferous aether. With this model in mind, he described the magnetic field created by a current through a conductor. However, he also knew that "electrical effects" were propagated across metal plates separated by a dielectric (a capacitor). In his paper "On Physical Lines of Force, Part III", Maxwell states:

"Bodies which do not permit a current of electricity to flow through them are called insulators. But though electricity does not flow through them, electrical effects are propagated through them, and the amount of these effects differs according to the nature of the body; so that equally good insulators may act differently as dielectrics."

To have a complete model of the electric and magnetic interactions, he needed to explain the "electrical effects" across a dielectric. Maxwell writes:

"Here then we have two independent qualities of bodies, one by which they allow of the passage of electricity through them, and the other by which they allow of electrical action being transmitted through them without any electricity being allowed to pass. A conducting body may be compared to a porous membrane which opposes more or less resistance to the passage of a fluid, while a dielectric is like an elastic membrane which may be impervious to the fluid, but transmits the pressure of the fluid on one side to that on the other."

Here he is describing a type of current that is transmitted across the capacitor as some type of pressure and not as a physical current. It is in this context that he came up with the concept of the ***displacement current.***

"The effect of this action on the whole dielectric mass is to produce a general displacement of the electricity in a certain direction. This displacement does not amount to a current, because when it has attained a certain value it remains constant, but it is the commencement of a current, and its variations constitute currents in the positive or negative direction, according as the displacement is increasing or diminishing. The amount of the displacement depends on the nature of the body, and on the electromotive force"

What Maxwell is describing here is the action of the electric field across a capacitor. When a varying current is applied to the plates of a capacitor, the electric charges on the plates vary with time, so the electric field between the plates also varies with time. So the displacement current is proportional to the time rate change in the electric field, and the displacement current is related to the electric field by

$$I_D = \varepsilon \frac{\partial}{\partial t} \iint_S E \bullet dA$$

The value under the integral sign is the sum of the electric field strength passing through a small area dA, which is just the electric flux, so we can write

$$I_D = \varepsilon \frac{\partial \Phi_E}{\partial t}$$ The displacement current

This states that the displacement current is equal to the permittivity of a substance times the time rate change of electric flux. This displacement current term was added to Ampere's law giving

$$\oint \mathbf{B} \bullet d\mathbf{l} = \mu(I + I_D)$$

Fully written as

$$\boxed{\oint_C \mathbf{B} \bullet d\mathbf{l} = \mu I_{enclosed} + \mu\varepsilon \frac{\partial \Phi_E}{\partial t}}$$ Ampere-Maxwell 6.3

Note that the displacement current and, therefore, the electric flux Φ_E only refers to an electric field across the plates of a capacitor and not the electric field inside the wire. When determining the magnetic field surrounding a wire, the displacement current is zero, and we get the original Ampere equation. When looking at the magnetic field surrounding a capacitor, the current between the plates is zero, and we are only left with the displacement current.

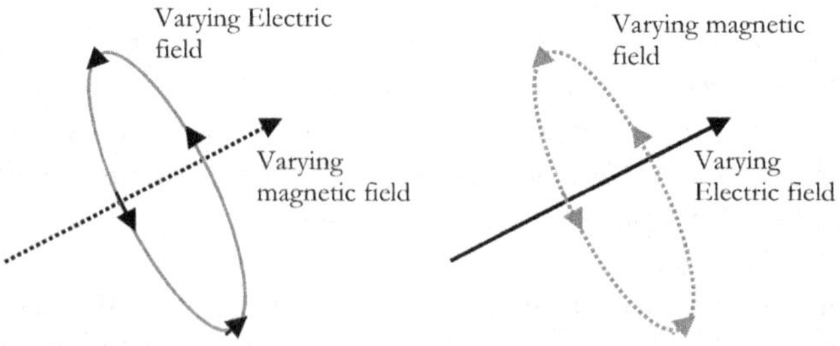

Figure 6.2
A varying magnetic field has an electric field, and a varying electric field has a magnetic field.

The Ampere-Maxwell equation in differential form is

$$\boxed{\nabla \times \mathbf{B} = \mu\mathbf{J} + \mu\varepsilon \frac{\partial \mathbf{E}}{\partial t}}$$ 6.4

When dealing with the displacement only, the current density J goes to zero and

$$\boxed{\nabla \times \mathbf{B} = \mu\varepsilon \frac{\partial \mathbf{E}}{\partial t}}$$

There is now a nice symmetry between this equation and the differential form of Faraday's law of induction. Equation 5.13

$$\nabla \times \mathbf{E} = -\frac{\partial \mathbf{B}}{\partial t}$$

From these two equations, we can see that a changing electric field has an associated magnetic field, and a changing magnetic field has an associated electric field. And neither field requires the existence of charged particles for them to exist. This was an enormous breakthrough and unified both the electric and magnetic fields into one field called the ***electromagnetic field***. Note that a time-varying electric field does not create a magnetic field; a varying electric field just has an associated magnetic field. Likewise, a time-varying magnetic field does not create an electric field; it just has an associated electric field. This means that both fields exist together in one field called the electromagnetic field and are tied together, see Figure 6.2.

6.2 THE WAVE EQUATION

Consider a cork floating on a perfectly calm flat sea. Ignoring any surface currents, wind or tides, if a wave passes by, the cork will stay in its current location and will bob up and down in sympathy with the crests and troughs, as shown in Figure 6.3. The water itself is displaced up and down, but there is no other displacement; that is, the water does not move to the right or left. What is moving along the x-axis is the energy required to displace the water (up and down). For the water and cork to be moved up and down, work must be done, and work is energy. The wave (and therefore the energy) travels in some direction, moving from one location to another. This is called the direction of propagation. A wave can be considered *a disturbance in the medium that transports energy from one location to another*. If the acceleration of the medium (the up and down movement) is in a direction perpendicular to the propagation direction, then the wave is called a ***transverse wave***, as in Figure 6.3.

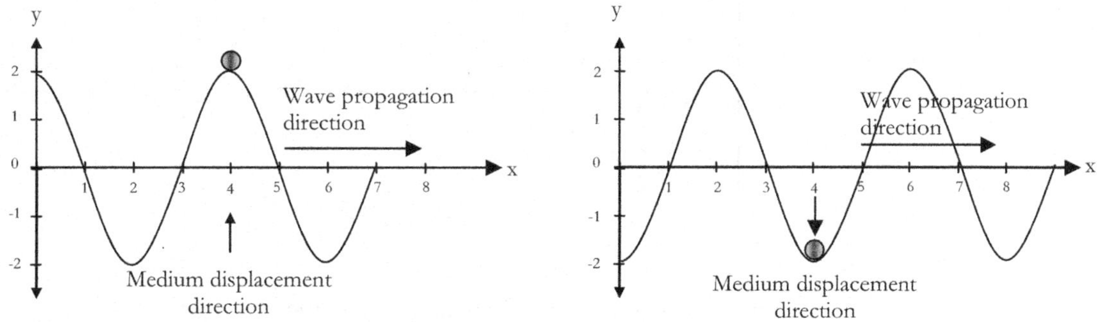

Figure 6.3
Transverse wave.

A wave where the displacement of the medium is in the same direction as the wave propagation is called a ***longitudinal wave***. An example of a longitudinal wave is a sound wave. Sound waves are caused by the displacement of air molecules. The air molecules bunch up, causing areas of high and low pressure known as pressure waves. Air molecules in a high-pressure region collide with other air molecules imparting kinetic energy causing these molecules to move in the direction of propagation until another high-pressure area is formed, and they collide with other molecules to keep the propagation going, as shown in Figure 6.4. These pressure waves are detected by the eardrum, and we hear sound.

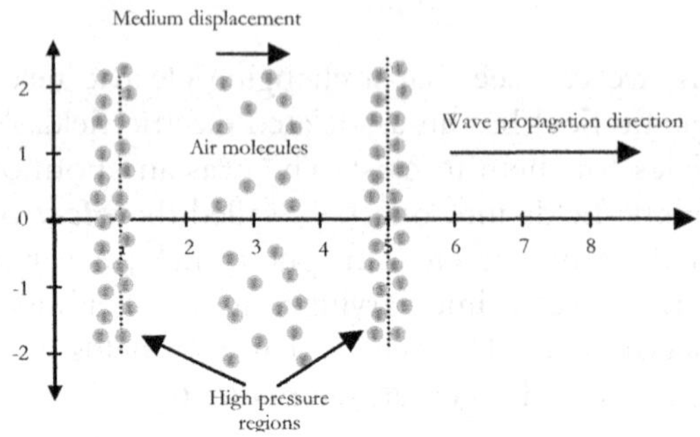

Figure 6.4
Longitudinal wave.

If we take a snapshot of a transverse wave at some time, we have a frozen wave in time that would appear something like the wave shown in Figure 6.5. In this particular case, the wave is shaped like a cosine. A wave has a maximum value called the ***peak amplitude*** (or just the amplitude), shown as A. Consider the x and y scales of the graph to be in metres, so in this example, the amplitude of the wave is 2 metres. The length at which a wave starts to repeat itself is called the ***wavelength***, denoted by lambda (λ). The wavelength is easiest to see if taken from peak-to-peak values. In this example, the peaks occur at x=0 and x=4, so the wavelength is (4 - 0) 4 metres.

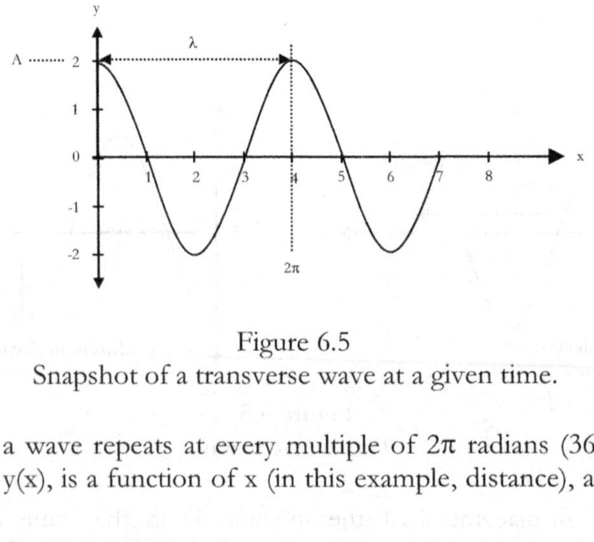

Figure 6.5
Snapshot of a transverse wave at a given time.

In terms of angles, a wave repeats at every multiple of 2π radians (360°). The instantaneous amplitude of the wave, y(x), is a function of x (in this example, distance), and we could determine this value as follows

$$y(x) = A\cos(\frac{2\pi}{\lambda}x)$$

where A is the peak amplitude, and the term inside the bracket translates the x (distance) value into degrees or radians as required by the cosine function. So when x is equal to a multiple of the wavelength λ, the wave repeats itself. The term 2π/λ is called the ***wave number*** and is denoted by beta β. Sometimes the letter k is used to define the wave number, but we will use β.

$$\beta = \frac{2\pi}{\lambda} \qquad \text{The wave number} \qquad 6.5$$

When x is one wavelength (x=λ), this term becomes 2π, and the value of the function y(x) is 2 metres, just as it was at the start.

This description is fine for a snapshot, but it does not tell us how the wave changes over time. In Figure 6.6, the wave has moved to the right at some time interval Δt, shown as a dotted line. Say the initial snapshot was taken at time t=0. At the origin (x=0), the height of the wave was 2 metres, and at a one-metre distance (x=1), the height of the wave was 0.

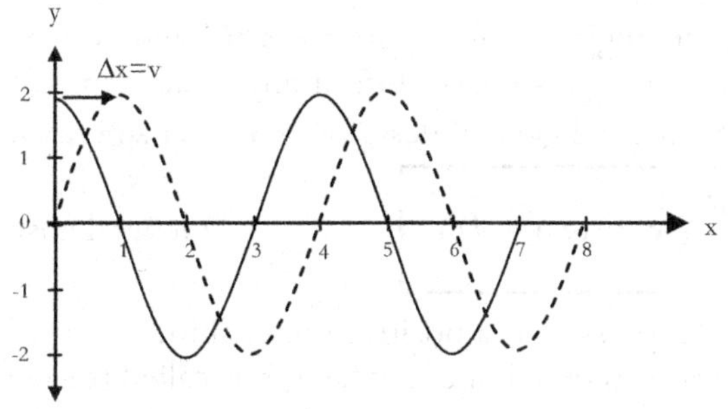

Figure 6.6
Transverse wave over time

However, now at time Δt at x=1, the height of the wave is 2 metres and not zero anymore. This could be described by saying the wave has been translated by Δx to the right. This translation is simply given by the velocity of the wave multiplied by time (the equation for distance)

$$\Delta x = vt$$

where v is the speed of the wave in the x direction. If we consider Δx to be the new origin, then at Δt, the wave looks like it did at t=0. To add this velocity component to our wave function, the x translation distance needs to be subtracted from the current value of x. Apply this to the wave function gives

$$y = A\cos(\frac{2\pi}{\lambda}(x - \Delta x))$$

Expressed as a velocity, gives

$$y = A\cos(\frac{2\pi}{\lambda}(x - vt)) \qquad \text{Forward travelling wave}$$

We now have a wave function that takes into account the wavelength of the wave and the velocity of the wave. Note if the wave travelled to the left, then the speed term would be added to x.

$$y = A\cos(\frac{2\pi}{\lambda}(x + vt)) \qquad \text{Backward travelling wave}$$

Use the negative sign in the cosine term if the wave moves to the right and the positive sign if the wave is moving to the left. Applying this equation to the above example, the wave moved 1 metre to the right ($\Delta x=1$) during some time interval. If we wanted to know the height of the wave (y value) at a distance of two metres (x=2) from the origin after some time interval Δt, then

$$y = A\cos(\frac{2\pi}{4}(2-1))$$
$$= 0$$

From the dotted curve, the value of y is indeed 0 at x =2.

Lastly, say during the time Δt, the wave met with some inference that caused the translation to be greater or smaller than would be expected than if the wave had not met this inference. This extra translation could be described as a change in angle from the original 0° angle the wave started with. This change in angle is called a ***phase shift*** and is denoted by theta (θ) and can be expressed in degrees or radians. This phase shift, if any, is accounted for by adding the phase shift value θ to the equation. So the full general description of a cosine wave would be

$$\boxed{y = A\cos(\frac{2\pi}{\lambda}(x \pm vt) + \theta)} \qquad \text{General cosine equation} \qquad 6.6$$

Note that θ can always be expressed as a positive phase shift.

The time taken for a wave to travel one wavelength is called the period (T) and is the inverse of the frequency f.

$$\boxed{f = \frac{1}{T}} \qquad \text{frequency} \qquad 6.7$$

We can write this wave function in several forms depending on what values are known. Each form can be derived from one another, and they describe the same wave function, just in different ways. For instance, the propagation speed of a wave is related to its wavelength and period, as speed is distance divided by the time taken to travel that distance. So the speed of a wave is given by the wavelength divided by the period.

$$\boxed{v = \frac{\lambda}{T} = \lambda f = \frac{2\pi f}{\beta}} \qquad \text{Velocity of propagation} \qquad 6.8$$

And wavelength can be calculated as

$$\boxed{\lambda = \frac{v}{f} = \frac{2\pi f}{\beta}} \qquad \text{Wavelength} \qquad 6.9$$

Using Equation 6.8 for the velocity in Equation 6.6, we can write

$$y = A\cos(\frac{2\pi}{\lambda}x \pm \frac{2\pi}{\lambda}\frac{\lambda}{T}t + \theta)$$

Giving

$$y = A\cos(\frac{2\pi}{\lambda}x \pm \frac{2\pi}{T}t + \theta)$$

Now, 2π/T is called the angular velocity, how quickly it revolves or changes angle in radians per second. This is designated by lowercase omega (ω).

$$\boxed{\omega = \frac{2\pi}{T} = 2\pi f}$$ Angular velocity (radians/s) 6.10

So we could write the wave function as

$$y = A\cos(\frac{2\pi}{\lambda}x - \omega t + \theta) = A\cos(\beta x - \omega t + \theta)$$ Forward travelling wave equation using angular velocity 6.11

$$y = A\cos(\frac{2\pi}{\lambda}x + \omega t + \theta) = A\cos(\beta x + \omega t + \theta)$$ Backward travelling wave equation using angular velocity 6.11b

In this example, we have used a wave described by the cosine function, but a wave could be a sine function or a combination (superposition) of waves. If a sine wave were used, the ωt term would be positive for forward travelling waves and negative for backward travelling waves.

Consider the cork bobbing up and down in the y direction on a forward travelling transverse wave as described above. The speed of this up and down motion can be calculated by taking the first partial derivative of the wave equation with respect to time. Remember, partial derivatives keep all other variables constant, so the value of x will be constant. Taking the first partial derivative of Equation 6.11 with respect to time and using β as the wave number, we have

$$v_y = \frac{\partial y}{\partial t} = \omega A\sin(\beta x - \omega t + \theta)$$

If we now take the partial derivative of speed with respect to time, we get the acceleration of the medium in the y direction (up and down).

$$a_y = \frac{\partial^2 y}{\partial t^2} = -\omega^2 A\cos(\beta x - \omega t + \theta)$$

Now consider the wave propagation in the x direction. If we take the partial derivative of the wave equation with respect to x rather than time, we determine how the amplitude changes for a change in x, and we get the slope of the wave for any given x (a change in y with respect to a change in x). If we take the second partial derivative with respect to x, we get how this slope changes for x. This indicates how rapidly the wave changes as a function of distance. As a wave is the propagation of energy, we can think of it as how the energy changes as the wave travels. So we have

$$\partial y / \partial x = -\beta A\sin(\beta x - \omega t + \theta)$$

And the second partial derivative is

$$\partial^2 y \Big/ \partial x^2 = -\beta^2 A\cos(\beta x - \omega t + \theta)$$

Taking the ratio of these two derivatives

$$\frac{\partial^2 y}{\partial t^2} \Big/ \frac{\partial^2 y}{\partial x^2} = \frac{-\omega^2 A\cos(\beta x - \omega t + \theta)}{-\beta^2 A\cos(\beta x - \omega t + \theta)}$$

$$= \omega^2/\beta^2$$

$$= \frac{(2\pi f)^2}{(\frac{2\pi}{\lambda})^2}$$

$$= (\lambda f)^2$$

From Equation 6.8, velocity is given by λf. So we can write

$$\frac{\partial^2 y}{\partial t^2} \Big/ \frac{\partial^2 y}{\partial x^2} = v^2$$

where v is the propagation velocity of the wave. Rearranging the above equation and using u rather than y as a function of x and t, we get

$$\boxed{\frac{\partial^2 u(x,t)}{\partial t^2} = v^2 \frac{\partial^2 u(x,t)}{\partial x^2}} \qquad \text{The wave equation} \qquad 6.12$$

This is known as the linear wave equation in one dimension. Linear because the original wave function is a function of x and t and not some non-linear value such as x^2. One dimension, because the direction of propagation (x) is in a single dimension. This equation can be extended to other spatial dimensions.

The second-order partial derivative on the left tells us the acceleration up and down for a transverse wave, or from left to right for a longitudinal wave. The second-order partial derivative on the right tells us how the energy changes over distance. In summary, this equation tells us that the acceleration (up and down) is related to the propagation velocity of the wave and how energy changes with distance.

6.3 ELECTROMAGNETIC WAVES

Consider the differential form of Maxwell's equations in a vacuum. As there is no charge in a vacuum, Gauss's law becomes

$$\nabla \bullet \mathbf{E} = 0$$

The divergence of the magnetic field is always zero.

$$\nabla \bullet \mathbf{B} = 0$$

A time-varying magnetic field can exist, so

$$\nabla \times \mathbf{E} = -\frac{\partial \mathbf{B}}{\partial t} \qquad \text{Faraday}$$

And a time-varying electric field can exist, and

$$\nabla \times \mathbf{B} = \mu_0 \varepsilon_0 \frac{\partial \mathbf{E}}{\partial t} \qquad \text{Ampere-Maxwell}$$

The first two equations say that the divergence of the electric and magnetic field in a vacuum is zero; there is just as much electric and magnetic field flowing into some tiny region as coming out. The third, Faraday's law, says that a changing electric field exists around a changing magnetic field, and the electric field is orthogonal to the magnetic field. The fourth, Ampere-Maxwell's law, says that a changing magnetic field exists around a changing electric field, and the magnetic field is orthogonal to the electric field. This means that the electric and magnetic fields are coupled together and do not exist independently. Therefore the electric and magnetic fields can propagate through space as waves without any electric charges or wires.

These travelling waves are known as ***electromagnetic waves***, and just as water waves are disturbances in the water medium, these electromagnetic waves are considered to be disturbances in the ***electromagnetic field.*** As waves are a mechanism for the transport of energy, these electromagnetic waves transport electromagnetic energy from one location to another. The electric and magnetic fields each contain equal amounts of energy. The energy density of an electromagnetic wave is

$$\begin{aligned} U &= U_E + U_B \\ &= \frac{\varepsilon_0 E^2}{2} + \frac{B^2}{2\mu_0} \end{aligned} \qquad \text{Energy of an electromagnetic wave} \qquad 6.13$$

Figure 6.7 shows the relationship between the electric and magnetic fields in an electromagnetic wave. The direction of propagation is the same for both fields, and the electric field is orthogonal to the magnetic field.

Say we have an electromagnetic wave travelling along the x-axis, as shown in Figure 6.8a. If we select some moment in time and graph the electromagnetic wave, we have a snapshot of the electric field strength, the magnetic field strength and the direction of propagation, as shown in Figure 6.8b. The electric field vector is a function of x (the direction of propagation) and time (t), but the field strength points only in the y direction. The electric vector in component form is

$$\begin{aligned} \mathbf{E} &= E_x\hat{\mathbf{i}} + E_y\hat{\mathbf{j}} + E_z\hat{\mathbf{k}} \\ &= 0\hat{\mathbf{i}} + E(x,t)\hat{\mathbf{j}} + 0\hat{\mathbf{k}} \\ &= E_y(x,t)\hat{\mathbf{j}} \end{aligned}$$

Solving for the curl of the electric field in Faraday's equation

$$\begin{aligned} \nabla \times \mathbf{E} &= \left(\frac{\partial E_z}{\partial_y} - \frac{\partial E_y}{\partial_z}\right)\hat{\mathbf{i}} - \left(\frac{\partial E_z}{\partial_x} - \frac{\partial E_x}{\partial_z}\right)\hat{\mathbf{j}} + \left(\frac{\partial E_y}{\partial_x} - \frac{\partial E_x}{\partial_y}\right)\hat{\mathbf{k}} \\ &= \left(0 - \frac{\partial E_y}{\partial_z}\right)\hat{\mathbf{i}} - (0-0)\hat{\mathbf{j}} + \left(\frac{\partial E_y}{\partial_x} - 0\right)\hat{\mathbf{k}} \end{aligned}$$

$$\nabla \times \mathbf{E} = -\frac{\partial E_y}{\partial_z}\hat{\mathbf{i}} + \frac{\partial E_y}{\partial_x}\hat{\mathbf{k}}$$

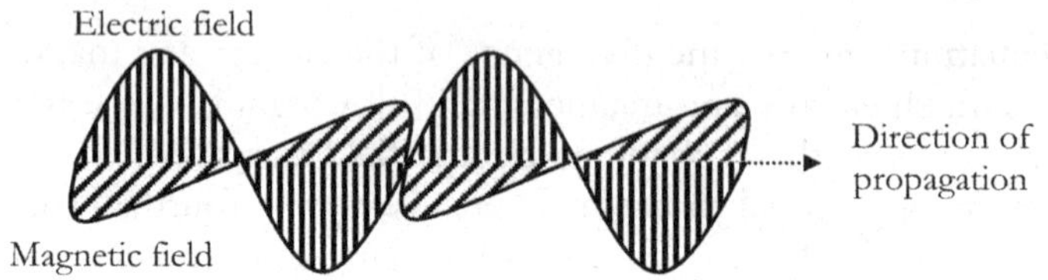

Figure 6.7
An electromagnetic wave.

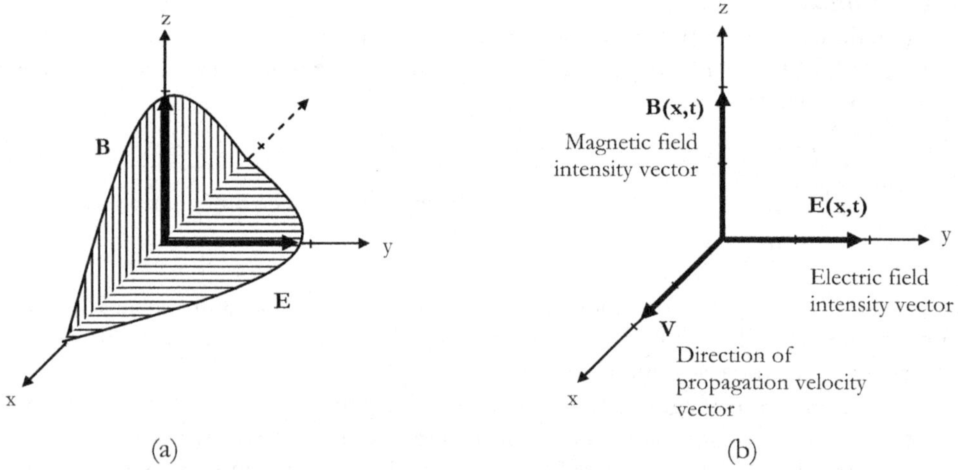

Figure 6.8
Vector components of an electromagnetic wave.

The electric field strength does not change with a change in the z-direction, only with a change in the x-direction. Therefore the partial derivative with respect to z is zero, and we can write

$$\nabla \times \mathbf{E} = \frac{\partial \mathbf{E}}{\partial x}$$

We can now write Faraday's equation as

$$\frac{\partial \mathbf{E}}{\partial_x} = -\frac{\partial \mathbf{B}}{\partial t} \qquad 6.14$$

Performing the same analysis on the magnetic field, the magnetic field strength only points in the z-direction, but its propagation direction is along the x-axis, just like the electric field. So the magnetic field is a function of x and t.

$$\mathbf{B} = B_x\hat{\mathbf{i}} + B_y\hat{\mathbf{j}} + B_z\hat{\mathbf{k}}$$
$$= 0\hat{\mathbf{i}} + 0\hat{\mathbf{j}} + B(x,t)\hat{\mathbf{k}}$$
$$= B_z(x,t)\hat{\mathbf{k}}$$

Evaluating the curl of the magnetic field in the Ampere-Maxwell equation in a vacuum

$$\nabla \times \mathbf{B} = (\frac{\partial B_z}{\partial_y} - \frac{\partial B_y}{\partial_z})\hat{\mathbf{i}} - (\frac{\partial B_z}{\partial_x} - \frac{\partial B_x}{\partial_z})\hat{\mathbf{j}} + (\frac{\partial B_y}{\partial_x} - \frac{\partial B_x}{\partial_y})\hat{\mathbf{k}}$$
$$= (\frac{\partial B_z}{\partial_y} - 0)\hat{\mathbf{i}} - (\frac{\partial B_z}{\partial_x} - 0)\hat{\mathbf{j}} + (0-0)\hat{\mathbf{k}}$$
$$= \frac{\partial B_z}{\partial_y}\hat{\mathbf{i}} - \frac{\partial B_z}{\partial_x}\hat{\mathbf{j}}$$

The magnetic field strength does not change with a change in the y direction, only with a change in the x direction. Therefore the partial derivative with respect to y is zero, and we can write

$$\nabla \times \mathbf{B} = -\frac{\partial \mathbf{B}}{\partial_x}$$

Now Ampere-Maxwell becomes

$$\frac{\partial \mathbf{B}}{\partial x} = -\mu_0 \varepsilon_0 \frac{\partial \mathbf{E}}{\partial t} \qquad 6.15$$

Now taking the partial derivative with respect to x for the electric field, Equation 6.14

$$\frac{1}{\partial x}\frac{\partial \mathbf{E}}{\partial x} = -\frac{1}{\partial x}\frac{\partial \mathbf{B}}{\partial t}$$
$$\frac{\partial^2 \mathbf{E}}{\partial x^2} = -\frac{1}{\partial x}\frac{\partial \mathbf{B}}{\partial t}$$

If we consider the second-order derivatives of the electric and magnetic field wave function to be continuous (and they are), then the order of partial differentiation does not matter, so we can swap the order and using Equation 6.15, we get

$$\frac{\partial^2 \mathbf{E}}{\partial x^2} = -\frac{1}{\partial t}\frac{\partial \mathbf{B}}{\partial x}$$
$$= -\frac{1}{\partial t}(-\mu_0 \varepsilon_0 \frac{\partial \mathbf{E}}{\partial t})$$

And

$$\frac{\partial^2 \mathbf{E}}{\partial x^2} = \mu_0 \varepsilon_0 \frac{\partial^2 \mathbf{E}}{\partial t^2}$$

Rearranging

$$\frac{\partial^2 \mathbf{E}}{\partial t^2} = \frac{1}{\mu_0 \varepsilon_0}\frac{\partial^2 \mathbf{E}}{\partial x^2} \qquad 6.16$$

This is the wave Equation 6.12 for the electric field, with a propagation speed of

$$v = \frac{1}{\sqrt{\mu_0 \varepsilon_0}}$$

Performing the same analysis for the magnetic field, and taking the partial derivative with respect to x for Equation 6.15

$$\frac{1}{\partial x}\frac{\partial \mathbf{B}}{\partial x} = \frac{1}{\partial x}(-\mu_0 \varepsilon_0 \frac{\partial \mathbf{E}}{\partial t})$$

$$\frac{\partial^2 \mathbf{B}}{\partial x^2} = -\mu_0 \varepsilon_0 \frac{1}{\partial t}(\frac{\partial \mathbf{E}}{\partial x})$$

Now from Equation 6.14

$$\frac{\partial \mathbf{E}}{\partial x} = -\frac{\partial \mathbf{B}}{\partial t}$$

So

$$\frac{\partial^2 \mathbf{B}}{\partial x^2} = -\mu_0 \varepsilon_0 \frac{1}{\partial t}(-\frac{\partial \mathbf{B}}{\partial t})$$

$$= \mu_0 \varepsilon_0 \frac{\partial^2 \mathbf{B}}{\partial t^2}$$

Rearranging

$$\frac{\partial^2 \mathbf{B}}{\partial t^2} = \frac{1}{\mu_0 \varepsilon_0}\frac{\partial^2 \mathbf{B}}{\partial x^2} \qquad 6.17$$

This is also a wave equation with a propagation speed of

$$\boxed{v = \frac{1}{\sqrt{\mu_0 \varepsilon_0}}} \qquad \text{The speed of light in a vacuum} \qquad 6.18$$

Performing dimensional analysis on Equation 6.18, we get

$$v = 1 \Big/ \sqrt{\frac{Henrys}{metre}\frac{Farads}{metre}}$$

$$= 1 \Big/ \sqrt{\frac{Kg.m^2}{s^2 A^2}\frac{s^4 A^2}{Kg.m^2}\frac{1}{m^2}}$$

$$= 1 \Big/ \sqrt{\frac{s^2}{m^2}}$$

$$= m/s$$

We get metres per second which is speed. Plugging the values of μ_0 and ε_0 into Equation 6.18, we get a propagation speed of 2.99×10^8 metres per second, which is the speed of light in a vacuum. Faraday believed that light was also an electromagnetic wave, and when the speed of light was eventually measured, it was found to be in excellent agreement with Maxwell's predicted value.

6.4 THE ELECTROMAGNETIC SPECTRUM

All electromagnetic waves travel at the same speed in a vacuum; that is, the speed of light. The speed of light is signified by the letter "C". Given that the speed of light is constant, the only difference between visible light and other electromagnetic waves is the wavelength and frequency;

Band	Abbreviation	Description	Upper Frequency	Upper Wave Length
Radiowaves <1GHz	ELF	extremely low frequency	30 Hz	10,000 km
	SLF	super low frequency	300 Hz	1,000 km
	ULF	ultra low frequency	3 kHz	100 km
	VLF	very low frequency	30k Hz	10 km
	LF	low frequency	300 kHz	1 km
	MF	medium frequency	3 MHz	100 m
	HF	high frequency	30 MHz	10 m
	VHF	very high frequency	300 MHz	1 m
	UHF	ultra high frequency	3000 MHz	10 cm
Microwaves >=1GHz	SHF	super high frequency	30 GHz	1 cm
	EHF	extremely	300 GHz	1 mm
Infrared		infrared	400×10^{12} Hz	750 nm
Visible light		visible light	780×10^{12} Hz	380 nm
Ultraviolet light		ultraviolet light	30×10^{15} Hz	10 nm
X-rays		X-rays	30×10^{18} Hz	10 pm
Gamma rays		gamma rays	$>30 \times 10^{18}$ Hz	<10 pm

Table 6.1
Electromagnetic bands.

the higher the frequency, the shorter the wavelength. Visible light forms just one small region of a much greater spread of frequencies known as the ***electromagnetic spectrum***. The electromagnetic spectrum is broadly split into sections, as shown in Figure 6.9.

Table 6.1 lists the designated wavelengths and frequencies of the electromagnetic spectrum in greater detail. Different authors give different frequencies for frequency bands. I have used the bands given by the International Telecommunications Union up to the EHF band.

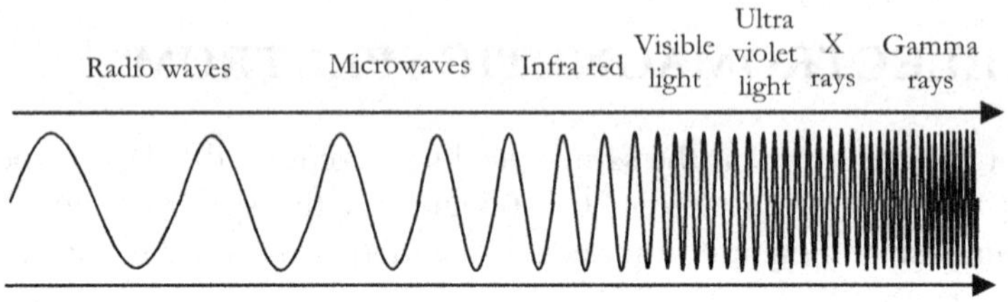

Figure 6.9
The electromagnetic spectrum.

7

Circuit Analysis

7.1 DIRECT AND ALTERNATING CURRENT

Connecting a constant voltage source, like a battery, to a circuit produces ***Direct Current*** (DC) that always flows in one direction. Circuit analysis can be performed by considering current as the flow of electrons, called *electron current* flow, where the direction of the current is from the negative terminal to the positive terminal. This reflects how actual current flows in a circuit. However, when the phenomenon of electricity was first discovered, it was not known whether the current was due to the flow of positive particles or negative particles. They had a fifty per cent chance of getting it right, and they chose wrong. It was believed that current was due to the movement of positive particles, and it stuck. Circuit analysis to this day is generally performed by choosing the positive flow of particles moving from the positive side of the voltage source to the negative side. This type of current flow is called *conventional current* flow. And unless otherwise stated, conventional current flow will be used in this book for circuit analysis.

When a voltage source is not constant, and the polarity of the terminals flips from positive to negative and from negative to positive, the current through the circuit follows suit and flips direction as well. This type of current is called ***Alternating Current*** (AC), as the direction of the current flow alternates over time, see Figure 7.1. The waveform of an AC source may be square, triangular or any infinite variety. The waveform of interest to us will be sinusoidal in nature. The mains AC voltage coming into your house is sinusoidal with a frequency of 50 or 60 Hz, depending on where you live. The carrier wave of AM radio is sinusoidal in nature, as well as audio signals. When combined, however, the signal is a superposition (mixing) of all these waves, and the resultant waveform looks like a mess. However, it still alternates in voltage and current.

Voltage potential difference is a relative measurement, and typically circuit voltages are measured with reference to some branch of the circuit known as ***common***. The branch of a circuit connected directly to the ground is called earth, or ground, and quite often is also chosen as common. As such, common is also quite often referred to as earth or ground.

Although most times the power source of a circuit is considered a voltage source, on occasions, the power source can be treated as a current source to make circuit analysis easier in certain circumstances. The circuit symbols for voltage and current sources are shown as follows.

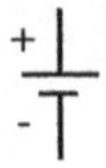

DC voltage source

AC voltage source

AC current source

7.2 RMS VOLTAGE AND CURRENT

AC voltage and current values may be given in peak-to-peak values, peak values or RMS values. See Figure 7.1. RMS stands for *root mean squared* and is derived by equating the average power of an AC signal to a DC equivalent value. We quite often express and analyse AC circuits using RMS values. The RMS value is the peak value divided by the square root of 2.

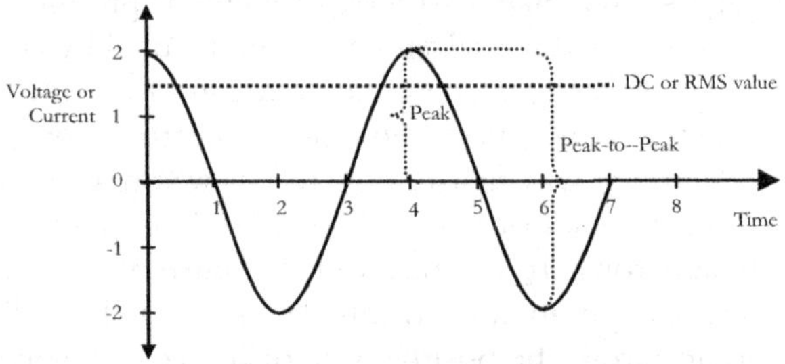

Figure 7.1
DC and AC voltage and current.

$$\boxed{V_{RMS} = \frac{V_{PEAK}}{\sqrt{2}}}$$ RMS voltage 7.1

$$\boxed{I_{RMS} = \frac{I_{PEAK}}{\sqrt{2}}}$$ RMS current 7.2

7.3 POWER

Power indicates the amount of energy expended per second, or the energy available to be released per second, and is given by

$$\boxed{P = \frac{E}{t}} \qquad \text{Power} \qquad 7.3$$

There are two types of power; *reactive power* and *real power*. Reactive power stores the energy and is available for release back into the circuit, and as such, does not result in a loss of energy from the system. Some devices, such as capacitors and inductors, store energy in the electric or magnetic field and give back most of this stored energy to the circuit. For ideal reactive components, the AC power loss over a full cycle is zero. For real-world components, however, there will generally be some small losses due to resistance.

Real power, on the other hand, results in a loss of energy from the circuit. Energy is used to perform work by converting the circuit's electrical energy into other forms of energy, such as heat, kinetic energy or sound. For instance, resistors will transform some of the electrical energy into heat. A speaker will convert electrical energy into mechanical energy (vibration of the speaker membrane).

For electrical circuits, power is determined by multiplying current by voltage.

$$\boxed{P = VI} \qquad \text{Electrical power(watts)} \qquad 7.4$$

For real power, the units are in *watts*. Reactive power is usually stated in *volt-amps reactive*. For power supplies and some devices, the total power is given in *volt-amps* that may include real and reactive components.

In AC systems with reactive components, there may be a phase angle between the voltage and current given by θ. Also, we generally talk about average power over a cycle for AC systems, as this gives the equivalent DC power loss, making circuit analysis a little easier. The average power is determined by multiplying the RMS current by the RMS voltage. The average power delivered to a load is

$$\begin{aligned} P_{AV} &= V_{RMS} I_{RMS} \cos(\theta) \\ &= \frac{V_{Peak}}{\sqrt{2}} \frac{I_{Peak}}{\sqrt{2}} \cos(\theta) \\ &= \frac{V_{Peak} I_{Peak}}{2} \cos(\theta) \end{aligned} \qquad \text{Average AC electrical power(watts)} \qquad 7.5$$

Using Ohm's Law, we can rearrange the power formula in the following ways

$$\boxed{P_{AV} = V_{RMS} I_{RMS}} \qquad 7.5a$$

And

$$\boxed{P_{AV} = I_{RMS}^2 R} \qquad 7.6$$

And

$$\boxed{P_{AV} = V_{RMS}^2 / R} \qquad 7.7$$

7.4 BASIC CIRCUIT COMPONENTS

The following section describes the fundamental electronic components. All other electronic components can be modelled using these basic components, whether they are diodes, transistors or integrated circuits. Other electronic components may change how the device reacts to changes in voltage, current or frequency, but fundamentally they all rely on these basic components. For instance, a diode changes its resistance depending on the polarity of the voltage applied, presenting a small resistance to current in one direction and a very high resistance in another direction. Diodes and coils are covered in their own Chapters.

Every electronic component is either made using wire or has wires attached as leads; therefore, every component has some tiny resistance, inductance and even capacitance depending on the geometry of the device. In this respect, every component can be considered as a collection of components made up of resistance, capacitance and inductance. This can lead to a maximum frequency limit in which a particular component can be used. Above a specific frequency, an inductor may start to look like a capacitor, and a capacitor may start to look like an inductor. This is particularly noticeable for physically large coils (inductors) we will wind ourselves and will be covered in the Chapter on coil design.

Resistors

Resistors are devices that resist the flow of current. The resistance is due to the collisions of electrons with atoms and other electrons in the material. These collisions result in energy loss from the circuit as electrical energy is converted into heat. Resistors are built to withstand a given continuous wattage (power) value. If the power dropped across the resistor goes above the stated value, you run the risk of damaging the resistor. Typical resistor wattage values used in electronics are ¼, ½ and 1 watt.

Large-wattage resistors will be physically big enough for the resistance value to be printed on the resistor body. Small wattage resistors of ¼ and ½ watt will have the resistance value shown as a colour code. The resistance value is not exact, and the last colour in the band will specify the tolerance value ranging from 0.1 to 10%. For instance, the colour code markings on the resistor shown in Figure 7.2 indicate a 24 thousand ohm resistor with a ±5% tolerance value. This means the resistance value could be between 22.8 kΩ and 25.2 kΩ.

Generally, there will be four or five bands of colours. The first two colours for four bands, or the first three colours for five bands, represent the first two or three digits of the resistance value. The next colour will be the multiplying value, and the last colour band will be the tolerance value. See Figure 7.2 for resistor colour code values.

Resistors allow both AC and DC currents to pass. The circuit symbol for a fixed resistor is

Fixed resistor

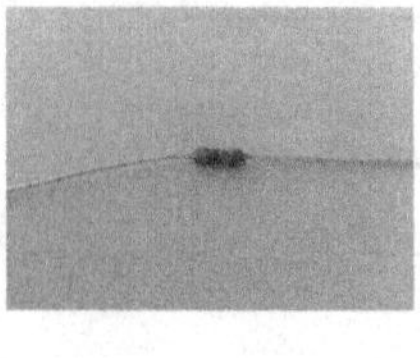

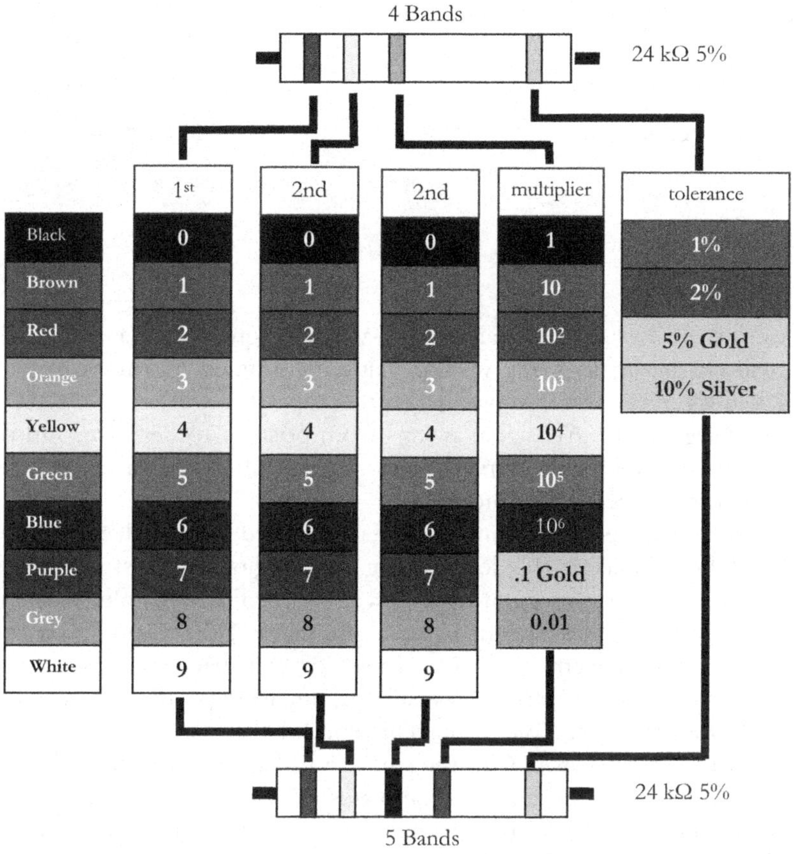

Figure 7.2
Resistor colour code chart.

Not all resistors are fixed in value. There are resistors that change the resistance value by turning a wiper arm that swipes across the resistive material; thus varying the resistance value. The resistance value ranges from 0 to some maximum resistance value. Variable resistors are physically large enough to have the maximum resistance value printed on the body. The circuit symbols for a variable resistor is

Variable resistor

A variable resistor with three terminals, with the middle terminal attached to a wiper arm, is called a potentiometer. It allows the resistance value to be split into three different portions; the total resistance across terminals 1 and 3, the resistance across terminals 1 and 2 and the resistance value across terminals 2 and 3. If the resistance changes in a linear fashion, the potentiometer is called a linear potentiometer. If the resistance changes in a logarithmic fashion, then the potentiometer is called a logarithmic potentiometer. Logarithmic potentiometers (log pots) are typically used as volume controls as they approximate the sensitivity of the human ear to changes in volume. The circuit symbol for a potentiometer is

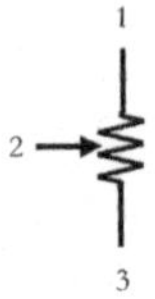

Potentiometer

Capacitors

Capacitors are devices that block DC and allow time-varying signals such as AC to pass. In a DC circuit, the plates will charge up to some voltage value determined by the design of the circuit. When fully charged, the direct current flow across the capacitor will cease. In AC circuits, the plates charge and discharge as the AC signal swings from positive to negative, and capacitors are said to allow AC signals to pass. Capacitors also store electrical energy in the electric field using the capacitance effect, as discussed in Chapter 4.

Capacitors are constructed using two metal plates separated by a dielectric. The capacitance value in picofarads will be printed on the device using three digits. The first two digits represent a numeric value, with the last digit representing a power of 10. For example, a ceramic capacitor with 102 printed in the body would represent 10×10^2 pF, that is 1 thousand pF or 1 nF.

The physical size and the dielectric used in a capacitor will determine the capacitance value. Even though a dielectric is considered to be an insulator, due to impurities, there will still be some small current that will flow across the dielectric. This current flow is called *leakage current*, and for our purposes, we want to use capacitors with the smallest leakage current possible. Current across the capacitor that does not add a charge across the plates is a loss of energy, and we want to store as much energy as we can, considering the small amount of energy crystal radios have to begin with.

There are many types of capacitors, but the most commonly used fixed capacitors are electrolytic, polyester and ceramic.

Electrolytic capacitors are constructed using a sandwich of aluminium with one side coated with aluminium oxide, an electrolytic past and another layer of aluminium. The sandwich is rolled up with the non-coated side being the positive side and the coated aluminium in direct contact with the paste as the negative side. They have large relative capacitance values in the order of one to several thousand microfarads. They are polarized, meaning they must be connected in the circuit the right way around. They are typically used when a large DC voltage component is applied across the capacitor, such as in a DC power supply circuit. The positive terminal of an electrolytic capacitor is connected to the positive side of the circuit, and the negative side of the capacitor is connected to the negative side of the circuit. Electrolytic capacitors have a relatively high leakage current. The circuit symbol for electrolytic capacitors is

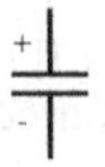

Electrolytic and polarized capacitors

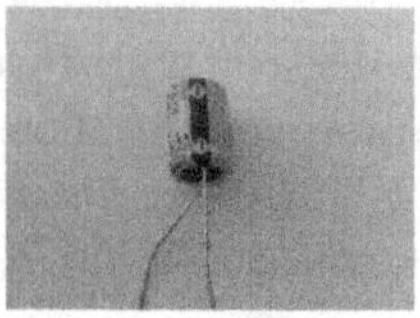

Polyester capacitors use polyester film as the dielectric. They typically have lower capacitance values than electrolytic capacitors, and range from 10 nanofarads to 10 microfarads and are not polarized. They are considered to be general-purpose capacitors and are commonly found in audio circuits.

Ceramic capacitors use ceramic material as the dielectric and have small capacitance values ranging typically from 1 picofarad to 470 nanofarads. Ceramic capacitors have very low leakage current and also have very good characteristics at high frequencies, meaning they still look like capacitors at high frequencies. These capacitors are often used in oscillators and RF tank circuits.

The circuit symbol for non-polarized fixed capacitors is

Fixed capacitor

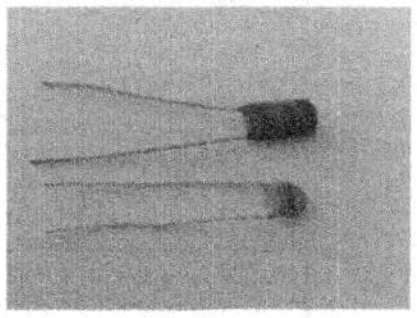

Variable capacitors are made using two sets of parallel plates with an air dielectric. Each set of plates consists of one or more flat conductors with an air gap between them. One set of parallel plates is fixed to the body, and the other is attached to a shaft that allows the plates to be intermeshed, changing the shared surface area between the plates. Changing the surface area between the plates changes the capacitance. Variable capacitors come in all shapes and sizes and typically range from a few picofarads to hundreds of picofarads. They have very low leakage currents. We will use a variable capacitor as the tuning capacitor in our crystal radios. The circuit symbol for the variable capacitor is

Variable capacitor

Inductors

Inductors are devices designed to have a high value of inductance. Commercially available inductors come in a wide range of values and sizes. Small inductors are colour coded to indicate the value of inductance in henries. The inductors we are interested in will be the inductors (coils) that we design and make ourselves. The circuit symbol for an inductor is

Inductor

7.5 VOLTAGE AND CURRENT RELATIONSHIPS

Resistor Voltage and Current

When a DC voltage is applied across a resistor, the voltage across the resistor is constant, as shown in Figure 7.3, and the voltage is given by Ohm's law.

$$V_R = I_R R$$

The DC current through the resistor is also constant, and using Ohm's law, is

$$I_R = \frac{V_R}{R}$$

When a sinusoidal AC voltage is applied across a resistor, the voltage and current across the resistor are in synch with the frequency of the voltage source, Figure 7.4.

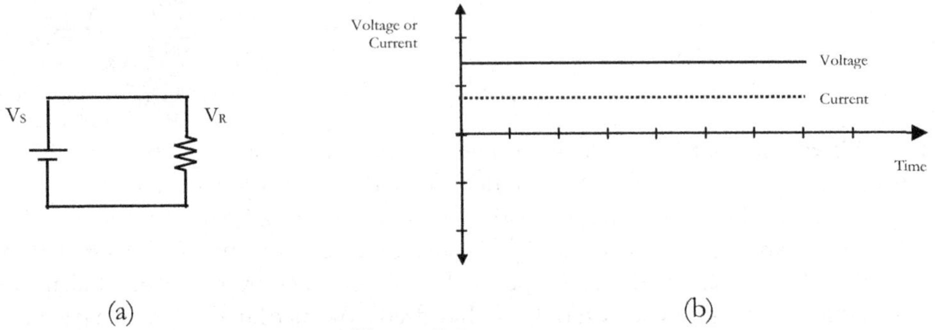

Figure 7.3
Resistor DC voltage and current response.

The instantaneous (meaning at any given time t) voltage across the resistor is given by

$$v_R(t) = i_R(t)R$$

and the instantaneous current across the resistor is given by

$$i_R(t) = \frac{v_R(t)}{R}$$

We could also rearrange Ohm's law to determine the resistance as

$$R = \frac{v_R(t)}{i_R(t)}$$

Resistance is an intrinsic property of a resistor and does not change. So the ratio of the voltage to current should be constant for any given time t. If we substitute the sinusoidal equation for voltage and current, we have

$$R = \frac{V_M \sin(\omega t + \theta)}{I_M \sin(\omega t + \theta)}$$

where V_M is the peak (amplitude) of the voltage, and I_M is the peak (amplitude) of the current. The sine values cancel in this instance, leaving

$$R = \frac{V_M}{I_M}$$

which is a constant value R.

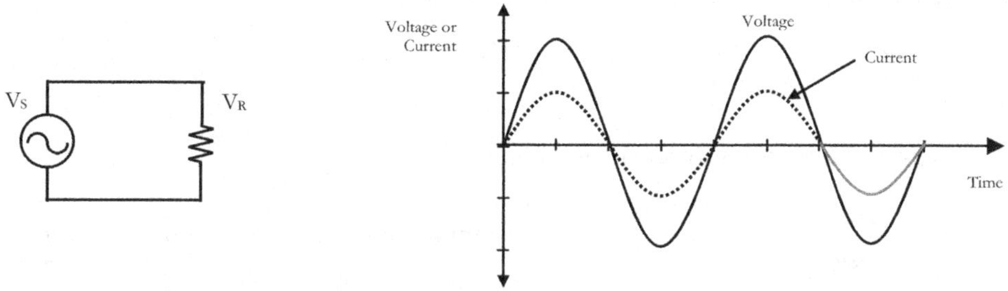

Figure 7.4
Resistor AC voltage and current response.

Capacitor Voltage and Current

Now let us apply a DC voltage across a capacitor, assuming there is no initial charge on the plates. When the DC voltage is first applied, there will be a surge of current as there is no charge and, therefore, no voltage across the capacitor plates to oppose the flow of current. Even though current does not flow across the dielectric, it does flow to and from the capacitor plates. As the charge accumulates on the plates, the voltage across the capacitor rises and the potential difference between the capacitor and the source voltage decreases. This decrease in potential difference causes the current to decrease. This continues until the voltage across the capacitor equals the voltage source and no more current flows, as shown in Figure 7.5. At the steady state condition, a capacitor blocks DC current.

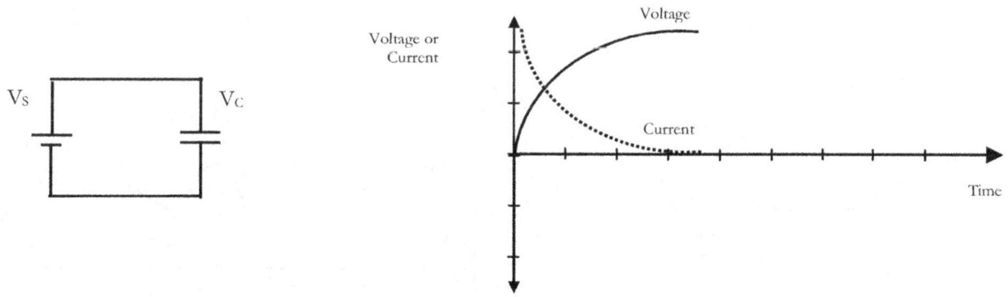

Figure 7.5
Capacitor DC voltage and current response.

When an AC voltage source is applied across a capacitor, from Equation 4.23, the instantaneous charge on a capacitor is

$$q_c(t) = Cv_c(t)$$

A changing charge on the plates produces a changing voltage across the capacitor so we can write

$$\frac{dq_c(t)}{dt} = C\frac{dv_c(t)}{dt}$$

Now a change in charge with respect to time is just current

$$i(t) = \frac{dq(t)}{dt}$$ Instantaneous current 7.8

So we can express the current in terms of a change in voltage

$$i_c(t) = C\frac{dv_c(t)}{dt}$$ Instantaneous current through a capacitor 7.9

We can find the voltage across the capacitor after some period of time from time=0 to Time=t by integrating 7.9 with respect to time.

$$\int_{t_0}^{t} dv_c(t) = \frac{1}{C}\int i_c(t)dt$$

And the voltage across a capacitor at a given time is given by

$$v_c(t) = \frac{1}{C}\int i_c(t)dt + v(t_0)$$ Instantaneous voltage across a capacitor 7.10

As the capacitor is in parallel with the voltage source, the voltage across the capacitor will follow the source.

$$v_S(t) = v_C(t) = V_M \sin(\omega t)$$

So solving for current using Equation 7.9, we have

$$i_c(t) = C\frac{d(V_M \sin(\omega t))}{dt}$$
$$= V_M \omega C \cos(\omega t)$$

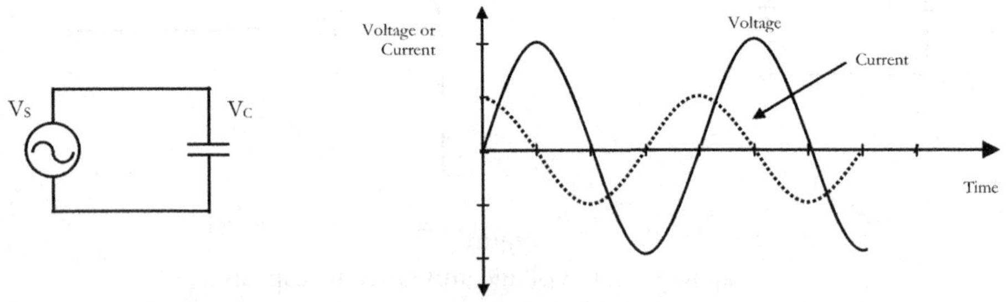

Figure 7.6
Capacitor AC voltage and current response.

The maximum value of a cosine function occurs at cos(0) = 1. Therefore, the value of $V_M\omega C$ must be the maximum current I_M.

$$I_M = V_M \omega C$$

And a cosine is the same as a sine function shifted by 90° to the right, so that we can write for the current

$$i_c(t) = I_M \sin(\omega t + \frac{\pi}{2})$$

Plotting voltage and current in Figure 7.6, we see there is a 90° phase difference between voltage and current. We say that voltage lags the current by 90° or alternatively, that current leads the voltage by 90° across a capacitor. Unlike in the DC circuit, where the current is blocked, in an AC circuit, the capacitor allows AC current to flow.

Inductor Voltage and Current

When a DC voltage is first applied across an inductor, the rate of change of current will be high, going from no current to some current in a very small time period. This rapid rate of change of current will produce a high back emf to counteract this change in current, so the current across the inductor will be small, and the voltage will be high. As the current increases, the rate of change of current decreases, meaning the opposition to the current, the emf, will also decrease. Eventually, the opposition to the current ceases, and the inductor appears as just a length of wire. The current across the inductor will be at a maximum, and the voltage across the inductor will be zero, see Figure 7.7. At the steady state condition, the inductor is a short at DC and allows DC current to pass.

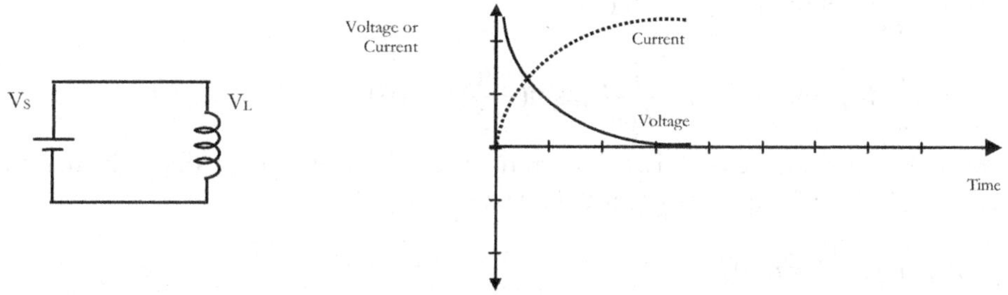

Figure 7.7
Inductor DC voltage and current response.

Applying an AC voltage across an inductor, the back emf will follow the change in current. Using Equation 5.15, the instantaneous voltage across an inductor is given by

$$v_L(t) = L\frac{\partial i_L(t)}{\partial t}$$

The voltage across the inductor will be in phase with the voltage source, that is

$$v_L(t) = v_S(t) = V_M \sin(\omega t)$$

Solving the differential equation

$$L\int di = \int v_L(t)dt$$

And

$$\int di = \frac{1}{L}\int v_L(t)dt$$

We are interested in the current between some initial time t_0 and onward. This will form the limits of integration so we can write

$$\int_{t_0}^{t} di = \frac{1}{L}\int v_L(t)dt$$

And

$$i_L(t) = \frac{1}{L}\int v_L(t)dt + i_L(t_0) \qquad \text{Instantaneous current through an inductor} \qquad 7.11$$

Taking the initial time $t_0 = 0$, that is when we first apply the voltage. Solving 7.11 for a sinusoidal voltage source gives

$$\begin{aligned} i_L(t) &= \frac{1}{L}\int_0^t V_M \sin(\omega t)dt + i_L(0) \\ &= -\frac{1}{L\omega}V_M \cos(\omega t)\Big|_0^t + i_L(0) \\ &= \frac{1}{L\omega}V_M \sin(\omega t - \frac{\pi}{2})\Big|_0^t + i_L(0) \\ &= \frac{1}{L\omega}V_M \sin(\omega t - \frac{\pi}{2}) - \frac{1}{L\omega}V_M \sin(-\frac{\pi}{2}) + i_L(0) \end{aligned}$$

As voltage divided by impedance is the current, the second term on the right-hand side of the equation is just the initial current $i_L(0)$ at t=0. That is

$$\frac{1}{L\omega}V_M \sin(-\frac{\pi}{2}) = i_L(0)$$

So we can write

$$\begin{aligned} i_L(t) &= \frac{1}{L\omega}V_M \sin(\omega t - \frac{\pi}{2}) - i_L(0) +_L (0) \\ &= \frac{V_M}{L\omega}\sin(\omega t - \frac{\pi}{2}) \end{aligned}$$

With the peak current given by

$$I_M = {}^{V_M}\!/_{L\omega}$$

We can write

$$i_L(t) = I_M \sin(\omega t - \frac{\pi}{2})$$

where I_M is the peak current. Plotting AC voltage and current in Figure 7.8, we see there is a -90° phase difference between voltage and current. We say that voltage leads the current by 90° or alternatively that current lags voltage by 90° across an inductor.

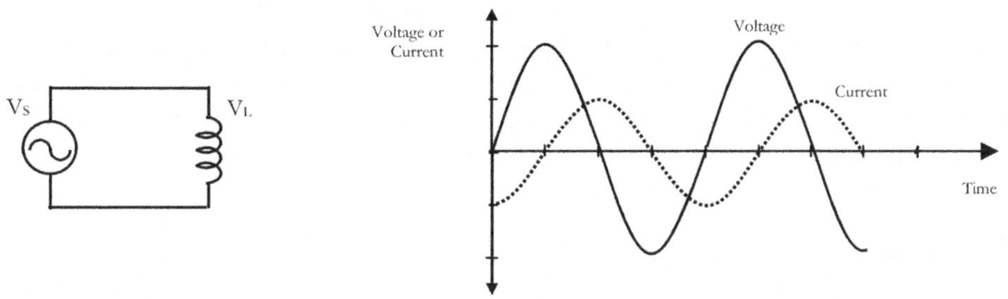

Figure 7.8
Inductor AC voltage and current response.

7.6 PHASORS

In the preceding section, we analysed the voltage and current for three basic electronic components using a simple circuit consisting of an AC power source and a single component. Real-world AC circuits will have many basic components resulting in phase shifts of voltage and current. We can analyse the circuit in the time domain using complicated differentiation and integration and then add up the sinusoidal values at some time t. This would make even the most basic of circuits difficult to analyse. Luckily for us, there is a way we can analyse circuits using Ohm's law and other well-established techniques for circuit analysis using something called phasors.

Consider the voltage and current relationship for the inductor, as shown in Figure 7.8. We can represent these sinusoidal waveforms in a different manner. Say we represent the peak voltage V_M and the peak current I_M as vectors starting at the origin, with the y-axis representing the instantaneous value of either the current or the voltage. And then represent the passage of time as an angle of rotation between the vectors and the x-axis, with the voltage angle given by ωt (the angular velocity times time) and the current angle given by $\omega t-\pi/2$ as given by the time domain equations for voltage and current. If we now project the instantaneous value onto another graph with ωt along the x-axis (rectangular coordinates) representing the passage of time. We can see that as time progresses, the vectors rotate around the origin, and the projection of both the voltage and current onto the rectangular graph (Figure 7.9) produces the same sinusoidal pattern as seen in Figure 7.8. Therefore both graphs represent the same phenomena over time. Note that time "t" is expressed in terms of a fraction of a period and

$$\omega t = \frac{2\pi}{T}(t)$$

For example, when t =0, ωt=0. When t=T/8

$$\omega t = \frac{2\pi}{T}\frac{T}{8} = \frac{\pi}{4}$$

When ωt = 2π, one complete cycle, the projection will look the same when ωt =0. Vectors with a constant vector length that rotate around the origin over time are called ***phasors***.

If we take a snapshot of the phasors at some time t, usually t = 0, and provided the frequency is the same through each component, we can treat these sinusoids as vectors with length and some phase angle θ with respect to the x positive axis. When analysing phasors, we select a frequency of interest and perform the circuit analysis for a given frequency. If the circuit frequency changes, we redo the analysis for the new frequency. Note that resistors, capacitors and inductors do not change the frequency; only the amplitudes and phase angles will change.

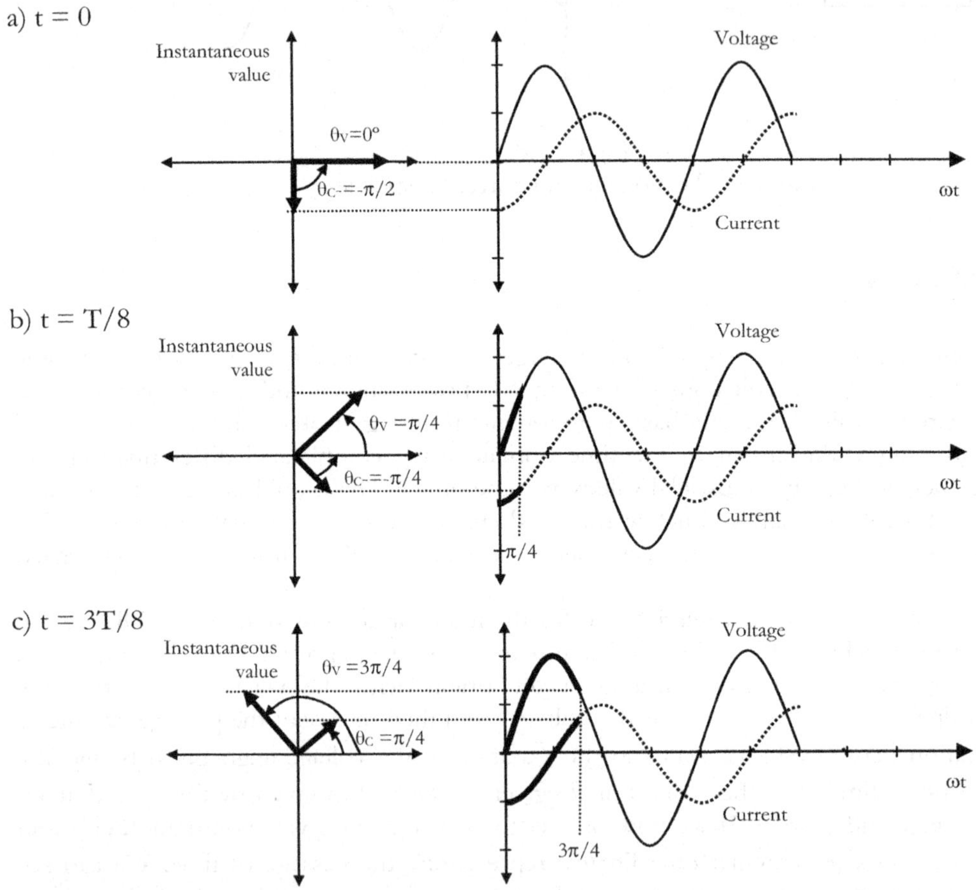

Figure 7.9
Inductor voltage and current response represented as phasors.

We can represent circuit values of voltage, current and impedance (yet to be discussed) in phasor notation using polar form or rectangular form. The polar form is

$$\mathbf{A} = A\angle\theta$$

Like vectors, phasors will be in bold letters and the magnitude in normal letters. When writing the phase angle θ, degrees will be used for clarity; however, typically, the calculations will be done in radians.

We can convert from the time domain to the phasor domain as follows

$$v(t) = V_M \sin(\omega t \pm \theta) \Rightarrow V_M \angle \pm \theta$$

For instance, we would write the time domain representation of the inductor voltage, at time t=0, in the above example as

$$v_L = V_M \sin(\omega t + 0)$$

Something in the time domain simply means that it is a function (dependent) of time. In the phasor domain, the time dependence disappears because we are only interested in the angle between two vectors or the angle between a vector and the x-axis. The above time domain voltage would be written in the phasor domain as

$$\mathbf{V_L} = V_M \angle 0^\circ$$

The inductor current in the time domain would be

$$i_L(t) = I_M \sin(\omega t - \frac{\pi}{2})$$

And in the phasor domain would be

$$\mathbf{I_L} = I_M \angle -90^\circ$$

The above representation of the phasor is in a format called polar form. We can express phasors in another form using a complex rectangular coordinates system. In Figure 7.10, the y-axis is represented by jY and –jY, where j represents the square root of minus 1, and the x-axis represents real numbers. The square root of minus one does not exist in the real world, and any number using j (Mathematicians use i) is called imaginary.

$$\boxed{j = \sqrt{-1}} \qquad 7.12$$

A number that has both real and imaginary components is called a ***complex number*** and has the form

$$A = X \pm jY$$

It is possible to use one form only, either polar or rectangular when manipulating phasors. However, as you will see, it makes life a lot easier to move between the two representations depending on the algebraic manipulation we are doing. We will use polar form when multiplying or dividing phasors and the complex rectangular form when adding or subtracting.

The phasor shown in Figure 7.9b is shown in the complex coordinate system in Figure 7.10. In polar form, the voltage is

$$\mathbf{V_L} = V_M \angle \theta_V$$

And in rectangular format, the voltage is

$$\mathbf{V_L} = X + jY$$

where X is the x (real) component of the vector, and Y is the y (imaginary) component of the vector. The current in polar form is

$$\mathbf{I_L} = I_M \angle \theta_I$$

And in rectangular format, is

$$\mathbf{I}_L = X - jY$$

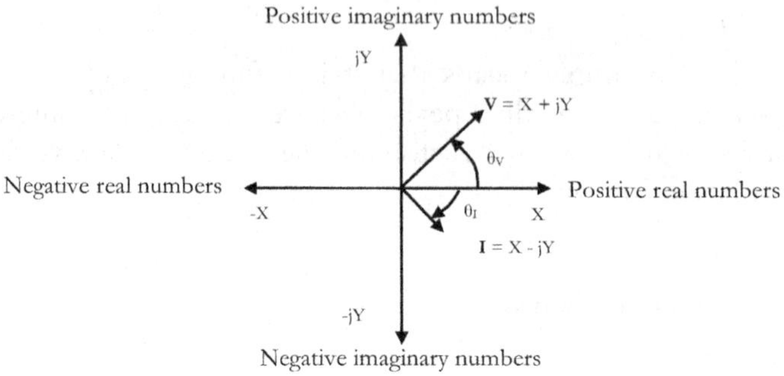

Figure 7.10
Phasor representation of voltage and current.

Phasor Algebraic Manipulation

Given two phasors in polar form

$$\mathbf{A} = A\angle\theta_A$$

$$\mathbf{B} = B\angle\theta_B$$

We multiply two phasors as follows

$$\begin{aligned}\mathbf{AB} &= (A\angle\theta_A)(B\angle\theta_B) \\ &= AB\angle(\theta_A + \theta_B)\end{aligned} \qquad 7.13$$

The magnitudes of the two phasors are multiplied, and the angles are added. To divide two polar numbers

$$\begin{aligned}\frac{\mathbf{A}}{\mathbf{B}} &= \frac{A\angle\theta_A}{B\angle\theta_B} \\ &= \frac{A}{B}(\angle\theta_A - \theta_B)\end{aligned} \qquad 7.14$$

The magnitudes of the two phasors are divided, and the denominator angle is subtracted from the numerator angle.

To add or subtract two phasors, we use complex numbers. The complex rectangular number format is the same for adding and subtracting vectors. The real and imaginary components of each phasor are added or subtracted. For example, given two phasors in rectangular format

$$\mathbf{A} = A_X + jA_Y$$

$$\mathbf{B} = B_X + jB_Y$$

Adding phasor **A** and **B**

$$\begin{aligned}\mathbf{A}+\mathbf{B} &= (A_X + jA_Y)+(B_X + jB_Y) \\ &= (A_X + B_X)+(jA_Y + jB_Y)\end{aligned} \qquad 7.15$$

And subtracting phasor **B** from phasor **A**

$$\begin{aligned}\mathbf{A}-\mathbf{B} &= (A_X + jA_Y)-(B_X + jB_Y) \\ &= (A_X - B_X)+(jA_Y - jB_Y)\end{aligned} \qquad 7.16$$

We also need to convert from one form to another. To convert from polar form to rectangular form, we need to split the magnitude of the phasor into its real and imaginary components. This is done by the following equations.

$$\begin{aligned}A_X &= A\cos(\theta_A) \\ A_Y &= A\sin(\theta_A)\end{aligned} \qquad 7.17$$

And the rectangular format of the phasor can now be written using the calculated components as

$$\mathbf{A} = A_X + jA_Y$$

To convert from rectangular form to polar form, we need to calculate the magnitude of the phasor and its phase angle. The magnitude is just the Pythagorean formula and is given by

$$A = \sqrt{(A_X)^2 + (A_Y)^2} \qquad 7.18$$

The phase angle is calculated using the arctangent of the imaginary component divided by the real component.

$$\theta_A = \tan^{-1}(\frac{A_Y}{A_X}) \qquad 7.19$$

And we can now write the phasor **A** in polar form

$$\mathbf{A} = A\angle\theta_A$$

To calculate the roots, such as a square root, of a complex number, first convert the number into polar form. Take the root of the magnitude and divide the angle by the root number.

$$\sqrt[n]{\mathbf{A}} = \sqrt[n]{A}\angle(\frac{\theta_A}{n}) \qquad 7.20$$

For example, the square root of the phasor **A** would be

$$\sqrt[2]{\mathbf{A}} = \sqrt[2]{A}\angle(\frac{\theta_A}{2})$$

Euler's Formula and Phasors

There is an important mathematical relationship between sinusoidal waves and complex numbers called Euler's formula.

$$e^{jx} = \cos(x) + j\sin(x) \qquad \text{Euler's formula} \qquad 7.21$$

This allows us to express phasors in another format. This format, like the others, has its advantages in certain circumstances. The symbol e, is called Euler's number and is the base number of the natural logarithm (ln) and is approximately 2.718281828459. That is,

$$\ln x = \log_e x$$

And

$$\ln e^x = x$$

A function using Euler's number e is called an exponential function. For example

$$f(x) = e^{ax+b}$$

where a and b are constants. If we differentiate the exponential function above with respect to x, we get

$$\frac{df(x)}{dx} = \frac{d(e^{ax+b})}{dx}$$
$$= ae^{ax+b}$$

The derivative of an exponential function is the original function times the derivative of the exponent (the top bit ax+b). If we were to differentiate it again, the second derivative with respect to x, we would get

$$\frac{d^2 f(x)}{dx^2} = a^2 e^{ax+b}$$

Expressing a phasor as an exponential is useful when differentiating or integrating. We can express phasors in exponential format as follows

$$\mathbf{A} = Ae^{j\theta} \qquad \text{A phasor in exponential form} \qquad 7.22$$

To summarize, a phasor **A** can be expressed in any of the following ways

$$\mathbf{A} = A\angle\theta^\circ = A_X + jA_Y = A\cos(\theta) + jA\sin(\theta) = Ae^{j\theta} \qquad 7.23$$

7.7 REACTANCE AND IMPEDANCE

If we know the voltage and current across a resistor in a circuit, we can use Ohm's law to calculate the resistance as follows.

$$R = \frac{V}{I}$$

This resistance is in opposition to the current and is the cause of the voltage drop across a resistor. Also, resistance represents a loss of energy from the system as the electrical energy is converted into heat. For pure capacitors and inductors, the energy is stored and is not lost. There is, however, as we have seen, an opposition to the flow of AC. This opposition, like resistance, is the cause of the voltage drop across a capacitor or inductor. We call this type of opposition to the flow of AC current ***reactance*** and use the letter X to represent reactance. For a capacitor, we have ***capacitive reactance***, X_C, and for an inductor, we have ***inductive reactance***, X_L.

Capacitive Reactance

For AC circuits, we want to know how a circuit responds at a particular frequency. Consider a sinusoidal voltage across a capacitor given by the following function

$$v_C(t) = V_M \sin(\omega t)$$

In polar phasor form, this is

$$\mathbf{V}_C = V_M \angle 0^\circ$$

The current across a capacitor leads the voltage by 90°; therefore, the current in the time domain is

$$i_c(t) = I_M \sin(\omega t + \frac{\pi}{2})$$

And in polar form, is

$$\mathbf{I}_C = I_M \angle 90^\circ$$

We can now use Ohm's law, in phasor form, to calculate the opposition to the current, which is the capacitive reactance.

$$\begin{aligned}\mathbf{X_C} &= \frac{\mathbf{V_C}}{\mathbf{I_C}} \\ &= \frac{V_M \angle 0^\circ}{I_M \angle 90^\circ} \\ &= \frac{V_M}{I_M} \angle (0^\circ - 90^\circ)\end{aligned}$$

In section 7.5, we determined that the maximum current is given by

$$I_M = V_M \omega C$$

Using this relationship, we can write

$$\begin{aligned}\mathbf{X_C} &= \frac{V_M}{V_M \omega C} \angle (-90^\circ) \\ &= \frac{1}{\omega C} \angle (-90^\circ) \\ &= \frac{1}{2\pi f C} \angle (-90^\circ)\end{aligned}$$

We now have an equation to determine the magnitude of the capacitive reactance

$$\boxed{X_C = \frac{1}{2\pi f C}} \qquad \text{Capacitive reactance magnitude (ohms)} \qquad 7.24$$

In phasor form, the capacitive reactance can be written in either polar or rectangular form and is

$$\boxed{\mathbf{X_C} = -jX_C = X_C\angle(-90^\circ)} \qquad \text{Phasor form of capacitive reactance} \qquad 7.25$$

Inductive Reactance

Using Ohm's law to calculate the inductive reactance and given that current lags voltage by 90° across an inductor, we can write

$$\mathbf{X_L} = \frac{\mathbf{V_L}}{\mathbf{I_L}}$$

$$= \frac{V_M\angle 0^\circ}{I_M\angle -90^\circ}$$

$$= \frac{V_M}{I_M}\angle(0^\circ + 90^\circ)$$

And the peak current is

$$I_M = \frac{V_M}{L\omega}$$

Substituting this relationship in the above equation, we get

$$\mathbf{X}_L = \frac{V_M}{\frac{V_M}{L\omega}}\angle 90^\circ$$

$$= L\omega\angle 90^\circ$$

We can now write an equation to determine the magnitude of the inductive reactance as

$$\boxed{X_L = 2\pi f L} \qquad \text{Inductive reactance magnitude (ohms)} \qquad 7.26$$

And in phasor form

$$\boxed{\mathbf{X_L} = jX_L = X_L\angle 90^\circ} \qquad \text{Phasor form of capacitive reactance} \qquad 7.27$$

Impedance

In the above analysis, the capacitor and the inductor were treated as if they had no resistance. In reality, a component will always have some resistance due to its physical construction. In most instances, the resistance in a capacitor is so low that it can safely be ignored. For large coil inductors, however, the wire resistance cannot be ignored. The total opposition to the current that includes both reactance and resistance is called ***impedance*** and is signified by the letter Z.

If resistance and reactance are treated as phasors and are plotted on a phasor diagram, the resistance will appear purely as real, and reactance will appear purely as imaginary, as shown in

Figure 7.11. These two phasors are at right angles to one another and will form a right-angle triangle. The magnitude of the hypotenuse of this right-angle triangle is given by Pythagoras

$$\boxed{Z = \sqrt{R^2 + X^2}} \qquad \text{Impedance magnitude (ohms)} \qquad 7.28$$

And it represents the magnitude of the impedance.

A pure inductor will always be represented by a positive imaginary number jY, Figure 7.11a. A pure capacitor will always be represented by a negative imaginary number -jY, Figure 7.11b. The phasor form of the impedance contains a real number (the resistance) and an imaginary number (the reactance) and is, therefore, a complex number. The impedance of an inductor in phasor form is

$$\boxed{\mathbf{Z}_{\mathrm{L}} = R_L + jX_L = Z_L\angle\theta^\circ} \qquad \text{Phasor inductor impedance} \qquad 7.29$$

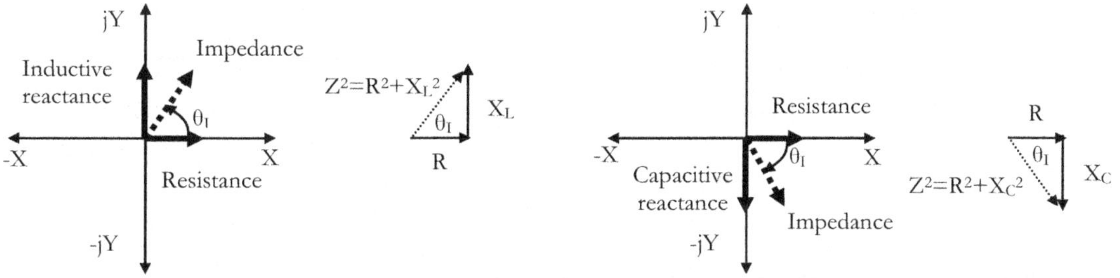

(a) Looks inductive. (b) Looks capacitive.

Figure 7.11
Phasor impedance diagram.

And for a capacitor is

$$\boxed{\mathbf{Z}_C = R_C - jX_C = Z_C\angle -\theta^\circ} \qquad \text{Phasor capacitor impedance} \qquad 7.30$$

Note that typically the resistance of a capacitor is so low it is normally ignored (R_C=0), and the impedance is just the capacitive reactance.

In this book, we will use the following circuit diagram component for impedance.

Impedance (maybe purely resistive, purely reactive or a combination of both)

7.8 SERIES CIRCUITS

Two components of a circuit are considered to be in series when one terminal of a component is connected to only one terminal of another component, and this connection point is not connected to any other current-carrying component. That is, the same current flows through each of the components.

Consider Figure 7.12a. The voltage source is connected to one end of resistor R_1, and the other end of R_1 is connected to only one component, namely R_2. All the components are said to be connected in series, and this circuit is called a ***series circuit***. The current flowing across each component in series is the same. Any part of a circuit with one or more components in series is called a ***branch***.

In Figure 7.12b, we added another resistor, R_3. Resistors R_1 and R_2 are now no longer in series as their connection point is connected to another current-carrying component (R_3). The current flowing across R_1 splits, and a portion of the current flows across R_2 and another portion of the current flows across R_3. Note that if only one terminal of resistor R_3 was connected to the circuit and the other end was not connected, then no current would flow across R_3, and R_1 and R_2 would still be in series.

(a) R1 and R2 in series　　(b) R1 and R2 not in series

Figure 7.12
Series and non-series circuits.

Total Series Resistance and Impedance

To find the total resistance of resistors in series, simply add the resistance values.

$$R_T = R_1 + R_2 + R_3 + \ldots R_n$$ Total resistance of resistors in series 7.31

For the circuit in figure7.12a, the total resistance of the circuit is

$$\begin{aligned} R_T &= 10 + 47 \\ &= 57\Omega \end{aligned}$$

And similarly, for the total impedance of a branch, add the individual component impedance phasors.

$$\mathbf{Z_T} = \mathbf{Z_1} + \mathbf{Z_2} + \mathbf{Z_3} + \ldots \mathbf{Z_n}$$ Total impedance of impedances in series 7.32

Total Series Capacitance

The total capacitance of capacitors in series is

$$\frac{1}{C_T} = \frac{1}{C_1} + \frac{1}{C_2} + \frac{1}{C_3} + \ldots \frac{1}{C_n}$$ Total capacitance of capacitors in series 7.33

If only two capacitors are in series, you can calculate the total capacitance using

$$C_T = \frac{C_1 C_2}{C_1 + C_2}$$ Total capacitance of two capacitors in series 7.34

Total Series Inductance

For inductors in series, the total inductance is the sum of the inductors, provided they are far enough apart that mutual inductance is not a factor. Mutual inductance will be discussed later.

$$L_T = L_1 + L_2 + L_3 + \ldots L_n$$ Total inductance of inductors in series 7.35

Examples of Series Circuit Analysis

Example 7.1

Given the circuit in Figure 7.13, find the total current drawn from the power supply.

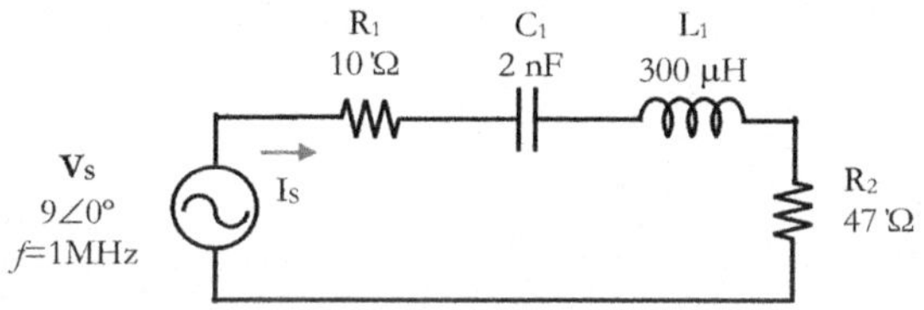

Figure 7.13

We are given an AC supply voltage of 9 volts RMS (assume RMS unless otherwise stated) at a frequency of 1 MHz. If we calculate the total impedance of the circuit, we can determine the current I_S using Ohm's law.

$$\mathbf{I_S} = \frac{\mathbf{V_S}}{\mathbf{Z_T}}$$

First, calculate the reactance values for the capacitor and inductor at the given frequency. The capacitive reactance is

$$X_C = \frac{1}{2\pi f C}$$

$$= \frac{1}{2\pi(10^6)(2x10^{-9})}$$

$$= 79.58$$

As this is capacitive reactance, we can write this in phasor form as

$$\mathbf{X_C} = -j79.58$$

And the inductive reactance is

$$X_L = 2\pi fL$$
$$= 2\pi(10^6)(300x10^{-6})$$
$$= 1884$$

And the phasor form of the inductive reactance is

$$\mathbf{X_L} = j1884$$

The total impedance for the series circuit is the sum of the impedances for each component (Equation 7.32). As this is phasor addition, we add the real components together and the imaginary components together.

$$Z_T = R_1 - jX_C + jX_L + R_2$$
$$= 10 - j79.58 + j1{,}884 + 47$$
$$= (10 + 47) + (-j79.58 + j1{,}884)$$
$$= 57 + j1{,}804.42$$

Converting from rectangular form to polar first, calculate the magnitude of the total impedance.

$$Z_T = \sqrt{(57)^2 + (1804.42)^2}$$
$$= 1805\Omega$$

Next, calculate the phase angle.

$$\theta_{Z_T} = \tan^{-1}(\frac{1804.42}{57})$$
$$= 88^\circ$$

In polar form, the total impedance is

$$\mathbf{Z_T} = 1805\angle 88^\circ$$

We can now plug this into Ohm's law to calculate the current

$$\mathbf{I_S} = \frac{\mathbf{V_S}}{\mathbf{Z_T}}$$

$$\mathbf{I_S} = \frac{9\angle 0^\circ}{1805\angle 88^\circ}$$
$$= 0.00499\angle -88^\circ$$

The magnitude of the current running through the circuit given in Figure 7.13 is 4.99 mA with a phase angle, with respect to the voltage source, of -88°. That is, the current lags the voltage by 88°, which is what we expect for an inductor. We could represent the total impedance as a single impedance value in polar form, as shown in Figure 7.14a, or in rectangular form, showing what the impedance looks like in terms of components, as in Figure 7.14b. Because the imaginary value is positive, it looks like an inductor; we say that the circuit "looks" inductive.

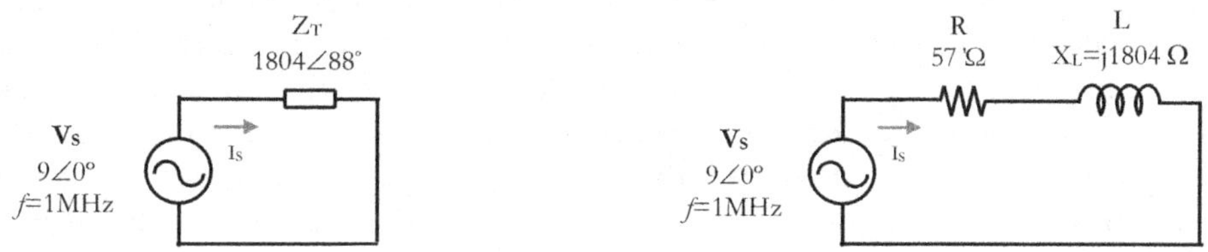

Figure 7.14
The total impedance of the circuit in Figure 17.13.

Example 7.2

Using the circuit in Figure 7.13, calculate the voltage drop across each of the components in the circuit. As this is a series circuit, the same current runs through each component. We now know the impedance for each component, and we know the current, so once again, we can use Ohm's law to calculate the voltages.

For the resistor R_1

$$\begin{aligned}\mathbf{V}_{R_1} &= \mathbf{I}_{R_1}\mathbf{Z}_{R_1}\\ &= (0.00499\angle-88^\circ)(10\angle0^\circ)\\ &= (0.00499x10)(\angle-88^\circ+0^\circ)\\ &= 0.0499\angle-88^\circ\,volts\end{aligned}$$

For the resistor R_2

$$\begin{aligned}\mathbf{V}_{R_2} &= \mathbf{I}_{R_2}\mathbf{Z}_{R_2}\\ &= (0.00499\angle-88^\circ)(47\angle0^\circ)\\ &= (0.00499x47)(\angle-88^\circ+0^\circ)\\ &= 0.2345\angle-88^\circ\,volts\end{aligned}$$

For capacitor C_1

$$\begin{aligned}\mathbf{V}_{C_1} &= \mathbf{I}_{C_1}\mathbf{Z}_{C_1}\\ &= (0.00499\angle-88^\circ)(79.58\angle-90^\circ)\\ &= (0.00499x79.58)(\angle-88^\circ-90^\circ)\\ &= 0.3971\angle-178^\circ\,volts\end{aligned}$$

For inductor L_1

$$\begin{aligned}\mathbf{V}_{L_1} &= \mathbf{I}_{L_1}\mathbf{Z}_{L_1}\\ &= (0.00499\angle-88^\circ)(1884\angle90^\circ)\\ &= (0.00499x1884)(\angle-88^\circ+90^\circ)\\ &= 9.4\angle2^\circ\,volts\end{aligned}$$

Notice that the voltage across the inductor is larger than the supply voltage. Has something gone wrong here? The answer is no. The voltage really will be higher across the inductor in this circuit than the supply voltage. It is due to the phase shifts of the capacitor and inductor. The voltage across the inductor and capacitor can be very much higher than the supply voltage as we approach resonance. However, if we sum the voltages as phasors across each of the components, we will end up with the magnitude of the supply voltage. Welcome to the confusing and wacky world of AC circuits. This is why phasors **must** be used to analyse AC circuits.

To convince ourselves that this is true, we will convert each voltage to rectangular form, as they are easier to add.

$$\begin{aligned}\mathbf{V_S} &= \mathbf{V_{R_1}} + \mathbf{V_{R_2}} + \mathbf{V_{C_1}} + \mathbf{V_{L_1}} \\ &= (0.0017414 - j0.049869) + (0.0081839 - j.234357) + (-.3968581 - j0.0138585) + \\ &\quad (9.394 + j.3280552) \\ &= 9.007 - j0.0299\end{aligned}$$

And the amplitude is

$$\begin{aligned}V_S &= \sqrt{9.007^2 + 0.0299^2} \\ &= 9.007\end{aligned}$$

So the amplitude of the total voltage drops across the circuit is 9.007 volts. The extra 7 mV and the imaginary term in the final result are due to rounding errors. Notice the real component of the voltage drop across the capacitor is negative as the phase angle is -178°.

Kirchhoff's Voltage Law

We see that taking the phasor algebraic sum of all the voltage drops equals the source voltage. This brings us to a very important law known as ***Kirchhoff's voltage law*** (KVL), which states *the algebraic sum of the potential rises and drops around a closed path is zero* (as we have demonstrated in this example).

$$\boxed{\mathbf{V_S} = \mathbf{V_1} + \mathbf{V_2} + \mathbf{V_3} + \ldots \mathbf{V_n}} \qquad \text{KVL}$$

The Voltage Divider Rule

Another important rule is the ***voltage divider rule.*** For two or more impedances in series, if you know the branch voltage and the total branch impedance, then you can calculate the voltage across individual components.

$$\boxed{\mathbf{V}_X = \mathbf{V_S}\frac{\mathbf{Z}_X}{\mathbf{Z_T}}} \qquad \text{The voltage divider rule} \qquad 7.36$$

Example 7.3

Using the voltage divider rule, calculate the voltage drop across the capacitor for the circuit in Figure 7.13. As there is only one branch in this circuit and from KVL, the sum of the component voltage drops must be the same as the source voltage, and V_S=9 volts. We have already calculated the total impedance of the circuit and the impedance of the capacitor. So we can write 7.36 as

$$\mathbf{V}_{C_1} = 9\angle 0^\circ \frac{79.58\angle -90^\circ}{1804\angle 88^\circ}$$
$$= 0.397\angle -178^\circ$$

This is the same as the voltage calculated for the capacitor in Example 7.2.

7.9 PARALLEL CIRCUITS

Two components or branches are in parallel when they have two points in common. In Figure 7.15a, all three components are in parallel as they each have two connection points in common. In Figure 7.15b, components R_1 and R_2 are in series, but this series branch is in parallel with the other two components, L_1 and C_1.

Notice that the current splits when it encounters a parallel branch. In Figure 7.15a, the source current I_S splits with some current flowing through inductor L_1 and the rest I_2 flowing into the next branch. Current I_2 then splits, with current I_3 flowing through resistor R_1 and current I_4 across the capacitor. The branch currents flow back to the source power supply via the common rail so that the total current going into the circuit is the same as the current going back to the supply voltage. The current flowing through the parallel branches may differ for each branch; however, the voltage across each parallel branch is the same.

In Figure 7.15a, the voltage across L_1 will be the same as the voltage across R_1, which will be the same as the voltage across C_1. In Figure 7.15b, L_1 and C_1 will have the same voltage drop, and the parallel branch consisting of R_1 and R_2 will have the same voltage drop across the entire branch as L_1 and C_1. Also, R_1 and R_2 are in series with each other, and the current I_3 will flow across each resistor. But the voltage drop across the individual resistors R_1 and R_2 will be determined by the resistance value and the current.

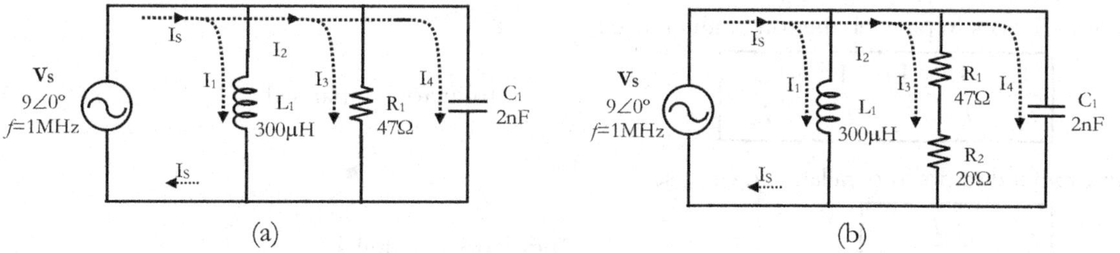

Figure 7.15
Parallel circuits.

In this example, each of the branches is in parallel with the voltage source. The amplitudes of the voltages for circuit 7.15a are

$$V_S = V_{L1} = V_{R1} = V_{C1}$$

And for circuit 7.15b are

$$V_S = V_{L1} = (V_{R1} + V_{R2}) = V_{C1}$$

Total Parallel Resistance and Impedance

The total resistance of resistors in parallel is given by

$$\frac{1}{R_T} = \frac{1}{R_1} + \frac{1}{R_2} + \frac{1}{R_3} + \ldots \frac{1}{R_n}$$

Total resistance of resistors in parallel 7.37

For two resistors in parallel, you can use

$$R_T = \frac{R_1 R_2}{R_1 + R_2}$$

Two resistors in parallel 7.38

Similarly, for impedance, the total parallel impedance is given by the phasor equation

$$\frac{\mathbf{1}}{\mathbf{Z_T}} = \frac{\mathbf{1}}{\mathbf{Z_1}} + \frac{\mathbf{1}}{\mathbf{Z_2}} + \frac{\mathbf{1}}{\mathbf{Z_3}} + \ldots \frac{\mathbf{1}}{\mathbf{Z_n}}$$

Total impedance in parallel 7.39

For two impedances in parallel, we can use

$$\mathbf{Z_T} = \frac{\mathbf{Z_1 Z_2}}{\mathbf{Z_1 + Z_2}}$$

Two impedances in parallel 7.40

Total Parallel Capacitance

The total capacitance of capacitors in parallel is the summation of the capacitance values

$$C_T = C_1 + C_2 + C_3 + \ldots C_n$$

Capacitors in parallel 7.41

Total Parallel Inductance

For inductors in parallel, the total inductance is

$$\frac{1}{L_T} = \frac{1}{L_1} + \frac{1}{L_2} + \frac{1}{L_3} + \ldots \frac{1}{L_n}$$

Inductors in parallel 7.42

For two inductors in parallel, you can use

$$L_T = \frac{L_1 L_2}{L_1 + L_2}$$

Inductors in parallel 7.43

The Current Divider Rule

For series components, we had the voltage divider rule. For parallel components, we have the ***current divider rule***. The current divider rule gives us the branch current for a particular branch for two or more branches in parallel, as is

$$\boxed{\mathbf{I_X} = \mathbf{I_S}\frac{\mathbf{Z_T}}{\mathbf{Z_X}}}$$ Current divider rule 7.44

where Z_T is the total parallel impedance as given by Equation 7.39, and Z_X is the impedance of the branch of interest. For circuits with two parallel branches only, Equation 7.44 simplifies to

$$\mathbf{I_1} = \mathbf{I_S}\frac{\mathbf{Z_2}}{\mathbf{Z_1}+\mathbf{Z_2}}$$

and 7.45

$$\mathbf{I_2} = \mathbf{I_S}\frac{\mathbf{Z_1}}{\mathbf{Z_1}+\mathbf{Z_2}}$$

Examples of Parallel Circuit Analysis

Example 7.4

Given the circuit in Figure 7.16, calculate the supply current I_S.

First, calculate the total impedance of the circuit and then calculate the supply current. The frequency of the power supply, the inductor L_1, and the capacitor C_1 are the same as in Example 7.1, so we will use the already calculated reactive values. That is

$$\mathbf{X_C} = -j79.58$$
$$\mathbf{X_L} = j1884$$

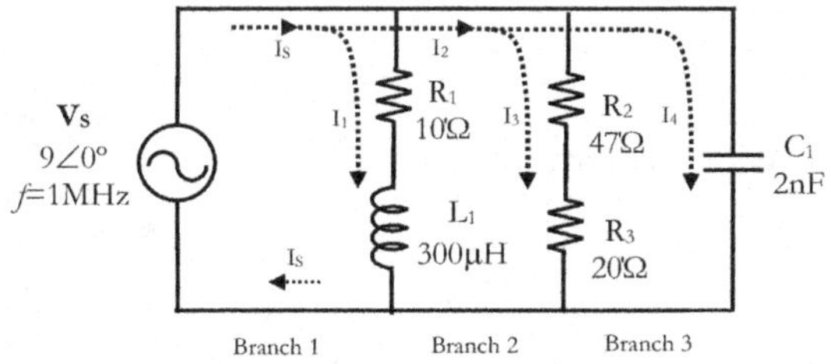

Figure 7.16

The total impedance is

$$\frac{\mathbf{1}}{\mathbf{Z_T}} = \frac{\mathbf{1}}{\mathbf{Z}_{B1}} + \frac{\mathbf{1}}{\mathbf{Z}_{B2}} + \frac{\mathbf{1}}{\mathbf{Z}_{B3}}$$

Where Z_{B1} is the impedance of branch 1, Z_{B2} is the impedance of branch 2, and Z_{B3} is the impedance of branch 3.

$$\mathbf{Z_{B1}} = R_1 + jX_{L_1}$$
$$= 10 + j1884$$
$$= 1884\angle 89.7^\circ$$

$$\mathbf{Z_{B2}} = R_2 + R_3$$
$$= 47 + 20$$
$$= 67\angle 0^\circ$$

$$\mathbf{Z_{B3}} = -jX_{C_1}$$
$$= -j79.58$$
$$= 79.58\angle -90^\circ$$

So the total impedance of the circuit is

$$\frac{\mathbf{1}}{\mathbf{Z_T}} = \frac{1}{1884\angle 89.7^\circ} + \frac{1}{67\angle 0^\circ} + \frac{1}{79.58\angle -90^\circ}$$
$$= 5.3x10^{-4}\angle -89.7^\circ + 149.2x10^{-4}\angle 0^\circ + 125.7x10^{-4}\angle 90^\circ$$
$$= 0.0278x10^{-4} - j5.299x10^{-4} + 149.2x10^{-4} + j125.7x10^{-4}$$
$$= 149.2x10^{-4} + j120.4x10^{-4}$$
$$= 191.7x10^{-4}\angle 38.9^\circ$$

And

$$\mathbf{Z_T} = \frac{1}{191.7x10^{-4}\angle 38.9^\circ}$$
$$= 52.2\angle -38.9^{\circ\circ}$$

Using Ohm's law to calculate the supply current

$$\mathbf{I_S} = \frac{\mathbf{V_S}}{\mathbf{Z_T}}$$

$$\mathbf{I_S} = \frac{9\angle 0^\circ}{52.2\angle -38.9^\circ}$$
$$= 0.1724\angle 38.9$$

Example 7.5

Using the current divider rule calculate the current through each branch.

$$\mathbf{I}_{B1} = \mathbf{I_S}\frac{\mathbf{Z_T}}{\mathbf{Z_{B1}}}$$

$$= 0.1724\angle 38.9^\circ \frac{52.2\angle -38.9^\circ}{1884\angle 89.7^\circ}$$

$$= 0.00478\angle -89.7^\circ$$

$$\mathbf{I}_{B2} = \mathbf{I_S}\frac{\mathbf{Z_T}}{\mathbf{Z_{B2}}}$$

$$= 0.1724\angle 38.9^\circ \frac{52.2\angle -38.9^\circ}{67\angle 0^\circ}$$

$$= 0.1343\angle 0^\circ$$

$$\mathbf{I}_{B3} = \mathbf{I_S}\frac{\mathbf{Z_T}}{\mathbf{Z_{B3}}}$$

$$= 0.1724\angle 38.9^\circ \frac{52.2\angle -38.9^\circ}{79.58\angle -90^\circ}$$

$$= 0.1131\angle 90^\circ$$

Example 7.6

Calculate the voltage across each component.

The voltage across each component in a branch is the branch current times the impedance of the component.

$$\mathbf{V_{R1}} = \mathbf{I_{B1}R_1}$$

$$= 0.00478\angle -89.7^\circ (10\angle 0^\circ)$$

$$= 0.0478\angle -89.7^\circ$$

$$\mathbf{V_{L_1}} = \mathbf{I_{B1}Z_{L_1}}$$

$$= 0.00478\angle -89.7^\circ (1884\angle 90^\circ)$$

$$= 9.006\angle 0.3^\circ$$

$$\mathbf{V_{R_2}} = \mathbf{I_{B2}Z_{R_2}}$$

$$= 0.1343\angle 0^\circ (47\angle 0^\circ)$$

$$= 6.3121\angle 0^\circ$$

$$\mathbf{V}_{\mathbf{R}_3} = \mathbf{I}_{\mathbf{B2}}\mathbf{Z}_{\mathbf{R}_3}$$
$$= 0.1343\angle 0^\circ (20\angle 0^\circ)$$
$$= 2.6860\angle 0^\circ$$

$$\mathbf{V}_{\mathbf{C}_1} = \mathbf{I}_{\mathbf{B3}}\mathbf{Z}_{\mathbf{C}_1}$$
$$= 0.1131\angle 90^\circ (79.58\angle -90^\circ)$$
$$= 9.0004\angle 0^\circ$$

Kirchhoff's Current Law

Kirchhoff's current law (KCL) states that *the algebraic sum of the current entering a circuit must equal the algebraic sum of the current leaving the circuit.* That is

$$\boxed{\mathbf{I}_S = \mathbf{I}_1 + \mathbf{I}_2 + \mathbf{I}_3 + \ldots \mathbf{I}_n} \qquad \text{KCL} \qquad 7.46$$

where I_1, I_2 etc., are the branch currents. Kirchhoff's current law applies to both series and parallel circuits. In the series circuit, the same current flows through each component, so there is only one branch, so KCL holds.

Example 7.7

Using the circuit in Figure 7.16, calculate the algebraic sum of the branch currents, and confirm that KVL holds for parallel circuits.

$$\mathbf{I}_\mathbf{S} = \mathbf{I}_{\mathbf{B1}} + \mathbf{I}_{\mathbf{B2}} + \mathbf{I}_{\mathbf{B3}}$$
$$0.1724\angle 38.9^\circ = (0.00478\angle -89.7^\circ) + (0.1343\angle 0^\circ) + 0.1131\angle 90^\circ$$
$$= 2.503x10^{-5} - j478x10^{-5} + 0.1343 + j0.1131$$
$$= 0.1343 + j0.10832$$
$$= 0.1725\angle 38.9^\circ$$

The difference in the last digit of the magnitude is due to rounding errors. Checking KVL, the source voltage should equal the voltage drop across the circuit.

$$\mathbf{V}_\mathbf{S} = \mathbf{I}_\mathbf{S}\mathbf{Z}_\mathbf{T}$$
$$9.0\angle 0^\circ = 0.1725\angle 38.9^\circ (52.2\angle -38.9^\circ)$$
$$= 9.0045\angle 0^\circ$$

7.10 MAGNETIC CIRCUITS

The electrical circuits described above deal with current and voltages due to the electric field. The analysis is based largely on Ohm's law relating impedance, current and voltage. We could write Ohm's law in a general term as follows:

$$Effect = \frac{cause}{opposition}$$

where the effect is the current that is caused by the voltage, and is constrained by an opposition to the current. That is

$$I = \frac{V}{Z}$$

Similarly, when dealing with magnetic fields, the ***magneto motive force*** (MMF), the thing producing the magnetic field or the cause, has a similar relationship. In this case, the effect is the magnetic lines of flux Φ_M, and the opposition is called the ***reluctance*** ($\mathfrak{R}$), the opposition to creating magnetic flux. Thus

$$\Phi_M = \frac{F_M}{\mathfrak{R}} \qquad 7.47$$

where F_M is the magneto motive force. When dealing with coils, the MMF is proportional to the number of turns around the core and the current through the turns of wire. That is

$$F_M = NI \qquad \text{MMF (amp-turns)} \qquad 7.48$$

where the units of magnetic force are amp-turns (At). The reluctance is determined by

$$\mathfrak{R} = \frac{\ell_M}{\mu A} \qquad \text{Reluctance(inverse henry } H^{-1}\text{)} \qquad 7.49$$

where ℓ_M is the length of the magnetic path (typically the length of the core), and A is the cross-sectional area. If we divide the MMF by the length of the magnetic path, we have the magnetic force per metre. This is known as the ***Magnetising Force*** (H). Sometimes H is also referred to as the magnetic field strength or magnetic field intensity just to make sure it is confused with B. In this book, we will refer to H as the Magnetising force. The units can be expressed as amp-turns per metre (**At/m** - obsolete) or in amps per metre (**A/m**); they mean the same thing.

$$H = \frac{F_M}{\ell_M} = \frac{NI}{\ell_M} \qquad \text{The magnetising force (At/m)} \qquad 7.50$$

The magnitude of the magnetic field strength B is related to the magnetising force and the permeability of the material by the following relationship

$$\boxed{B = \mu_R \mu_0 H} \qquad \text{Magnetic field strength} \qquad 7.51$$

We will deal with magnetic circuits and these relationships when examining coils using ferromagnetic cores in a later chapter.

7.11 DECIBELS

The logarithm of a ratio of any two numbers is called a bel and is defined as

$$B = \log_{10} \frac{P_2}{P_1} \qquad \text{bel} \qquad 7.52$$

One-tenth of a bel is called a ***decibel***, so there are 10 decibels to a bel. This is defined as

$$dB = 10\log_{10}\frac{P_2}{P_1} \qquad \text{decibel} \qquad 7.53$$

At first, it may appear that there are ten bels per decibel, but we are expressing the bel in one-tenth of the units - just the same way 1-metre can be expressed as 1,000 millimetres. If we measure some object to be 2.4 metres, we multiply by one-thousand to give a length of 2,400 millimetres. In the same way, the bel is multiplied by 10 to express the units in decibels.

Engineers often express power as a gain in relation to a set value, such as a milliwatt. In the case of an antenna, the power output is often expressed (in decibels) as a gain above that of a dipole or isotropic antenna.

Voltage and current can also be expressed in decibels. Similarly to power, the voltage gain ratio can be with respect to some other fixed value such as a millivolt, or when the voltage drops across a device by some ratio, say by a half. If we write the ratio of power in terms of voltage across a device, we get

$$dB = 10\log_{10}\frac{V_2^2/R}{V_1^2/R}$$

$$= 10\log_{10} V_2^2/V_1^2$$

And the voltage gain, expressed in decibels, becomes

$$dB_V = 20\log_{10} V_2/V_1 \qquad \text{Volts in decibels} \qquad 7.54$$

8

Resonant Circuits

8.1 SERIES RLC CIRCUITS

An RLC circuit is simply a circuit containing resistance (R), inductance (L) and capacitance (C). A series RLC means that each of these components is in series. Consider the circuit shown in Figure 8.1. The current I_{IN} flows in series through each of the components. The reactance of the capacitor C and the inductor L is frequency-dependent.

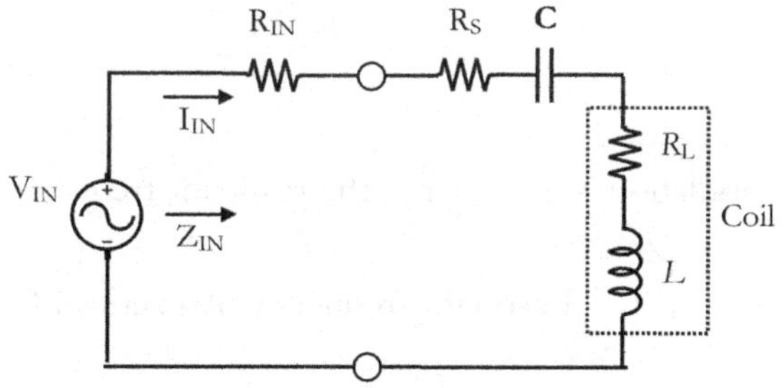

Figure 8.1
Series RLC circuit.

The capacitive reactance is given by

$$X_C = \frac{1}{2\pi fC}$$

and the inductive reactance is given by

$$X_L = 2\pi fL$$

Resonance

Figure 8.2 shows a graph of the impedance over a range of frequencies for each component. The impedance of the resistor is constant, ignoring any skin effect. The impedance of the inductor is linear with respect to frequency, and the capacitor's impedance is high at low frequencies and drops as the frequency increases. There is a frequency in which both the capacitive and inductive reactances are equal in magnitude, and the total impedance of the circuit is simply the total resistance value of the resistors. This represents minimum circuit impedance, and the highest possible current will flow. The frequency in which this occurs is called the ***resonant frequency*** and is defined for series RLC circuits as the frequency at which the capacitive reactance and the inductive reactance are equal in magnitude.

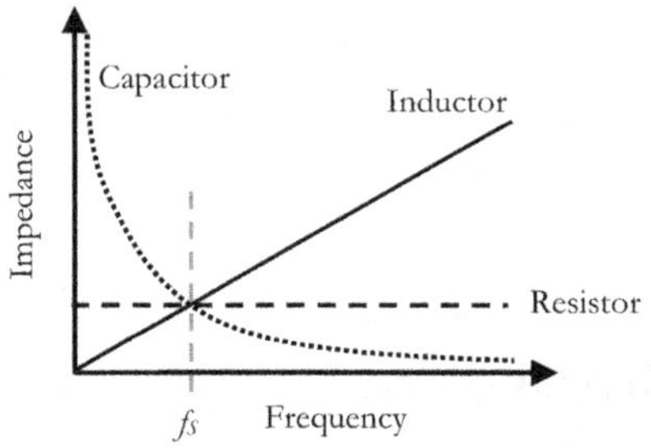

Figure 8.2
Individual component impedance.

Setting the inductive reactance equal to the capacitive reactance, we get

$$X_L = X_C$$

$$2\pi f_S L = \frac{1}{2\pi f_S C}$$

With a bit of algebraic manipulation, we can write the resonant frequency f_S as

$$f_S = \frac{1}{2\pi\sqrt{LC}}$$

Resonant frequency of a series LC circuit 8.1

Given an inductor, we can calculate the capacitor required to form a resonant circuit for a given frequency by rearranging Equation 8.1.

$$C = \frac{1}{4\pi^2 f^2 L} \quad 8.2$$

And similarly, given a capacitor, we can calculate the inductor required to form a resonant circuit for a given frequency

$$L = \frac{1}{4\pi^2 f^2 C} \quad 8.3$$

Impedance

The impedance of the circuit shown in Figure 8.1 is the sum of the component impedances.

$$\mathbf{Z_{IN}} = R_{IN} + R_S + R_L + jX_L - jX_C$$

Note that the resistance R_L is the resistance of the coil. In general, the impedance of a series RLC circuit is

$$\mathbf{Z_{IN}} = R_{TOT} + jX_L - jX_C$$ Impedance of a series RLC circuit 8.4

where R_{TOT} is the sum of all the resistance values. In this example

$$R_{TOT} = R_{IN} + R_S + R_L$$

At resonance X_L and X_C have the same magnitude, that is

$$jX_L - jX_C = 0$$ Inductive and capacitive reactance at resonance

Therefore the impedance of a series RLC circuit at resonance Z_0 is just the total series resistance

$$\boxed{\mathbf{Z_0} = R_{TOT} = R_1 + R_2 + ...R_N}$$ Impedance of a series RLC circuit at resonance 8.5

This represents the minimum impedance state of the circuit resulting in the highest current. Figure 8.3 shows the magnitude of the input impedance $\mathbf{Z_{IN}}$ of the series RLC circuit as a function of frequency. At frequencies below the resonant frequency, the circuit "looks" capacitive as the capacitive reactance is greater than the inductive reactance. At resonance, the circuit looks resistive as the capacitive and inductive reactance cancel. At frequencies above resonance, the circuit looks inductive.

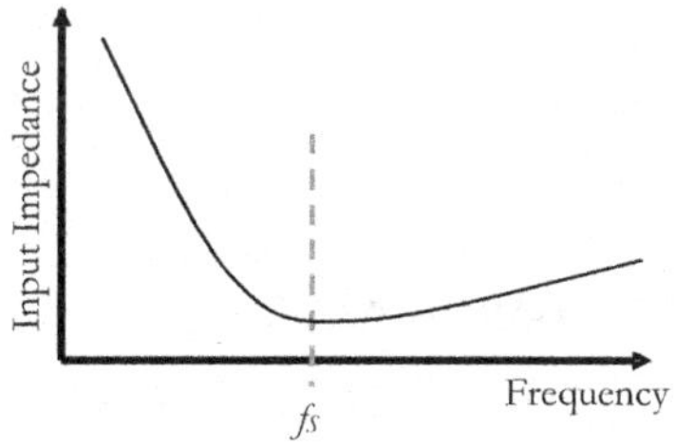

Figure 8.3
Impedance as a function of frequency for series RLC circuit.

Current and Voltage Response

The line current I_{IN}, the current flowing directly out of the power supply and into the circuit, is given by

$$\mathbf{I_{IN}} = \frac{\mathbf{V_S}}{\mathbf{Z_{IN}}}$$

where Z_{IN} is the input impedance of the circuit at a given frequency. As this is a series RLC circuit, the same current will flow through each component. Figure 8.4a is a graph of the input current as a function of frequency. The magnitude of the voltage across a component is simply the line current multiplied by the component's impedance at a given frequency. For circuit 8.1 is

$$\mathbf{V}_{\mathbf{R_S}} = \mathbf{I}_{\mathbf{IN}}\mathbf{R}_{\mathbf{S}}$$
$$\mathbf{V}_{\mathbf{L}} = \mathbf{I}_{\mathbf{IN}}\mathbf{Z}_{\mathbf{L}}$$
$$\mathbf{V}_{\mathbf{C}} = \mathbf{I}_{\mathbf{IN}}\mathbf{Z}_{\mathbf{C}}$$

Figure 8.4b shows the graph of the coil (inductor) voltage as a function of frequency. The current and voltage rise slowly until we approach resonance, and then there is a rapid increase in current and voltage followed by a rapid decline. At the resonant frequency, both the current and the voltage across the inductor reach a peak. Note that the graph is not quite symmetrical about the resonant frequency, and the current increases rapidly as we approach resonance and then trails off a little slower after resonance.

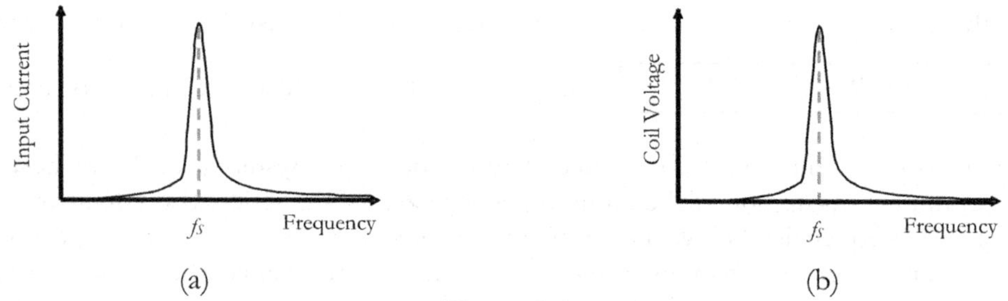

Figure 8.4
Current and coil voltage as a function of frequency for series RLC circuit.

At resonance, the line current is

$$\mathbf{I}_0 = \frac{\mathbf{V}_{\mathbf{IN}}}{\mathbf{Z}_0}$$

And at resonance, the voltage across a resistor R_S is

$$\mathbf{V}_{\mathbf{R_S}} = \mathbf{I}_0\mathbf{R}_{\mathbf{S}}$$

Treating the coil as an ideal inductor and the capacitor as an ideal capacitor (that is, they have no internal resistance), at resonance, the capacitive and inductive reactances are of the same magnitude, and since the same current flows through both, the voltages are the same.

$$V_{L0} = V_{C0}$$
$$I_0 X_L = I_0 X_C$$

Applying the voltage divider rule to circuit 8.1, the voltage across the capacitor (or inductor) at resonance is

$$V_{L0} = \frac{X_L}{Z_0} V_{IN} = V_{C0} = \frac{X_C}{Z_0} V_{IN}$$

The Quality Factor (Q)

If the reactance X is larger than the circuit resistance R_{TOT}, and it usually is, we get a voltage across the capacitor and inductor larger than the supply voltage. The ratio of the reactance to the resistance can be considered as a factor determining the voltage across a reactive component. Therefore the voltage across an inductor or capacitor can be calculated by multiplying the source voltage by some factor. This factor is called the ***Quality factor*** or just Q and is generally greater than one and can be thought of as a voltage multiplier. In general, we can write the voltage across an inductor or capacitor at resonance as

$$V_{L0} = V_{C0} = \frac{X_L}{R_{TOT}} V_{IN} = QV_{IN}$$ Voltage across an inductor or capacitor at resonance 8.6

Where the Q of the circuit is

$$Q = \frac{X_L}{R_{TOT}} = \frac{X_C}{R_{TOT}}$$ The Q of an RLC circuit 8.7

Q is an extremely important factor when designing crystal radios and will be constantly referenced throughout this book. It is a major factor in determining the sensitivity and selectivity of a crystal radio, and as such, requires a deep understanding.

Consider the circuit shown in Figure 8.5. Assume there is no initial charge on the capacitor. Power is applied to the circuit, and current starts to flow from the power supply through the resistor across the capacitor and inductor. The current flowing through the resistor represents a real power loss (heat); however, the current flowing through the inductor and capacitor represents reactive power and is available to be returned to the circuit. The capacitor stores energy in the electric field, and the inductor stores energy in the magnetic field; this is referred to as power gain in Figure 8.5a, as it represents an increase of stored energy above what the component had some time before.

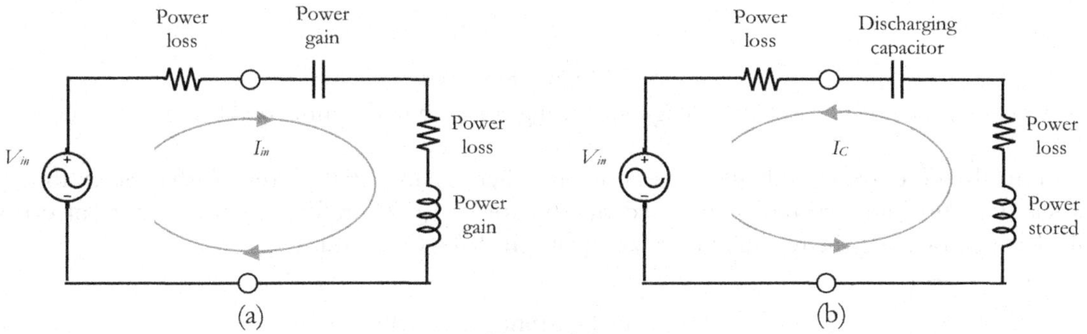

Figure 8.5
Series RLC circuit power loss and gain.

When the capacitor discharges, the discharge current flows through the entire circuit loop, through each of the elements and the power supply, see Figure 8.5b. The current from the capacitor flowing through the resistor represents a real power loss, and the current flowing through the inductor is stored in the magnetic field. So only some of the reactive power of the capacitor is lost, and the rest is retained. However, while this is occurring, the voltage supply is still providing the same reactive power as it did for the last cycle, and as only some of the previously gained reactive power was lost due to resistance, this represents an overall reactive power gain

$$P_{GAIN} = i_{IN}^{2} X$$

This means that the voltage across the capacitor will increase with each cycle, and as the voltage increases, the capacitor discharge current increases. This increase in discharge current results in a greater power loss through the resistor, as given by

$$P_{LOSS} = i_{C}^{2} R$$

This charge and discharge cycle is also true for the inductor just 180° out of phase with the capacitor. When the total power loss across the total resistance due to the inductor and capacitor returning energy to the circuit equals the total reactive power gained from the power supply for a single cycle, equilibrium is reached, and the circuit attains resonance. As much energy is provided by the power supply as is lost in the circuit.

Figure 8.6 shows a screen capture of an oscilloscope showing the voltage across a capacitor increasing over time. An RLC circuit was tuned to resonance at 1 MHz. An oscilloscope probe was attached across the capacitor, and power was applied. This image clearly shows the voltage increasing for each cycle until equilibrium is reached.

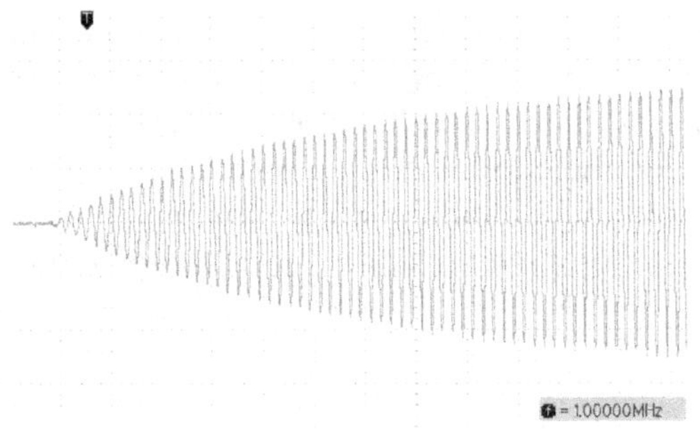

Figure 8.6
Oscilloscope image showing voltage increasing for each cycle until equilibrium is reached.

From the description and oscilloscope image, there is an initial current I_I that flows through the circuit. This initial current is boosted by a factor of Q. Therefore, the initial current can be calculated by dividing the resonant circulating current I_0 by Q, as follows

$$I_I = \frac{I_0}{Q}$$ Initial current for an RLC circuit 8.8

This stored energy results in a high circulating (line current) current through the circuit. Due to Ohm's law, a high current across an impedance results in a high voltage, thus the voltage across the inductor and capacitor can be very much higher than the power supply voltage.

This is an example of a negative feedback mechanism. Negative feedback is a way systems have of ensuring stability. Negative feedback occurs in many natural systems, including biological systems. If conditions change, either externally or internally, a system will re-adjust itself to a new Q or stability point. If the conditions are changed so that not enough negative feedback occurs, then the system becomes unstable and usually ceases to exist in its current form.

We are now in a position to give an exact definition of Q.

Q is defined as 2π times the peak energy stored divided by the average energy lost, at resonance, during a single cycle.

$$\boxed{Q = \frac{2\pi(\text{Peak\_Stored\_Enegy\_Per\_Cycle})}{\text{Average\_Real\_Energy\_Loss\_Per\_Cycle}}}$$

Definition of Q

Q can be defined for coils, capacitors or entire circuits. The same definition applies. Q is frequency dependent, and as we are dealing with the AM broadcast band, the standard resonant frequency will be set at 1 MHz unless otherwise stated.

Let's examine the Q of a coil. For a coil, the peak energy stored is in the magnetic field.

$$E_{PEAK} = \frac{1}{2} I^2_{(PEAK)}\, L$$

where $I_{S(PEAK)}$ is the peak current value, and L is inductance. The average energy loss across some resistor over one cycle is

$$E_{LOSS} = I^2_{RMS} T R_{TOTAL}$$

where I_{RMS} is the RMS current value (average DC equivalent), T is the period of time for one cycle, and R_{TOTAL} is the total resistance. For a coil, the loss of energy is due to the AC resistance of the coil. Therefore

$$E_{LOSS} = \left(\frac{I_{(PEAK)}}{\sqrt{2}}\right)^2 R_{coil(ac)} T$$

$$= \left(\frac{I_{(PEAK)}}{\sqrt{2}}\right)^2 R_{coil(ac)} \frac{1}{f}$$

where $R_{coil(ac)}$ is the AC resistance of the coil at the given frequency. Using the definition of Q, the Q of a coil is

$$Q_{COIL} = \frac{2\pi(\frac{1}{2} I^2_{(PEAK)} L)}{\left(\frac{I_{(PEAK)}}{\sqrt{2}}\right)^2 R_{coil(ac)} \frac{1}{f}}$$

$$= \frac{\frac{I^2_{(PEAK)}}{2} 2\pi L}{\frac{I^2_{(PEAK)}}{2} R_{coil(ac)} \frac{1}{f}}$$

And

$$Q_{COIL} = \frac{2\pi fL}{R_{coil(ac)}}$$

Apart from current resistive losses, there may be other losses such as dielectric losses R_ε and magnetic losses R_M. The dielectric and magnetic loss will be discussed in the Chapter on coils. Therefore the Q of an isolated coil (not used in a circuit) is,

$$Q_{COIL} = \frac{X_L}{R_{coil(ac)} + R_\varepsilon + R_M} \qquad \text{The Q of an isolated coil} \qquad 8.9$$

The internal resistance of a capacitor is generally very small, a fraction of an ohm; therefore, the Q of a capacitor is typically many times higher than for coils. The Q of a capacitor is

$$Q_{CAP} = \frac{X_C}{R_{CAP(ac)}} \qquad \text{The Q of a capacitor} \qquad 8.10$$

However, in a circuit such as the series RLC circuit shown above, at resonance, the capacitive reactance is equal to the inductive reactance, and the peak energy stored is the same for each element. Note, the peak is not the addition of both the peak energy of the capacitor and inductor

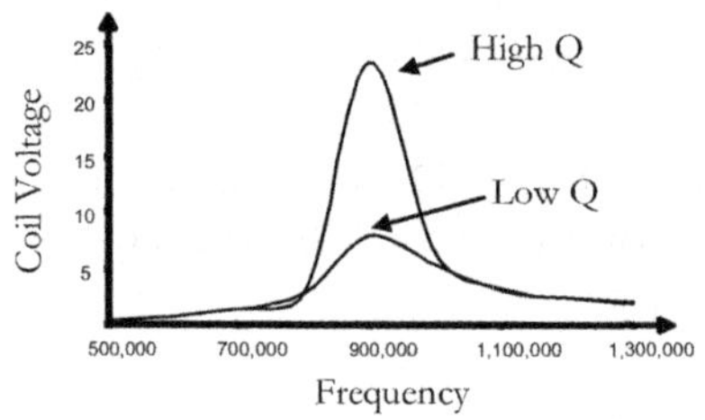

Figure 8.7
High vs. low Q coil voltage.

as the voltage is phase shifted by 180°, and only one component will have peak energy at any one time. The resistance is the total resistance of the circuit, including the power supply resistance, coil resistance, capacitor resistance and any other loss of electrical energy from other devices such as earphones. The Q of the circuit is

$$Q_{Circuit} = \frac{X_L}{R_{TOTAL}} = \frac{X_C}{R_{TOTAL}} \qquad \text{The Q of a series RLC circuit} \qquad 8.11$$

A high ratio of reactance to resistance results in a high Q, and a low ratio results in a low Q. For a given inductor, if the resistance or power losses increase, the Q will drop. See Figure 8.7. So low Q coils result in lower voltages at resonance than high Q coils. A high Q coil in a crystal radio will have greater sensitivity than a low Q coil. Sensitivity is the ability to detect low-power signals.

Selectivity

Apart from voltage boost and sensitivity, another important property of concern to crystal radio makers is selectivity. The selectivity of a circuit is its ability to select a range of frequencies and reject all others. In terms of power, this range is defined by the frequencies at which the power drops by half; that is 3dB down from the maximum. There will be two frequencies where the power drops by half, one below the resonant frequency at f_l and one above the resonant frequency at f_h. See Figure 8.8. The difference between these two frequencies is called the ***bandwidth***.

$$BW = f_h - f_l$$

The narrower the bandwidth, the more selective a circuit is. Since power is the product of voltage and current, at half power, the voltage be $1/\sqrt{2}$ of the maximum voltage ($0.707V_{MAX}$), and the current will be $1/\sqrt{2}$ of the maximum current ($0.707I_{MAX}$).

$$\begin{aligned} \frac{1}{2}P_{MAX} &= \frac{1}{2}V_{MAX}I_{MAX} \\ &= (\frac{1}{\sqrt{2}}V_{MAX})(\frac{1}{\sqrt{2}}I_{MAX}) \\ &= (0.707V_{MAX})(0.707I_{MAX}) \end{aligned}$$

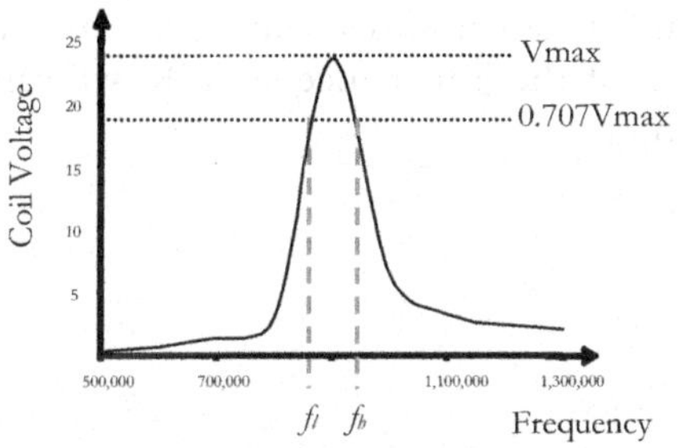

Figure 8.8
The upper and low frequencies where the power drops by half.

Consider Figure 8.7; if we express the voltage as a ratio of the maximum voltage for each curve (called normalising), each maximum will occur at a value of 1 regardless of the curve's actual voltage. Normalising the curves allows us to see the effect of different Q values on the bandwidth, as shown in Figure 8.9. The bandwidth for the high Q curve is smaller than the low Q curve. So not only are high Q coils desirable for sensitivity, but they are also important for selectivity.

At resonance, the impedance is just the total resistance R of the circuit, and the maximum current is given by

$$I_{MAX} = \frac{V_{MAX}}{R}$$

At the upper and lower bandwidth frequencies, the current is 0.707 (or $^1/\sqrt{2}$) of the current at resonance. That is

$$\frac{1}{\sqrt{2}} I_{MAX} = \frac{V_{MAX}}{\sqrt{2}R}$$

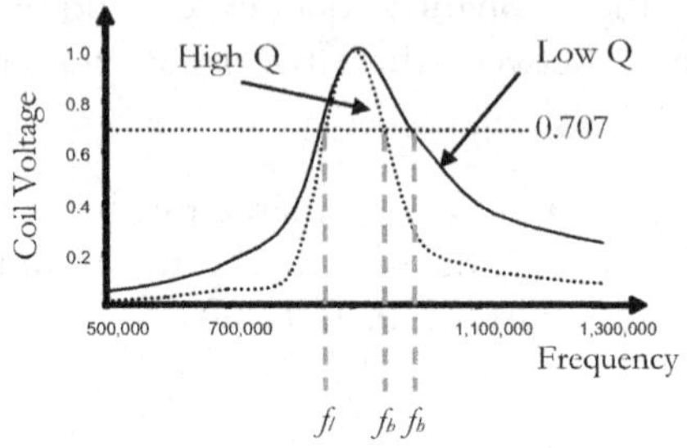

Figure 8.9
Graph of Q versus normalised coil voltage as a function of frequency for series RLC circuit.

This means the impedances at the upper and lower frequencies are √2 times the impedance at resonance. The magnitude of the impedance at resonance is R, so the magnitude of the impedance at these frequencies is just √2R. Note, that we are only talking about the magnitude, the absolute value. The impedance of a circuit, when non-resonant, will have a reactive component and be complex. If we set the magnitude of the impedance as √2R, we can put this in the equation for calculating the impedance magnitude

$$\sqrt{2}R = \sqrt{R^2 + (X_L - X_C)^2}$$

$$2R^2 = R^2 + (X_L - X_C)^2$$

$$R^2 = (X_L - X_C)^2$$

If we first consider the case of the upper bandwidth frequency f_h, where $X_L > X_C$, by taking the square root of the equation, we can write

$$R = X_L - X_C$$

$$R - X_L + X_C = 0$$

And setting $\omega_h = 2\pi f_h$,

$$R - \omega_h L + \frac{1}{\omega_h C} = 0$$

Multiply both sides by ω_h

$$\omega_h R - \omega_h^2 L + \frac{1}{C} = 0$$

Dividing both sides by -L and rearranging, we arrive at

$$\omega_h^2 - \omega_h \frac{R}{L} - \frac{1}{LC} = 0$$

This is simply a quadratic, and we can find the roots of a quadratic

$$\omega_h = \frac{-(-R/L)+\sqrt{[(-R/L)]^2-[-4/LC]}}{2}$$

By simplifying and writing and expressing ω_h as a frequency, we finally get

$$f_h = \frac{1}{2\pi}\left[\frac{R}{2L}+\frac{1}{2}\sqrt{(\frac{R}{L})^2+\frac{4}{LC}}\right]$$

Repeating the procedure for f_l, where $X_L < X_C$, we can write

$$R = X_C - X_L$$

$$R + X_L - X_C = 0$$

Giving

$$\omega_l^2 + \omega_l\frac{R}{L} - \frac{1}{LC} = 0$$

And f_l is

$$f_l = \frac{1}{2\pi}\left[\frac{-R}{2L}+\frac{1}{2}\sqrt{(\frac{R}{L})^2+\frac{4}{LC}}\right]$$

Using the definition of bandwidth, we can write

$$BW = f_h - f_l$$

$$= \frac{1}{2\pi}\left[\frac{R}{2L}+\frac{1}{2}\sqrt{(\frac{R}{L})^2+\frac{4}{LC}}\right] - \frac{1}{2\pi}\left[\frac{-R}{2L}+\frac{1}{2}\sqrt{(\frac{R}{L})^2+\frac{4}{LC}}\right]$$

$$= \frac{R}{2\pi L}$$

The ratio of the resonant frequency f_0 to the bandwidth is

$$\frac{f_0}{f_h - f_l} = \frac{f_0}{\frac{R}{2\pi L}}$$

$$= \frac{2\pi L f_0}{R}$$

From Equation 8.7, this is simply the Q of the coil, and we can write the Q in terms of bandwidth.

$$\boxed{Q_{coil} = \frac{f_Q}{f_h - f_l}}$$ The Q in terms of bandwidth 8.12

where f_Q is the selected Q resonant frequency.

8.2 PARALLEL RLC CIRCUITS

Consider the circuit in Figure 8.10a, showing a voltage source with a series resistor connected to an inductor and capacitor in parallel. This is not a parallel RLC circuit. To transform this circuit so that all the components are in parallel, we can use a technique called ***source transformation*** to replace the voltage source with its series resistor for a current source with a resistor in parallel, where the current source is given by

$$I_{IN} = \frac{V_{In}}{R_{IN}}$$

In Figure 8.10b, we have a parallel RLC circuit.

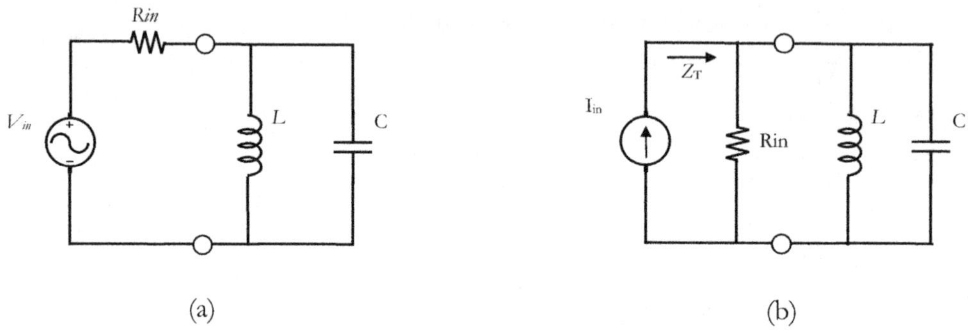

Figure 8.10
Transformed voltage source circuit to a current source circuit.

Resonance

For this circuit to be resonant, the capacitive and inductive reactance must be equal, just as it was for the series RLC circuit. However, in the real world, inductors will have some series resistance due to the wire. Figure 8.11a shows a circuit with a coil and a capacitor having internal resistance. In this case, the presence of these resistances changes the conditions for resonance, as it is no longer in the form given by circuit 8.10b. The input impedance of circuit 8.11a is

$$\frac{1}{Z_T} = \frac{1}{R_{IN}} + \frac{1}{R_L + jX_L} + \frac{1}{R_C - jX_C} \qquad 8.13$$

However, we want it to look like a parallel RLC circuit, as shown in Figure 8.11b.
In circuit 8.11b, resonance occurs when

$$X'_L = X'_C$$

We can transform the circuit in Figure 8.11a into the 8.11b form by multiplying the imaginary terms in Equation 8.13 by their complex conjugates. We then collect the real and imaginary terms as follows

$$\frac{1}{\mathbf{Z_T}} = \frac{1}{R_{IN}} + \frac{1}{R_L + jX_L} + \frac{1}{R_C - jX_C}$$

$$= \frac{1}{R_{IN}} + \frac{1}{R_L + jX_L}\left(\frac{R_L - jX_L}{R_L - jX_L}\right) + \frac{1}{R_C - jX_C}\left(\frac{R_C + jX_C}{R_C + jX_C}\right)$$

$$\frac{1}{\mathbf{Z_T}} = \frac{1}{R_{IN}} + \frac{R_L - jX_L}{R^2{}_L + X^2{}_L} + \frac{R_C + jX_C}{R^2{}_C + X^2{}_C}$$

$$= \frac{1}{R_{IN}} + \frac{R_L}{R^2{}_L + X^2{}_L} + \frac{-jX_L}{R^2{}_L + X^2{}_L} + \frac{R_C}{R^2{}_C + X^2{}_C} + \frac{jX_C}{R^2{}_C + X^2{}_C}$$

$$= \left(\frac{1}{R_{IN}} + \frac{R_L}{R^2{}_L + X^2{}_L} + \frac{R_C}{R^2{}_C + X^2{}_C}\right) + \left(\frac{-jX_L}{R^2{}_L + X^2{}_L} + \frac{jX_C}{R^2{}_C + X^2{}_C}\right)$$

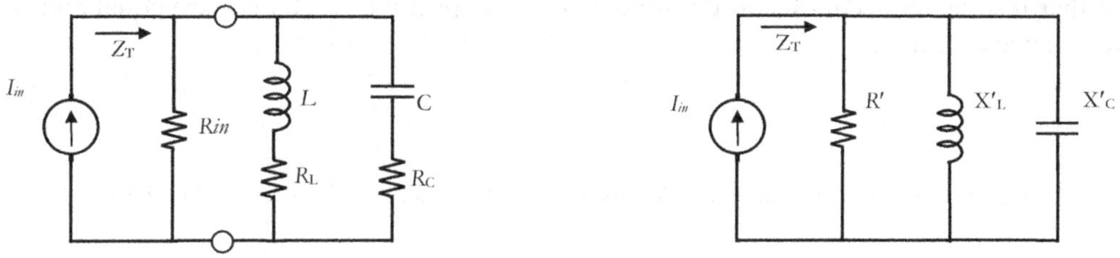

(a) (b)

Figure 8.11
A transformed RLC circuit required for resonance.

Equating the real and imaginary parts to circuit 8.11b, we have

$$\frac{1}{R'} = \left(\frac{1}{R_{IN}} + \frac{R_L}{R^2{}_L + X^2{}_L} + \frac{R_C}{R^2{}_C + X^2{}_C}\right)$$

$$\frac{1}{\mathbf{X'_L}} = \frac{-jX_L}{R^2{}_L + X^2{}_L}$$

$$\frac{1}{\mathbf{X'_C}} = \frac{jX_C}{R^2{}_C + X^2{}_C}$$

And resonance occurs when

$$\left(\frac{-jX_L}{R^2{}_L + X^2{}_L} + \frac{jX_C}{R^2{}_C + X^2{}_C}\right) = 0 \qquad 8.14$$

After some lengthy algebraically manipulating of Equation 8.14, we finally arrive at an equation for the parallel resonant frequency

$$\boxed{f_P = \frac{1}{2\pi\sqrt{LC}}\sqrt{\frac{L - CR_L^2}{L - CR_C^2}}}$$ Resonant frequency of a parallel RLC circuit 8.15

where R_L is the AC resistance of the coil, and R_C is the AC resistance of the capacitor. Typically the series capacitance resistance is so small it is usually neglected, and we can write 8.15 as

$$\boxed{f_P = f_S\sqrt{1 - \frac{R_L^2 C}{L}}} \qquad 8.16$$

where f_S is the series resonant frequency. The square root term is always less than one, so the parallel resonant frequency will always be less than the series resonant frequency for any given inductance and capacitance.

For the component values used in a crystal radio and the frequencies involved, the value of the inductor L will be many orders of magnitude greater than the capacitance value C and the ratio will tend to zero. Therefore, the difference between the series and parallel resonant frequencies is very small indeed, in the order of a fraction of a per cent. As such, the formula for the series resonance is often used. For example, given a 300 uH coil with 10-Ω resistance and a capacitor of 100 pF, the series resonant frequency is 918.881 kHz. The parallel frequency is 918.866 kHz; that is a 0.0016% difference. For this reason, when dealing with parallel RLC circuits, we will adopt the position that resonance occurs when the inductive and capacitive reactances are equal and use the series resonance equation.

Impedance

An inductor in parallel with a capacitor is known as a tank circuit. The impedance of the tank

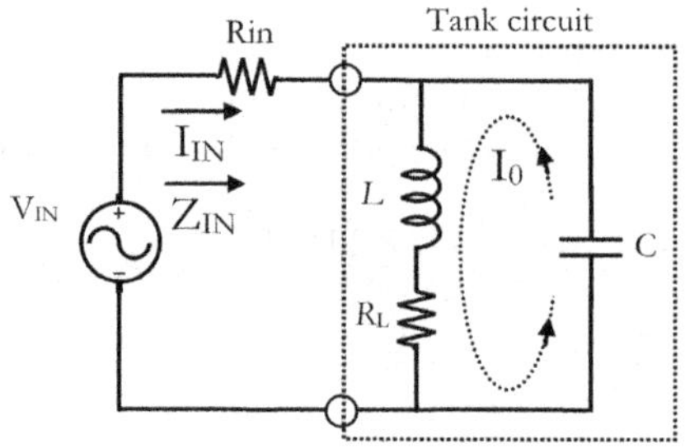

Figure 8.12
Tank circuit.

circuit is the parallel impendence of the inductor and capacitor (see Equation 7.40).

$$Z_{TANK} = \frac{Z_L Z_C}{Z_L + Z_C} \tag{8.17}$$

The graph of the tank circuit impedance as a function of frequency is shown in Figure 8.13.

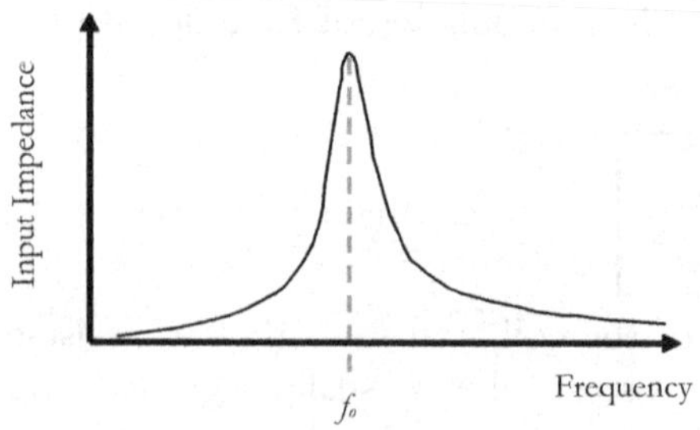

Figure 8.13
Tank circuit impedance as a function of frequency.

At resonance, the input impedance is high, and the line current I_{IN} will be at its smallest. This is opposite to the series circuit, where at resonance, the input impedance is the smallest, and the line current is at its greatest. At resonance, the input impedance of the tank circuit will be

$$Z_0 = \frac{Z_L Z_C}{R_L}$$

Current and Voltage Response

The input impedance for the circuit in 8.12 is

$$Z_{IN} = R_{IN} + \frac{Z_L Z_C}{Z_L + Z_C}$$

And the input current (the line current) is by Ohm's law

$$I_{IN} = \frac{V_{IN}}{Z_{IN}}$$

As the tank circuit impedance approaches resonance, the input impedance Z_{IN} becomes larger. And at resonance, the tank circuit impedance is at its maximum. Therefore, the line current will be at its minimum, as shown in Figure 8.14a.

The voltage across the tank circuit, however, will be at its greatest, just like in the series circuit, Figure 8.14b. The voltage across the inductor and the capacitor can be many orders of magnitude higher than the source voltage due to the circulating current I_0, see Figure 8.12. The energy coming from the power supply is stored in the inductor and the capacitor, just like the series circuit. And just like the series circuit, the energy accumulates as charge on the capacitor plates. At resonance, the power supply adds enough energy in the form of the current to keep the tank circuit oscillating.

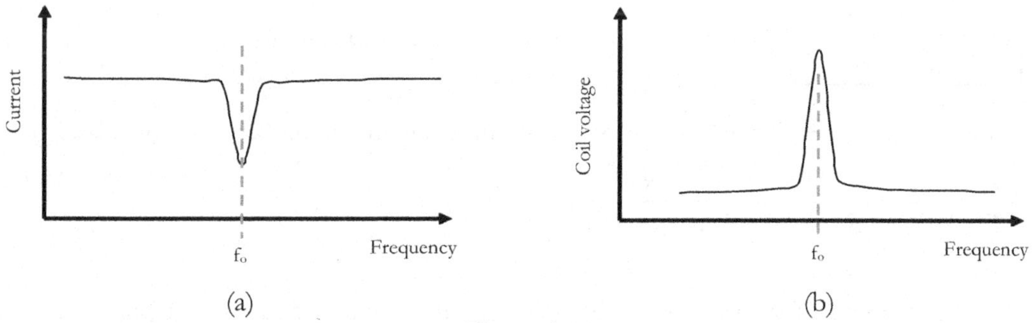

Figure 8.14
Current and coil voltage as a function of frequency for a parallel RLC circuit.

The Quality Factor (Q)

Consider the transformed parallel RLC circuit in Figure 8.11b. At resonance

$$\mathbf{X'_L} = \mathbf{X'_C}$$

The peak energy stored in the inductor equals the peak energy stored in the capacitor

$$\frac{1}{2}LI^2_{Peak} = \frac{1}{2}CV^2_{Peak}$$

As the components are in parallel, the voltage will be the same across each component. In determining the series Q, the current was common, and the Q was determined using the current. For the parallel circuit, the voltage is common across each element, so voltage will be used to determine the Q of the circuit. The power dropped across the resistance Ŕ using the power Equation 7.7 is

$$P_{R'} = \frac{\left(\frac{V_{(PEAK)}}{\sqrt{2}}\right)^2}{R'}\frac{1}{f}$$

Using this in our definition of Q, we can write

$$\begin{aligned} Q_P &= \frac{2\pi\frac{1}{2}CV^2_{Peak}}{\frac{\left(\frac{V_{PEAK}}{\sqrt{2}}\right)^2}{R'}\frac{1}{f}} \\ &= \frac{2\pi fC}{\frac{1}{R'}} \\ &= 2\pi fCR' \end{aligned}$$

That is

$$\boxed{Q_P = \frac{R'}{X_C} = \frac{R'}{X_L}} \qquad \text{Q of a parallel RLC circuit} \qquad 8.18$$

The apparent resistance R´ is the transformed resistance, and for circuit 8.11a, is given by

$$\frac{1}{R'} = \left(\frac{1}{R_{IN}} + \frac{R_L}{R^2{}_L + X^2{}_L} + \frac{R_C}{R^2{}_C + X^2{}_C}\right)$$

Note that the definition for Q is the same for both parallel and series circuits (the ratio of the energy stored to the energy lost) and that R´ is the transformed or apparent resistance calculated for a given resonant frequency and is not simply just a resistance value. As we shall see in a later Chapter, the tank circuit used for tuning crystal radios can be modelled as a series circuit, and we will use a modified version of Equation 8.11 to calculate the Q of a crystal radio with the total resistance, including all the losses.

8.3 THE CRYSTAL RADIO CIRCUIT

It is time we introduced the crystal radio circuit diagram, shown in Figure 8.15. This diagram will appear in Chapter headings when the Chapter has a direct bearing on the development of the model and the construction of the crystal radio. The relevant section of the circuit diagram will be highlighted.

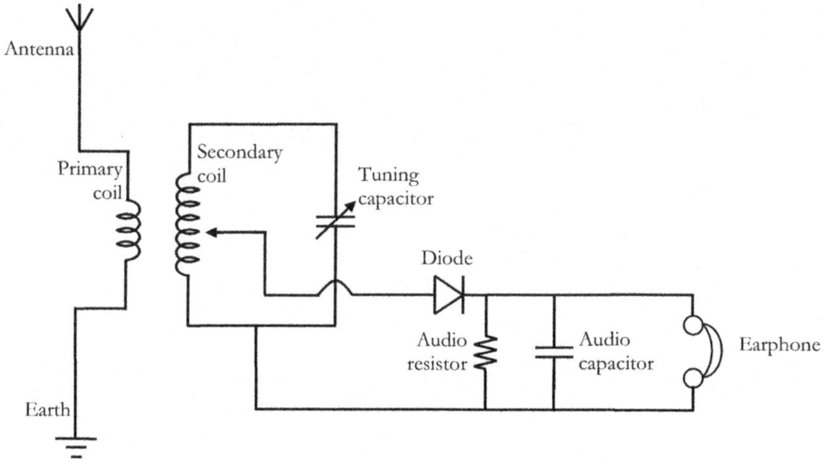

Figure 8.15
Circuit diagram of a crystal radio.

9

Transmission Line Theory

9.1 OVERVIEW

A transmission line is simply a pair of wires used to transmit voltage and current from one location to another. A pair of wires running from an audio amplifier to a speaker is a transmission line. A coaxial cable running from an antenna into a short-wave receiver is a transmission line. The power generated from a power station is transmitted to your home via a series of transmission lines. These lengths of wire form a completed circuit from the power source to the load at the other end. Typically transmission lines are constructed using two wires in close proximity to one another. Sometimes single wires are used, relying on the conductivity of the ground to provide the return path.

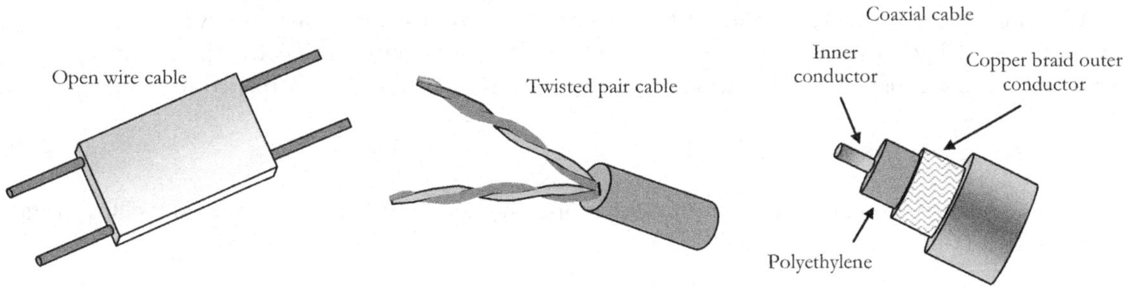

Figure 9.1
Common transmission line cables.

Figure 9.1 shows some commonly used transmission line cables. The simplest is the open wire cable which consists of two wires spaced by an insulator. These were widely used in the days of analogue TV to connect an indoor antenna to a TV set. The twisted pair cable consists of two wires twisted around one another, with a single cable having one or more sets of twisted pairs.

These are used as telephone or computer network cables.

Coaxial cables are typically used with RF equipment. Commercially available RF equipment is designed to connect to coaxial cable transmission lines of known electrical characteristics, and the transmission line is then connected to an antenna of known electrical characteristics. The antenna and transmission line are matched to the device to provide optimal performance, and the operator simply buys the correct transmission line and antenna and connects them. Crystal radio operators, on the other hand, will quite often grab some random length of wire and string it up to use as an antenna. Crystal radio antennas may range from only a few metres to thirty metres in length or more, depending on the availability of space and your living situation. This antenna forms part of the crystal radio circuit, and we must determine the electrical characteristics in order to design a crystal radio that will operate at optimal (or something close to it) performance. The antenna is our only source of power, and we want to extract as much power as we can.

Transmission line theory will be used to analyse antennas, transmission lines and even some properties of the coils we are going to wind. Therefore we need to spend some time analysing and gaining an understanding of this topic.

9.2 CURRENT AND VOLTAGE RESPONSE

In our analysis of transmission lines, we will consider them to be uniform in construction. That is, the materials used in the construction, wire and dielectric, do not change, and the separation distance between the wires does not change.

Consider a sinusoidal voltage source attached to a transmission line, with a length greater than one wavelength of the sinusoidal input signal. At the end of the transmission line, we attach a load of some impedance Z_{LOAD}, as shown in Figure 9.2. The voltage waveform takes some period of time (a short time but some time) to travel from the source voltage to a load at the end of the cable. If we freeze time at the instant the waveform reaches the load and measure the voltage at all points along the transmission line, we would have a voltage waveform, shown as a solid line. The voltage waveform represents the voltage value on the top wire with respect to the bottom wire (treating the bottom wire as common). This voltage signal travelling from the source to the load is called the forward travelling wave. By convention, the location of the load is taken as the origin x=0. Using Equation 6.11, we can write a function for the forward travelling wave. The distance from the load back to the voltage source will result in a negative value for the distance, so the value of x in the equation will be negative. Also, the angular velocity ωt is related to the distance by

$$\varpi t = \frac{2\pi}{\lambda}\Delta x \qquad 9.1$$

And this term will appear positive in our equation. The equation for the forward travelling voltage wave thus becomes

$$v^{+}(x,t) = V_{M}^{+}\cos(\omega t - \beta x + \theta^{+})$$

where V^{+}_{M} represents the peak magnitude of the forward travelling voltage wave, θ^{+} is the initial phase angle and β is the wave number.

In the above example, we assumed there was no loss of signal as the voltage wave propagated along the transmission line. This is not true, as there will always be some loss along the transmission line from wire resistance. Whether these losses are great enough to be of interest depends on the system under study. However, for completeness, we will incorporate a loss factor in the equation.

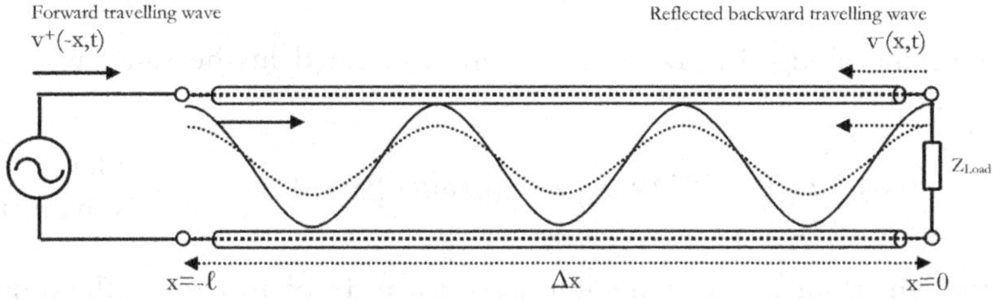

Figure 9.2
Voltage waveform on a transmission line.

We can represent the loss as a dampening of the signal. As energy is lost, the maximum amplitude of the wave will decrease, and it takes the form of an exponential decay term as a function of distance x, $e^{-\alpha x}$, and the wave equation becomes

$$v^+(x,t) = V_M^+ e^{-\alpha x} \cos(\omega t - \beta x + \theta^+) \qquad \text{Voltage forward travelling wave} \qquad 9.2$$

where V^+_M represents the peak magnitude of the forward travelling voltage wave, θ^+ is the initial phase angle, and β is the wave number.

If we now resume time and wait for the same period, and then freeze time again, we would now have a forward travelling wave and a backward travelling wave (dotted). This backward travelling wave is the result of reflection at the load. Due to the conservation of energy, any energy not absorbed by the load is returned to the system and travels back along the transmission line. This is similar to how light, an electromagnetic wave, may reflect from an object.

This backward travelling wave is the reflection of the wave given by Equation 9.2. Both the forward and backward waves will have the same angular velocity, so the ωt term will not change. The value of x will be flipped, however, and any phase change will be given by θ^-. Therefore the backward travelling wave is described by the following equation

$$v^-(x,t) = V_M^- e^{\alpha x} \cos(\omega t + \beta x + \theta^-) \qquad \text{Voltage backward travelling wave} \qquad 9.3$$

where V^-_M represents the peak magnitude of the backward travelling voltage wave, θ^- is any phase change due to the reflection and β is the wave number. The backward travelling wave may have a different peak value depending on the reflection, and there may be some phase shift in relation to the forward wave, particularly if the load impedance has any reactance, capacitive or inductive.

The state of the transmission line at this time is called the ***steady state***, meaning from this time onward, the voltage and current on the transmission line will resemble this picture with both a forward and backward wave. The voltage continues to be sinusoidal and changes over time, however. The period before this state occurs is called the ***transitional state***. Both the transitional state and the steady state can be analysed. But it is the steady state that is of most interest to us.

During steady state, the forward and backward travelling waves will interfere with on-another and the resultant waveform is the superposition (addition) of the forward and reflected wave. The complete time domain voltage function of the transmission line is, therefore

$$v(x,t) = V_M^+ e^{-\alpha x}\cos(\omega t - \beta x + \theta^+) + V_M^- e^{\alpha x}\cos(\omega t + \beta x + \theta^-)$$ Resultant voltage waveform 9.4

The current travelling along the transmission line is treated in the same way, and the time domain current function is

$$i(x,t) = I_M^+ e^{-\alpha x}\cos(\omega t - \beta x + \theta^+) + I_M^- e^{\alpha x}\cos(\omega t + \beta x + \theta^-)$$ Resultant current waveform 9.5

We can write the time domain equations in phasor form. In phasor form, the time dependence disappears, and as phase angles are constant and are not a function of x, we can incorporate phase shifts into the peak values and express the peak values as complex numbers. The voltage equation in phasor form becomes

$$\begin{aligned}\mathbf{V(x)} &= V_M^+ e^{-\alpha x} e^{-j\beta x + j\theta^+} + V_M^- e^{\alpha x} e^{+j\beta x + j\theta^-} \\ &= (V_M^+ e^{j\theta^+}) e^{-(\alpha + j\beta)x} + (V_M^- e^{j\theta^-}) e^{(\alpha + j\beta)x} \\ &= \tilde{V}_M^+ e^{-(\alpha + j\beta)x} + \tilde{V}_M^- e^{(\alpha + j\beta)x}\end{aligned}$$

where $\tilde{V}_M$ is a complex number containing the phase angle. The loss factor α and the wavenumber β in the exponent form a complex number, and we refer to this complex number as gamma, the ***propagation constant*** and

$$\gamma = \alpha + j\beta$$ 9.6

We can now write the phasor equations for the voltage as

$$\boxed{\mathbf{V(x)} = \tilde{V}_M^+ e^{-\gamma x} + \tilde{V}_M^- e^{\gamma x}}$$ Phasor form of voltage 9.7

And the phasor form for current as

$$\boxed{\mathbf{I(x)} = \tilde{I}_M^+ e^{-\gamma x} + \tilde{I}_M^- e^{\gamma x}}$$ Phasor form of current 9.8

9.3 CHARACTERISTIC IMPEDANCE

At the steady state, both the forward and backward travelling waves exist independently of each other. The net waveform, however, is the supersession of both waves; that is, we take into account how they interfere with one another. It is the net waveform that is observed and the independent waves that are hidden from view. For example, if we were to observe the audio waveform of someone speaking, we would not see the individual frequencies but the superposition of all the frequencies creating a complicated wave pattern. These individual frequencies still exist and, by using filters, can be extracted if desired.

Say that the superposition of both the forward and backward waves produces an unacceptably high voltage spike on the transmission line such that it causes a problem, or that it is even so high it melts the wire. Perhaps the load is a speaker, and we want all the power dropped across the speaker and not wasted in the line. We, therefore, want to reduce the amplitude of the backward travelling wave due to reflection. If all the energy is absorbed by the load, then no reflection will occur. For purely resistive impedances, maximum power transfer occurs when the load impedance matches the output impedance (for complex impedances, this occurs when the load is the complex conjugate of the output impedance). If both impedances are purely resistive and are matched, then all the energy will be absorbed by the load, and no reflection will occur. Therefore we need to determine the impedance of the line where it contacts the load (the output impedance). We can use Ohm's law to determine the impedance of the line at any given position (x) along the line as follows,

$$\mathbf{Z_0} = \frac{\mathbf{V^+(x)}}{\mathbf{I^+(x)}}$$
$$= \frac{\tilde{V}_M^+ e^{-\gamma x}}{\tilde{I}_M^+ e^{-\gamma x}}$$
$$= \frac{\tilde{V}_M^+}{\tilde{I}_M^+}$$

Notice that the distance (x) component cancels, and therefore the impedance is not a function of distance, meaning it will be constant throughout the length of the transmission line. If you increase the length of the transmission line, this impedance will not change and is a property of the construction of the transmission line itself. The ratio of the maximum voltage and the maximum current of the forward (or backward) travelling wave is called the ***characteristic impedance*** Z_0. This is the impedance at every infinitesimally small section along the transmission line, including the end of the line. If we were to place a resistor at the load end with this value Z_0, all the energy would be absorbed, and no reflection would occur.

Characteristic impedance is an important factor for ham radio operators as the reduction in reflections results in smaller standing waves (to be discussed later), resulting in a lowering of the loss of RF energy. However, a more important factor for crystal radio enthusiasts is the input impedance of the transmission line (and antenna), as seen by the primary coil. This is not the same as characteristic impedance. Energy is transferred to the secondary coil via magnetic induction. As magnetic induction is directly proportional to the current, we wish to maximise the current. This is achieved by making the primary circuit resonant. We will discuss input impedance and achieving resonance in later Chapters.

Also, note that characteristic impedance cannot be measured directly. The losses on a transmission line are represented by the total AC resistance of the length of the wire and the parallel resistance (conductance) distributed along the entire length of the cable. If you were to measure the resistance along the length of a cable, it would increase as you go along.

The reflected wave also has a characteristic impedance that is the same as the forward travelling wave's characteristic impedance. However, as we shall see later, that is given by the negative of the voltage to current ratio.

$$\boxed{\mathbf{Z_0} = \frac{\tilde{V}_M^+}{\tilde{I}_M^+} = -\frac{\tilde{V}_M^-}{\tilde{I}_M^-}} \qquad \text{Characteristic impedance} \qquad 9.9$$

Rearranging Equation 9.9 for current, we can write the time domain transmission line current Equation 9.5 as

$$i(x,t) = \frac{V_M^+}{Z_0} e^{-\alpha x} \cos(\omega t - \beta x + \theta^+) - \frac{V_M^-}{Z_0} e^{\alpha x} \cos(\omega t + \beta x + \theta^-) \qquad 9.10$$

And in phasor form, Equation 9.8 is written as

$$\mathbf{I(x)} = \frac{\tilde{V}_M^+}{Z_0} e^{-\gamma x} - \frac{\tilde{V}_M^-}{Z_0} e^{\gamma x} \qquad 9.11$$

9.4 THE TELEGRAPH EQUATIONS

In Chapter 6, we developed the wave Equation 6.12 by taking the second partial derivative with respect to distance (x) and with respect to time (t). If we take the second partial derivative with respect to distance (x) of the phasor voltage Equation 9.7, we get

$$\frac{\partial^2 \mathbf{V(x)}}{\partial x^2} = \gamma^2 \tilde{V}_M^+ e^{-\gamma x} + \gamma^2 \tilde{V}_M^- e^{\gamma x} \qquad 9.12$$

We see that gamma (the propagation constant) squared becomes a coefficient of the wave equation. We will use this relationship to determine the loss factor (α), the wave number (β) and ultimately, the characteristic impedance and other useful parameters needed to analyse a transmission line. To determine gamma, we need to express our voltage and current waveforms in the form given by Equation 9.12.

Any given length of wire will have both resistance and self-inductance, no matter how small. A two-wire transmission line will also have some capacitance between the wires depending on the proximity and dielectric properties of the material separating the wires. As no dielectric is perfect, there will also be some conductance across the wires (the inverse of resistance). The values of these properties will depend on the construction of the transmission line and will differ between transmission line types. These values of resistance, inductance, capacitance and conductance are distributed along the entire length of the transmission line. If we select some infinitesimally short length of line, we can model a two-wire transmission line as a set of lumped components, as shown in Figure 9.3. The values of resistance ($\acute{R}$), inductance ($\acute{L}$), capacitance ($\acute{C}$) and conductance ($\acute{G}$) are values per metre.

Consider a transmission line with a DC voltage source connected. Power is turned on for a short time, sending a pulse down the line. When the power is first applied, current will flow across the inductor and start to rapidly charge up the first capacitor. The charge on the capacitor will then start to discharge through the next segment of the transmission line through the next inductor and onto the next capacitor, and so on down the line. In this way, the signal is propagated down the transmission line, with the velocity of propagation depending on the values of the inductive and capacitive components in the infinitesimally small section. This velocity is always less than the speed of light in a vacuum. The ratio of the velocity of propagation to the speed of light in a vacuum is called the ***velocity factor*** (V_F)

$$\boxed{V_F = \frac{V_P}{C}} \qquad \text{Velocity factor} \qquad 9.13$$

where V_P is the velocity of propagation, and C is the speed of light in a vacuum. This charging mechanism along the transmission line is the same for AC as for DC currents.

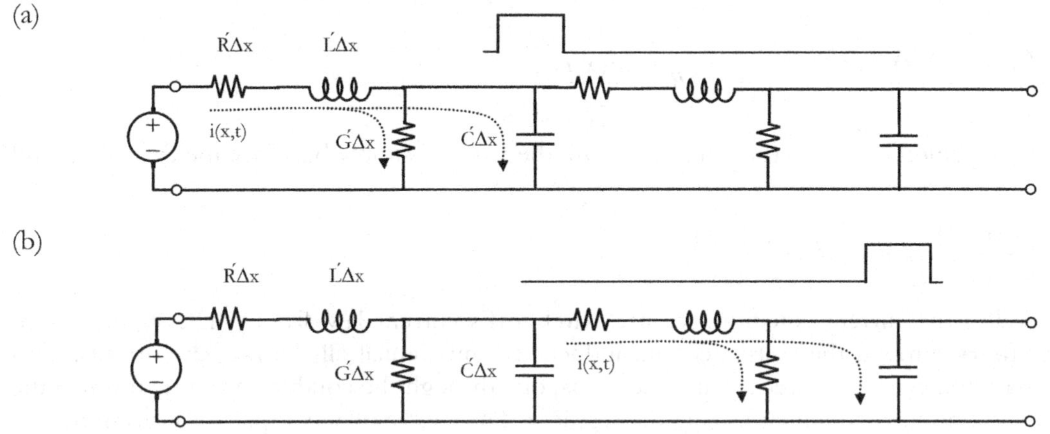

Figure 9.3
Propagation of signal down a transmission line.

Consider Figure 9.4, showing some small segment of a transmission line. Using Kirchhoff's

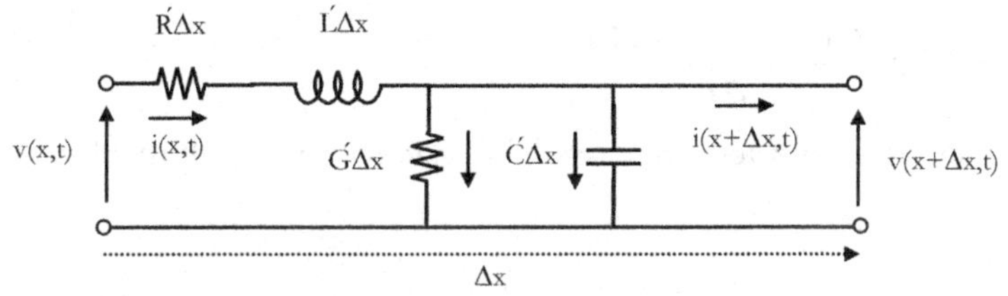

Figure 9.4
Component model of a transmission line segment.

voltage law (KVL), we can write an equation that describes the voltage relationships along this small length of cable. From KVL the sum of the voltage drops around the loop equals the voltage source. The resistor and inductor are in series and the same current will pass through both components. The conductance and capacitance are in parallel so the voltage drop across both components will be the same, that is, v(x+Δx,t). Using Equation 5.15, we can write voltage across the inductor as

$$v_L = L'\Delta x \frac{\partial i(x,t)}{\partial t}$$

And the KVL time domain equation for the loop becomes

$$v(x,t) = i(x,t)R'\Delta x + \frac{\partial i(x,t)}{\partial t} L'\Delta x + v(x+\Delta x, t)$$

Grouping the voltages and dividing both sides by Δx, we can write

$$-\left(\frac{v(x+\Delta x, t) - v(x,t)}{\Delta x} \right) = \frac{i(x,t)R'\Delta x}{\Delta x} + \frac{\partial i(x,t)}{\partial t} \frac{L'\Delta x}{\Delta x}$$

And

$$-\left(\frac{v(x+\Delta x,t)-v(x,t)}{\Delta x}\right)=i(x,t)R'+\frac{\partial i(x,t)}{\partial t}L'$$

Solving this equation by taking the limit of Δx→0, the left side simply becomes the derivative, and we get

$$-\left(\frac{\partial v(x,t)}{\partial x}\right)=i(x,t)R'+\frac{\partial i(x,t)}{\partial t}L' \qquad 9.14$$

To develop the current equation, we use Kirchhoff's current law (KCL). The input current i(x,t) that flows through the resistance and inductance must equal all the branch currents. This input current branches into three separate sections; one through the conductance, one through the capacitor and one current branch leaving the segment. The charge on any capacitor is given by

$$Q=CV$$

Taking the partial derivative of Q and V with respect to time, we will get the current through the line capacitance as

$$\frac{\partial Q}{\partial t}=C'\Delta x\frac{\partial V}{\partial t}$$

$$i=C'\Delta x\frac{\partial v(x+\Delta x,t)}{\partial t}$$

The current through the conductance (parallel resistance) using Ohm's law is

$$\begin{aligned}i&=\frac{V}{R}\\&=G'\Delta x(V)\\&=G'\Delta x(v(x+\Delta x,t))\end{aligned}$$

Putting this all together, for the three branch currents, the KCL equation becomes

$$i(x,t)=G'\Delta x(v(x+\Delta x,t))+C'\Delta x\frac{\partial v(x+\Delta x,t)}{\partial t}+i(x+\Delta x,t)$$

Grouping the currents and dividing by Δx, we get

$$\begin{aligned}-\left(\frac{i(x+\Delta x,t)-i(x,t)}{\Delta x}\right)&=\frac{G'\Delta x(v(x+\Delta x,t))}{\Delta x}+\frac{C'\Delta x}{\Delta x}\frac{\partial v(x+\Delta x,t)}{\partial t}\\&=G'v(x+\Delta x,t)+C'\frac{\partial v(x+\Delta x,t)}{\partial t}\end{aligned}$$

And taking the limit of Δx→0, we arrive at

$$-\left(\frac{\partial i(x,t)}{\partial x}\right)=G'v(x,t)+C'\frac{\partial v(x,t)}{\partial t} \qquad 9.15$$

If we convert Equations 9.14 and 9.15 from the time domain into the phasor domain, we effectively eliminate the time variable, and the voltage and current equations can be expressed in terms of the distance along the transmission line. Considering the voltage Equation 9.14 first, the voltage across the inductor is given by

$$v_L = L'\frac{\partial i(x,t)}{\partial t}$$

The inductor is in series with the resistance and will share the same current. So using Ohm's law, we can write the voltage across the inductor in the phasor domain as

$$\mathbf{V_L} = \mathbf{Z'_L I(x)}$$

We can write Equation 9.14 in phasor form as

$$\begin{aligned}\frac{\partial \mathbf{V(x)}}{\partial x} &= -R'\mathbf{I(x)} - Z'_L\mathbf{I(x)} \\ &= -(R' + Z'_L)\mathbf{I(x)} \\ &= -(R' + jX'_L)\mathbf{I(x)} \\ &= -(R' + j2\pi fL')\mathbf{I(x)}\end{aligned} \qquad 9.16$$

Now considering the current Equation 9.15, the current across the capacitor is

$$i_C = C'\frac{\partial v(x,t)}{\partial t}$$

In phasor form, using Ohm's law, we can write the current through the capacitor as

$$\mathbf{I_C} = \frac{\mathbf{V}(x)}{\mathbf{Z'_C}}$$

So in phasor form 9.15 becomes

$$\begin{aligned}\frac{\partial \mathbf{I(x)}}{\partial x} &= -\left(G'\mathbf{V(x)} + \frac{\mathbf{V(x)}}{Z'_C}\right) \\ &= -\left(G'\mathbf{V(x)} + \mathbf{V(x)}\Big/\frac{-j}{2\pi fC'}\right) \\ &= -\left(G'\mathbf{V(x)} + j2\pi fC'\mathbf{V(x)}\right) \\ &= -(G' + j2\pi fC')\mathbf{V(x)}\end{aligned} \qquad 9.17$$

We have two equations with two unknown functions, one for voltage V(x) and one for current I(x). We would like to express these equations in terms of a single function in order to solve for V(x) and I(x). That is, we wish to find a function V(x) that describes the voltage along the transmission line and a function I(x) that describes the current along the transmission line.

Taking the partial derivative with respect to x for Equation 9.16, we get

$$\frac{\partial^2 \mathbf{V(x)}}{\partial x^2} = -(R' + j2\pi fL')\frac{\partial \mathbf{I(x)}}{\partial x}$$

Now substituting Equation 9.17 for the partial derivative current term in the above equation, we get

$$\frac{\partial^2 \mathbf{V(x)}}{\partial x^2} = -(R' + j2\pi fL')(-(G' + j2\pi fC'))\mathbf{V(x)}$$

Simplifying, we finally arrive at

$$\boxed{\frac{\partial^2 \mathbf{V(x)}}{\partial x^2} = (R' + j2\pi fL')(G' + j2\pi fC')\mathbf{V(x)}}$$ Voltage telegraph equation 9.18

For the current Equation 9.17, if we take the partial derivative with respect to x, we get

$$\frac{\partial \mathbf{I}^2\mathbf{(x)}}{\partial x^2} = -\left((G' + j2\pi fC')\frac{\partial \mathbf{V(x)}}{\partial x}\right)$$

Substituting Equation 9.16 into the partial derivative term of the voltage, we get

$$\frac{\partial \mathbf{I}^2\mathbf{(x)}}{\partial x^2} = -(G' + j2\pi fC')(-(R' + j2\pi fL')\mathbf{I(x)}))$$

And finally,

$$\boxed{\frac{\partial \mathbf{I}^2\mathbf{(x)}}{\partial x^2} = (R' + j2\pi fL')(G' + j2\pi fC')\mathbf{I(x)}}$$ Current telegraph equation 9.19

Equations 9.18 and 9.19 are called the ***telegraph equations*** (in phasor form) as they were first derived when the electrical characteristics of telegraph transmission lines needed to be understood.

9.5 DETERMINING THE PROPAGATION CONSTANT

The wave Equation 9.12 can be written as

$$\begin{aligned}\frac{\partial^2 \mathbf{V(x)}}{\partial x^2} &= \gamma^2 \tilde{V}_M^+ e^{-\gamma x} + \gamma^2 \tilde{V}_M^- e^{\gamma x} \\ &= \gamma^2 (\tilde{V}_M^+ e^{-\gamma x} + \tilde{V}_M^- e^{\gamma x}) \\ &= \gamma^2 \mathbf{V(x)}\end{aligned}$$

Relating Equation 19.18 to this wave equation, we can see that the propagation constant gamma (γ) is given by the electrical characteristics of the cable

$$\gamma^2 = (\alpha + j\beta)^2 = (R' + j2\pi fL')(G' + j2\pi fC')$$

And

$$\boxed{\gamma = (\alpha + j\beta) = \sqrt{(R' + j\omega L')(G' + j\omega C')}}$$ Propagation constant 9.20

Applying a lot of algebraic manipulation and using the quadratic formula, we can solve for α and β individually. The loss factor becomes

$$\alpha = \frac{\sqrt{(R'G' - \omega^2 L'C') + \sqrt{(R'^2 + \omega^2 L'^2)(G'^2 + \omega^2 C'^2)}}}{2} \qquad 9.21$$

And the wave number becomes

$$\beta = \frac{\sqrt{-(R'G' - \omega^2 L'C') + \sqrt{(R'^2 + \omega^2 L'^2)(G'^2 + \omega^2 C'^2)}}}{2} \qquad 9.22$$

To determine the loss factor and wave number, enter the cable's electrical characteristics in Equation 9.20. The real value will be α, and the imaginary term will be β. Or you can solve Equations 9.21 and 9.22 separately to get α and β.

9.6 DETERMINING THE CHARACTERISTIC IMPEDANCE

We now need to determine the characteristic impedance given the transmission line values. Substituting Equations 9.7 and 9.8 into Equation 9.16, we get

$$\frac{\partial \mathbf{V(x)}}{\partial x} = -(R' + j\omega L')\mathbf{I(x)}$$

$$\frac{\partial(\tilde{V}_M^+ e^{-\gamma x} + \tilde{V}_M^- e^{\gamma x})}{\partial x} = -(R' + j\omega L')(\tilde{I}_M^+ e^{-\gamma x} + \tilde{I}_M^- e^{\gamma x})$$

$$\gamma \tilde{V}_M^+ e^{-\gamma x} - \gamma \tilde{V}_M^- e^{\gamma x} = (R' + j\omega L')\tilde{I}_M^+ e^{-\gamma x} + (R' + j\omega L')\tilde{I}_M^- e^{\gamma x}$$

The forward and backward travelling waves are independent, so we can treat them as separate equations. For the forward wave

$$\gamma \tilde{V}_M^+ e^{-\gamma x} = (R' + j\omega L')\tilde{I}_M^+ e^{-\gamma x}$$

$$\frac{\gamma \tilde{V}_M^+ e^{-\gamma x}}{\tilde{I}_M^+ e^{-\gamma x}} = (R' + j\omega L')$$

$$\frac{\tilde{V}_M^+}{\tilde{I}_M^+} = \frac{(R' + j\omega L')}{\gamma}$$

And the backward wave

$$-\tilde{V}_M^- e^{\gamma x} = (R' + j\omega L')\tilde{I}_M^- e^{\gamma x}$$

$$-\frac{\gamma \tilde{V}_M^- e^{\gamma x}}{\tilde{I}_M^- e^{\gamma x}} = (R' + j\omega L')$$

$$-\frac{\tilde{V}_M^-}{\tilde{I}_M^-} = \frac{(R' + j\omega L')}{\gamma}$$

We see that the characteristic impedance for the forward and backward waves are the same. Substituting Equation 9.20 for gamma, we get

$$\frac{\tilde{V}_M^+}{\tilde{I}_M^+} = -\frac{\tilde{V}_M^-}{\tilde{I}_M^-}$$

$$= \frac{(R' + j\omega L')}{\sqrt{(R' + j\omega L')(G' + j\omega C')}}$$

and

$$\frac{\tilde{V}_M^+}{\tilde{I}_M^+} = \sqrt{\frac{R' + j\omega L'}{G' + j\omega C'}}$$

So the characteristic impedance of a transmission line is

$$\mathbf{Z_0} = \frac{\tilde{V}_M^+}{\tilde{I}_M^+} = -\frac{\tilde{V}_M^-}{\tilde{I}_M^-} = \sqrt{\frac{R' + j\omega L'}{G' + j\omega C'}} \qquad 9.23$$

Characteristic impedance of a lossy transmission line

The above equation is the characteristic impedance of a lossy transmission line. The inductive and capacitive reactances are frequency-dependent. As frequency increases, so do the reactance values and when the resistance per metre and the conductance per metre are small compared to the reactance values, then the characteristic impedance looks quite constant, as shown in Figure 9.5. For RF frequencies, this will be generally true so that we can treat the characteristic impedance as constant.

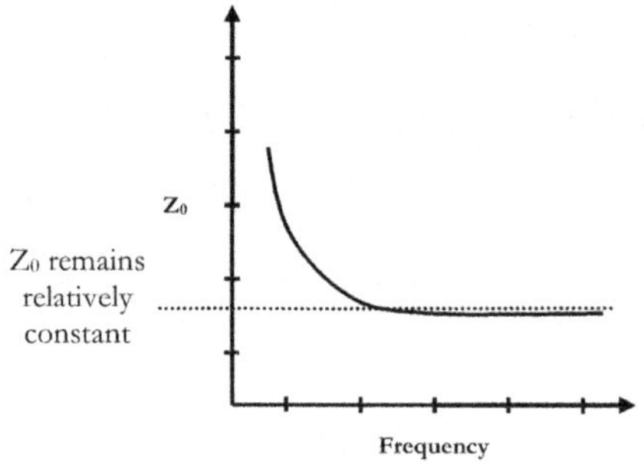

Figure 9.5
Typical characteristic impedance curve as a function of frequency.

9.7 THE REFLECTION COEFFICIENT

When a wave is reflected, the maximum amplitude of the reflected wave V^-_M will be some portion of the maximum amplitude of the forward travelling wave V^+_M. We can express this portion as a ratio of the reflected wave amplitude to the forward amplitude. This ratio is called the ***reflection coefficient***, capital gamma (Γ).

$$\mathbf{\Gamma(x)} = \frac{\tilde{V}_M^- e^{\gamma x}}{\tilde{V}_M^+ e^{-\gamma x}}$$

The reflection occurs when the forward wave reaches the load when x=0. At x=0, the exponential terms become equal to one. So we can write,

$$\mathbf{\Gamma} = \frac{\tilde{V}_M^-}{\tilde{V}_M^+}$$

From Equation 9.11, the maximum current can be written in terms of voltage and characteristic impedance. At x=0, the voltage and current equations become

$$\mathbf{V}(0) = \widetilde{V}_M^+ e^{-\gamma 0} + \widetilde{V}_M^- e^{\gamma 0} = \widetilde{V}_M^+ + \widetilde{V}_M^-$$

$$\mathbf{I}(0) = \frac{\widetilde{V}_M^+}{\mathbf{Z_0}} e^{-\gamma 0} - \frac{\widetilde{V}_M^-}{\mathbf{Z_0}} e^{\gamma 0} = \frac{\widetilde{V}_M^+}{\mathbf{Z_0}} - \frac{\widetilde{V}_M^-}{\mathbf{Z_0}}$$

The voltage of the resultant wave, when it meets the load, must be the same as the voltage across the load due to KVL. So we can write,

$$V(0) = V_L$$

$$\widetilde{V}_M^+ + \widetilde{V}_M^- = \widetilde{V}_L \qquad 9.24$$

The value of the resultant current due to the travelling waves must be the same as the current across the load, and we can write,

$$I(0) = I_L$$

$$\frac{\widetilde{V}_M^+}{\mathbf{Z_0}} - \frac{\widetilde{V}_M^-}{\mathbf{Z_0}} = \frac{\widetilde{V}_L}{\mathbf{Z}_L} \qquad 9.25$$

Substituting Equation 9.24 into 9.25 for V_L, we get

$$\frac{\widetilde{V}_M^+}{\mathbf{Z_0}} - \frac{\widetilde{V}_M^-}{\mathbf{Z_0}} = \frac{\widetilde{V}_M^+ + \widetilde{V}_M^-}{\mathbf{Z}_L}$$

$$\frac{\widetilde{V}_M^+}{\mathbf{Z_0}} - \frac{\widetilde{V}_M^-}{\mathbf{Z_0}} = \frac{\widetilde{V}_M^+}{\mathbf{Z}_L} + \frac{\widetilde{V}_M^-}{\mathbf{Z}_L}$$

$$\frac{\widetilde{V}_M^+}{\mathbf{Z_0}} - \frac{\widetilde{V}_M^+}{\mathbf{Z}_L} = \frac{\widetilde{V}_M^-}{\mathbf{Z}_L} + \frac{\widetilde{V}_M^-}{\mathbf{Z}_0}$$

$$\widetilde{V}_M^+ \left(\frac{\mathbf{Z}_L - \mathbf{Z_0}}{\mathbf{Z_0}\mathbf{Z}_L}\right) = \widetilde{V}_M^- \left(\frac{\mathbf{Z}_0 + \mathbf{Z}_L}{\mathbf{Z}_0\mathbf{Z}_L}\right)$$

$$\left(\frac{\mathbf{Z}_L - \mathbf{Z_0}}{\mathbf{Z_0}\mathbf{Z}_L}\right)\left(\frac{\mathbf{Z}_0\mathbf{Z}_L}{\mathbf{Z}_0 + \mathbf{Z}_L}\right) = \frac{\widetilde{V}_M^-}{\widetilde{V}_M^+}$$

And

$$\frac{\mathbf{Z}_L - \mathbf{Z_0}}{\mathbf{Z}_0 + \mathbf{Z}_L} = \frac{\widetilde{V}_M^-}{\widetilde{V}_M^+}$$

That is

$$\mathbf{\Gamma} = \frac{\mathbf{Z}_L - \mathbf{Z_0}}{\mathbf{Z}_0 + \mathbf{Z}_L}$$

The reflection coefficient for the current is

$$\mathbf{\Gamma}_C = \frac{\widetilde{I}_M^-}{\widetilde{I}_M^+}$$

And

$$\boldsymbol{\Gamma}_C = \frac{-\dfrac{\tilde{V}_M^-}{Z_0}}{\dfrac{\tilde{V}_M^+}{Z_0}} = -\frac{\tilde{V}_M^-}{\tilde{V}_M^+} = -\boldsymbol{\Gamma}$$

Therefore the reflection coefficient for the current is 180° out of phase with the voltage reflection coefficient. More specifically, the reflected current wave is 180° out of phase with the forward voltage wave. Finally, we can write the reflection coefficient as

$$\boxed{\boldsymbol{\Gamma} = \frac{\tilde{V}_M^-}{\tilde{V}_M^+} = -\frac{\tilde{I}_M^-}{\tilde{I}_M^+} = \frac{\mathbf{Z}_L - \mathbf{Z}_0}{\mathbf{Z}_L + \mathbf{Z}_0}} \qquad \text{Reflection coefficient} \qquad 9.26$$

Note, if the load were another length of transmission line with a different characteristic impedance value, then we would use the characteristic impedance of the attached line as the load impedance. That is

$$\boxed{\boldsymbol{\Gamma} = \frac{\mathbf{Z}(2)_0 - \mathbf{Z}(1)_0}{\mathbf{Z}(2)_0 + \mathbf{Z}(1)_0}} \qquad \text{Reflection coefficient at the join of two transmission lines} \qquad 9.27$$

Thus giving the reflection coefficient at the join of the two transmission lines.

9.8 INPUT IMPEDANCE

The characteristic impedance is the impedance of an infinitely small section of the line and is not the impedance for the entire length of the line. For a given transmission line, the characteristic impedance is constant no matter the length. However, the total impedance of a line is a function of the length of the line and will be different depending on the line length.

Ohm's law tells us that impedance is the ratio of voltage to current, and the phase angle between the voltage and current determines how resistive, capacitive or inductive a circuit looks. The same is true for transmission lines. The phase angle between the voltage and current will determine what the transmission line "looks" like to our circuit. Distilling a circuit down to a single impedance value as viewed from the input or output side is called the ***Thevenin equivalent impedance***.

For lumped components such as resistors, capacitors and inductors (for a given frequency), this ratio (impedance) will not change. However, in the case of transmission lines, the impedance not only varies with frequency but also varies with position along the line. The input impedance is the impedance as "seen" from the source side looking towards the load as a function of distance along the line, as shown in Figure 9.6.

Once determined, however, the impedance of the transmission line can be modelled as any other lumped component and circuit analysis performed. In phasor form, the impedance of the resultant waveform using Ohm's law is,

$$\mathbf{Z(x)} = \frac{\mathbf{V(x)}}{\mathbf{I(x)}}$$

Note impedance, voltage and current are functions of distance x. Using Equations 9.7 for voltage **V(x)** and Equation 9.11 for the current **I(x)** and using the reflection coefficient to express the backward travelling maximum amplitude, we can write

$$\mathbf{Z(x)} = \frac{\tilde{V}_M^+ e^{-\gamma x} + \Gamma \tilde{V}_M^+ e^{\gamma x}}{\frac{\tilde{V}_M^+}{\mathbf{Z_0}} e^{-\gamma x} - \frac{\Gamma \tilde{V}_M^+}{\mathbf{Z_0}} e^{\gamma x}}$$

Simplifying, we get an expression for the input impedance as a function of distance.

$$\boxed{\mathbf{Z(x)} = \mathbf{Z_0} \frac{e^{-\gamma x} + \Gamma e^{\gamma x}}{e^{-\gamma x} - \Gamma e^{\gamma x}}} \qquad \text{Input impedance of a transmission line} \qquad 9.28$$

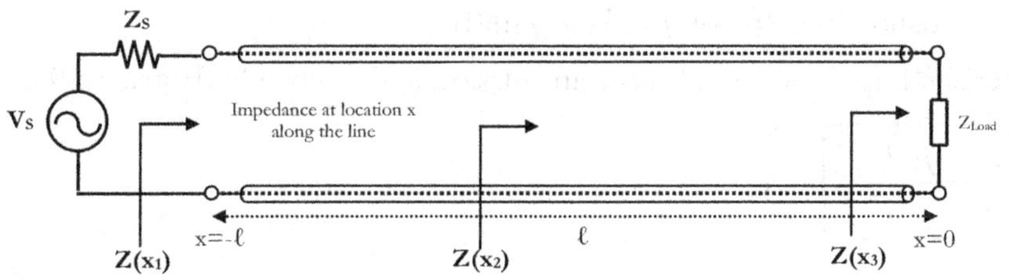

Figure 9.6
Input impedance as seen at a distance x from the load.

Lossy Transmission Line

Given a lossy (α>0) transmission line of length ℓ with some load impedance Z_L the input impedance of the line at x =- ℓ is

$$\mathbf{Z}(-\ell) = \mathbf{Z_0} \frac{e^{-\gamma(-\ell)} + \Gamma e^{\gamma(-\ell)}}{e^{-\gamma(-\ell)} - \Gamma e^{\gamma(-\ell)}}$$

And

$$\mathbf{Z}(-\ell) = \mathbf{Z_0} \frac{e^{\gamma\ell} + \Gamma e^{-\gamma\ell}}{e^{\gamma\ell} - \Gamma e^{-\gamma\ell}}$$

Substituting the equation for the reflection coefficient Γ we get,

$$\mathbf{Z}(-\ell) = \mathbf{Z_0} \left. e^{\gamma\ell} + \left(\frac{\mathbf{Z}_L - \mathbf{Z_0}}{\mathbf{Z}_L + \mathbf{Z_0}}\right) e^{-\gamma\ell} \middle/ e^{\gamma\ell} - \left(\frac{\mathbf{Z}_L - \mathbf{Z_0}}{\mathbf{Z}_L + \mathbf{Z_0}}\right) e^{-\gamma\ell} \right.$$

And

$$\mathbf{Z}(-\ell) = \mathbf{Z_0}\frac{\mathbf{Z}_L(e^{\gamma\ell} + e^{-\gamma\ell}) + \mathbf{Z}_0(e^{\gamma\ell} - e^{-\gamma\ell})}{\mathbf{Z}_L(e^{\gamma\ell} - e^{-\gamma\ell}) + \mathbf{Z}_0(e^{\gamma\ell} + e^{-\gamma\ell})}$$

Using some algebraic and trigonometric substitutions, this equation simplifies to

$$\boxed{\mathbf{Z}(-\ell) = \mathbf{Z_0}\frac{\mathbf{Z}_L + \mathbf{Z}_0 \tanh(\gamma\ell)}{\mathbf{Z}_0 + \mathbf{Z}_L \tanh(\gamma\ell)}} \quad \text{Input impedance of a lossy transmission line} \qquad 9.29$$

where tanh is the hyperbolic tangent function. This is a general equation for the input impedance of a transmission line of length ℓ. Note that gamma (γ) is complex, and the tanh function takes real arguments only, so this function needs to be calculated using the following formula,

$$\begin{aligned}\tanh(\gamma\ell) &= \tanh((\alpha + j\beta)\ell)) \\ &= \tanh(\alpha\ell + j\beta\ell) \\ &= \frac{\sinh(\alpha\ell)\cos(\beta\ell) + j\cosh(\alpha\ell)\sin(\beta\ell)}{\cosh(\alpha\ell)\cos(\beta\ell) + j\sinh(\alpha\ell)\sin(\beta\ell)}\end{aligned} \qquad 9.30$$

The characteristic impedance of a lossy transmission line is given by Equation 9.23

$$\boxed{\mathbf{Z_0} = \sqrt{\frac{R' + j\omega L'}{G' + j\omega C'}}}$$

The electrical wavelength of the transmission line is given by Equation 6.9

$$\boxed{\lambda = \frac{2\pi}{\beta}}$$

where β is determined by Equation 9.20 or calculated using Equation 9.22. And the velocity of propagation of the transmission line is given by Equation 6.8

$$\boxed{v_p = \lambda f = \frac{2\pi f}{\beta}}$$

Lossless Transmission Line

If the resistance of the line is very much smaller than the inductive reactance (R<<$2\pi f$L), and if the conductance across the line is very much smaller than the capacitive susceptance (the reciprocal of reactance) (G<<$2\pi f$C), then we can treat the line as lossless. That is α=0 and R´=0, and G´=0.

Treating a line as lossless simplifies the analysis greatly. We get a bunch of simplified equations. The characteristic impedance becomes

$$\boxed{Z_0 = \sqrt{\frac{L'}{C'}}} \qquad 9.31$$

Note that the characteristic impedance for lossless lines is purely real and is, therefore, resistive.

The propagation constant gamma becomes

$$\boxed{\gamma = j\beta} \tag{9.32}$$

Beta, the wave number, becomes

$$\boxed{\beta = \omega\sqrt{L'C'}} \tag{9.33}$$

And the input impedance becomes

$$\boxed{\mathbf{Z}(-\ell) = Z_0 \frac{\mathbf{Z}_L + jZ_0 \tan(\beta\ell)}{Z_0 + j\mathbf{Z}_L \tan(\beta\ell)}} \quad \text{Input impedance of a lossless transmission line} \tag{9.34}$$

The wavelength can be written in terms of the properties of the transmission line. Given β, we can write

$$\beta = \frac{2\pi}{\lambda}$$

$$\beta = \omega\sqrt{L'C'}$$

And

$$\frac{2\pi}{\lambda} = 2\pi f\sqrt{L'C'}$$

$$\lambda = \frac{2\pi}{2\pi f\sqrt{L'C'}}$$

Finally

$$\boxed{\lambda = \frac{1}{f\sqrt{L'C'}}} \tag{9.35}$$

The wavelength is given by

$$\lambda = \frac{v}{f}$$

$$\frac{v}{f} = \frac{1}{f\sqrt{L'C'}}$$

$$v = \frac{f}{f\sqrt{L'C'}}$$

And the velocity of propagation can be written as

$$\boxed{v_p = \frac{1}{\sqrt{L'C'}} = \frac{2\pi f}{\beta}} \tag{9.36}$$

The velocity of propagation of a signal through the transmission line, expressed as a ratio of the propagation of electromagnetic waves through a vacuum, is called the ***velocity factor***.

$$\boxed{v_f = \frac{v_p}{C} = \frac{2\pi f}{\beta C}} \qquad 9.37$$

Where C is the speed of light through a vacuum.

Short-Circuit as a Load

For a short-circuited transmission line, the load impedance is zero (Z_L=0), and the reflection coefficient is

$$\mathbf{\Gamma} = \frac{0 - \mathbf{Z_0}}{0 + \mathbf{Z_0}}$$

$$= -1\angle 0^{\circ\circ}$$

Or

$$\mathbf{\Gamma} = 1\angle 180^\circ$$

This means the forward wave is reflected 180 out of phase. The input impedance for a lossy line is

$$\mathbf{Z}(-\ell) = \mathbf{Z_0}\frac{0 + \mathbf{Z}_0 \tanh(\gamma\ell)}{\mathbf{Z}_0 + (0)\tanh(\gamma\ell)}$$

$$= \mathbf{Z_0}\frac{\mathbf{Z}_0 \tanh(\gamma\ell)}{\mathbf{Z}_0}$$

And

$$\boxed{\mathbf{Z}(-\ell) = \mathbf{Z_0}\tanh(\gamma\ell)}$$ Short circuit as load – lossy line 9.38

The input impedance for a lossless line is

$$\boxed{\mathbf{Z}(-\ell) = jZ_0 \tan(\beta\ell)}$$ Short circuit as load – lossless line 9.39

Open-Circuit as a Load

For an open-circuit transmission line (no load attached), the load impedance is infinite (Z_L=∞), and the reflection coefficient is

$$\mathbf{\Gamma} = \lim(Z_L \to \infty)\frac{\mathbf{Z}_L - \mathbf{Z_0}}{\mathbf{Z}_L + \mathbf{Z_0}}$$

$$= 1\angle 0^\circ$$

Strictly speaking, infinities in equations are not nice; however, we can use L'Hôpital's rule and take the derivative with respect to Z_L of the numerator and denominator, and we get 1. And the input impedance for a lossy line is

$$\mathbf{Z}(-\ell) = \mathbf{Z_0}\frac{e^{-\gamma(-\ell)} + \Gamma e^{\gamma(-\ell)}}{e^{-\gamma(-\ell)} - \Gamma e^{\gamma(-\ell)}}$$

And

$$\mathbf{Z}(-\ell)=\mathbf{Z_0}\frac{e^{\gamma\ell}+e^{-\gamma\ell}}{e^{\gamma\ell}-e^{-\gamma\ell}}$$

Or

$$\boxed{\mathbf{Z}(-\ell)=\mathbf{Z_0}\coth(\gamma\ell)=\frac{\mathbf{Z_0}}{\tanh(\gamma\ell)}}$$ Open circuit as load – lossy line 9.40

where coth is the hyperbolic cotangent function. And for a lossless line is

$$\boxed{\mathbf{Z}(-\ell)=-jZ_0\cot(\beta\ell)=-\frac{jZ_0}{\tan(\beta\ell)}}$$ Open circuit as load – lossless line 9.41

Note that the characteristic impedance Z_0 for a lossless line is purely resistive. Therefore the input impedance is purely reactive due to the –j term.

Matched Load (Z_0=Z_L)

When the load impedance is equal to the characteristic impedance (Z_L=Z_0), the reflection coefficient becomes

$$\begin{aligned}\boldsymbol{\Gamma}&=\frac{\mathbf{Z}_0-\mathbf{Z_0}}{\mathbf{Z}_0+\mathbf{Z_0}}\\&=0\angle 0^\circ\end{aligned}$$

The input impedance for a lossy and lossless line is

$$\begin{aligned}\mathbf{Z}(-\ell)&=\mathbf{Z_0}\frac{e^{-\gamma(-\ell)}+\Gamma e^{\gamma(-\ell)}}{e^{-\gamma(-\ell)}-\Gamma e^{\gamma(-\ell)}}\\&=\mathbf{Z_0}\frac{e^{\gamma\ell}}{e^{\gamma\ell}}\end{aligned}$$

And

$$\boxed{\mathbf{Z}(-\ell)=\mathbf{Z_0}}$$ Matched load – lossy and lossless line 9.42

9.9 STANDING WAVES

If the load impedance matches the transmission line's characteristic impedance, all the energy will be absorbed by the load, and reflection will not occur. Whenever an unmatched load is attached, some of the forward travelling wave will be reflected, and the transmission line is said to be ***resonant***.

This reflected voltage would mix with the forward wave producing a superposition (addition and subtraction) of the two travelling waves. The effect is to produce a resultant wave that appears to stay in a single position with only the amplitude changing in time. See Figure 9.7. This stationary wave is called a ***standing wave***. The position where this standing wave appears along the line is dependent on the line characteristics, the length of the line, the frequency of the wave and the load impedance.

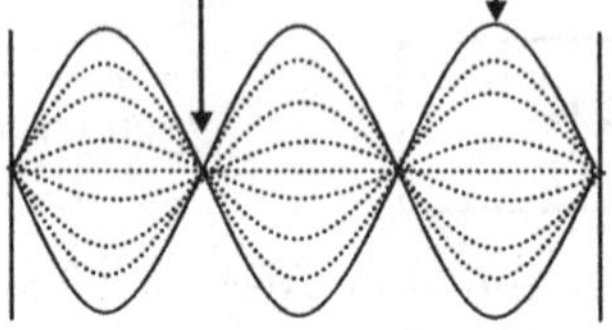

Figure 9.7
Standing waves.

The purpose of a transmission line is to deliver power from one location to another, and a reflected wave means that some of this signal is not being absorbed by the load and, in this respect, may be considered a waste of energy. It is generally not possible to perfectly match a transmission line to a load, and there will always be some reflection taking place. A measure of this reflected wave can be used to indicate the efficiency of power transfer by the transmission line. One such measure is called the ***voltage standing wave ratio*** (VSWR). It is defined as the absolute maximum peak voltage (anti-node) of the standing wave to the absolute minimum voltage (node) of the standing wave.

$$VSWR = \frac{\left|V^S_{MAX}\right|}{\left|V^S_{MIN}\right|}$$

where V^S_{MAX} is the maximum standing wave voltage, and V^S_{MIN} is the minimum standing wave voltage. There will be some point along the transmission line where the addition of both the forward and reflected wave will give you a maximum resultant value. We can write this as

$$\begin{aligned} V^S_{MAX} &= \left|V^+_M\right| + \left|V^-_M\right| \\ &= \left|V^+_M\right| + \left|\Gamma V^+_M\right| \\ &= (1 + |\Gamma|)\left|V^+_M\right| \end{aligned} \qquad 9.43$$

Note that V^+_M is the magnitude of the complex value and that the reflection coefficient $\mathbf{\Gamma}$ is a phasor. Unbolded Γ represents the magnitude of the reflection coefficient. And $|\Gamma|$ is the absolute value of the magnitude. So we are dealing with the absolute value of real numbers.

Similarly, there will be a minimum value for the resultant standing wave where the reflected wave subtracts from the forward wave, and

$$\begin{aligned} V^S_{MIN} &= \left|V^+_M\right| - \left|V^-_M\right| \\ &= \left|V^+_M\right| - \left|\Gamma V^+_M\right| \\ &= (1 - |\Gamma|)\left|V^+_M\right| \end{aligned} \qquad 9.44$$

We can now rewrite the voltage standing wave ratio in terms of the reflection coefficient and the peak value of the forward travelling wave as

$$\frac{|V^S_{MAX}|}{|V^S_{MIN}|} = \frac{(1+|\Gamma|)|V^+_M|}{(1-|\Gamma|)|V^+_M|}$$

And

$$\boxed{VSWR = \frac{|V^S_{MAX}|}{|V^S_{MIN}|} = \frac{(1+|\Gamma|)}{(1-|\Gamma|)}} \qquad \text{Voltage standing wave ratio} \qquad 9.45$$

There is also a standing wave ratio for the current CSWR. The VSWR and the CSWR values are equal, but there could be a phase angle between the two, and they may appear on the transmission line in a different location. We can use Ohm's law to calculate the maximum and minimum current standing wave amplitudes.

$$|I^S_{MAX}| = \frac{|V^S_{MAX}|}{Z_0} \; and$$
$$|I^S_{MIN}| = \frac{|V^S_{MIN}|}{Z_0} \qquad 9.46$$

And

$$CSWR = VSWR = \frac{(1+|\Gamma|)}{(1-|\Gamma|)} \qquad \text{Current standing wave ratio} \qquad 9.47$$

Figure 9.8 shows how to draw a reflected wave when the reflection coefficient is -1. First, continue to draw the forward wave beyond the boundary of the transmission line, shown as dotted, 9.8a. Fold this wave along the x-axis 9.8b. Then fold back across the y-axis 9.8c. The reflected wave is shown as a dotted line. Draw the rest of the sinusoid backwards. The points where the forward and reflected wave intersect form the peaks (positive or negative) of the standing wave, shown as a bold line in Figure 9.8c. If the coefficient is negative but not -1, then the amplitude of the reflected wave will be less than the forward wave.

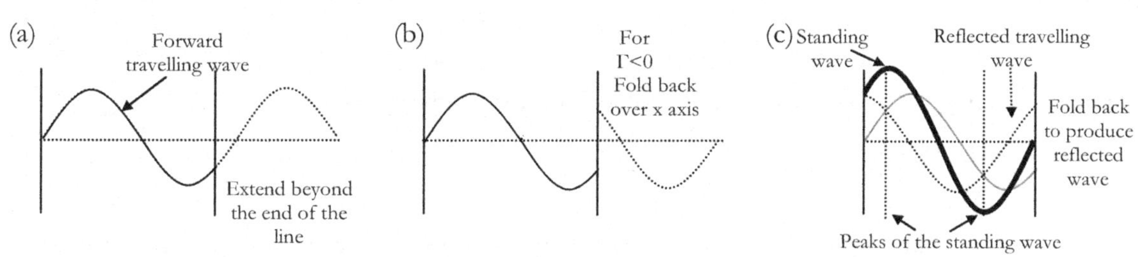

Figure 9.8
Drawing a reflected wave for Γ=-1.

Figure 9.9 shows how to draw a reflected wave when the reflection coefficient is greater than zero. Continue to draw the forward wave beyond the boundary of the transmission line, then fold back across the y-axis, as shown in Figure 9.9b. Once again, the standing wave peaks are at the intersection points.

Figure 9.10 shows the development of a standing wave over time for a lossless transmission line. Figure 9.10a shows the development of a standing wave when the reflection coefficient is -1 and 9.10b when the reflection coefficient is 1. The wave can be a voltage waveform or a current waveform. The forward wave is shown as a solid line, the reflected wave as a dotted line and the standing wave as a thick line.

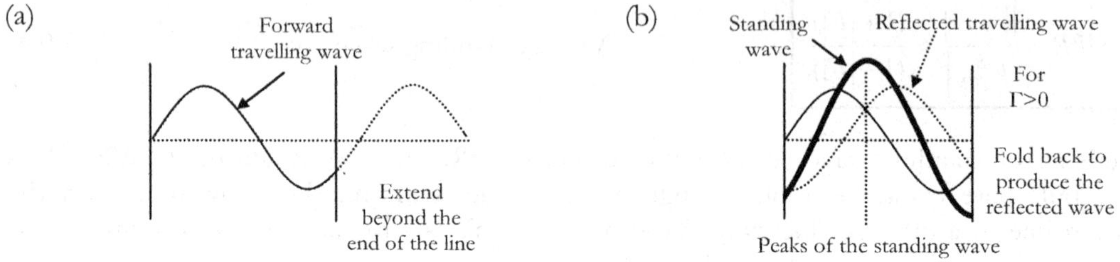

Figure 9.9
Dawing reflected wave for Γ=1.

Standing Wave on a Shorted Transmission Line

For a transmission line with a short at the load, the complex magnitude of the reflection coefficient is -1, and the absolute value is 1. From Equation 9.45, the VSWR is

$$VSWR = \frac{(1+1)}{(1-1)}$$
$$= \frac{2}{0}$$
$$= \infty$$

From Equation 9.43, the maximum standing voltage wave peak value somewhere along the line is

$$V_{MAX}^{S} = (1+1)\left|V_{M}^{+}\right|$$
$$= 2\left|V_{M}^{+}\right|$$

And the standing voltage wave minimum value from Equation 9.44 is

$$V_{MIN}^{S} = (1-1)\left|V_{M}^{+}\right|$$
$$= 0$$

We can see that the maximum amplitude of the standing wave is twice the forward wave amplitude, and the minimum value is zero, giving a VSWR of infinity. At the load, the voltage is always zero as the reflected wave is always 180° out of phase with the forward wave (at this location), and the reflected wave will cancel the forward wave. As the reflected wave moves back from the load, the forward wave is also moving, and the phase relationship between the forward and reflected wave will change, resulting in a standing wave with a node (zero point) at the load, see Figure 9.10a.

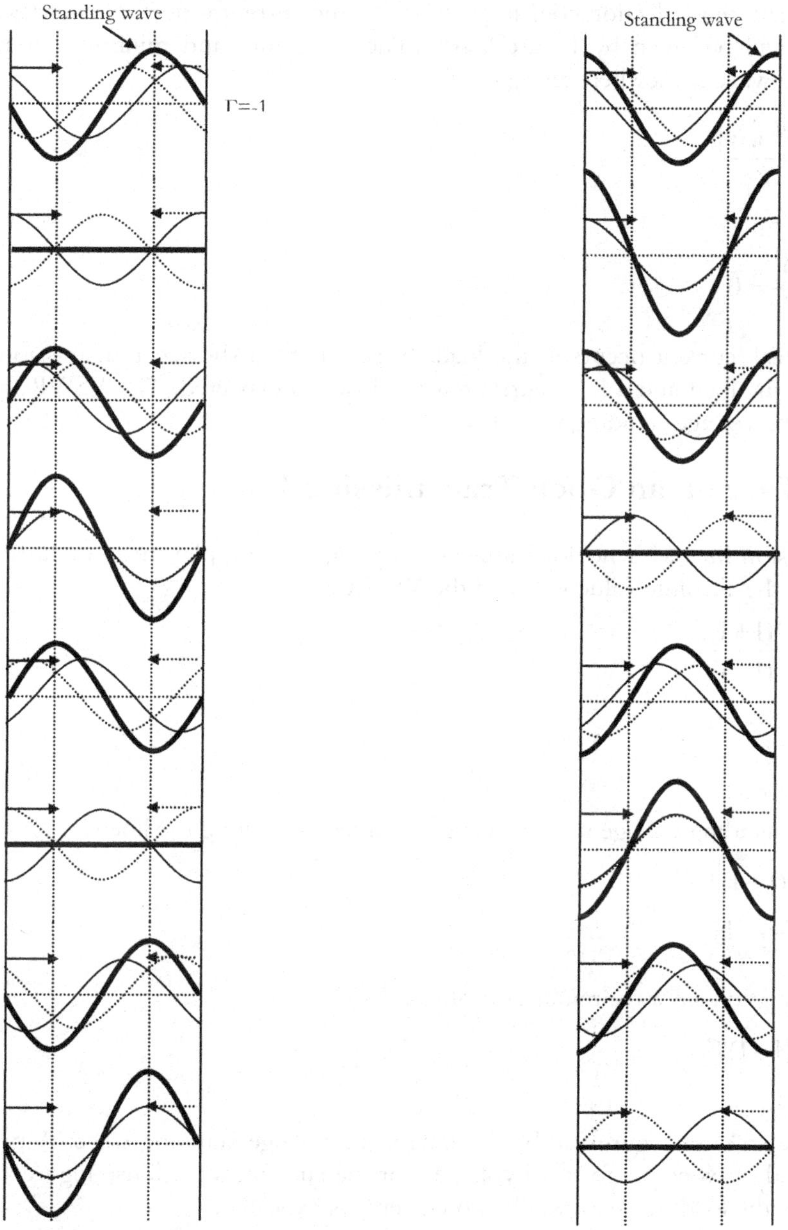

(a)
Standing wave with a negative reflection coefficient.

(b)
Standing wave with a positive reflection coefficient.

Figure 9.10

For the current, the reflection coefficient is $-\Gamma$, so the current reflection coefficient is 1. The standing wave will be given by Figure 9.10b. The maximum and minimum amplitude for the current standing wave somewhere along the line is

$$\left|I^S_{MAX}\right| = \frac{2\left|V^S_{MAX}\right|}{Z_0}$$

and

$$\left|I^S_{MIN}\right| = \frac{0}{Z_0} = 0$$

Notice that a peak current occurs at the load. In summary, a shorted transmission line will have zero voltage at the load and a high current at the load, as expected. The VSWR is infinite as the node value of the voltage standing wave is zero.

Standing Wave on an Open Transmission Line

For a transmission line with no load attached (open), the complex magnitude of the reflection coefficient is 1, the absolute value is 1, and the VSWR is

$$\begin{aligned} VSWR &= \frac{(1+1)}{(1-1)} \\ &= \frac{2}{0} \\ &= \infty \end{aligned}$$

The maximum standing voltage wave peak value somewhere along the line is

$$\begin{aligned} V^S_{MAX} &= (1+1)\left|V^+_M\right| \\ &= 2\left|V^+_M\right| \end{aligned}$$

And the standing voltage wave minimum value is

$$\begin{aligned} V^S_{MIN} &= (1-1)\left|V^+_M\right| \\ &= 0 \end{aligned}$$

As Γ=1, we will use Figure 9.10b to examine the voltage standing wave. A peak voltage will occur at the load of twice the forward voltage. For the current, we will use Figure 9.10a. A current node appears at the load, so there will be no current, as expected.

Standing Wave on a Matched Transmission Line

If we match the load impedance with the characteristic impedance of a transmission line, no reflection occurs and Γ=0. As the forward voltage wave moves along the line, each point along the transmission line will attain a maximum voltage of V^+_M, including the load. The voltage nodes equal the voltage anti-nodes, and the line will have a VSWR of 1. And similarly, for the current wave, the maximum across the load will be I^+_M.

Summary of Standing Wave Properties

Table 9.1 summarises the transmission line standing wave conditions for different loads.

Load	Max load voltage	Max load current	Current phase angle	Γ	VSWR
Short	0	$2I_M$	-90°	-1	∞
Open	$2V_M$	0	+90°	1	∞
Matched	V_M	I_M	0°	0	1

Table 9.1
Transmission line standing wave properties for different loads.

9.10 DETERMINING THE ELECTRICAL CHARACTERISTICS

To calculate the input impedance of a transmission line, we need to know some basic characteristics of the cable. These characteristics can be measured, calculated or given in a speciation sheet. The values needed are:

1. The length of the cable.
2. The inductance per metre L'.
3. The capacitance per metre C'.
4. The resistance per metre R'.
5. The conductance per metre G'.

In most instances, G´ is so small that it can be ignored. If we are treating the line as lossless, then R and G and not required. However, for energy loss, we will need to know the total resistance of the cable. We can determine the inductance and capacitance per unit length values in two ways. The first way is to measure the inductance and capacitance of the length of line (see measuring techniques Chapter) and calculate the per metre values by dividing through by the length of the transmission line.

The second way is by using the relative permittivity of the dielectric, the relative permeability of the wire and the spacing between the wires in the cable; that is the geometry of the cable. We can calculate the inductance per metre of a two-wire transmission line by using the following equation.

$$L' = \frac{\mu_0 \mu_R}{\pi} \cosh^{-1}\left(\frac{s}{W_d}\right)$$ Inductance per metre for a two-wire transmission line 9.48

where Ĺ is the inductance per metre, s is the distance between the centres of the conductors in the transmission line, and W_d is the wire diameter. Note that the hyperbolic function $\cosh^{-1}$ (arcosh) can be calculated using

$$\cosh^{-1}(x) = \ln(x + \sqrt{x^2 - 1})$$ Arcosh calculation 9.49

We can calculate the per unit length capacitance using

$$C' = \frac{\pi\varepsilon_0\varepsilon_R}{\cosh^{-1}(\frac{S}{W_d})}$$

Capacitance per metre for a two-wire transmission line 9.50

We can use these values to calculate the characteristic impedance using Equation 9.23 and the propagation constant using Equation 9.20.

For low-loss transmission lines (and they generally are), we can use Equation 9.36 to determine the velocity of propagation. Using the inductance per metre and the capacitance per metre in Equation 9.36 for the velocity of propagation, we get

$$V_P = \frac{1}{\sqrt{L'C'}}$$

$$V_P = 1 \bigg/ \sqrt{\left(\frac{\mu_0\mu_R}{\pi}\cosh^{-1}(\frac{S}{W_d})\right)\left(\frac{\pi\varepsilon_0\varepsilon_R}{\cosh^{-1}(\frac{S}{W_d})}\right)}$$

Simplifying

$$V_P = \frac{1}{\sqrt{\mu_0\mu_R\varepsilon_0\varepsilon_R}}$$

$$= \frac{1}{\sqrt{\mu_0\varepsilon_0}}\frac{1}{\sqrt{\mu_R\varepsilon_R}}$$

$$= C\frac{1}{\sqrt{\mu_R\varepsilon_R}}$$

The velocity of propagation can be written as

$$V_P = \frac{1}{\sqrt{L'C'}} = \frac{C}{\sqrt{\mu_R\varepsilon_R}} = CV_f$$

Velocity of propagation for a transmission line 9.51

We can either use the inductance and capacitance per metre or the relative permeability and the relative permittivity to determine the velocity of propagation for low-loss transmission line. The velocity factor can be written as,

$$V_f = \frac{V_P}{C} = \frac{1}{C\sqrt{L'C'}} = \frac{1}{\sqrt{\mu_R\varepsilon_R}}$$

Velocity factor a transmission line 9.52

The electrical wavelength becomes

$$\lambda = \frac{V_P}{f}$$

Electrical wavelength for a transmission line 9.53

If we multiply the velocity of propagation by the inductance per metre, we get

$$L'V_P = \frac{L'}{\sqrt{L'C'}}$$
$$= \frac{\sqrt{L'}}{\sqrt{L'}}\frac{\sqrt{L'}}{\sqrt{C'}}$$
$$= \frac{\sqrt{L'}}{\sqrt{C'}}$$
$$= Z_0$$

So we can calculate the characteristic impedance by multiplying the propagation velocity by the inductance per metre.

$$\boxed{Z_0 = L'V_P}$$ Lossless transmission line characteristic impedance 9.54

9.11 WORKED EXAMPLES

The following transmission line will be referred to as the standard transmission line for this book. This line is a length of relatively cheap and readily available standard 4-wire telephone cable with two pairs of cores connected to form a two-wire transmission line. With two pairs of wires connected, the inductance per metre needs to be reduced by a factor of two (half) as the wire inductance is in parallel, and the capacitance per metre needs to be doubled as the capacitance is also in parallel. Using Equation 9.48 to calculate the inductance per metre and divide by two.

$$L' = 0.5\left[\frac{\mu_0\mu_R}{\pi}\cosh^{-1}\left(\frac{s}{W_d}\right)\right]$$
$$= 0.5\left[\frac{\mu_0 1.0}{\pi}\cosh^{-1}\left(\frac{1.0x10^{-3}}{0.5x10^{-3}}\right)\right]$$
$$= 0.5\left[0.52678x10^{-6}\right]$$
$$= 0.26339x10^{-6}$$

Description	Value
Transmission line length	10 m
Wire diameter	0.5 mm (x2)
Line relative permeability	$\mu_R = 1.0$
Line relative permittivity (using PVC)	$\varepsilon_R = 3.0$
Distance between centres of conductors	1.0 mm

Table 9.2
The standard two-wire transmission line.

Calculate the capacitance per metre using Equation 9.50

$$C' = 2.0\left[\frac{\pi\varepsilon_0\varepsilon_R}{\cosh^{-1}(\frac{s}{W_d})}\right]$$

$$= 2.0\left[\frac{\pi\varepsilon_0(3)}{\cosh^{-1}(\frac{1.0x10^{-3}}{0.5x10^{-3}})}\right]$$

$$= 2.0\left[6.3365x10^{-11}\right]$$

$$= 126.73x10^{-12}$$

Calculate the total wire AC resistance at 1 MHz using Equation 5.24. As we have two wires in the transmission line, the length needs to be multiplied by 2. However, we have also connected wires in pairs reducing the resistance by half (the resistances are in parallel).

$$R_W = 0.5\left[2.0\frac{\rho\ell_A}{\pi(2r-\delta)\delta}\right]$$

$$= \frac{(1.68x10^{-8})10}{\pi(2(0.25x10^{-3}) - 6.5234x10^{-5})6.5234x10^{-5}}$$

$$= 1.8855$$

The resistance per metre is the total AC resistance divided by the length of the cable.

$$R' = \frac{R_W}{\ell_T}$$

$$= \frac{1.8855}{10}$$

$$= 0.18855$$

Now we can calculate the characteristic impedance of the transmission line, setting G´ to zero

$$\mathbf{Z_0} = \sqrt{\frac{R' + j\omega L'}{G' + j\omega C'}}$$

$$= \sqrt{\frac{0.18855 + j(6.280x10^6)(0.26339x10^{-6})}{j(6.280x10^6)(126.73x10^{-12})}}$$

$$= 45.73\angle -3.2^\circ$$

$$= 45.66 - j2.59$$

And calculate gamma.

$$\gamma = \sqrt{(R' + j\omega L')(G' + j\omega C')}$$

$$= \sqrt{(0.18855 + j(6.280x10^6)(0.26339x10^{-6}))(j(6.280x10^6)(126.73x10^{-12}))}$$

$$= 0.002065 + j0.03636$$

Note that α = 0.002065 and β = 0.03636. Having determined the electrical characteristics of the transmission line, we can calculate the input impedance for a given load. We will use a short as a load for this exercise, and calculate the input impedance of the transmission line using Equation 9.29. Note we could have used Equation 9.38.

$$\mathbf{Z}_{IN} = \mathbf{Z}_0 \frac{\mathbf{Z}_L + \mathbf{Z}_0 \tanh(\gamma\ell)}{\mathbf{Z}_0 + \mathbf{Z}_L \tanh(\gamma\ell)}$$

$$= (45.73\angle -3.2^\circ)\frac{0.0 + (45.73\angle -3.2^\circ)\tanh((0.002065 + j0.03636)10.0)}{(45.73\angle -3.2^\circ) + 0.0((0.002065 + j0.03636)}$$

$$= (45.73\angle -3.2^\circ)\frac{(45.73\angle -3.2^\circ)\tanh((0.002065 + j0.03636)10.0)}{(45.73\angle -3.2^\circ)}$$

$$= 0.6398 - j120.02$$

$$= 120.0\angle -89.69^\circ$$

Calculate the velocity factor using Equation 9.52

$$V_f = \frac{1}{\sqrt{\mu_R \varepsilon_R}}$$

$$= \frac{1}{\sqrt{1.0(3.0)}}$$

$$= 0.58$$

The waveform travels down the transmission line at 0.58 the speed of light. We can calculate the velocity of propagation using Equation 9.51

$$V_P = CV_f$$

$$= 2.99792458x10^8(0.58)$$

$$= 1.739x10^8$$

And the electrical wavelength using Equation 9.53 at 1MHz is

$$\lambda = \frac{V_P}{f}$$

$$= \frac{1.739x10^8}{1x10^6}$$

$$= 1.739x10^2$$

That is 173.9 metres.

10

Antennas

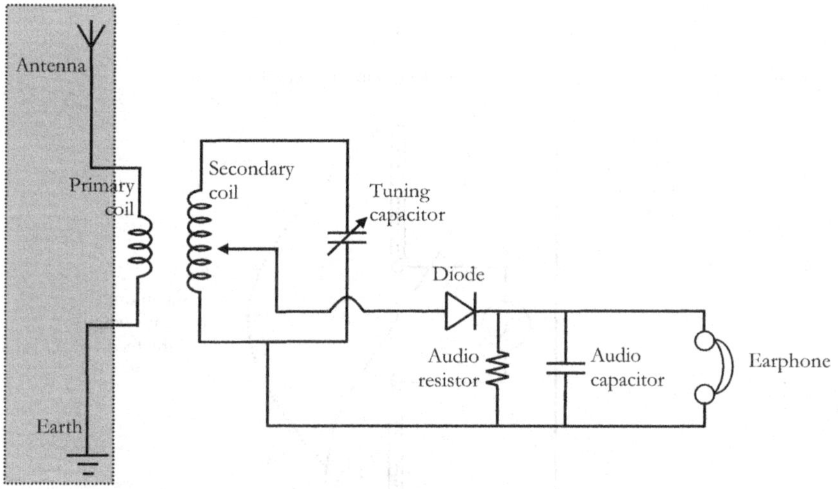

10.1 HALF-WAVE DIPOLE ANTENNA

Consider a half-wavelength open-ended (no load $\Gamma=1$) transmission line, as shown in Figure 10.1. A voltage standing wave and a current standing wave will develop across the line (refer to Figure 9.10 for the shape of the standing waves). If the same signal were now injected into the middle of two lengths of wire 180° apart, each one-quarter of a wavelength long (one-half a wavelength in total), voltage and current standing waves would be created across the wires matching the half-wavelength transmission line pattern.

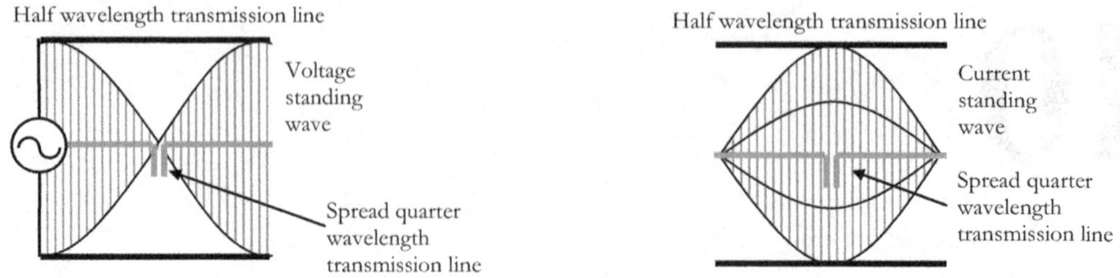

Figure 10.1
Open-end half-wave transmission line voltage and current standing waves.

Figure 10.2 shows such a transmission line, with two lengths of wire separated at the feed point, each length of wire being one-quarter wavelength long. A transmission line constructed using this geometry is called a ***half-wave dipole antenna***. Dipole means it has two active antenna elements and a half-wave because the total length of the active elements is one-half a wavelength long.

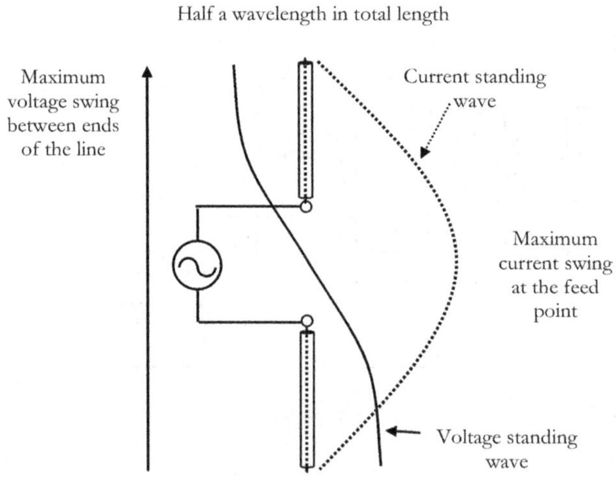

Figure 10.2
Half-wavelength dipole antenna.

10.2 HOW AN ANTENNA CREATES ELECTROMAGNETIC WAVES

Consider a transmitting dipole antenna, shown in Figure 10.3. When a voltage is applied to the antenna, the electric field inside the wire causes charges (electrons) to migrate to one end. However, the circuit is not complete, and the charges congregate, causing a charge separation at both ends (an electric dipole). One end being overall positive and the other negative. This dipole produces an electric field around the antenna, as shown in Figure 10.3a. The voltage across the antenna is at a maximum, as shown in Figure 10.2. The velocity of the electrons at the ends of the antenna is zero. Therefore, the current is zero, as shown in Figure 10.2.

When the direction of the applied voltage flips, the electrons accelerate in the opposite direction with the maximum current at the feed point. This acceleration causes a rapid change in the electric field lines surrounding the antenna. As nothing in the universe happens instantaneously, this change in the electric field takes time to propagate out to the field lines far away. So the shape of the field lines closest to the antenna will be different than the shape of the field lines at some distance, as shown in Figure 10.3b.

When the two charges meet at the feed point, the electric field disappears. However, the outer field lines are not aware that the field has gone and, as such find themselves on their own. They become disconnected from the antenna and are free to move in space. This is called pinch-off; see Figure 10.3c. The distance the electric field lines "pinch-off" is vaguely defined, but is approximately given by the following equation

$$D_f \geq \frac{2\ell^2}{\lambda} \qquad \text{Pinch-off distance} \qquad 10.1$$

where ℓ is the length of the antenna, and λ is the wavelength of the frequency in free space. For a half-wave dipole at 1 MHz, this distance is about 37.5 metres.

The electric field occupying a distance greater than D_f is known as the ***far field,*** and the electric field closer to the antenna than D_f is called the ***near field.*** Pinch-off occurs for both the electric and magnetic fields, and they are free to travel as electromagnetic waves. These electromagnetic, or radio waves, can now be picked up by a distant antenna.

An interesting property of an antenna is that good transmitting antennas also make good receiving antennas for the same frequency. This is known as ***antenna reciprocity.*** A given antenna will have the same characteristics whether it is used for transmitting or receiving.

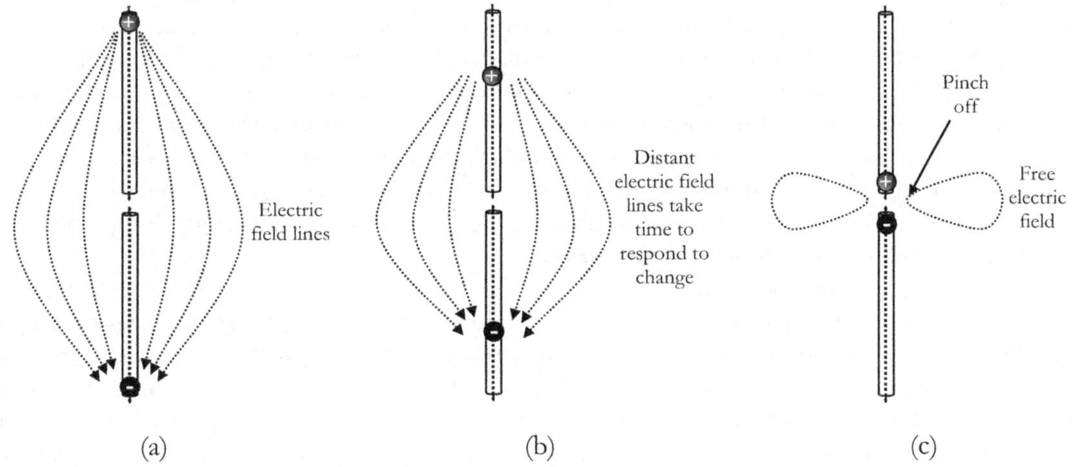

Figure 10.3
Antenna radiation.

10.3 THE MONOPOLE ANTENNA AND GROUND

The ***polarization*** of an electromagnetic wave is determined by the orientation of the electric field. If the electric field points vertically with respect to the ground, it is ***vertically polarized***. An electromagnetic wave with an electric field horizontal to the ground is ***horizontally polarized***. Broadcast band AM signals are vertically polarized. The electromagnetic field created by an antenna will be polarised in the orientation of the length of the wire. An antenna pointing upward will create vertically polarized electromagnetic waves. And due to antenna reciprocity, a vertically oriented antenna will receive vertically polarized radio waves.

A 1 MHz radio wave in free space will have a wavelength of about 300 metres. If we were to create a half-wave dipole to use in the AM broadcast band, we would need a 150 metre (492 feet) antenna; which is not very realistic. Fortunately, radio waves are electromagnetic waves, just like light, but with longer wavelengths. Just as light waves can be reflected by polished metal surfaces, radio waves can be reflected by metal surfaces or surfaces with high conductivity. Note that reflection and refraction are statistical in nature. Electromagnetic waves penetrate into a substance given by its penetration depth. This is defined similarly to the skin depth, where the intensity of the incident electric field drops off by 1/e (36%) from that at the surface. So RF signals still penetrate metals, with 36% of the electric field penetrating right into the bulk of the metal, dropping off exponentially towards the core. How deep most of the signal penetrates the material depends on the frequency and the material in question - the higher the frequency, the smaller the penetration depth.

RF waves are wavefronts containing vast numbers of individual EM waves known as photons. The more photons a wavefront has, the greater the intensity. Photons not 100% reflected will give up their energy to the antenna to cause current to flow. This must be the case; otherwise, we would be violating the conservation of energy. Work is done on the electrons, work requires energy, and this energy comes from the RF wave.

Consider one arm of a half-wave dipole; it has a length of one-quarter of a wavelength. We place a solid metal disk underneath, but not touching the arm, with a radius of at least one-quarter wavelength, as shown in Figure 10.4a. Radio waves coming in from an angle will strike the metal disk; some will be absorbed, and some will be reflected. The reflected waves from the metal disk impacting the antenna will contribute to the overall strength of the electric field along the length of the antenna. An increase in the maximum electric field will increase the magnitude of the charge separation and will increase the maximum voltage difference between the ends. The reflective material underneath the antenna is called a ***ground plane*** and acts like a mirror for radio waves. For perfectly reflective ground planes, the voltage across the monopole antenna will approach voltages normally seen across a dipole. An antenna one-quarter of a wavelength long that uses a ground plane is called a ***quarter-wave monopole***.

Of course, there is no such thing as a perfect mirror or ground plane, but we will treat it as such to make the analysis simpler. There is a technique called the ***method of images*** used to study the ground effect. If you look into a mirror, from your point of view, images behind you look like they appear deeper in the mirror and can be thought of as extending into some virtual environment behind the mirror. Using this technique, we can consider a mirror image of the charge distribution on the antenna at any given time exists behind the ground plane, as shown in Figure 10.4b. The ground plane acts as a virtual arm of the antenna and, for analytical purposes, can be treated as such.

For high-frequency RF, using solid metal objects as a ground is feasible. However, for AM frequencies, we would require a solid metal disk with a diameter of 300 metres. Sometimes AM broadcast antennas use metal strips or lengths of wire placed under the ground to aid in the ground effect. For us, however, this is still impractical. As it turns out, the Earth is quite conductive to varying degrees, depending on the soil type and water content, and can be used as a ground plane. A true earth ground won't be as effective as a metal ground, but RF waves will reflect to some degree, depending on the soil type.

What if we now drove a metal rod into the ground giving us another real arm of a dipole antenna? Using the method of images, we would have an antenna appearing, as shown in Figure 10.5.

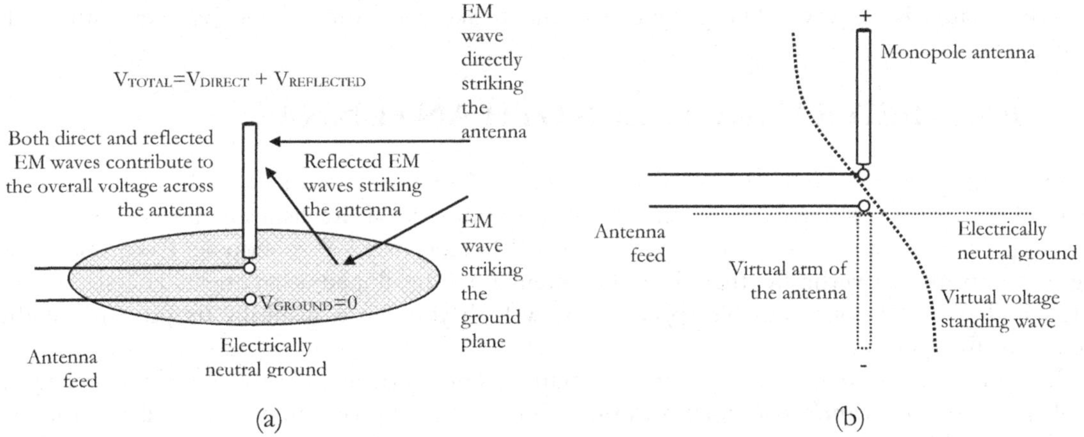

Figure 10.4
Monopole antenna.

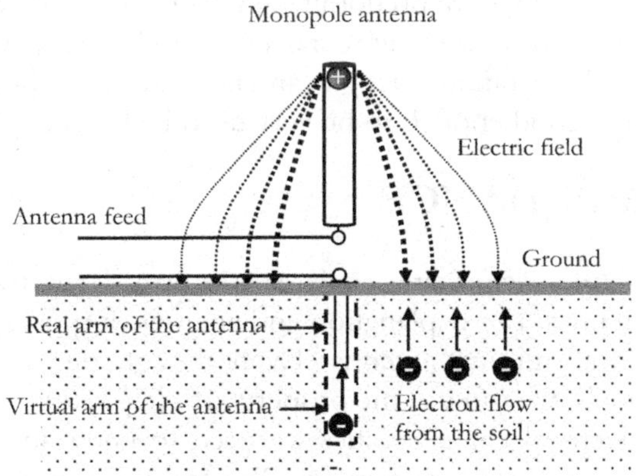

Figure 10.5
Monopole using a ground rod.

Electrons will flow through the metal to provide current, and also through the earth. The metal rod will have very little resistance. However, the resistivity of the soil will be much greater, resulting in a much larger resistance. Think of the soil as being a resistive extension to the metal rod. This resistance will result in a real power loss. The more soil the electrons need to travel through before reaching the metal, the greater the ground resistance will be. If we use a longer rod and or a rod with a greater surface area, then we should reduce the ground resistance. This is, in fact, the case. This ground resistance can be calculated, and it will be discussed when we talk about the impedance of the antenna. Also, placing the ground rod directly underneath the antenna should result in a high voltage. However, for practical antennas, there will be a large horizontal length component, and the charge distribution will be different. Therefore, practicalities will determine where the ground rod is placed.

Also, the charge in the antenna creates an electric field that extends down to the ground (like a capacitor) and is called self-capacitance. The electrons in the ground will feel the electric force and move accordingly. Toward the surface when the charge is positive, and away from the surface when the charge is negative. The ground acts much like a conductor and has very little self-capacitance.

10.4 END-FED RANDOM-LENGTH ANTENNAS

An AM band quarter-wave monopole would be about 75 metres (246 feet) high. Although one-half the height of a half-wave dipole, this is still very impractical. Realistically, unless you have a large property, the limit will be around 37.5 metres (123 feet) or a ⅛th monopole. Even this is too long to be positioned vertically; therefore, the antenna will be sloped at an angle. The size of the angle will depend on your available space and how high you can realistically fix one end of the antenna off the ground.

This type of monopole, whose antenna orientation is not vertical and is not ¼ of a wavelength, is called an ***end-fed random-length antenna***. This is the type of antenna we will use for our crystal radios. These types of antennas can be difficult to analyse, especially when we use any available object that might act like an antenna and a ground that may be nothing more than a water pipe running for how long and who knows where. Note, I consider water pipes to be the ground of last resort. They typically do not penetrate very deep into the earth and may even be mostly PVC piping. If your water pipes are predominately PVC then they will be practically useless as an earth. Also, the characteristics of an antenna can be changed by the objects around it. Nevertheless, we can analyse these random-length antennas sufficiently to get characteristics of interest to us, with results being good enough to be very useful when designing crystal radios.

10.5 ANTENNA IMPEDANCE

Due to the principle of antenna reciprocity, analyses performed for transmitting antennas are equally valid for receiving antennas. For explanation and analysis purposes, sometimes the antenna will be considered to be a transmitting antenna. Even though our crystal radios will not be transmitting but receiving signals, we can use the property of reciprocity.

The important antenna electrical characteristics we need to know in order to design a crystal radio are the real resistance and the input capacitance. The real resistance represents a loss and will affect the Q of the radio. The input capacitance will determine the design of the primary coil needed to achieve resonance. Antennas tend to be far more lossy than transmission lines, so we shall consider the antenna to be a lossy transmission line.

Characteristic Impedance of an Antenna

To determine the input capacitance, we need to calculate the characteristic impedance Z_0 of the antenna. The characteristic impedance of a transmission line is given by Equation 9.23. In studying transmission lines, we used Equations 9.48 and 9.50 to determine the inductance-per-metre and capacitance-per-metre to be used in Equation 9.23. However, Equations 9.48 and 9.50 assume the line has two parallel conductors, and the electric field inhabits the material between the two conductors. For an antenna, the surrounding material is essentially free-space, and there is no parallel conductor to modify the magnetic and electric fields. There is only self-capacitance to the ground, as shown in Figure 10.5. Therefore the ground permittivity will only have a slight effect on the total surrounding antenna permittivity. Inside the ground, there exists very little capacitance as the ground essentially acts like a conductor, and so any internal ground self-capacitance can be ignored. The relative permittivity surrounding the antenna will be referred to as the antenna's relative permittivity ε_A.

The energy stored in the magnetic field surrounding the antenna and ground rod is due to the permeability of the surrounding material. Assuming the soil does not contain a high concentration of materials such as iron oxide, we can treat the relative permeability of the soil as 1. Therefore, we will treat the permeability as the inductance-per-metre and the permittivity as the capacitance-per-metre. The characteristic impedance of a monopole antenna becomes

$$\mathbf{Z_0} = \sqrt{\frac{R' + j\omega\mu_0\mu_R}{G' + j\omega\varepsilon_0\varepsilon_A}}$$ Characteristic impedance of a monopole antenna 10.2

And setting

$$L' = \mu_0\mu_R$$
$$and$$
$$C' = \varepsilon_0\varepsilon_A$$

we can analyse our antenna using the lossy transmission-line equations.

From experience, it is known that the velocity factors of antennas range from 0.9 to just under 1.0. That is

$$0.9 \leq \frac{1}{C\sqrt{L'C'}} < 1$$

For a velocity factor of 0.95, the apparent antenna permittivity ε_A will be 1.1. The permittivity of the soil increases with water content, so if you have soggy soil, select an ε_A of 1.2.

Radiation Resistance

When an antenna is used to transmit signals, the energy contained in the radiated electromagnetic wave represents a loss of energy from the system. Due to the conservation of energy, the energy of the electromagnetic wave must come from somewhere. In this case, the energy contained in the electromagnetic wave is taken from the kinetic energy of the electrons. This results in a change in the electron's velocity, slowing them down. This can be thought of as another type of resistance to the flow of current. Just as normal Ohmic losses (resistance) result in a loss of electrical energy as heat, this loss represents a loss of electrical energy in the form of electromagnetic radiation. This type of resistance is called ***radiation resistance*** and can form the major resistive component of the input impedance of an antenna. For a centre-fed half-wave dipole, the radiation resistance is about 73 Ω which is the often quoted input resistance.

The actual radiation resistance depends on many factors, such as antenna length, feed point, wire, frequency and surrounding objects etc. Table 10.1 can be used to arrive at an approximate value depending on the antenna type.

Antenna type	Radiation resistance
Centre-fed half-wave dipole	73 Ω
Short dipole $\lambda/50 < \ell < \lambda/10$	$20\pi^2\left(\frac{\ell}{\lambda}\right)^2$
Quarter-wave monopole (good ground)	36.5
Short monopole dipole $\ell << \lambda/4$ (good ground)	$40\pi^2\left(\frac{\ell}{\lambda}\right)^2$

Table 10.1
Approximate radiation resistance of different antenna types.

Ground Resistance

The ground is a conductor, but compared to copper and other metals, not a very good one. Like all conductors, the ground has resistance, which varies greatly between soil types. The ground or soil resistance is approximated by the following equation.

$$R_S = \frac{\rho_G}{2\pi\ell_G}\left[\ln(\frac{8\ell_G}{d}) - 1\right]$$ Soil resistance 10.3

where ρ_G is the resistivity of the ground, ℓ_G is the length of the ground rod in the earth, and "d" is the diameter of the ground rod. Table 10.2 gives some typical ground resistivity values for various soil types.

The antenna ground resistance is dependent on the antenna geometry and frequency and is given by

$$R_G = \frac{R_S}{2\pi}\left[\ln(\frac{\lambda}{2a})\sin^2(\beta\ell) - 4\left(\sin(\sqrt{\frac{\pi^2}{4} + (\beta\ell)^2} - \cos(\beta\ell)\right)\cos(\beta\ell)\right]$$ Antenna ground resistance 10.4

where R_S is the soil resistance as calculated by Equation 10.3, "a" is the diameter of the antenna, ℓ is the length of the monopole, β is the wave number, and λ is the wavelength.

Let us look at the effect of connecting a single-coil crystal radio directly to an antenna. Say we have a 300 uH coil with an AC resistance of 9 Ω. At 1 MHz, the coil Q is,

$$Q_{Coil} = \frac{(2\pi 10^6)300x10^{-6}}{9}$$

$$= 209$$

Say we have an antenna ground resistance of 21.08 Ω at 1 MHz, a reasonable value. The Q of the coil connected to an antenna, accounting for ground losses, will be

$$Q_{Circuit} = \frac{(2\pi 10^6)300x10^{-6}}{9+21.08}$$
$$= 63$$

That is nearly a 70% loss in Q due to the ground resistance only. This is the reason you should never make a crystal radio where the tank circuit connects directly to the antenna. The only crystal radio to consider building is one where the primary coil connects to the antenna with a separate tuning (tank circuit) coil. This is called a ***loose-coupler***.

Soil type	Resistivity (Ω-m)
Clay, compacted	100 – 200
Clay sand	50 – 500
Siliceous sand	200 – 300
Soil, chalky	100 – 300
Stony ground	1,500 – 3,000
Stony sub-soil, grass-covered	300 - 500
Granite	1,500 – 10,000
Limestone	500 – 1,000
Mica schist	800
Sandstone	1,500 – 10,000
Schist, shale	50 – 300
Peat, turf	5 – 100
Soil, swampy	1 – 30

Table 10.2
Ground resistivity for various soil types.

Antenna Reactance

Determining the reactance of an antenna is quite a complicated process, as the value of the reactance is heavily dependent on the current distribution along the antenna (as is the radiation resistance). The current flowing in some small length of the antenna affects the current distribution in neighbouring wire segments along the entire length of the antenna. The presence of

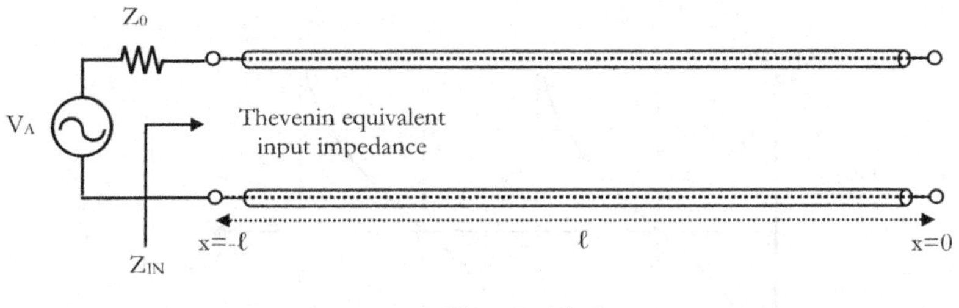

Figure 10.6
Transmission line model for an antenna.

external objects also affect the current distribution. High-end analysis would require the current distribution and the input impedance of the antenna to be split into small sections. The current would then be analysed in each section with respect to all the other sections, both internally and externally. However, for simple wire antennas, a reasonable approximation can be made by assuming a sinusoidal current distribution and determining the reactive component of the input impedance by transmission line theory. Figure 10.6 shows the equivalent transmission line model for an antenna.

The voltage V_A is the induced voltage across the antenna. We can treat the source voltage as though it is appearing across the antenna's input, and the source impedance will be the impedance of some small infinitesimally small section of the line; that is, the characteristic impedance Z_0.

This model can be used for an end-fed monopole antenna of length ℓ, with the earth treated as the return path of the transmission line. Treating the antenna as a lossy open-ended transmission line allows us to calculate the input impedance using Equation 9.40. And is given by

$$\mathbf{X_A} = imag\{\mathbf{Z_0}\coth(\gamma\ell)\} = imag\left\{\frac{\mathbf{Z_0}}{\tanh(\gamma\ell)}\right\}$$ Reactance of an antenna 10.5

This equation returns the Thevenin equivalent impedance, with the equivalent resistance given by the real number and the equivalent reactance given by the imaginary part of the complex number. This equation returns the equivalent resistance and reactance required to give the phase relationship between the voltage and current on a transmission line, as seen externally. However, it does not tell us what is going on inside the antenna. The returned antenna reactance value will be used to match the primary coil's reactance as it is the reactance seen by the primary coil. However, when determining power loss in the antenna, we need to know the actual resistance across the antenna and the Thevenin equivalent resistance will not be used. We need to know the actual internal resistance of the antenna, and it is this resistance value we shall use.

Treating the antenna as lossless and using Equation 9.41 to plot a graph for different relative antenna lengths ℓ in terms of the frequency, we can get a feel of the antenna reactance as the length changes, as shown in Figure 10.7. The cotangent function repeats (has a period) of π radians, so the impedance of the antenna will repeat every half wavelength ($\ell=\pi/2$). That is

$$\beta\ell = \left(\frac{2\pi}{\lambda}\right)n\frac{\lambda}{2} = n\pi$$

From the graph, we can see that when an antenna is less than ¼ of a wavelength, the antenna looks like a capacitor. If the length is over ¼ of the wavelength, then the antenna looks like an inductor. At exactly ¼ of a wavelength, the antenna looks resistive and is, therefore, resonant. This switching between capacitance and inductance repeats every ½ a wavelength.

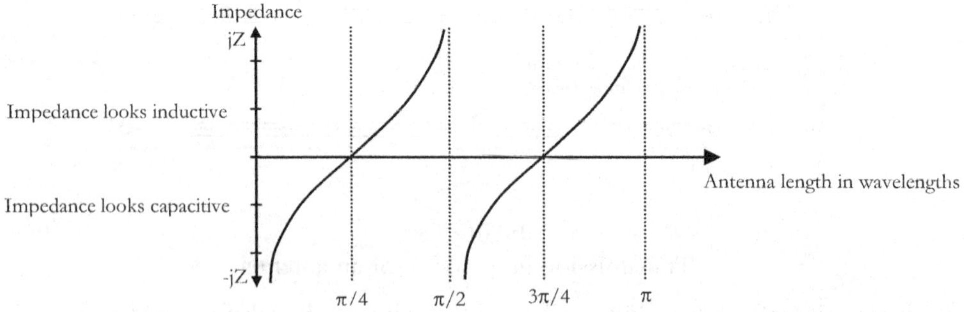

Figure 10.7
Antenna input impedance.

For a 1 MHz signal, a ¼ wavelength will be 75 metres (246 feet), so any antenna less than 75 metres will look capacitive at this frequency. The antenna lengths we will use will be much shorter than this and look capacitive for all AM frequencies.

Total Antenna Input Impedance

The total input impedance of an antenna is complex, consisting of a real component and an imaginary component.

$$Z_A = R_A + jX_A$$

The real component contains all of the real resistances in the antenna. These include wire resistance, radiation resistance and ground resistance. The imaginary component will be either inductive or capacitive. Our antennas will be shorter than a quarter of a wavelength, look capacitive, and be modelled as such.

Figure 10.8 shows the circuit model of an antenna with a load attached. Where V_A is the voltage induced across the antenna, R_A is the total antenna resistive losses. C_A is the apparent capacitance value, and Z_L is the load impedance. The impedance of our antennas will look like

$$\boxed{\mathbf{Z_A} = R_A - jX_A} \qquad \text{Impedance form of our antennas} \qquad 10.6$$

The total restive losses are

$$\boxed{R_A = R_W + R_G + R_{RAD}} \qquad \text{Total antenna resistance} \qquad 10.7$$

where R_W is the AC resistance of the antenna wire, R_G is the ground resistance, R_{RAD} is the radiation resistance. Therefore, the total antenna input impedance is

$$\boxed{\mathbf{Z_A} = (R_W + R_G + R_{RAD}) - imag\left\{\frac{\mathbf{Z_0}}{\tanh(\gamma\ell)}\right\}} \qquad \text{Total antenna input impedance} \qquad 10.8$$

For normal half-wave dipoles, the radiation resistance R_{RAD} is the major contributor to resistance. For the monopoles using the earth, the radiation resistance will be quite small. The AC wire resistance will be a contributor, but the greatest resistance will be the ground resistance.

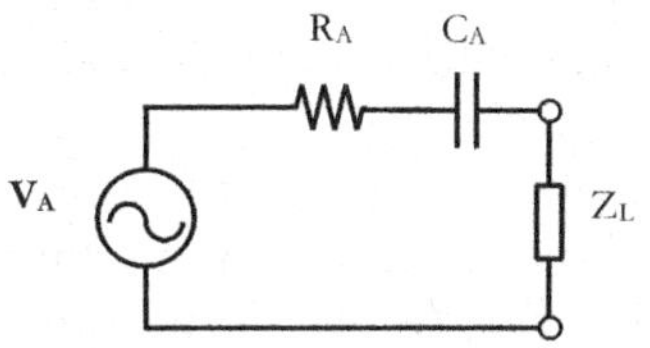

Figure 10.8
Circuit model of a receiving antenna with a load impedance.

The equivalent capacitor value can be calculated by using the capacitive reactance formula

$$C_A = \frac{1}{2\pi f X_A}$$ Antenna equivalent capacitance value 10.9

Making the Antenna Resonant

By making the antenna circuit in our crystal radio series resonant, we will receive the greatest amount of current across our primary coil. A large current across the primary will produce a large magnetic field which in turn will induce a large current and voltage in the secondary coil.

We can achieve resonance in an antenna by effectively removing all the capacitive reactance by adding a load inductor with the same reactance value as our antenna. So for our antenna, which looks capacitive, we would add an inductive load such that the load impedance is

$$\mathbf{Z}_\mathrm{L} = R_L + jX_A$$

As all the components are in series, the total impedance of circuit 10.8 becomes

$$\begin{aligned}\mathbf{Z}_T &= \mathbf{Z}_A + \mathbf{Z}_L \\ &= (R_A - jX_A) + (R_L + jX_A) \\ &= (R_A + R_L) + (jX_A - jX_A) \\ &= R_A + R_L\end{aligned}$$

And now looks totally resistive and thus resonant. You can use Equation 8.3 to determine the value of the inductor required to make the circuit resonant

10.6 ATTACHING A TRANSMISSION LINE

Unless you are using a short indoor antenna or want to sit outside next to your antenna, you will need long leads attached to your crystal radio. We can use a two-wire transmission line and use the capacitance of the transmission line to our advantage. When we are dealing with transmission lines, we treat them as a cascade, applying the result of the first transmission line to the second and so on. The transmission line equations take into account the electrical characteristics of the entire transmission line resulting in an equivalent single-value capacitance in series with a voltage source.

Note that adding an actual capacitor in series would result in a reduction of the input capacitance of our antenna (see Equation 7.33), the opposite of what we want. And adding a capacitor in parallel across the antenna will result in a path to ground, reducing our signal strength and will not result in resonance. The only time you would add a capacitor in series in the antenna circuit is to reduce capacitance.

When adding a transmission line to an antenna, treat the input impedance of the antenna as the load, as shown in Figure 10.9a, and calculate the input impedance at the beginning of the transmission line using Equation 10.10. This results in the equivalent circuit 10.9b. You can then calculate the inductance required to make the circuit series resonant. You can use Equation 10.10 to calculate the input impedance for a lossy line or Equation 10.11 for the lossless case.

$$\mathbf{Z}_{IN} = \mathbf{Z}_0 \frac{\mathbf{Z}_A + \mathbf{Z}_0 \tanh(\gamma\ell)}{\mathbf{Z}_0 + \mathbf{Z}_A \tanh(\gamma\ell)}$$ Input impedance with transmission line - lossy 10.10

$$\mathbf{Z}_{IN} = Z_0 \frac{\mathbf{Z}_A + jZ_0 \tan(\beta\ell)}{Z_0 + j\mathbf{Z}_A \tan(\beta\ell)}$$ Input impedance with transmission line - lossless 10.11

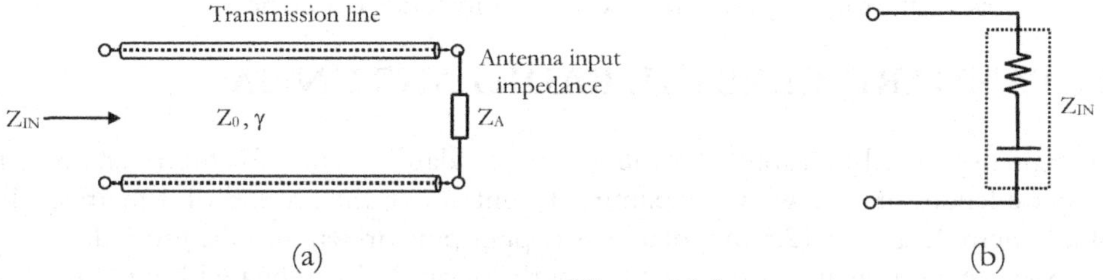

Figure 10.9
Input impedance of a transmission line connected to the antenna.

Figure 10.10a plots the coil inductance required to make the antenna circuit resonant as a function of frequency using our standard 25-metre wire antenna (described below) and a 10-metre two-wire transmission line described in Chapter 9. The primary coil of our crystal radio will require fewer windings if we use a transmission line, particularly for the lower frequencies. The inductance values required when using an antenna only range from 28-396 uH. When the two-wire transmission line is attached, the range reduces to 4.5-59 uH.

Figure 10.10b plots the voltage across the primary coil at resonance as a function of frequency for both the antenna only and with the transmission line. From the graph, you can see the voltage is much greater for primary coils of high inductance values. However, this voltage advantage is largely lost due to mutual inductance losses (described in a later Chapter), and the station volume will not be noticeably different (at least when the primary is at resonance). This will become clearer in later Chapters when we build both small inductance coils and large inductance coils to achieve resonance and why adding a transmission line to increase capacitance is, in most instances, a good thing.

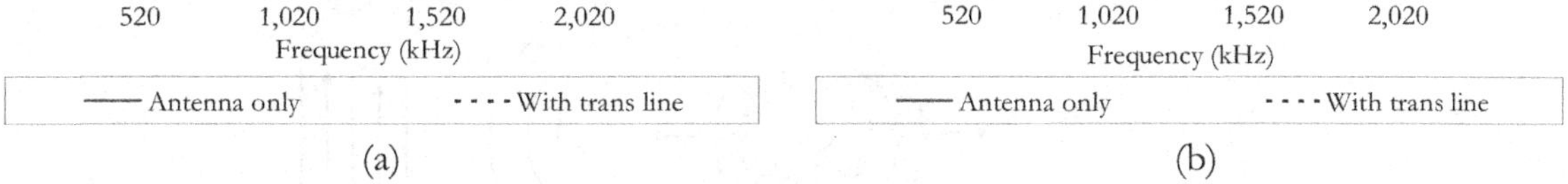

Figure 10.10
Load inductance values and primary coil voltages at resonance.

In summary, here are some rules of thumb, provided the total length of the antenna is less than ¼ of a wavelength.

- The longer the antenna, the greater the input series capacitance.
- Adding a transmission line will increase the series input capacitance to the antenna, and the longer the transmission line, the greater the capacitance.
- An inductor with the same capacitive reactance value as the antenna input capacitance, placed in series with the antenna input, will result in series resonance.

10.7 OUR STANDARD CRYSTAL RADIO ANTENNA

The monopole antenna specified in Table 10.3 will be the standard antenna for future analysis. It consists of a 25-metre length of wire with a maximum height above the ground of 4 metres. The earth consists of a 2-metre length of 12.5 mm diameter copper pipe driven into the ground.

The wire length was chosen as it is a reasonable length for an AM antenna without being too short or excessively long for a standard house yard. Secondly, my local electronics store sells this wire in spools of 25 metres, so it is convenient and cheap. This ground rod was chosen because 12.5 mm copper piping is readily available and will give you a good ground at two metres of depth. I have selected a ground resistivity of 100 ohm-metres as this is a typical value for common clay and sand soil types.

Description	**Value**
Monopole length	25 m
Monopole diameter	0.5 mm
Angle to the ground plane	9.2°
Vertical height =length x sin(θ)	4.0 m
Ground rod length	2 m
Ground rod diameter	12.5 mm
Ground resistivity	100 Ω-m

Table 10.3
Our standard crystal radio antenna.

10.8 ANTENNA GAIN

Consider a theoretically very small dipole antenna, so small it can be considered as a point source. This type of theoretical antenna is called an ***isotropic antenna***. Electromagnetic waves radiate spherically from this point source equally in all directions, as shown in Figure 10.11. As the radio

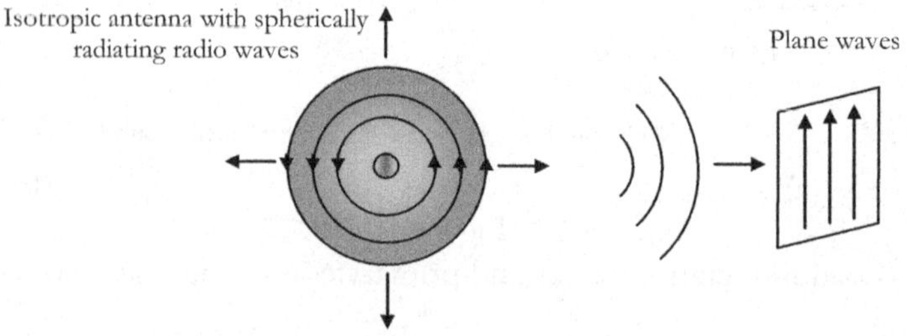

Figure 10.11
Isotropic antenna radiation pattern.

waves radiate outward, the surface of the sphere gets larger and larger. At some distance from the antenna, the surface of the sphere will be so large that it can be considered flat, and the electric field is pointing up or down in a plane, the same way a local region on the earth's surface is considered flat. This type of wave is called a plane wave.

Now consider the radiation pattern of a half-wave dipole, as shown in Figure 10.12a. There are lobes where the field intensity is greater than in other regions. The field strength intensity inside these lobes can be expressed as an increase with respect to an equivalent isotropic antenna. This multiplication factor is called the antenna gain G_A and has directionality due to the shape of the radiation pattern. Note that both antennas radiate the same amount of total power, but for dipoles, more power is radiated in a particular direction rather than equally spherically. There is no amplification of total power. This gain is often expressed as a decibel value with respect to the isotropic antenna. A typical gain for a half-wave dipole is 2.15 dBi (with the i meaning with respect to an isotropic antenna) at right angles to the direction of the antenna. To express this as a decimal value, take the anti-log.

$$V_{decimal} = 10^{(V_{LOG})/10}$$

For a half-wave dipole, the gain expressed as a decimal value is

$$\begin{aligned} G &= 10^{2.15/10} \\ &= 10^{0.215} \\ &= 1.64 \end{aligned}$$

The shape and gain of an antenna's radiation pattern depend on the signal wavelength and surrounding material. Long antennas can produce multiple lobes.

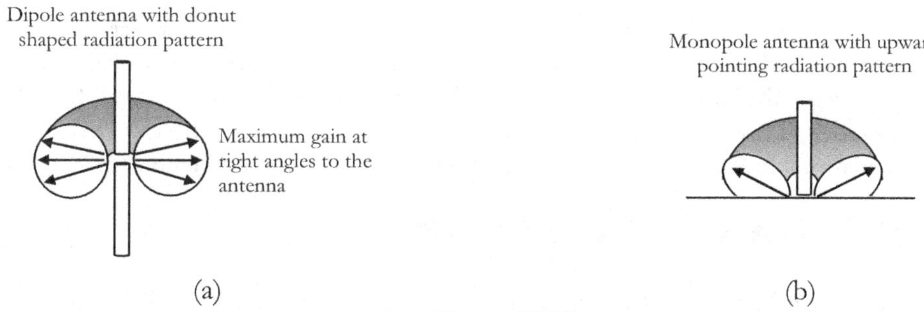

Figure 10.12
Dipole and monopole antenna radiation patterns.

For a ¼ wave monopole, the doughnut is sliced in half, and the effect of the ground produces a radiation pattern slightly tilted upward, resulting in an increase of 3dB gain over the ½ wave dipole, with an isotropic gain of 5.2 dBi (decimal 3.3), Figure 10.12b. The gain increases as the length of the monopole increases, up to a maximum of 6.6 dBi (decimal 4.6) for a ⅝th monopole over a perfectly reflective earth. Table 10.4 lists the isotropic gain for different length monopole antennas.

Commercial AM radio stations broadcast on antennas over ¼ of a wavelength long. Often a single transmission tower will be shared amongst broadcasters to lower costs. In my location, a single 180-metre tower is shared by three radio stations at 585 kHz, 720 kHz and 810 kHz. This results in the single tower acting as different length antennas of 0.35λ, 0.43λ and 0.49λ,

Monopole length	Gain (dBi)	Gain (decimal)
1/4λ (0.25λ)	5.2	3.3
1/2λ (0.50λ)	6.2	4.1
5/8λ (0.625λ)	6.6	4.6

Table 10.4
Typical monopole antenna gains.

respectively. This results in a different gain for each radio station. We can approximate the decimal value isotropic gain of a monopole antenna as

$$G_T = (\frac{H}{\lambda} + .75)3.3 \qquad 10.12$$

where H is the length of the monopole in metres, and λ is the wavelength for the frequency of interest. The results for the three radio stations sharing the same tower are shown in Table 10.5.

Station frequency	Wavelength	Monopole length	Perfect Gain (dBi)	Perfect Gain (decimal)	Realistic Gain (dBi)	Realistic Gain (decimal)
585 kHz	512	0.35λ	5.6	3.6	4.6	2.9
720 kHz	416	0.43λ	5.9	3.9	4.9	3.0
810 kHz	370	0.49λ	6.1	4.1	5.1	3.2

Table 10.5
Local radio stations sharing a single antenna.

The results aren't an exact match but are good enough to be useful. These results are for a perfectly reflecting earth. However, the earth is not a perfect reflector and radials of wire are generally buried beneath the antenna to increase reflectivity. Even so, we can safely reduce the gain by at least one dBi. This would result in a realistic gain, as shown in the last two columns of the table. For calculations in this book, we shall use a transmission antenna gain of 3.0 decimal.

10.9 RECEIVED ANTENNA POWER

The power received by an antenna is a function of the signal's incident (arriving) power and the receiving antenna's ability to capture this power. Consider a transmitting isotropic antenna. The signal radiates outward as a sphere. As this sphere expands, the power per unit area decreases and represents a decrease in power available for our antenna to capture. The power density at any given point some distance d from the transmitter is the total power radiated by the transmitter divided by the surface area of the sphere at distance d. This gives us power in watts per square metre and is called the ***power density*** of the transmitting antenna $P^D{}_T$ arriving at some distance d. For a non-isotropic antenna, there will be an associated power gain G_T in some direction. Assuming the distance d is great enough so the radio signal arrives as a plane wave, and our receiving antenna is in line with the directionality of the transmitting antenna, we can arrive at a formula to determine the power density arriving at a receiving antenna.

$$P_T^D = \frac{P_T G_T}{4\pi d^2} \qquad \text{Antenna arriving power density} \qquad 10.13$$

Where P_T is the power fed to the terminals of the transmitting antenna, d is the distance between the transmitting and receiving antennas, and G_T is the power gain of the transmitting antenna in some direction. The incident electric field captured by the receiving antenna creates voltage and current standing waves. The shape and distribution of these standing waves along the antenna depend on many factors, including frequency, the length of the antenna, the angle of the antenna to the incident wave and other power losses. These factors are lumped into a term with units of metres squared and is called the ***effective area*** A_e or ***effective aperture***, see Figure 10.13.

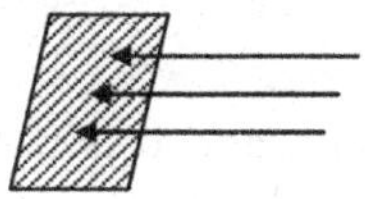

Figure 10.13
The ability of an antenna to receive power can be treated as an area.

An antenna's effective aperture also depends on the load attached to the antenna. When a load is included in the effective aperture, the incident received power density multiplied by the effective aperture is the power dropped across the load. We can write this as,

$$P_R = (P_T^D A_e) \qquad \text{Received power across an antenna load} \qquad 10.14$$

Using Equation 10.13, we can write this as

$$\boxed{P_R = \frac{P_T G_T A_e}{4\pi d^2}} \qquad \text{Received power across an antenna load} \qquad 10.15$$

If we know the effective aperture of the receiving antenna, the gain of the receiving antenna can be calculated as

$$G_R = {4\pi A_e}/{\lambda^2}$$

And

$$A_e = {G_R \lambda^2}/{4\pi}$$

where λ is the free space wavelength. Inserting this into Equation 10.15, we arrive at another form to calculate the received power. This is called the Friis equation.

$$\boxed{P_R = \frac{P_T G_T G_R \lambda^2}{(4\pi d)^2}} \qquad \text{Received power across an antenna the Friis equation} \qquad 10.16$$

10.10 RADIO WAVE PROPAGATION AND LOSS

Low-frequency radio waves (below 3 MHz) travelling horizontally along the ground tend to hug the ground as they travel. These, unsurprisingly, are called ground waves and can travel great distances depending on the terrain. As these ground waves propagate outward from the antenna, they interact with the ground, causing some of the power to be absorbed by the ground and some to be reflected. This interaction is dependent on the ground over which the wave is travelling, the wavelength of the radio wave, and the ground reflection coefficient. We can calculate the reflection coefficient for a given ground using the following formula.

$$\Gamma = \frac{\sin(\theta) - \dfrac{\sqrt{\varepsilon_g - \cos^2(\theta)}}{\varepsilon_g}}{\sin(\theta) + \dfrac{\sqrt{\varepsilon_g - \cos^2(\theta)}}{\varepsilon_g}}$$

Ground reflection coefficient 10.17

where ε_g is the relative permittivity of the ground, and θ is the angle of incidence with respect to the ground. The relative permittivity of the ground greatly depends on the water content. For dry ground, the relative permittivity could be as low as 2. For water-saturated ground, the relative permittivity could be as high as 28. So unless you know the relative permittivity of your soil, you will need to make an educated guess. For example, a radio wave with an angle of incidence of 45° and a ground with ε_g=2 will give a reflection coefficient of 0.07. For water-soaked ground, with an ε_g=28, the reflection coefficient will be 0.59. So that the ground, even at its best, is a poor reflector of radio waves. A typical ground relative permittivity will be around 3.

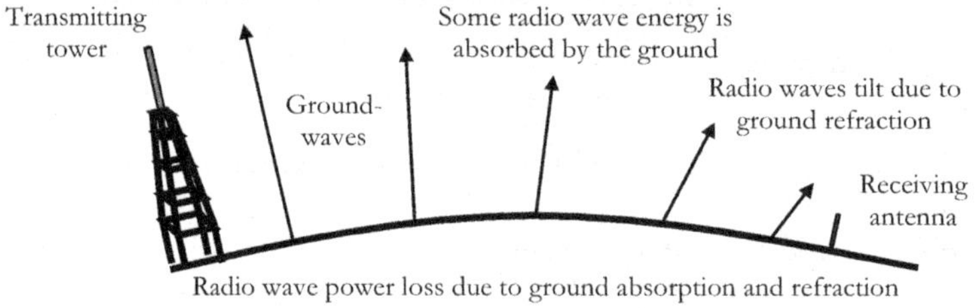

Figure 10.14
Ground wave propagation.

As well as absorption and reflection, radio waves are refracted (bent) due to ground interaction. Higher-frequency radio waves bend more than low-frequency radio waves. This causes the propagating wave to tip over a bit as it travels. The further it travels, the more it tips over. When the wave reaches your antenna, some of the power will be due to the incident wave hitting your antenna directly head-on, some will be at an angle to your antenna due to refraction, and some power will come from reflected waves. This will result in a reduction of power received by your antenna with respect to the power it could have received if all the incident waves were striking your antenna at a zero-degree angle. This power loss increases with frequency and distance. Also, objects such as hills and buildings will affect the radio wave as well. Ultimately all these losses result in an upper distance limit to which the wave will travel along the ground.

We need to take into account these ground and object collision losses, but coming up with an accurate model is virtually impossible. However, we can take into account some of these losses by adding a ground loss factor. We know that the longer the wave travels, the greater the losses, so just like the reduction in power density was a function of distance, we can include this loss as a function of distance by adjusting the exponent of Equation 10.13 as follows,

$$P_T^D = \frac{P_T G_T}{4\pi d^{(2+n)}}$$ Arriving power density at an antenna 10.18

where n is the ground loss factor. The value for n can be empirically derived, and will vary from one location to another. If you are close to your radio station, say less than 10 km, you may wish to ignore it and set n to 0. You may also want to vary the ground loss factor depending on frequency, increasing it slightly for higher frequencies. For my environment, n is between 0.1 and 0.3. For our analysis, we will choose a ground loss factor of 0.2.

Sky-wave Propagation

Not all waves emitted from a radio station will travel along the ground. Some will be directed upward. These are called ***sky-waves***. These sky-waves can be refracted back to the ground by a band of the atmosphere known as the ionosphere. How well the ionosphere refracts radio waves depends on the frequency of the wave and the density of ions (electrically charged atoms) in the band at any given time. The ionosphere extends from about 40 km to about 400 km above the surface of the Earth. The actual range varies depending on the season and how active the sun is.

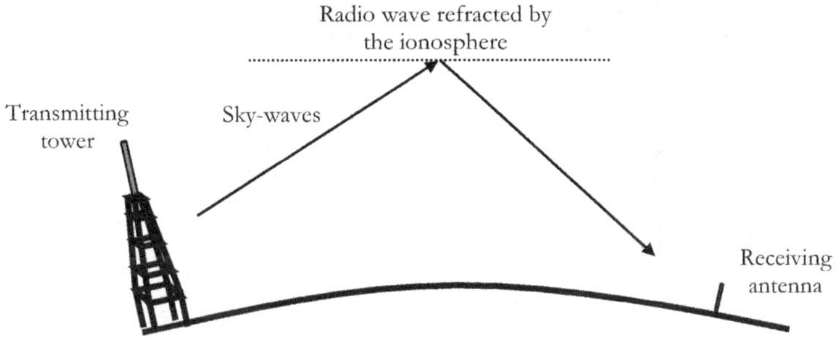

Figure 10.15
Sky-wave refracted by the ionosphere.

During the day, energetic charged particles from the sun collide with air molecules in the upper atmosphere causing electrons to be stripped from the atoms. The once neutral molecule now becomes positively charged. As the day progresses, more atoms are ionised, lowering the altitude of the start of the ionisation layer. Radio waves striking this ionised region are refracted, and some will arrive back at the surface of the Earth some distance (skip distance) from the transmission tower.

At night the reverse occurs, with the higher air density regions in the lower atmosphere recombining first to form neutral atoms. As the night progresses, the start of the ionised layer increases in altitude, resulting in longer skip distances. Typically the ability to hear these long distance radio stations increases to a maximum at around midnight. Then as more ions recombine, the ability of the ionosphere to refract waves decreases. Listening to these long distance radio stations is called ***DXing*** and is very popular amongst crystal radio and ham radio operators.

10.11 EFFECTIVE LENGTH OF AN ANTENNA

Consider a dipole antenna of length L. At some frequency, the length of the antenna will be one full wavelength and have a standing wave current distribution, as shown in 10.16a. As the frequency decreases, the current distribution will change along the length of the antenna. When the wavelength is L/2, we have a resonant half-wave dipole, Figure 10.16b. Decreasing the frequency further will result in the current standing wave shown in Figure 10.16c. The area under the curve represents a geometrical representation of the current distribution. We can express the same area under the curve as a rectangular region by multiplying the current amplitude I_0 by some length ℓ_e. This length is called the ***effective length*** of the antenna and is directly related to the ability of an antenna to induce a current or voltage across the antenna. Notice that as the wavelength increases (frequency decreases) with respect to a given full wave antenna length, the total area under the curve decreases, and the effective length decreases.

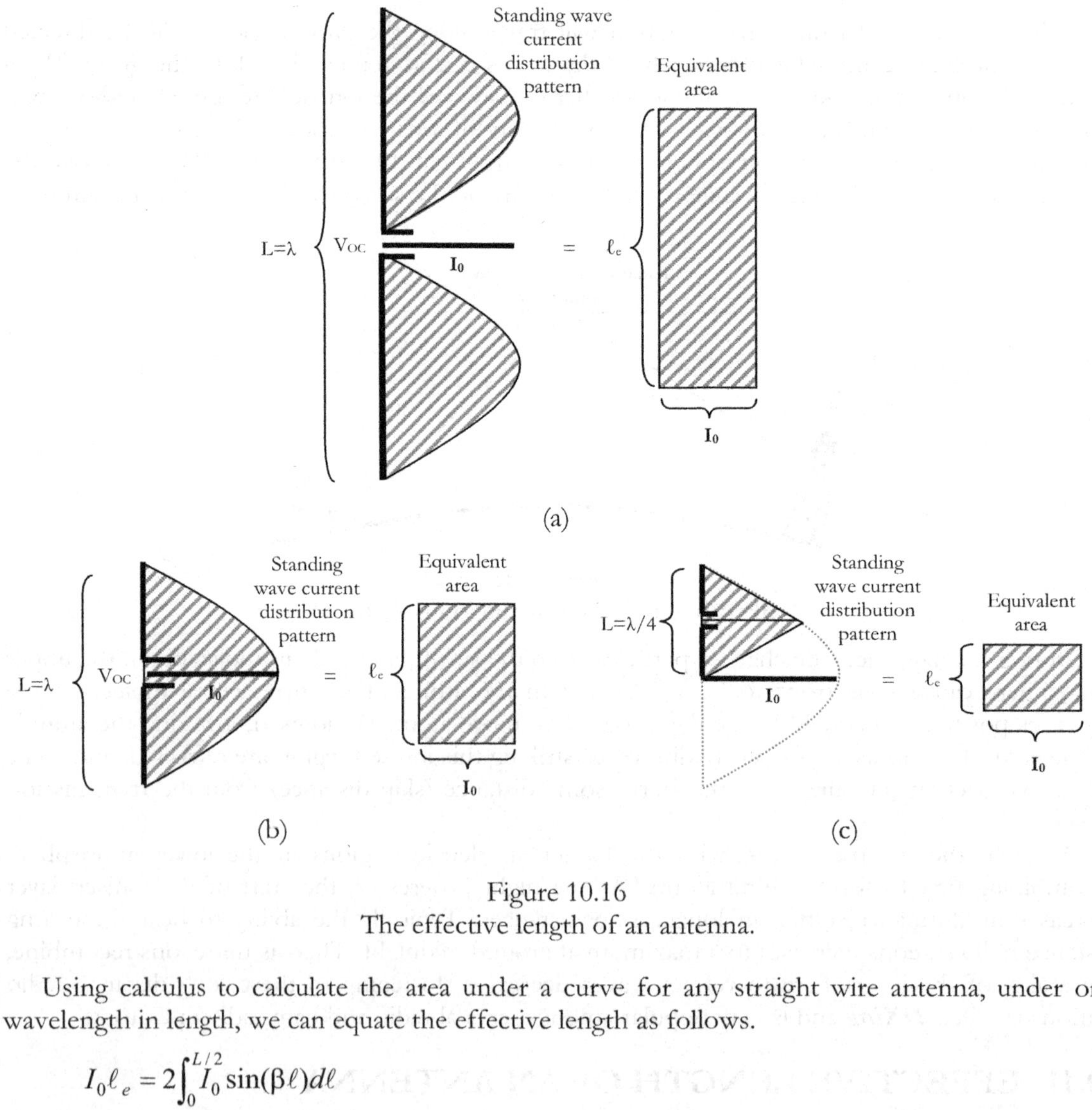

Figure 10.16
The effective length of an antenna.

Using calculus to calculate the area under a curve for any straight wire antenna, under one wavelength in length, we can equate the effective length as follows.

$$I_0\ell_e = 2\int_0^{L/2} I_0 \sin(\beta\ell)d\ell$$

Evaluating the integral where β is the wave number, we get,

$$I_0\ell_e = 2I_0\int_0^{L/2} \sin(\frac{2\pi}{\lambda}\ell)d\ell$$

$$= -\frac{2\lambda}{2\pi}I_0\left(\cos(\frac{2\pi}{\lambda}\ell)\right)\Bigg|_0^{L/2}$$

$$I_0\ell_e = -\frac{\lambda}{\pi}I_0\left(\cos(\frac{2\pi}{\lambda}\frac{L}{2}) - \cos(0)\right)$$

$$= \frac{\lambda}{\pi}I_0\left(1 - \cos(\frac{L}{\lambda}\pi)\right)$$

So the effective length of a dipole becomes

$$\boxed{\ell_e = \frac{\lambda_A}{\pi}\left(1 - \cos(\frac{L}{\lambda_A}\pi)\right)}$$ Effective length of a dipole antenna 10.19

where λ_A is the electrical wavelength of the antenna, and L is the length of the antenna in metres.

The effective length is used to determine the open circuit voltage induced across an antenna due to the incident electric field. From Equation 4.11, we can write the voltage across an antenna due to the incident electric field as

$$V_{OC} = E\ell_e$$

For a dipole, there are two active elements, and the incident wave impinges across both arms. For a monopole, however, there is only one active element, and the effective length is reduced by one-half. The shape of the current sine wave distribution will be similar to the dipole, however, and so the length of a monopole must be multiplied by 2 (the mirror effect). Therefore the effective length of a monopole is given by

$$\boxed{\ell_e = \frac{\lambda_A}{2\pi}\left(1 - \cos(\frac{2H}{\lambda_A}\pi)\right)}$$ Effective length of a monopole 10.20

where λ_A is the electrical wavelength of the antenna, and H is the height (length) of the monopole.

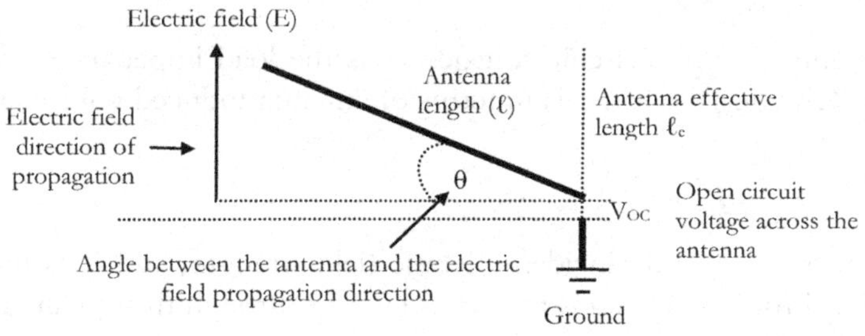

Figure 10.17
Incident electric field striking a monopole antenna at an angle.

In Figure 10.17, we have a radio wave arriving at a receiving antenna at some angle θ to the horizontal axis. We can write the maximum voltage induced across the antenna as

$$V_{OC} = E^{I} \ell_{e} \sin(\theta) \quad \text{Maximum induced open circuit voltage across an antenna} \qquad 10.21$$

where E^I is the magnitude of the incident electric field strength, ℓ_e is the effective length of the antenna, and θ is the incident angle (the angle between the direction of propagation of the electric field and the antenna).

10.12 CALCULATING THE EFFECTIVE APERTURE

Consider a radio wave striking an antenna with a load attached, as shown in Figure 10.18a. A voltage is induced across a monopole antenna between points A and B and is shown as the open circuit voltage V_{OC}. This is redrawn as an equivalent circuit in diagram 10.18b, as seen by the open

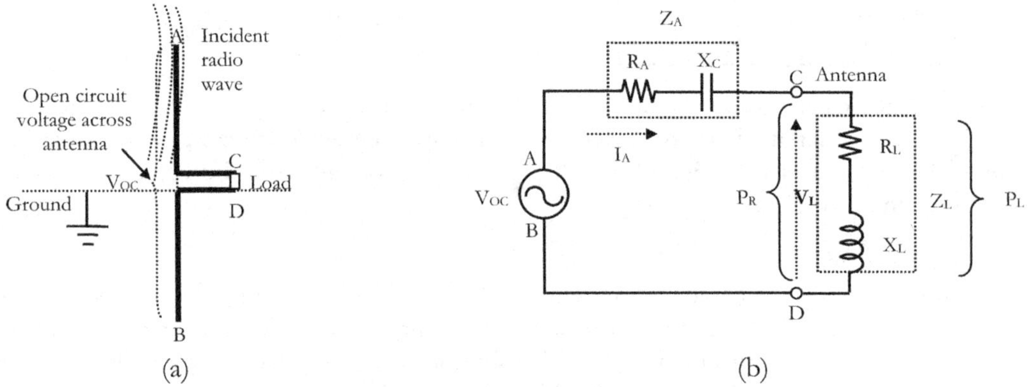

Figure 10.18
Power delivered to the antenna load.

circuit voltage. The voltage between the feed line at points C and D is dropped across the load. Using the voltage divider rule, we can write,

$$\mathbf{V_L} = \mathbf{V_A} = \mathbf{V_{OC}} \frac{\mathbf{Z_L}}{\mathbf{Z_A} + \mathbf{Z_L}} \quad \text{Voltage across an antenna load} \qquad 10.22$$

where V_{OC} is the antenna open circuit voltage, Z_L is the load impedance, and Z_A is the antenna input impedance. Rewriting the formula in terms of antenna-induced voltage, we get

$$\mathbf{V}_L = \mathbf{E^I} \bullet \ell_{\mathbf{e}} \frac{\mathbf{Z_L}}{\mathbf{Z_A} + \mathbf{Z_L}} \qquad 10.23$$

where E^I is the magnitude (peak) incident electric field strength, and ℓ_e is the antenna's effective length. The current through the load is the series current through the circuit. That is,

$$\mathbf{I_A} = \mathbf{I_L} = \frac{\mathbf{V_{OC}}}{\mathbf{Z_A} + \mathbf{Z_L}} = \frac{\mathbf{E^I} \bullet \ell_{\mathbf{e}}}{\mathbf{Z_A} + \mathbf{Z_L}} \qquad 10.24$$

The average power across the load is,

$$\begin{aligned}\mathbf{P_L} &= \frac{1}{2}\mathbf{V}_L\mathbf{I}_L \\ &= \frac{1}{2}\left(\mathbf{E^I} \bullet \ell_e \frac{\mathbf{Z_L}}{\mathbf{Z_A}+\mathbf{Z_L}}\right)\left(\frac{\mathbf{E^I} \bullet \ell_e}{\mathbf{Z_A}+\mathbf{Z_L}}\right) \\ &= \frac{(\mathbf{E^I} \bullet \ell_e)^2\mathbf{Z_L}}{2(\mathbf{Z_A}+\mathbf{Z_L})^2}\end{aligned} \qquad 10.25$$

The average power contained in the incident electric field in free space for the same effective length of the antenna is

$$\begin{aligned}\mathbf{P_I} &= \frac{1}{2}\mathbf{V_I}\mathbf{I_I} \\ &= \frac{1}{2}(\mathbf{E^I}\ell_\mathbf{e})\left(\frac{\mathbf{E^I}\ell_\mathbf{e}}{\mathbf{Z_o}}\right) \\ &= \frac{\mathbf{E^{I}}^2\ell_\mathbf{e}^2}{2\mathbf{Z_0}}\end{aligned}$$

where Z_0 is the characteristic impedance of free space. If we divide the incident power P_I by the area represented by ℓ_e^2, we get a power density.

$$\mathbf{P_T^D} = \frac{\mathbf{E^{I}}^2\ell_\mathbf{e}^2}{2\ell_\mathbf{e}^2\mathbf{Z_0}}$$

And

$$\mathbf{P_T^D} = \frac{\mathbf{E^{I}}^2}{2\mathbf{Z_0}} \qquad 10.26$$

The average received power across points A and B is the power density times the effective aperture (effective area) of the antenna, and by Equation 10.14, we can write

$$\begin{aligned}\mathbf{P_R} &= \mathbf{P_T^D}A_e \\ &= \frac{\mathbf{E^{I}}^2 A_e}{2\mathbf{Z_0}}\end{aligned} \qquad 10.27$$

This power P_R is the power dropped across the load. And equating 10.27 to 10.25, we get

$$\frac{1}{2}\frac{\mathbf{E^{I}}^2}{\mathbf{Z_0}}A_e = \frac{(\mathbf{E^I} \bullet \ell_\mathbf{e})^2\mathbf{Z_L}}{2(\mathbf{Z_A}+\mathbf{Z_L})^2}$$

Rearranging to get an expression for the effective aperture

$$\boxed{A_e = \frac{Z_0 Z_L \ell_e^{\,2}}{(Z_A+Z_L)^2}} \qquad \text{Effective aperture of a loaded antenna} \qquad 10.28$$

where Z_0 is the characteristic impedance of free space (377 Ω). We can use Equation 10.28 as the effective aperture in Equation 10.14 to calculate the power dropped across the load. Note that the power consists of both resistive (real power) and reactive power.

10.13 CALCULATING THE PEAK INCIDENT ELECTRIC FIELD VALUE

Equating the two incident power density Equations 10.27 and 10.19

$$\frac{E^{I^2}}{2Z_0} = \frac{P_T G_T}{4\pi d^{(2+n)}}$$

And rearranging to give an expression for the peak incident electric field value, we get

$$\boxed{E^I = \sqrt{\frac{2Z_0 P_T G_T}{4\pi d^{(2+n)}}}} \quad \text{Peak value of incident electric field} \quad 10.29$$

where P_T is the average (RMS) power of the transmitter. This value can be used in Equation 10.21 to determine the peak antenna open circuit voltage V_{OC}.

10.14 TYPICAL RANGE OF ANTENNA VOLTAGES

Table 10.6 lists the typical range of peak open circuit voltages V_{OC} induced across our standard antenna for the AM broadcast band at a frequency of 1 MHz. These are the antenna open circuit voltages fed into the primary coil. The audibility depends on several factors, but for a well-designed radio, the audibility is what would be reasonably expected when the antenna circuit is resonant at the station frequency.

Antenna induced voltage (peak)	Audibility	Broadcast station
5 mV	Just audible.	Typical of long distance (DX) radio stations.
10 mV	Low level.	1 kilowatt at 15 kilometres distance.
15 mV	Medium-Low level.	2 kilowatt station at 15 kilometres distance.
25 mV	Medium level.	5 kilowatt station at 15 kilometres distance.
45 mV	Loud.	15 kilowatt at 15 kilometres distance.
250 mV	Very loud.	50 kilowatt station at 5 kilometres distance.

Table 10.6
Typical peak induced antenna voltages.

A longer and more effective antenna is capable of inducing voltages of well over a volt for close and high power stations. Stations inducing voltages of 100 mV or more can be a real problem when trying to listen to nearby low-power satiations. High Q coils and a greater separation between the primary and secondary coil will help. But if the induced voltage is so high, you may be unable to remove the signal. Other strategies, covered in a later Chapter, can be employed to reduce the strength of these stations known as “Boomers”.

10.15 SETTING UP YOUR ANTENNA AND GROUND

The 25-metre Wire Antenna

When setting up your antenna, you want to make it as ***high as possible for as long as possible***. So don't elevate one end and leave the rest trailing along the ground. The part trailing along the ground will have a large capacitance to ground and act like one wire in a transmission line, with the ground making up the other wire with a very large resistance resulting in a loss of energy. Your antenna should look something like Figure 10.19.

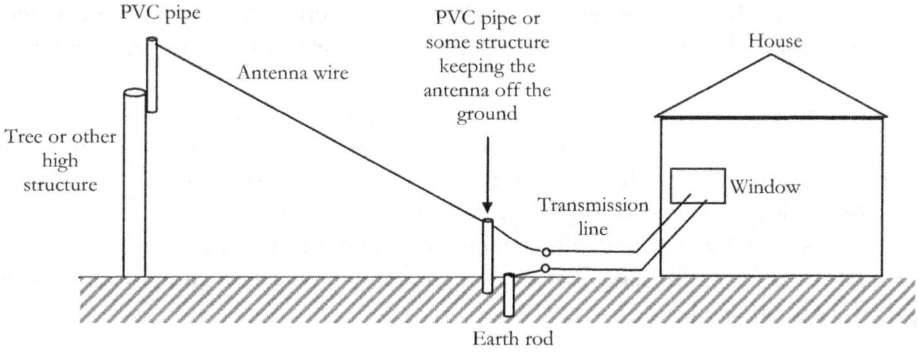

Figure 10.19
The standard 25-metre wire antenna and ground.

To set up a wire antenna, you will need 25 metres of wire or more. Drill a 2 mm hole at one end of a length of PVC pipe. The length is up to you, so select a length that is manageable. I chose a 2-metre length of pipe. Thread one end of your wire through the hole and tie a knot to keep the wire from sliding out. Attach this pipe to a tree, a downpipe or any suitable high structure. Use a ladder to get it as high as you can. Use string, wire or brackets to affix the pipe to the structure.

Using another length of PVC pipe, again choose a length that is manageable, drill two holes 90° apart and thread the wire through both holes. Pull the wire out, equalling the length of the pipe. At a distance where the wire is reasonably taut, bury the pipe sufficiently far into the ground to make a post. Of course, if you already have a post or tree at a suitable distance, use that. Also, if you don't have the room to run the antenna in a straight line, it can be taken off at different angles. You can use trees and other objects to keep the antenna wire off the ground. However, metal objects should be avoided as they might interfere with the electric field produced by the antenna.

Using a Downpipe Antenna

Downpipes and gutters around your house make for suitable ready-made antennas. The downpipe is the active element in your antenna and should be at least three metres in height; the higher, the better. Provided the downpipe is directly connected to your gutters, and they usually are, the total length of directly connected gutters adds to the overall length of your antenna. You want at least

25 metres of total antenna length (downpipe plus gutters) to make a good antenna. The longer the antenna, the more capacitive the input impedance will look, until the total length reaches about 45 metres. Above 45 metres in length, the input impedance will start to look inductive at higher frequencies, and your primary coil load will no longer make the antenna circuit resonant. In this case, you will need to add the 10-metre two-wire transmission line to make it look capacitive again. I suggest you add the line in any case, as the crystal radios we are going to build are designed for an antenna with high input capacitance. High antenna input capacitance values are desirable for crystal radios.

As long as no metal objects are grounding your gutters or downpipes, and your downpipes are directly connected to the gutters, making a continuous electrical connection, this should make a good antenna. This is the antenna I use, and I have found it to be even better than the 25-metre wire antenna described above. How well it works as an antenna depends on the height of the downpipe and the total electrically connected length of your gutters. You can always try both methods to see which works best for you.

When connecting the transmission line to the downpipe, you can use small test leads with alligator clips, or you can solder an alligator clip onto the transmission line. Note that downpipes will be coated with paint. Generally, this paint is only on the outside, but some may be on the inside. Ensure that at least one end of the alligator clip touches bare metal.

The only downside I have found when using this antenna is when it rains heavily; it can degrade the performance a bit. The water increases the capacitance to ground, reducing the voltage across the primary coil.

The Ground Rod

Before driving in the ground rod, ensure no electrical or plumbing pipes are running underneath. Check to see where the street wiring is coming into your house. Check wiring paths to sheds or any other electrically connected structures. Check for sewerage and other plumbing pipes. It is up to you to ensure you are safely driving the rod into the ground.

A metal rod driven into the ground makes the best earth, and the deeper the rod goes the better. First, drill a small hole about 1 mm near the top of one end to connect a ground wire. Or, if you are using an alligator clip, file around the end to remove any oxidation to make a good electrical connection. Drive a 2-metre 12.5 mm diameter copper pipe into the ground as far as you can get it, and then connect a ground wire of 30 cm or so.

Generally, earth rods are situated next to the feed end of the antenna; however, they don't have to be. The earth will still work very well if it is situated away from the antenna. The actual position depends on what is reasonable for your situation. For instance, I have an old metal irrigation pipe that goes all the way down to the water table about 20 metres from my house. I use this pipe as it is an excellent earth.

Using Taps and Water Pipes as Ground

While a tap or water pipe can be used as a ground, I have found them to be quite poor grounds. Apart from the difficulty in attaching a wire to a tap or pipe, they usually have some film build-up that prevents a good connection and must be removed. Even after making good metal contact, they still don't make good earths. Water pipes don't go that far into the ground. Also, if some or all of the underground piping is made of a non-conductive material, such as PVC, then it won't work well at all.

An alternative to water pipes is a long wire, 25 metres or more, spread out on the ground. It will capacitively couple to the ground, and although not a very good earth, it can be used in a pinch, especially if you have powerful radio stations nearby.

Using Electrical Earth as a Ground

Your housing electric system has a connection to earth; a rod outside your house driven into the ground. In the event of an electrical fault, it offers an electrical path back to the street transformer (through the ground). If the active wire touches a device's chassis, the current will follow the path through the ground rather than through you. Although this makes a good radio earth, I do not recommend using it as a potential exists for a shock (even fatal) if there is an electrical fault.

10.16 WORKED EXAMPLES

Given a station broadcasting on 1 MHz with an RMS (average) output power of 10 kW at a distance of 10 km, determine the voltage across the primary coil at resonance using our standard antenna and transmission line.

Determine the total resistance of our antenna

The wavelength at 1 MHz in a vacuum is

$$\begin{aligned} \lambda &= \frac{C}{f} \\ &= \frac{2.99792458x10^8}{1x10^6} \qquad \text{Wavelength} \\ &= 2.9979x10^3 \end{aligned}$$

From Table 10.1, the radiation resistance for a short monopole is given by

$$\begin{aligned} R_{RAD} &= 40\pi^2\left(\frac{\ell}{\lambda}\right)^2 \\ &= 40\pi^2\left(\frac{25}{2.9979x10^3}\right)^2 \qquad \text{Antenna radiation resistance} \\ &= 2.7453\Omega \end{aligned}$$

Next, determine the soil resistance using Equation 10.3, given we are using a 2-metre pole, 12.5mm in diameter and a ground resistivity of 100 Ω-m.

$$\begin{aligned} R_S &= \frac{\rho_G}{2\pi\ell_G}\left[\ln(\frac{8\ell_G}{d}) - 1\right] \\ &= \frac{100}{2\pi 2.0}\left[\ln(\frac{8(2.0)}{0.0125}) - 1\right] \qquad \text{Soil resistance} \\ &= 48.9769\Omega \end{aligned}$$

Now determine the antenna ground resistance using Equation 10.4. The diameter of the antenna is

0.5mm, and the wave number β is

$$\beta = \frac{2\pi}{\lambda}$$

$$= \frac{2\pi}{2.9979x10^3}$$ Wave number

$$= 0.02096$$

And,

$$\beta\ell = 0.02096(25)$$

$$= 0.5240$$

Using Equation 10.4, the ground resistance is,

$$R_G = \frac{R_S}{2\pi}\left[\ln(\frac{\lambda}{2a})\sin^2(\beta\ell) - 4\left(\sin(\sqrt{\frac{\pi^2}{4} + (\beta\ell)^2} - \cos(\beta\ell)\right)\cos(\beta\ell)\right]$$

$$= \frac{48.9769}{2\pi}\left[\ln(\frac{300}{2(5x10^{-4})})\sin^2(0.5240) - 4\left(\sin(\sqrt{\frac{\pi^2}{4} + (0.5240)^2} - \cos(0.5240)\right)\cos(0.5240)\right]$$ Antenna ground resistance

$$= 21.0819\Omega$$

Determine the skin depth using Equation 5.23. The resistivity of copper is 1.67999x10^{-8} Ω-m. The relative permeability of wire is 1.

$$\delta = \sqrt{\frac{2\rho}{\omega\mu}}$$

$$= \sqrt{\frac{2(1.6799x10^{-8})}{2\pi(1x10^6)(1.2566x10^{-6})}}$$ Skin depth

And

$$\delta = 6.5234x10^{-5}$$

Determine the AC resistance of the antenna using Equation 5.24

$$R_{AC} = \frac{\rho\ell}{\pi(2r - \delta)\delta}$$

$$= \frac{(1.6799x10^{-8})25}{\pi(2(0.00025) - 6.5234x10^{-5})6.5234x10^{-5}}$$ Antenna wire AC resistance

$$= 4.7138\Omega$$

The total antenna resistance is,

$$R_A = R_{AC} + R_G + R_{RAD}$$

$$= 4.7138 + 21.0819 + 2.7453$$ Total antenna resistance

$$= 28.541\Omega$$

Determine the input impedance of our antenna

Now calculate the no-load (open-circuit) input reactance of the antenna using Equation 10.5. First, we need to calculate the propagation constant gamma using Equation 9.20. The resistance per metre, inductance per metre and capacitance per metre is,

$$R' = \frac{28.541}{25.0}$$
$$= 1.14164$$

$$j\omega L' = j2\pi f\mu_0$$
$$= 7.8597$$

$$G' = 0.0$$

Using $\varepsilon_A = 1.0$

$$j\omega C' = j2\pi f\varepsilon_0\varepsilon_A$$
$$= 5.5633x10^{-5}$$

And,

$$\gamma = \sqrt{(R' + j\omega L')(G' + j\omega C')}$$
$$= \sqrt{(1.14164 + j7.8597)(0.0 + j5.5633x10^{-5})}$$
$$= 0.001511 + j0.021013$$

Antenna propagation constant

Calculate the characteristic impedance Z_0 using Equation 9.23.

$$\mathbf{Z_0} = \sqrt{\frac{R' + j\omega L'}{G' + j\omega C'}}$$

Antenna characteristic impedance

$$\mathbf{Z_0} = \sqrt{\frac{(1.14164 + j7.8597)}{(0.0 + j5.5633x10^{-5})}}$$
$$= 377.7085 - j27.1653$$
$$= 378.6841\angle -4.1137^\circ$$

The antenna Thevenin equivalent input impedance is,

$$\mathbf{Z_{IN}} = \frac{\mathbf{Z_0}}{\tanh(\gamma\ell)}$$
$$= \frac{378.6841\angle -4.1137^\circ}{\tanh((0.001511 + j0.021013)25.0)}$$
$$= 9.8759 - j651.9974$$

Antenna Thevenin input impedance

The capacitance of the antenna is,

$$C_A = \frac{1}{2\pi f X_A} = \frac{1}{2\pi(1x10^6)652} = 244pF$$

Antenna Thevenin capacitance

The input impedance of the antenna at 1 MHz is the real resistance and the reactance.

$$\mathbf{Z}_A = 28.541 - j651.9974 = 651\angle -87.5^\circ$$

Antenna input impedance

Determine the input impedance with the transmission line

Using the input impedance of our antenna as the load, we can calculate the Thevenin input impedance to our antenna using our standard transmission line. Using Equation 10.10 and the values for the transmission line calculated in Chapter 9, we can write,

$$\mathbf{Z}_{IN} = \mathbf{Z}_0 \frac{\mathbf{Z}_A + \mathbf{Z}_0 \tanh(\gamma \ell)}{\mathbf{Z}_0 + \mathbf{Z}_A \tanh(\gamma \ell)}$$

$$= (45.73\angle -3.2^\circ)\frac{(651\angle -87.5^\circ)+(45.73\angle -3.2^\circ)\tanh((0.002065+j0.03636)10.0)}{(45.73\angle -3.2^\circ)+(651\angle -87.5^\circ)\tanh((0.002065+j0.03636)10.0)}$$

$$= 1.5533 - j98.7053$$

$$= 98.7175\angle -89.1^\circ$$

Thevenin Input impedance to the antenna

The total resistance is the antenna resistance and the transmission line resistance.

$$R_T = R_A + R_{TRANS} = 28.541 + 1.8855 = 30.4265\Omega$$

Total input resistance

And the input impedance to our antenna is,

$$Z_A = 30.4265 - j98.7053 = 98.7175\angle -89.09^\circ$$

Input impedance to the antenna

The input capacitance is,

$$C_A = \frac{1}{2\pi f X_A} = \frac{1}{2\pi(1x10^6)98.7053} = 1{,}612pF$$

Antenna Thevenin capacitance

Determine the antenna open circuit voltage

Using Equation 10.29, determine the peak incident electric field value. Given the station power is 10 kW RMS and is 10 km away, and assuming an antenna transmission antenna gain of 3.0 decimal, and a ground loss factor of 0.2, we can write

$$E^I = \sqrt{\frac{2Z_0 P_T G_T}{4\pi d^{(2+n)}}}$$

And

$$E^I = \sqrt{\frac{2(377)(10x10^3)3.0}{4\pi(10x10^3)^{(2.2)}}}$$
$$= 0.05341$$

Peak incident electric field value

Now calculate the effective length of the antenna at 1 MHz using Equation 10.20.

$$\ell_e = \frac{\lambda_A}{2\pi}\left(1-\cos(\frac{2H}{\lambda_A}\pi)\right)$$
$$= \frac{299.01}{2\pi}\left(1-\cos(\frac{2(25.0)}{299.01}\pi)\right)$$
$$= 6.41689$$

Effective length of the antenna

Using Equation 10.21, we can calculate the open circuit voltage across the antenna.

$$V_{OC} = E^I \ell_e \sin(\theta)$$
$$= 0.05341(6.41689)\sin(9.2^\circ)$$
$$= 0.0548$$

Open circuit voltage across the antenna

Determine the voltage across the primary coil

The input impedance to our antenna is,

$$Z_A = 30.4265 - j98.7053$$
$$= 98.7175\angle -89.09^\circ$$

Input impedance to the antenna

For the circuit to become resonant, the primary coil must have an inductive reactance of j98.7053. Calculating the inductor required for resonance.

$$L = {X_L}/{2\pi f}$$
$$= {98.7053}/{2\pi(1x10^6)}$$
$$= 15.71uH$$

Inductance required for resonance

This coil will have a resistance at 1 MHz of 1.67 Ω (coil resistance will be discussed in a later Chapter). Using Equation 10.22, we can write

$$\mathbf{V_L} = \mathbf{V_{OC}}\frac{\mathbf{Z_L}}{\mathbf{Z_A}+\mathbf{Z_L}}$$

$$= 0.0548\frac{(1.67 + j98.7053)}{(30.4265 - j98.7053) + (1.67 + j98.7053)}$$

$$= 0.0548\frac{98.7175\angle 89.09^\circ}{32.0965\angle 0}$$

$$= 0.1685\angle 89.09^\circ$$

Voltage across the primary coil

11

Cylindrical Coils

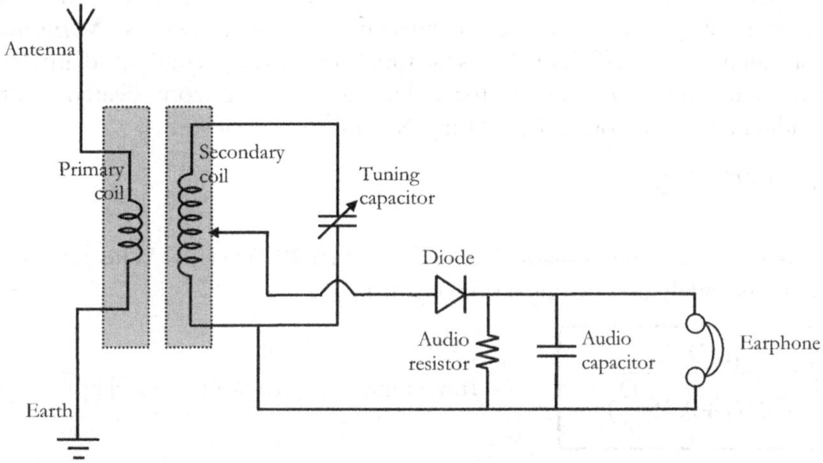

11.1 SELF-INDUCTANCE

The magnetic flux Φ going through a single loop of wire is given by the magnetic field strength multiplied by the area of the loop, Equation 5.3. If we add another loop, then the total flux will be 2Φ. If there are N loops in a coil, then the total magnetic flux through the coil will be NΦ. For a coil of N loops, the emf given by Faraday's Equation 5.12 becomes,

$$emf = -N\frac{\partial \Phi}{\partial t} \qquad 11.1$$

The self-inductance given for a loop by Equation 5.16, for a coil of N loops, becomes,

$$L = N\frac{\Phi}{I} \qquad 11.2$$

Using Equation 5.8 for the magnetic field inside a coil and Equation 11.2, we can write the inductance as,

$$\begin{aligned} L &= N\Phi/I \\ &= N(BA)/I \\ &= NA/I \; \mu NI/\ell \\ &= \mu N^2 A/\ell \end{aligned} \quad 11.3$$

where L is the inductance in henries, N is the number of loops in a coil, A is the area of a single loop, and ℓ is the overall length of the coil. This equation, however, ignores the effect that each loop has on other loops in the coil. It also assumes a uniform current distribution along the coil and ignores the skin effect, the proximity effect and the helicity of the coil. A more sophisticated equation is needed to determine the inductance of a coil to a reasonable accuracy.

Developing a more realistic real-world equation has been a long process spanning the last hundred years. Early modifications to the formula by Edward B. Rosa[1] included a table of correction factors that could be applied to cylindrical coils of given wire diameter to coil pitch ratios. Hantaro Nagaoka introduced a correction factor known as ***Nagaoka's coefficient*** that applied a calculated coefficient K as a function of the length-to-diameter ratio of the coil. However, calculating the coefficient for a given coil was very complicated and time-consuming.

The inductance equation incorporating Nagaoka's coefficient is,

$$L = \frac{\mu\pi r^2 N^2}{\ell} K$$

where K is the calculated Nagaoka's coefficient. In 1925, H. A. Wheeler[2] introduced a simplified formula that is widely used even to this day. It is,

$$\boxed{L = \frac{\mu\pi D^2 N^2}{4\ell(1+0.45\frac{D}{\ell})}} \quad \text{Inductance of a coil – very good approximation} \quad 11.4$$

where D is the diameter of the coil in metres. It is reasonably accurate for coils with a length-to-diameter ratio (ℓ/D) of 0.4 or greater. This is similar to the Nagaoka formula, with **a** Nagaoka coefficient K_W of,

$$K_W = \frac{1}{(1+0.45\frac{D}{\ell})}$$

In 1982 Wheeler revisited this subject and derived a formula with a different coefficient that removed the restriction on the length-to-diameter ratio. The coefficient for the 1982 Wheeler formula was modified by David W. Knight[3], resulting in a form giving a reported accuracy of less than ±0.03%. This modified coefficient is a function of the ℓ/D ratio and is given by

---

[1] Rosa, Edward B. (1906) Calculations of the self-inductance of single layer coils. Bulletin of the Bureau of Standards - volume 2, number 2.

[2] Wheeler, Harold A. (1928) Simple Inductance Formulas for Radio Coils. Proceedings of the I.R.E., October 1928, pp. 1398-1400

[3] Knight, David W. (2016) An introduction to the art of Solenoid Inductance Calculation.

$$K = \frac{2}{\pi}\left(\frac{\ell}{D}\right)\left[\ln(1+\frac{\pi D}{2\ell}) + \frac{1}{2.3004 + 3.2219\left(\frac{\ell}{D}\right) + 1.7793\left(\frac{\ell}{D}\right)^2}\right]$$ Nagaoka coefficient 11.5

where ℓ is the length of the coil in metres and D is the diameter of the coil in metres. The final form of a coil's inductance becomes,

$$L = \frac{\mu\pi D^2 N^2}{4\ell}K$$ Inductance of a coil 11.6

where L is the inductance in henries, N is the number of loops in a coil, ℓ is the length of the coil in metres, D is the diameter of the coil in metres, and K is the Nagaoka coefficient given by Equation 11.5. Refer to Figure 11.1 for coil dimensions. Pitch is measured from the centre of one loop to the centre of the next loop. The length of the coil is measured from the centre of the wire of the first loop to the centre of the wire in the last loop. The diameter is measured from the centre of the wires and is equal to the former diameter plus one wire diameter and insulation.

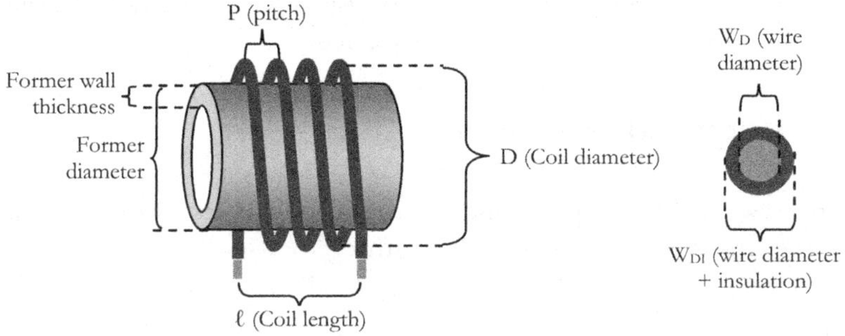

Figure 11.1
Coil dimensions.

Determine the Required Number of Turns

When winding a coil, we often know the inductance required and have selected the former, the wire and the pitch. However, we need to figure out how many turns to use. We can rearrange Equation 11.4 to get a good estimate of the length of the coil and then determine the number of turns. The coil length equals the number of turns multiplied by the pitch. That is,

$$\ell = NP$$ Coil length 11.7

as we measure from centre-to-centre of the wires. Note, however, that the physical length of the coil when measured across the leads, is longer by a wire diameter W_{DI}. That is,

$$\ell = NP + W_{DI}$$ Physical length of the coil

When physically measuring the length of the coil, take off one wire diameter for the calculations. In reality, using the physical length of the coil results in less than a 1% difference in the calculated inductance values for the coils we are making. Rearranging for the number of turns, we get

$$N = \frac{\ell}{P}$$ Number of turns 11.8

Substituting 11.8 into Equation 11.4, we get

$$L = \frac{\mu\pi D^2 \frac{\ell^2}{P^2}}{4\ell(1+0.45\frac{D}{\ell})}$$

Solving for the length of the coil, we get a quadratic equation in the form

$$\ell = \frac{-b+\sqrt{b^2-4ac}}{2a}$$
where
$$a = \mu\pi D^2, b = -4P^2L, c = -4(0.45)P^2LD$$
Very good estimate for the length of the coil 11.9

Using the length of the coil ℓ in Equation 11.8, we can determine the number of turns to use in Equation 11.6. Usually, the number of turns is not a whole number. When winding our coils, we will use complete windings, so round the number up or down to a whole number and plug this value into Equation 11.6 to determine the final inductance.

Determine the Required Diameter

Given an inductance value, the coil pitch and a length-to-diameter ratio, you may wish to know the number of turns and the diameter of the former to use. The number of turns can be calculated as follows,

$$N = \left[\frac{\ell}{D}\right]\frac{D}{P}$$ 11.10

Substituting 11.10 for the N, and 11.7 for ℓ in Equation 11.6 and rearranging for D, we get

$$D^2 = 4\ell L \Big/ \mu\pi N^2 K$$
$$= 4NPL \Big/ \mu\pi N^2 K$$
$$= 4PL \Big/ \mu\pi NK$$
$$= 4PL \Big/ \mu\pi\left(\frac{\ell}{D}\right)\left(\frac{D}{P}\right)K$$
$$D^3 = 4P^2L \Big/ \mu\pi\left(\frac{\ell}{D}\right)K$$

And

$$D = \sqrt[3]{\frac{4P^2 L}{\mu\pi\left(\frac{\ell}{D}\right)K}}$$ Calculating the coil diameter 11.11

The coil length is

$$\ell = \left[\frac{\ell}{D}\right]D \qquad 11.12$$

And the number of turns can be calculated using Equation 11.8.

11.2 SELF-CAPACITANCE

A coil not only has inductance but has internal resistance and capacitance. A real-world coil can be represented as a set of lumped components, as shown in Figure 11.2. From electrostatic theory, we know that a charge on a plate separated by an insulator (dielectric) will form capacitance. Initial models of coil self-capacitance assumed capacitance between the loops of the coil, treating them as series capacitance to arrive at a total coil capacitance. This was the accepted model for many years until someone measured the capacitance of various coils and found the model to be very inaccurate. Subsequently, there have been several equations developed to determine the coil's self-capacitance. The best models for coil self-capacitance are based on a combination of axial capacitance across the ends of the coil and transmission line theory based on the physical length of the wire. The combination of both the axial capacitance and the transmission line capacitance results in the total coil capacitance at self-resonance. The axial capacitance is greatly dependent on the geometry of the coil, and the transmission line capacitance is dependent on frequency and wire length. We can summarise the axial capacitance and the transmission line capacitance as follows.

$$C_L = C_A + C_T$$

where C_L is the total coil capacitance, C_A is the axial capacitance across the ends of the coil, and C_T is the contribution to the overall capacitance due to the wire length (transmission line theory).

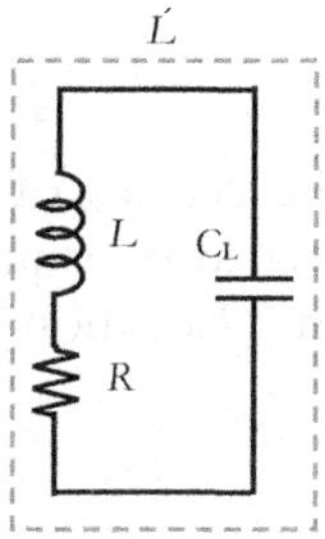

Figure 11.2
Circuit representation of a real-world coil.

An axial electric field has been shown to exist, as demonstrated by David W. Knight in his paper titled "self-resonance and self-capacitance of solenoid coils". His formula for the capacitance due to the axial electric fields is

$$C_A = \frac{2D\varepsilon_0(\varepsilon_X + \varepsilon_i)}{Ln(1+\pi^2 \frac{\ell}{D})} \quad \text{Coil axial (real) capacitance} \quad 11.13$$

where ε_x is the external relative permittivity (usually just 1 unless the coil is immersed in some material). ε_i is the relative permittivity of the coil former. D is the diameter of the coil in metres, and ℓ is the length of the coil in metres. Typically the coil former is not a solid object but a hollow cylinder such as a length of PVC piping, so the relative permittivity will be some combination of the relative permittivity of air (1.0) and the relative permittivity of the former material. David Knight gives this as

$$\varepsilon_i = \varepsilon_X\left(\frac{r_i}{r_o}\right)^2 + \varepsilon_f\left(1-\left(\frac{r_i}{r_o}\right)^2\right) \quad 11.14$$

where r_i is the inside radius of the former, r_o is the outside radius of the former ε_x is the external relative permittivity (usually just 1) and ε_f the relative permittivity of the former material.

The formula for total apparent capacitance at coil self-resonance is referred to by David Knight as the "doubly-asymptotic, empirically corrected formula", and is,

$$C_{DAE} = \frac{4\varepsilon_0\varepsilon_X\ell}{\pi}[1+K_C(\frac{1+\frac{\varepsilon_i}{\varepsilon_X}}{2})]\frac{1}{\cos^2\psi} \quad \text{Coil self-capacitance} \quad 11.15$$

where K_C is the capacitance coefficient given by

$$K_C = 0.717439\left(\frac{D}{\ell}\right) + 0.933048\left(\frac{D}{\ell}\right)^{3/2} + 0.106\left(\frac{D}{\ell}\right)^2 \quad 11.16$$

ψ is the pitch angle given by

$$\psi = \arctan(\frac{p}{\pi D}) \quad 11.17$$

where p is the pitch (distance between the centres of the wires between loops), and D is the diameter of the coil. The capacitance calculated by Equation 11.15 can be used to determine the self-resonant frequency of the coil. Using Equation 8.1, we can calculate the self-resonant frequency as follows

$$f_0 = \frac{1}{2\pi\sqrt{LC_{DAE}}}$$

Let us now examine the claim that self-capacitance at resonance is a combination of both real (axial capacitance) and apparent capacitance due to the transmission line effect. If we treat the capacitance shown in Figure 11.2 as real axial capacitance and unravel the coil, we end up with a load attached to a short-circuited transmission line, as shown in Figure 11.3. The length of the transmission line will be half the total wire length used in winding the coil. Using transmission line theory, the input impedance, as seen by the load, is the input impedance of a short-circuited transmission line. We can use Equation 9.39 to determine the input impedance as seen by the axial capacitance, and is given by

$$X_{LT} = jZ_0 \tan(\frac{2\pi\ell_T}{\lambda_T})$$

where λ_T is the electrical wavelength of the coil, $\ell_T = \ell_W/2$ is half the length of wire used in the coil, and Z_0 is the characteristic impedance of the coil. The characteristic impedance of the coil is the coil's self-inductance per metre multiplied by the velocity of propagation, Equation 9.54

$$Z_0 = L'V_P$$

The velocity of propagation is given by Equation 9.51

$$V_P = \frac{C}{\sqrt{\mu_R \varepsilon_i}}$$

where ε_i is given by Equation 11.14. And the electrical wavelength of the coil is given by

$$\lambda_T = \frac{V_P}{f}$$

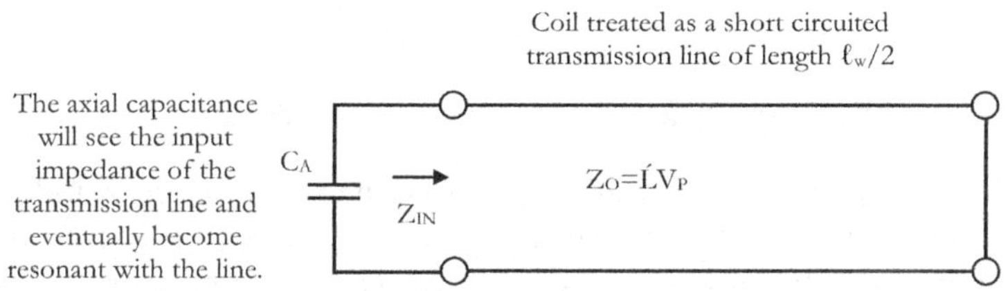

Figure 11.3
Coil treated as a short-circuited transmission line.

Graphing the transmission line input impedance as a function of wavelength (Figure 11.4), we see that the coil looks inductive until half a wavelength is reached, and then it looks capacitive. Below half a wavelength, as the frequency increases, the coil's inductive reactance increases.

If the self-resonant frequency is due to a combination of real axial capacitance and the transmission line effect, then at some frequency, the transmission line's inductive reactance will match the axial capacitance reactance, and the circuit will become resonant. The axial capacitance will see the Thevenin equivalent transmission line inductance. That is

$$|X_{C_A}| = |X_{LT}| \text{ at } f_0$$

If the frequency at which this occurs f_0 is predicted by C_{DAE}, and agrees with the measured self-resonant frequency, then this will confirm that the self-resonant capacitance is a combination of real capacitance and transmission line effect, and this apparent capacitance can be calculated using C_{DAE}.

To calculate the equivalent "apparent" self-capacitance when the coil becomes self-resonant, we can use the following formula.

$$C_{TRANS} = \frac{1}{4\pi^2 f_0^{\,2} L}$$

By measuring the self-resonant frequencies of various coils and calculating the transmission line input impedance for increasing frequencies and stopping when $X_{CA} = X_{LT}$, we will compare the results with the values predicted by C_{DAE}. These results are shown in Table 11.1. The first four coils were wound on PVC formers. The last coil was on a cardboard former. Each coil's relative permittivity was set to 3.0 for the calculations. The resonant frequency was measured for each coil (see measurement techniques Chapter).

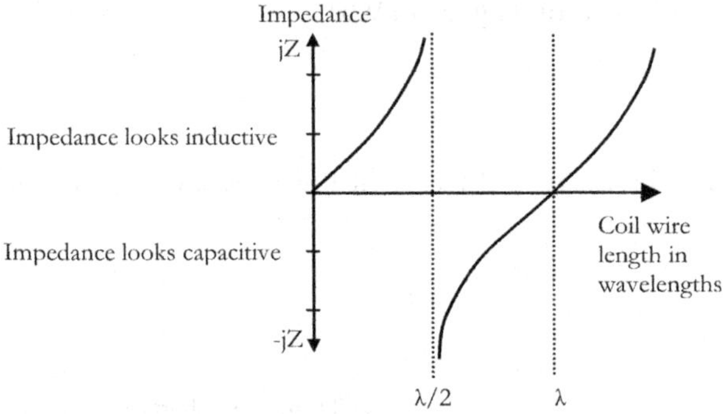

Figure 11.4
Coil impedance as a function of wavelength.

These results clearly show that the apparent self-capacitance of a coil is indeed a combination of the transmission line effect and some axial capacitance across the coil and that the axial capacitance formula is accurate and C_{DAE} is also an accurate predictor of self-capacitance at resonance.

Coil length (ℓ) mm	Coil Diam (D) mm	ℓ/D	Wire length (ℓ_W) m	L (uH)	f measured	f Trans (Xca=Xlt)	f DAE	C measured	C Trans	C_{DAE}
8.0	90.0	0.0889	3.68	32.5	11.74	11.58	11.74	5.65	5.81	5.65
28.0	103.0	0.2718	15.53	332.0	4.42	4.48	4.40	3.91	3.80	3.94
67.0	90.0	0.7444	15.83	228.0	5.99	6.12	6.19	3.10	2.97	2.90
65.0	75.0	0.8670	13.19	174.0	7.74	7.57	7.71	2.43	2.54	2.45
58.0	50.0	1.1640	18.85	441.5	5.55	5.50	5.74	1.86	1.90	1.74

Table 11.1
Comparison of measured and predicted self-resonant frequencies of coils and self-capacitance.

11.3 SELF-RESONANCE EFFECT ON Q

The self-capacitance of a coil determines the frequency at which a coil becomes ***self-resonant***. This self-resonance places an upper limit on the useful frequency. At frequencies higher than self-resonance, the inductor will start to look like a capacitor and cannot be used to tune to stations above this frequency.

Consider the tank circuit shown in Figure 11.5, a capacitor in parallel to a coil. Inductance L, the coil's internal resistance R and the parasitic capacitance C_P are part of one component; the inductor Ĺ Say we wish to determine the Q of the coil at some frequency f_Q. From our discussion on self-capacitance, we know the apparent inductance of the coil will change depending on the frequency. As Q is related to inductance, we can let α be some multiplication factor representing the effect of self-resonance on coil's Q. We can write,

$$Q_{eff} = Q_{coil}\alpha$$

where Q_{coil} is the Q of the coil, not taking into account the effect of self-capacitance. Q_{eff} is the Q of the coil taking into account this self-capacitance, which will be referred to as the ***effective Q***. Expressing the Q of the circuit using capacitance rather than inductance, we can write

$$\frac{1}{2\pi f_Q (C + C_P) R} = \frac{1}{2\pi f_Q (C) R}\alpha$$

Manipulating the equation, we can express α as

$$\alpha = \frac{2\pi f_Q (C) R}{2\pi f_Q (C + C_P) R}$$

And

$$\alpha = \frac{C}{C + C_P} \qquad 11.18$$

The frequency at which the circuit becomes resonant, assuming that the Q of the coil is much greater than 10, is given by the series resonant frequency equation. The self-resonant frequency of the coils is,

$$f_0 = \frac{1}{2\pi\sqrt{LC_P}}$$

And the resonant frequency of the tank circuit is

$$f_Q = \frac{1}{2\pi\sqrt{L(C + C_P)}}$$

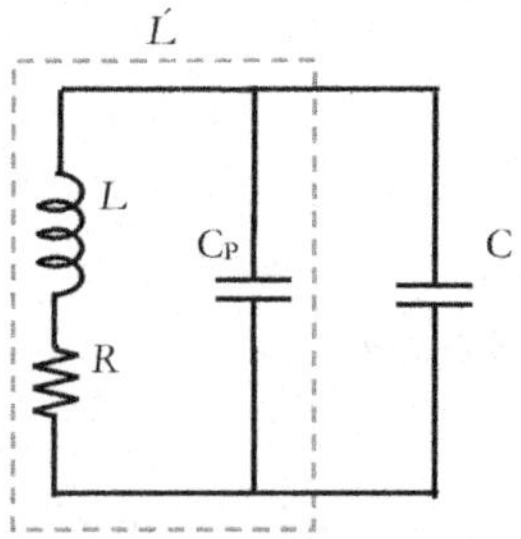

Figure 11.5
Tank circuit.

Expressing the self-resonant frequency of the coil and the frequency of the tank circuit as a ratio, we can write,

$$\frac{f_0}{f_Q} = \frac{2\pi\sqrt{L(C + C_P)}}{2\pi\sqrt{LC_P}}$$

Simplifying

$$\frac{f_0}{f_Q} = \frac{\sqrt{C + C_P}}{\sqrt{C_P}}$$

And

$$\frac{f_0^2}{f_Q^2} = \frac{C + C_P}{C_P}$$

Rearranging for the tank circuit capacitance C,

$$C = C_P \frac{f_0^2}{f_Q^2} - C_P$$

And

$$C = C_P(\frac{f_0^2}{f_Q^2} - 1)$$

Substituting the above value for C in Equation 11.18 for α, we get

$$\alpha = \frac{C}{C + C_P}$$

And

$$\alpha = \frac{C_P(\frac{f_0^2}{f_Q^2} - 1)}{C_P \frac{f_0^2}{f_Q^2}}$$

$$= \frac{\frac{f_0^2 - f_Q^2}{f_Q^2}}{\frac{f_0^2}{f_Q^2}}$$

And α, the self-resonant coefficient, becomes

$$\boxed{\alpha = \frac{f_0^2 - f_Q^2}{f_0^2}}$$ Self-resonant coefficient 11.19

The effective Q_{eff} becomes,

$$Q_{eff} = Q_{coil}(\frac{f_0^2 - f_Q^2}{f_0^2}) = Q_{coil}\alpha$$

Effective Q of a coil 11.20

This equation tells us that as an RLC circuit (either series or parallel) approaches the self-resonant frequency of the coil, the Q of the coil will drop. And when the resonant frequency is at the coil's self-resonant frequency, the effective Q of the coil is zero, and as such, the circuit is no longer tuneable. As this is intrinsic to the construction of a coil, the effective Q will be used as the measure of the coil's Q when not included in a circuit rather than Q_{coil}.

11.4 THE PROXIMITY EFFECT

Figure 11.6 shows the cross-section of two wires close to one another. In both wires, the current is travelling out of the page, and the magnetic field around wire-2 cuts wire-1 (use the right-hand rule for the direction of the magnetic field). The magnetic force felt by the conventional current in wire-1 forces the current initially to move to the right (using the right-hand rule). The magnetic force felt by a charge is always at right angles to the direction of travel, so as the direction of the current changes, the direction of the magnetic force felt by a charge changes, forcing the current to continually change direction, resulting in a current travelling in a loop (an eddy current).

These eddy currents change the distribution of current along the wire, causing more current to flow along one side of the wire than the other. In this instance, the eddy current travels in the same direction as the power supply current on the left side of wire-1 and travels in the opposite direction on the right side of wire-1. This causes a lopsided current distribution density, as shown in Figure 11.6b, with more current on the left side than on the right. The shape of the current distribution along the wire will depend on the position of the wire within the coil. However, the resultant current distribution will always be lopsided. This effect is called the ***proximity effect*** and will increase the resistance of the wire in a coil quite substantially.

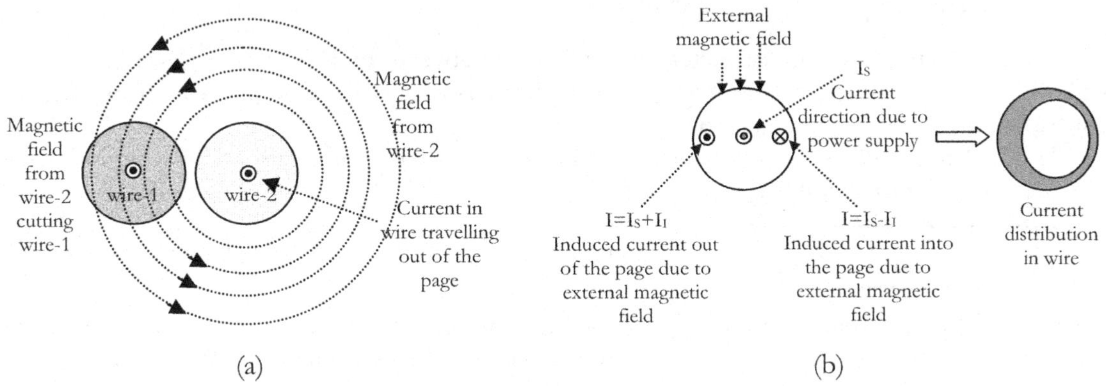

Figure 11.6
Proximity effect caused by another wire's magnetic field.

The interaction of the magnetic fields in a coil is extremely complicated. However, in 1947 R.G. Medhurst created a table combining both experimental and theoretical values of the ratio of a coil's AC resistance to that of the AC resistance of the same length of straight wire. That is,

$$\beta = \frac{R_{COIL}}{R_{AC}}$$

We will refer to these ratios as the ***proximity resistance factors***, which are given in Table 11.2. Once the ratio has been determined, we can use it to calculate the coil's total AC resistance.

$$R_{COIL} = R_{AC}\beta$$ Total AC resistance of a coil including the proximity effect 11.21

In Table 11.2, the shaded values represent the experimental values, with the theoretical values in clear cells. ℓ/D is the ratio of the length of the coil to the diameter of the coil. P/W_D is the pitch length to wire diameter ratio, where the pitch length is the length between the centre of one winding to the centre of the next winding. For values of ℓ/D and P/W_D that lay between the table values, we can use extrapolation to arrive at a proximity resistance factor as follows,

$$\beta = \frac{(Y_2 - \frac{\ell}{D})}{(Y_2 - Y_1)} f_1 + \frac{(\frac{\ell}{D} - Y_1)}{(Y_2 - Y_1)} f_2$$ Using extrapolation to determine a coil's proximity resistance factor 11.22

where the coil's ℓ/D ratio is between the ℓ/D ratios given in the table, with Y_1 being the smallest ratio and Y_2 the highest, and f_1 and f_2 are given by,

$$f_1 = \frac{(X_2 - \frac{P}{W_D})}{(X_2 - X_1)} \beta_{11} + \frac{(\frac{P}{W_D} - X_1)}{(X_2 - X_1)} \beta_{21}$$ 11.23a

$$f_2 = \frac{(X_2 - \frac{P}{W_D})}{(X_2 - X_1)} \beta_{12} + \frac{(\frac{P}{W_D} - X_1)}{(X_2 - X_1)} \beta_{22}$$ 11.23b

where the coil's pitch-to-wire diameter ratio P/W_D, is between the P/W_D ratios given in the table, with X_1 being the smallest ratio and X_2 the highest. $B_{col,row}$ is the β value read from the table corresponding to the column and row.

To calculate the coil's AC resistance, we first need to know the length of the wire in the coil. We can calculate this as follows,

$$\ell_W = \sqrt{(\pi ND)^2 + \ell^2}$$ Length of wire in a cylindrical coil 11.24

where ℓ_W is the wire length in metres, N is the number of turns in the coil, D is the diameter of the coil in metres, and ℓ is the coil length in metres. We then calculate the AC resistance due to the skin effect and multiply this by a proximity effect factor β.

The pitch of a coil is given by the length of the coil divided by the number of turns.

$$P = \frac{\ell}{N}$$ Pitch of a coil of a cylindrical coil 11.25

ℓ/D	P/W_D									
	1	1.111	1.25	1.429	1.667	2	2.5	3.333	5	10
0	5.31	3.73	2.74	2.12	1.74	1.44	1.2	1.16	1.07	1.02
0.2	5.45	3.84	2.83	2.2	1.77	1.48	1.29	1.19	1.08	1.02
0.4	5.65	3.99	2.97	2.28	1.83	1.54	1.33	1.21	1.08	1.03
0.6	5.8	4.11	3.1	2.38	1.89	1.6	1.38	1.22	1.1	1.03
0.8	5.8	4.17	3.2	2.44	1.92	1.64	1.42	1.23	1.1	1.03
1	5.55	4.1	3.17	2.47	1.94	1.67	1.45	1.24	1.1	1.03
2	4.1	3.36	2.74	2.32	1.98	1.74	1.5	1.28	1.13	1.04
4	3.54	3.05	2.6	2.27	2.01	1.78	1.54	1.32	1.15	1.04
6	3.31	2.92	2.6	2.29	2.03	1.8	1.56	1.34	1.16	1.04
8	3.2	2.9	2.62	2.34	2.08	1.81	1.57	1.34	1.165	1.04
10	3.23	2.93	2.65	2.27	2.1	1.83	1.58	1.35	1.17	1.04
∞	3.41	3.11	2.815	2.51	2.22	1.93	1.65	1.395	1.19	1.05

Table 11.2
Medhurst's proximity resistance factors.

11.5 OPTIMAL COIL GEOMETRY FOR INDUCTANCE

From Equation 11.6, the inductance of a coil is related to the number of turns, the length, and the diameter of the coil. The length, diameter, pitch and number of turns are dependent on one another, and each determines the total length of the wire used to wind the coil. This begs the question as to what is the least amount of wire required to wind a coil for a given inductance. Using the least amount of wire will not only save time and money but will affect Q. We will define "optimal", in this case, to mean the least amount of wire used to achieve a given inductance.

If we restrict ourselves to cylindrical coils with reasonably small pitch angles, that is, the distance between the loops is relatively small compared to the diameter of the coil, and for the type of coils we will be making, this will generally be the case. We can rewrite Equation 11.24 as,

$$\ell_W = \pi N D \qquad 11.26$$

The number of loops in a coil is given by,

$$N = \frac{\ell_W}{\pi D} \qquad 11.27$$

And the coil length is

$$\begin{aligned} \ell &= NP \\ &= \frac{\ell_W P}{\pi D} \end{aligned} \qquad 11.28$$

Using the Wheeler formula 11.4 for the inductance in this discussion as it is easy to analyse analytically. We rearrange the formula to include wire length.

$$L = \mu\pi D^2 N^2 \Big/ 4\ell(1+0.45\frac{D}{\ell})$$

$$L = \frac{\mu\pi}{4}\frac{D^2N^2}{(\ell + 0.45D)}$$

$$= \frac{\mu\pi}{4}\frac{D^2(\frac{\ell_W}{\pi D})^2}{\frac{\ell_W P}{\pi D} + 0.45D}$$

And

$$L = \frac{\mu \ell_W{}^2}{4\pi}\left[\frac{1}{\frac{\ell_W P}{\pi D} + 0.45D}\right] \qquad 11.29$$

If we treat the wire length and the pitch as constant, we can use this formula to determine how the inductance changes with respect to changes in the coil diameter. Figure 11.8 is a graph of

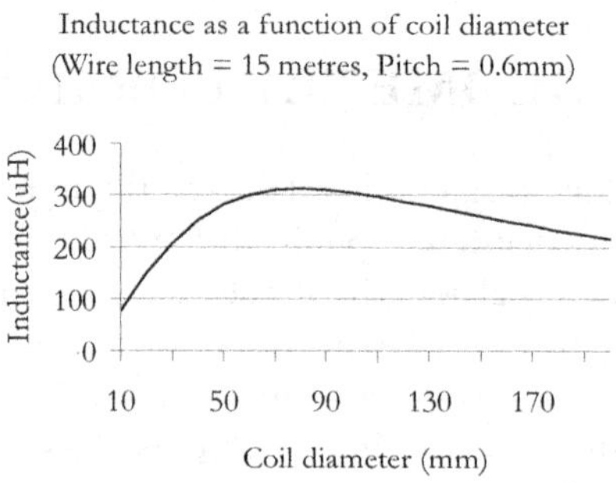

Figure 11.8
Inductance as a function of coil diameter for a given length of wire and pitch.

inductance as a function of coil diameter using Equation 11.29. The wire length was set to some arbitrary length of 15 metres, and the pitch was set to the wire diameter (conductor diameter plus enamel coating) of 0.6 mm.

From this graph, we can see there is a maximum inductance value obtained for a given wire length and pitch. We can use calculus to analyse this curve. Taking a derivative of a function gives us an equation for the tangential slope of the curve at each point. The slope of the curve at the peak inductance is zero. So if we take the partial derivative with respect to the diameter, keeping wire length and pitch constant, and set this value of the partial derivative to zero, we will find the diameter of the coil that gives us the peak inductance.

$$\frac{\partial L}{\partial D} = \frac{\mu \ell_W{}^2}{4\pi}\frac{\frac{\ell_W P}{\pi D^2} - 0.45}{\left(\frac{\ell_W P}{\pi D} + 0.45D\right)^2}$$

And setting this partial derivative to zero, we get

$$\frac{\mu \ell_W^{\ 2}}{4\pi}\left[\frac{\dfrac{\ell_W P}{\pi D^2}-0.45}{\left(\dfrac{\ell_W P}{\pi D}+0.45D\right)^2}\right]=0$$

The equation goes to zero when the numerator (the top term) goes to zero. So

$$\frac{\ell_W P}{\pi D^2}-0.45=0$$

$$\frac{\ell_W P}{\pi D^2}=0.45 \qquad 11.30$$

$$\frac{1}{D}\left|\frac{\ell_W P}{\pi D}\right|=0.45$$

Substituting the length of the coil given by Equation 11.28 into 11.30, we arrive at

$$\frac{\ell}{D}=0.45$$

Using the same analysis on Equation 11.6, we arrive at a more accurate value of the optimal ℓ/D ratio.

$$\boxed{\left.\frac{\ell}{D}\right|_{OPTIMAL}=0.407}$$ Optimal coil length to diameter ratio to achieve a given inductance 11.31

This means, that for a given inductance value, a coil with a length-to-diameter ratio of 0.407 requires the least amount of wire to wind the coil.

11.6 OPTIMAL COIL GEOMETRY FOR Q

In the above section, we determined that the inductance of a coil can be achieved using the least amount of wire when the length-to-diameter ratio is 0.407. One might think that using the least amount of wire will result in the minimum AC resistance of the coil and, therefore, the maximum Q. For solid core wires, this is not true, as we know the proximity effect is heavily dependent on the pitch-to-wire-diameter ratio P/W_D.

In Figure 11.9, the effective Q of a 300 uH coil is plotted as a function of the ℓ/D ratio for various pitch sizes. Each curve represents a different pitch-to-wire-diameter ratio. Figure 11.9a is for a wire diameter of 0.6 mm (conductor diameter 0.5 mm), and Figure 11.9b is for a wire diameter of 1.2 mm (conductor diameter 1.0 mm). The thicker wire produces coils of higher Q values, as expected. However, the overall shape of graphs 11.9a and 11.9b are very similar. A large increase in Q is achieved by increasing the distance between the loops by as little as half a wire diameter. As the P/W_D ratio increases, the Q increases up to a maximum value, and then the Q starts to decrease again. The maximum Q value is achieved at around one-and-a-half wire diameter spacing between the loops, a P/W_D ratio of 2.4.

The ℓ/D ratio at which the maximum Q, for different pitch sizes, ranges from about 0.3 to 0.5, with the maximum Q occurring at an ℓ/D ratio of 0.4 for a P/W_D ratio of 2.4. The change in Q can be explained as follows. As the P/W_D ratio increases, more wire is required to wind the coil and AC wire resistance increases. At small distances between the loops, the proximity effect is dominant in determining the total AC resistance of the coil, and the change in Q is initially sensitive to small changes in the gaps between the wire loops. This reduction in proximity effect is more than enough to counteract the increase in wire resistance, and the Q dramatically increases. As the gap increases, the proximity effect ceases to dominate the total resistance, and the AC wire resistance becomes the dominant factor. Also, increasing the length of the coil increases the self-capacitance, and the self-resonant frequency decreases. This increase in both wire resistance, and the reduction in self-resonant frequency, start to dominate, and the effective Q decreases.

This leads us to a rule of thumb:

"The maximum effective Q for a single layer cylindrical coil occurs when the pitch is one-and-a-half wire diameters between the loops and when the length-to-diameter ratio is around 0.4."

This rule of thumb is valid for solid core wires with diameters 1 mm or less and for inductance ranges between 200 to 500 uH. For Litz wire, the reduction in proximity effect by increasing the spacing between loops will be marginal. However, the optimal ℓ/D ratio of 0.4 still applies as this is the ratio required for the minimum length of wire required for a given inductance. For tightly wound coils, the maximum Q is reached at an ℓ/D ratio of around 0.3.

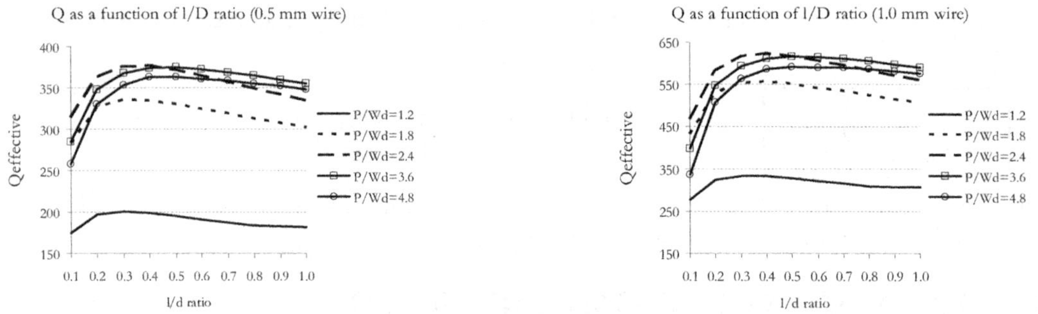

Figure 11.9
Q as a function of ℓ/D ratio for various pitch values – 300 uH inductor at 1 MHz.

11.7 COIL FORMERS AND DIELECTRIC LOSSES

The inductance values needed for AM crystal radio coils require lengths of wire in the tens of metres (at least for air coils). To wind a 300 uH coil, with an ℓ/D ratio of 0.4 and a spacing of one wire diameter, using a wire diameter of 1.25mm would require a wire length of 23 metres, with a coil diameter of 214 mm and a length of 84 mm. This coil would act like a slinky and needs some structural support. Unless you are using an extremely thick wire diameter of 2 mm or more, these coils will need to be wound on a former, and as such, we need to know what effect the former has on the Q.

From our discussion on self-capacitance, we know the greater the relative permittivity of a former, the lower the self-resonant frequency of the coil will be. For thin-walled formers, this reduction in self-resonant frequency is quite small, in the order of less than 1% for material with a relative permittivity of around 3.

As well as a reduction in Q due to self-resonance, there are also dielectric losses. These are due to the interaction with the electric field with molecules in the former. Energy may be lost in the form of heat or in the deformation of molecules. The dielectric loss is specified by the ***dissipation factor*** (or the ***loss tangent***) and is defined as the energy loss divided by the energy stored in the electric field within the material. That is,

$$D = \tan\sigma = \frac{\varepsilon''}{\varepsilon'} \quad \text{Dielectric dissipation factor} \quad 11.32$$

where ε'' represents the energy loss and ε' represents the energy stored in the electric field. Notice that this is the inverse of the definition of Q.

$$Q = \frac{1}{D} \quad 11.33$$

Both Q and the dissipation factor are frequency-dependent. We know there is some axial capacitance along the coil, and the dielectric of the former will affect this capacitance. We can treat this axial capacitance as a discrete capacitor, with some portion of the former acting as a dielectric. We can model this loss as an ***equivalent series resistance*** (ESR), and we can write,

$$D_F = \frac{R_{ESR}}{X_C} = 2\pi fC(R_{ESR}) \quad \text{ESR} \quad 11.34$$

The energy stored in a capacitor is,

$$E_{cap} = \frac{CV^2}{2}$$

The energy stored in the axial capacitance in the electric field will only be a fraction of the energy stored in the magnetic field of a coil. The energy stored in the magnetic field of an inductor is given by,

$$E_L = \frac{LI^2}{2}$$

Using Ohm's law, we can rewrite this as

$$E_L = \frac{LV^2}{2Z^2}$$

where V is the voltage across the inductor, and Z is the impedance across the inductor. Expressing the ratio of the energy stored in the electric field to the energy stored in the magnetic field, we can write

$$\frac{E_{CAP}}{E_L} = \frac{CV^2}{2}\frac{2Z^2}{LV^2}$$

$$= \frac{CZ^2}{L}$$

$$= \frac{C(2\pi fL)^2}{L}$$

And

$$\frac{E_{CAP}}{E_L} = 4\pi^2 f^2 CL \qquad 11.35$$

Multiplying Equation 11.35 by the dissipation factor will give us the proportion of the energy lost by the electric field in relation to the energy stored in the magnetic field. This proportion of energy loss is directly proportional to the reduction in Q. We can write this as,

$$\frac{E_{CAP}(D_F F_{ratio})}{E_L} = 4\pi^2 f^2 CL(D_F F_{ratio})$$

$$\frac{E_{CAP(LOSS)}}{E_L} = E_{LOSS}$$

$$E_{CAP(LOSS)} = E_L E_{LOSS}$$

And

$$E_{CAP(LOSS)} = E_L E_{LOSS}$$

Where

$$\boxed{E_{LOSS} = (4\pi^2 f^2 C_A L) D_F F_{ratio}}$$ Proportion of energy lost due to the dielectric 11.36

where C_A is the axial capacitance, D_F is the dissipation factor, and F_{ratio} is the proportion of the electric field interacting with the dielectric. E_{LOSS} is a dimensionless number. If we treat the axial capacitance as existing across the entire cross-section area of the coil, then only a fraction of the electric field will interact with the dielectric, and we will estimate this as follows

$$\boxed{F_{ratio} = 1 - \left(\frac{r_i}{r_o}\right)^2}$$ Proportion of the electric field interacting with the dielectric 11.37

where r_i is the inside radius of the former with no dielectric material, and r_O is the outside radius of the former.

The proportion of energy loss due to the dielectric can be expressed as an equivalent resistance, which can be used in Equation 8.9 to determine the Q of a coil. In Equation 8.9, the dielectric loss is given as R_ε. From the definition of Q, the peak energy is represented by the reactance of the coil X_L. And since $E_{CAP(LOSS)} = E_L E_{LOSS}$ we can write R_ε as,

$$\boxed{R_\varepsilon = X_L E_{LOSS}}$$ Dielectric loss represented as an equivalent resistance 11.38

Reduction in Q Due to the Former

Let us examine the effect formers have on the effective Q. We will compare Styrene based formers with PVC-based formers using the relative permittivity and the dissipation factors given in Table 11.3.

Given a typical crystal radio coil with an inductance of 300 uH, using a wire diameter of 0.5 mm with a pitch of .6 mm on a 75 mm former with an ℓ/D ratio of 0.4872. Let's assume the coil is wound on a PVC former with a wall thickness of 2 mm. From Equation 11.37

$$F_{ratio} = 1 - \left(\frac{r_i}{r_o} \right)^2$$
$$= 1 - \left(\frac{35.5}{37.5} \right)^2$$

And

$$F_{ratio} = 0.104$$

From Table 11.3 (using the worst case), the relative permittivity of PVC is 3.0, and the dissipation factor is 0.017. Using Equation 11.13 to calculate the axial capacitance, we get C_A=1.65 pF. Using Equation 11.36, we get a loss ratio at 1 MHz of

$$E_{LOSS} = (4\pi^2)(1.0x10^6)^2(1.65x10^{-12})(300x10^{-6})(0.017)(0.104)$$
$$= 3.45x10^{-5}$$

And Rε is

$$R_\varepsilon = X_L E_{LOSS}$$
$$= 1885(3.45x10^{-5})$$
$$= 0.0656$$

The effective Q at 1 MHz, ignoring the dielectric loss, is 196.05 with α=0.9698. Adding the dielectric loss, the effective Q becomes

$$Q_{eff} = \frac{X_L \alpha}{R_{coil(ac)} + R_\varepsilon}$$
$$= \frac{1885(0.9698)}{9.3244 + 0.065}$$
$$= 194.69$$

Now using a Styrene-based former with the same dimensions. The relative permittivity of Polystyrene is 2.4, and the dissipation factor is 0.0005. We get an axial capacitance C_A=1.61 pF, and

$$E_{LOSS} = (4\pi^2)(1.0x10^6)^2(1.61x10^{-12})(300x10^{-6})(0.0005)(0.104)$$
$$= 9.91x10^{-7}$$

Rε is

$$R_\varepsilon = X_L E_{LOSS}$$
$$= 1885(9.91x10^{-7})$$
$$= 0.0019$$

The effective Q at 1 MHz, ignoring the dielectric loss, is 196.19, and α=0.9705. Including the dielectric loss, the effective Q becomes

$$Q_{eff} = \frac{X_L \alpha}{R_{coil(ac)} + R_\varepsilon}$$

$$
\begin{aligned}
Q_{eff} &= \frac{1885(0.9705)}{9.3244 + 0.0019} \\
&= 196.15
\end{aligned}
$$

That is a 0.75% difference in effective Q between the formers. For high-Q coils, such as Litz wire coils, the effect is greater as the AC resistance of the coil is smaller, and R_ε will have a greater effect. For instance, a 300 uH coil using 175/46 Litz wire wound on a 75 mn Styrene former has an effective Q of 580.50, and on the PVC former, the Q is 571.67; a 1.52% difference. At 1,700 kHz, the difference is 3.98 % (Qs of 504.83 and 484.74, respectively).

When placed in a circuit, however, this drop is largely swamped by other loss factors yet to be explored. For instance, it is not unusual to have an equivalent resistive loss of 10 Ω or more above the intrinsic losses in the coil when operating a crystal radio. Using this operating resistive value, the Litz wire coil Qs at 1,700 kHz become 203 and 200, a 1.5% difference. This is at the upper range of differences and is the worst one would expect. For most coils and for lower frequencies, the difference would be a lot less. For the above copper coils, the operating Q at 1 MHz would be 94.23 and 94.53, respectively, a difference of 0.32%.

In summary, thin-walled formers with low relative permittivity and loss factors will have a noticeable effect only on isolated high-Q coils at high frequencies. For coils operating in crystal radios, the effect is dramatically reduced and is negligible in most instances. If thin-walled formers with low dissipation factors are readily and cheaply available to you, then you should use them. However, do not expect a noticeable improvement in the performance of a crystal radio. Piping with low dissipation factors are not readily and easily available for me. So in this book, I will use PVC.

Former type	**Abbreviation**	**Typical relative permittivity**	**Dissipation factor (1Mhz)**	**Recycle code**
Air		1.0		
Polytetrafluoroethylene	PTFE (Teflon)	2.1	<0.0001	
Low-density polyethylene	LDPE	2.2	0.0002	4
Polypropylene	PP	2.2	0.0003	5
High-density polyethylene	HDPE	2.3	0.0005	2
Cardboard		2.4		
Polystyrene	PS	2.4	0.0005	6
Polyvinyl chloride	PVC	2.7 - 3.0	0.017-0.006	3
Polycarbonate	PC	2.9	0.010	
Polyethylene terephthalate	PET (Mylar)	3.0	0.016	1
Acrylonitrile butadiene styrene	ABS	3.2	0.02	
Nylon		3.2	0.02	

Table 11.3
Relative permittivity of common plastic material.

11.8 COIL FORMERS AND MAGNETIC LOSSES

Just as the electric field can lose energy due to intervening material in a former, so can the magnetic field. This loss of energy from the magnetic field is called ***Magnetic Loss***. Magnetic loss can occur due to the formation of magnetic domains in the material or from the formation of eddy currents. As the materials we are using for formers are thin-walled plastics that do not form magnetic domains and are insulators with extremely high resistance that do not form any eddy currents, we can safely ignore any magnetic losses due to PVC formers or any insulating plastic formers. Magnetic losses will be discussed further in the Chapter on ferrite core coils.

11.9 COIL TAPS AND Q

Taps have a reputation for being Q killers. How true is this? The following is a discussion of experiments carried out using a Litz wire coil to determine the validity of this statement. Table 11.4 gives the specifications of the coil used in the experiments.

Description	Value
Coil inductance	217 uH
Coil diameter	75.9 mm
Coil length	55 mm
Number of windings	60
Wire	175/46 Litz wire (ID = 0.53 mm equiv, OD = 0.90)
Q_{eff} (at 1 MHz)	547

Table 11.4
Litz wire test coil specifications.

All Q measurements were taken at 1 MHz frequency using the Q measurement stand described in the Chapter on measurement techniques. A one-MHz five-volt peak-to-peak signal was injected into a primary coil at a distance of 150 mm from the Litz wire coil under test (the secondary coil). The primary coil consisted of three windings of 1.25 mm diameter copper wire with a coil diameter of 103 mm. The separation distance and the construction of the primary coil were chosen to minimise mutual inductance losses (described in a later Chapter). This setup results in an unusually high crystal radio operating Q, allowing small changes in Q to be measured.

A tap consists of a 70 mm length of Litz wire (the same wire used in the construction of the secondary coil). A tap was soldered to winding number 9 on the secondary coil. This tap was connected to the audio output circuit of the crystal radio, a diode in series with a piezoelectric earphone. With a tuning capacitor in parallel with the secondary coil, the secondary was set to resonance at one MHz, and the Q of the secondary coil was measured. This setup is equivalent to using a wiper-arm to select the audio rather than taps. This is a functional crystal radio without using taps. This is the base configuration for which other Q measurements will be compared, with the only changes being the addition and removal of taps. For each of the experiments, the tap at winding 9 was connected to the audio output. The operating Q was measured as 490.

Next, four extra taps were added to the secondary coil at windings 20, 30, 40 and 50, but they were not connected; they just hung down from the secondary coil. The Q was measured at 473, a 3.47% drop in Q.

Next, taps at winding 9, 20 and 30 were connected to the audio output circuit via a rotary switch, with the tap at winding 9 used for the output. The taps at winding 40 and 50 and the end lead of the coil were connected to a rotary switch to be used as a band selection switch. The output of the band selection rotary switch was connected to the tuning capacitor. The Q was measured as 450, an 8.16% drop.

The taps and rotary switch used for band selection were removed, leaving only the audio taps.

The Q was 503, an increase of 2.65%. There is obviously experimental error in this result; however, it clearly shows that the Q increases from the previous configuration. And clearly shows taps used for band selection near the top end of the coil, result in a noticeable drop in Q.

Lastly, the audio rotary switch was removed, and the two extra taps were left hanging. The Q was 482, a 1.63% decrease. The summary of these results is given in Table 11.5.

Experiment number	Description	Measured Q	% difference (from 490)
1	Tap 1 connected to the audio output. No other taps connected to the secondary coil.	490	N/A
2	Tap 1 connected to the audio output and 4 extra taps not connected	473	-3.47
3	3 lower taps connected for audio selection using a rotary switch, and 2 higher taps connected for band selection using a rotary switch	450	-8.16
4	3 taps connected to audio using rotary switch only	503	+2.65
5	Tap 1 connected to the audio output and 2 extra taps not connected	482	-1.63

Table 11.5
Operating Q values with different tap configurations.

The results of the experiments can be explained as follows.

1. Unconnected taps act like small sections of open-circuited transmission lines, looking slightly capacitive. This results in an overall reduction in the apparent inductance of the coil resulting in a small drop in Q.
2. A tap connected to a switch will have a complete circuit path due to some small capacitance across the terminals of the switch. This current will result in wire and possibly dielectric resistive losses through the circuit resulting in a noticeable reduction in Q when used for band selection.
3. Taps connected to a switch for audio selection will have a diode and earphone in circuit. The charge on the earphone raises the voltage level at which current will flow across the diode. Therefore, current flows for only a small portion of the total signal period, resulting in a small loss of energy during one cycle and a small drop in Q. This will be discussed in greater detail in a later Chapter.

We can draw the following conclusions from the results of these experiments.

1. Using taps (lengths of wire) not connected in a circuit results in a small reduction of the operating Q of approximately 0.87% for each added tap. So don't use taps in this fashion.
2. Taps connected as part of a band selection circuit result in a significant reduction of Q. If designed properly, these taps are not required (to be covered later), so don't use them.
3. Taps connected as part of the audio selection circuit result in very little, if any, measurable reduction in Q.

The reputation of taps as Q killers appears to be somewhat overstated. The only time there is a significant reduction in Q is when they are used for band selection. For a properly designed crystal radio, band selection is not required, and we will not be using taps for such purposes. When winding coils using copper enamelled wire, we will wind small loops on the coil. These small loops have no measurable effect on Q. We will either select a loop for audio takeoff by using a single wire connection, or by using a rotary switch.

The alternative to this approach is to use a wiper-arm. A wiper-arm is moved across the exposed wires of the secondary coil to select the audio takeoff. Let us examine the pros and cons of using a wiper-arm.

The pros:

1.0 Allows you to potentially select the loudest volume for a given signal by adjusting the position along the coil for audio detection. We shall see in later Chapters that even this is a dubious claim.
2.0 Saves using taps and a potential very small drop in Q.

Now for the cons:

1.0 More difficult to construct than using taps.
2.0 Need to remove all the enamel from a strip along the length of the coil without disturbing or having adjacent loops shorting one another.
3.0 Much harder to remove the enamel from coils with a pitch greater than a wire diameter, without disturbing the wire loops.
4.0 Due to oxidation and corrosion, the exposed copper wire and wiper contact must be re-sanded periodically; otherwise your Q will drop.
5.0 Getting the wiper-arm to make a good electrical contact with a single winding is very problematic. There needs to be sufficient pressure to make a decent connection; otherwise, resistance will be introduced, resulting in a loss of Q. If the pressure is too great, you will continually move the position of the loops with large pitch values. If the pitch angle is small, there is also a good probability the wiper arm will be touching more than one winding at a time.
6.0 It makes a lot of noise through the earphone when you adjust the wiper.
7.0 You need to remember the positions that gave you the best signal.
8.0 Difficult, if not impossible, to use with high Q Litz wire coils. When using Litz wire, you can't simply sand off the Litz wire coating. You need to remove the enamel coating on each of the tiny strands of wire as well, best achieved by using a hot soldering iron. As Litz wire is very flexible, each winding will most likely short with each other as you move the wiper.

For these reasons, we will only use the tap method of building Crystal radios in this book.

11.10 LITZ WIRE COILS

Litz wire is designed to reduce the AC resistance and the proximity effect in coils. At 1 MHz frequency, the skin depth due to the skin effect is about 65 micrometres. If the wire you use has a radius less than the skin depth, then the majority of the current will flow through the entire cross-sectional area of the wire, and the resistance at these frequencies will not be much greater than the resistance at DC. However, since the wire diameter is very small, it will have a very large DC resistance. We know that resistances in parallel reduce the total resistance. And the total resistance of resistors with the same value in parallel is given by $R_T=R/N$, where R is the resistance value, and N is the number of resistors in parallel. So to reduce the AC resistance, Litz wire is made up of very many small conductors in parallel. The Litz wire I use has 175 wires, each with a diameter of 0.0406 mm (46 Gauge); see Figure 11.10.

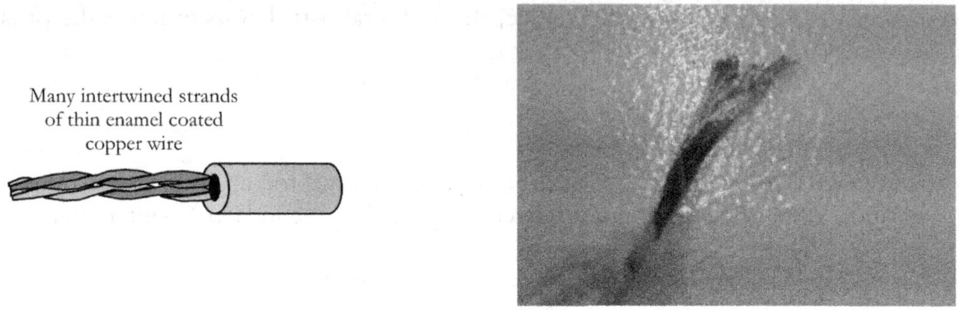

(a) (b)

Figure 11.10
Litz wire.

As Litz wire is constructed using many thin strands of wire, there will be a proximity effect acting on the strands inside the wire cable itself that does not occur when using single solid core wires. To reduce this internal proximity effect, each conductor is woven in such a way that the current density distribution caused by the proximity effect is reduced. The conductor will spend some time on the right side of the cable, then the left, then the middle, etcetera. Changing the orientation of the conductor with respect to the magnetic field inside the cable changes the shape of the current density as the conductor winds its way along the cable. This prevents the current density from becoming too great in one area. The reduction in AC resistance and proximity effect allows for very high-value Q coils to be wound.

Modifications to the Model

When calculating the inductance of a coil, the actual wire diameter is unimportant. However, the wire diameter and pitch are important when determining the proximity effect between coil loops. We will determine an equivalent wire diameter based on the total conductor cross-sectional area within the cable. The total cross-sectional surface area of the Litz cable is given by the cross-sectional area of a single strand multiplied by the number of strands. That is

$$\boxed{A_{LITZ} = N_S \pi (\frac{W_D}{2})^2}$$ Equivalent Litz wire core area 11.39

where N_S is the number of strands in the Litz cable, and W_D is the diameter of a single strand. The equivalent single core wire diameter with this area is thus given by

$$\boxed{W'_D = 2\sqrt{\frac{A_{LITZ}}{\pi}}}$$ Equivalent single-core diameter 11.40

Use the equivalent diameter of the bundle of conductors and treat this diameter as a single wire and carry out the calculations for inductance, wire length, self-resonant frequency and capacitance.

When calculating the Q of the coil, we need to make some modifications to the model to take into account the difference in resistance and the internal proximity effect of the Litz wire. Due to the small size of the strands in Litz wire, the skin depth is often a large proportion of the wire radius or even greater than the wire radius itself. Skin depth is defined as depth below the surface of the conductor where the current density falls to 0.37 (1/e) of the density at the surface. So even if the wire radius is smaller than the skin depth, there will be a small change in the current density. We must take this change in current density into account.

First, determine the DC resistance of a single strand of Litz wire using Equation 4.19. The length of the strand is the length of the wire used to wind a coil. To determine the cable DC resistance, divide the single-strand DC resistance by the number of strands in the cable (N_{Litz}).

$$R_{DC(Litz)} = \left[\rho \frac{\ell}{A}\right] \Big/ N_{Litz}$$ Litz wire cable DC resistance 11.41

Next, determine the AC resistance of a single strand. Typically this will not be much greater than the DC resistance of a single strand. The following procedure for determining the AC resistance of a wire where the diameter is small compared to the skin depth was given by David Knight[4]. First, determine the skin depth and DC resistance for a single strand. Calculate a modification factor Y as follows

$$Y = \frac{0.189774}{\left(1 + 0.272481\left[Z^{1.82938} - Z^{-.99457}\right]^2\right)^{1.0941}} \qquad 11.42$$

where Z is

$$Z = 0.62006\frac{r}{\delta} \qquad 11.43$$

and r is the radius of the wire, and δ is the skin depth. Next, modify the skin depth value.

$$\delta' = \delta\left[1 - e^{-\left(\frac{r}{\delta}\right)}\right] \qquad 11.44$$

Now calculate the AC resistance of the length of wire for the given frequency using the DC resistance, the modified skin depth value and the modification factor Y.

$$R_{AC} = R_{DC}\left(\frac{r^2}{\left(2r\delta' - \delta'^2\right)\left(1 + Y\right)}\right) \qquad 11.45$$

We now have the AC resistance due to skin depth only. The ratio of the total AC resistance of a single strand due to the skin effect, the proximity effect and capacitance between single strands is called the ***AC resistance factor*** and is given by

$$F_R = \frac{R_{ST}}{R_{AC}} \approx \frac{R_{ST}}{R_{DC}}$$

where R_{AC} is the AC resistance of a single strand of wire at a given frequency due to the skin effect

[4] Knight, David. (2016). Practical continuous functions for the internal impedance of solid cylindrical conductors.

(note for properly designed Litz wire for a given frequency, this will be close to the DC resistance of the wire). R_{ST} is the total AC resistance of a single strand, including all the resistive losses. The AC resistance factor is frequency dependent, and when Engineers design a Litz wire cable, they design it for a specific range of frequencies. And the higher the frequency, the smaller the single-strand diameter is. For AM broadcast frequencies, you should choose Litz wire with a single strand diameter of around 46 gauge (0.0406 mm). The AC resistance factor chosen takes into account manufacturing considerations such as complexity and cost. Table 11.6 lists typical AC resistance factors F_R for given wire strand diameters. For our interests, the 46 gauge F_R is 1.6. This should be the F_R at around 1 MHz.

Strand AWG Size	FR	Strand AWG Size	FR	Strand AWG Size	FR	Strand AWG Size	FR
32	1.06	37	1.15	42	1.35	47	1.64
33	1.07	38	1.18	43	1.41	48	1.68
34	1.09	39	1.22	44	1.47		
35	1.11	40	1.25	45	1.54		
36	1.13	41	1.30	46	1.60		

Table 11.6
General AC resistance factors for strand diameters.

How F_R changes with frequency is cable specific, and generally, we only know the number of strands and the strand size. In the absence of any other information, use the Fr given in Table 11.6 and the following formula to calculate the Fr for a given frequency.

$$F_R' = (F_R)^f \qquad 11.46$$

where F_R is the given AC resistance factor from Table 11.6, and f is the frequency in megahertz. If you are lucky enough to know how your purchased Litz wire F_R changes with frequency, then use that. Equation 11.46 is an empirical formula that fits the 175/46 Litz wire I bought. The total AC resistance of the Litz cable for a given frequency is then

$$R_{LITZ} = \frac{F_R' R_{AC}}{N_S} \qquad \text{Total AC resistance of a Litz wire cable} \qquad 11.47$$

where N_S is the number of strands of wire in the Litz cable.

Having determined the internal AC resistance of a Litz cable, we now need to determine the proximity effect due to the loops in a coil. We can use the technique described previously using Medhurst's table. This approach assumes that the wire consists of a single conductor, so we need to modify the proximity resistance factor β to take into account the reduction in proximity effect due to the Litz wire cable winding pattern. Treating the Litz wire as a single conductor with a diameter given by Equation 11.40, use the single core equivalent value as the conductor size and calculate the proximity resistance factor (β). And R_{COIL} becomes

$$R_{COIL} = \beta R_{LITZ} \qquad 11.48$$

Now we need to make an adjustment to β. Given

$$\beta = \frac{R_{COIL}}{R_{LITZ}}$$

$$= \frac{R_{LITZ} + R_{\Pr oximity}}{R_{LITZ}}$$

$$= \frac{R_{LITZ} + (R_{COIL} - R_{LITZ})}{R_{LITZ}}$$

$$= 1 + \frac{(R_{COIL} - R_{LITZ})}{R_{LITZ}}$$

The second term gives us the proportion by which the resistance is increased due to the proximity effect (the external magnetic field) for a solid core wire. Similarly, the Litz wire AC resistance factor can be written as

$$F'_R = 1 + \gamma$$

where γ is the factor by which the Litz wire total resistance increases due to the proximity effect. The proportion by which the Litz wire modifies the proximity resistance is therefore given by subtracting 1 from F'_R. If we multiply the second term by this value, we should get the proportion increase due to the proximity effect between loops. That is

$$\boxed{\beta' = 1 + (F'_R - 1)\frac{(R_{COIL} - R_{LITZ})}{R_{LITZ}}}$$

Litz wire proximity effect factor 11.49

And the total coil resistance, including the proximity effect, now becomes

$$\boxed{R'_{coil} = \beta' R_{LITZ}}$$

Litz wire coil total AC resistance 11.50

We can now use this Litz wire AC coil resistance in determining the Q of the coil.

Table 11.7 shows the difference in measured and calculated values for a 175/46 Litz wire coil wound on a 75 mm PVC former with 60 turns. Using ε_r = 2.7 and D_F = 0.006 (the lower values for PVC).

	Calculated	Measured	% difference
Inductance	228.78 uH	229.00	0.096
Effective Q (1 MHz)	544.47	546.59	0.39
Effective Q (1.7 MHz)	468.39	475.0	1.39

Table 11.7
Difference between calculated and measured value for a Litz wire coil.

11.11 WORKED EXAMPLE-1

Given we required a coil with an inductance of 300 uH, and we have chosen to use copper enamel wire with a core diameter of 0.63 mm and a wire diameter (including enamel) of 0.68 mm. We wish to maximise the Q of the coil. What will be the geometry of the coil (the diameter and length of the coil), and what is the effective Q at 1 MHz? Also, assume the coil former will be PVC with a wall thickness of 2 mm.

Determine the coil diameter

The maximum effective Q of a coil occurs when the length-to-diameter ratio is about 0.407 and the pitch is about 2.5 times the wire diameter. Therefore,

$$\frac{\ell}{D} = 0.407$$

and

$$P = 0.63x10^{-3} * 2.5$$
$$= 1.575x10^{-3}$$

Before using Equation 11.11 to determine the diameter, first, calculate the inductance coefficient K using Equation 11.5.

$$K = \frac{2}{\pi}\left(\frac{\ell}{D}\right)\left[\ln(1+\frac{\pi D}{2\ell}) + \frac{1}{2.3004 + 3.2219\left(\frac{\ell}{D}\right) + 1.7793\left(\frac{\ell}{D}\right)^2}\right]$$

$$= \frac{2}{\pi}(0.407)\left[\ln(1+\frac{\pi}{2}2.5) + \frac{1}{2.3004 + 3.2219(0.4) + 1.7793(0.4)^2}\right]$$

$$= 0.47595$$

Calculating the diameter using Equation 11.11

$$D = \sqrt[3]{\frac{4P^2L}{\mu\pi\left(\frac{\ell}{D}\right)K}}$$

$$= \sqrt[3]{\frac{4(1.575x10^{-3})^2 300x10^{-6}}{\mu_0\pi(0.407)0.47595}}$$

$$= 0.1573$$

Using Equation 11.12, the coil length is

$$\ell = [0.407]0.1573$$
$$= 0.0640$$

And the number of turns calculated using Equation 11.8 is

$$N = \frac{\ell}{P}$$

$$= \frac{64.0x10^{-3}}{1.575x10^{-3}}$$

$$= 40.63$$

The coil is 157.3 mm in diameter, 64.0 mm in length and requires 40.63 turns of wire.

Determine the resistive losses

To determine the Q of a coil, we need to calculate all the resistive losses. First, determine the length of the wire used in winding the coil. Using Equation 11.24

$$\ell_W = \sqrt{(\pi ND)^2 + \ell^2}$$
$$= \sqrt{(\pi 40.63(157.3x10^{-3})^2 + (64.0x10^{-3})^2}$$
$$= 20.09$$

The skin depth at this frequency is 6.5234 x 10^{-5} metres. And the AC wire resistance is 2.9156 Ω. Next, determine the proximity resistance factor. The coil's ℓ/D ratio is 0.407, and the coil's pitch-to-wire-diameter ratio is

$$\frac{P}{W_D} = 2.5$$

Using Table 11.2, the coil's ℓ/D ratio is between Y_1=0.4 and Y_2=0.6. The coil's P/W_D ratio is between X_1=2.5 and X_2=3.333. This cell range is highlighted in the following table.

ℓ/D	P/W_D									
	1	1.111	1.25	1.429	1.667	2	2.5	3.333	5	10
0	5.31	3.73	2.74	2.12	1.74	1.44	1.2	1.16	1.07	1.02
0.2	5.45	3.84	2.83	2.2	1.77	1.48	1.29	1.19	1.08	1.02
0.4	5.65	3.99	2.97	2.28	1.83	1.54	1.33	1.21	1.08	1.03
0.6	5.8	4.11	3.1	2.38	1.89	1.6	1.38	1.22	1.1	1.03
0.8	5.8	4.17	3.2	2.44	1.92	1.64	1.42	1.23	1.1	1.03
1	5.55	4.1	3.17	2.47	1.94	1.67	1.45	1.24	1.1	1.03
2	4.1	3.36	2.74	2.32	1.98	1.74	1.5	1.28	1.13	1.04
4	3.54	3.05	2.6	2.27	2.01	1.78	1.54	1.32	1.15	1.04
6	3.31	2.92	2.6	2.29	2.03	1.8	1.56	1.34	1.16	1.04
8	3.2	2.9	2.62	2.34	2.08	1.81	1.57	1.34	1.165	1.04
10	3.23	2.93	2.65	2.27	2.1	1.83	1.58	1.35	1.17	1.04
∞	3.41	3.11	2.815	2.51	2.22	1.93	1.65	1.395	1.19	1.05

Medhurst's proximity resistance factors.

B_{11}=1.33, B_{12}=1.38, B_{21}=1.21, B_{22}=1.22. Using these values in Equations 11.23a and 11.23b.

$$f_1 = \frac{(X_2 - \frac{P}{W_D})}{(X_2 - X_1)}\beta_{11} + \frac{(\frac{P}{W_D} - X_1)}{(X_2 - X_1)}\beta_{21}$$
$$= \frac{(3.333 - 2.5)}{(3.333 - 2.5)}1.33 + \frac{(2.5 - 2.5)}{(3.333 - 2.5)}1.38$$
$$= 1.33$$

$$f_2 = \frac{(X_2 - \frac{P}{W_D})}{(X_2 - X_1)}\beta_{12} + \frac{(\frac{P}{W_D} - X_1)}{(X_2 - X_1)}\beta_{22}$$

$$= \frac{(3.333 - 2.5)}{(3.333 - 2.5)}1.38 + \frac{(2.5 - 2.5)}{(3.333 - 2.5)}1.22$$

$$= 1.38$$

Using Equation 11.22

$$\beta = \frac{(Y_2 - \frac{\ell}{D})}{(Y_2 - Y_1)}f_1 + \frac{(\frac{\ell}{D} - Y_1)}{(Y_2 - Y_1)}f_2$$

$$= \frac{(0.6 - 0.407)}{(0.6 - 0.4)}1.33 + \frac{(0.0407 - 0.4)}{(0.6 - 0.4)}1.38$$

$$= 1.3317$$

And the total coil AC resistance using Equation 11.21 is

$$R_{COIL} = R_{AC}\beta$$

$$= (2.9156)1.3317$$

$$= 3.883$$

Determine the self-resonant frequency

Determine the self-capacitance at resonance using Equation 11.15. The material external to the coil is air and has a relative permittivity of 1. The relative permittivity of the PVC former is 3.0. We will use the diameter of the coil as the diameter of the former and with a wall thickness of 2 mm. Using Equation 11.14, we have

$$\varepsilon_i = \varepsilon_X\left(\frac{r_i}{r_o}\right)^2 + \varepsilon_f\left(1 - \left(\frac{r_i}{r_o}\right)^2\right)$$

$$= 1.0\left(\frac{0.07665}{0.07865}\right)^2 + 3.0\left(1 - \left(\frac{0.07665}{0.07865}\right)^2\right)$$

$$= 1.1$$

And ψ is the pitch angle in radians given by Equation 11.17

$$\psi = \arctan(\frac{p}{\pi D})$$

$$= \arctan(\frac{1.575x10^{-3}}{\pi 157.3x10^{-3}})$$

$$= 0.003187$$

The coefficient K_C is given by Equation 11.16

$$K_C = 0.717439\left(\frac{D}{\ell}\right) + 0.933048\left(\frac{D}{\ell}\right)^{3/2} + 0.106\left(\frac{D}{\ell}\right)^2$$

$$= 0.717439(2.457) + 0.933048(2.457)^{3/2} + 0.106(2.457)^2$$

$$= 5.9961$$

And

$$C_{DAE} = \frac{4\varepsilon_0\varepsilon_X\ell}{\pi}[1 + K_C(\frac{1+\frac{\varepsilon_i}{\varepsilon_X}}{2})]\frac{1}{\cos^2\psi}$$

$$= \frac{4\varepsilon_0 0.0640}{\pi}[1 + 5.9961(\frac{1+\frac{1.1}{1.0}}{2})]\frac{1}{\cos^2(0.003187)}$$

$$= 5.27x10^{-12}$$

The self-resonant frequency is then

$$f_0 = \frac{1}{2\pi\sqrt{300x10^{-6}(5.27x10^{-12})}}$$

$$= 4.0x10^6$$

Determine the dielectric losses

Calculate the axial capacitance given by Equation 11.13 using the ε_i of 1.1 calculated above.

$$C_A = \frac{2D\varepsilon_0(\varepsilon_X + \varepsilon_i)}{\ln(1+\pi^2\frac{\ell}{D})}$$

$$= \frac{2(0.1573)\varepsilon_0(1.0+1.1)}{\ln(1+\pi^2(0.407))}$$

$$= 3.6278x10^{-12}$$

Calculate the loss equivalent resistance $R\varepsilon$

$$F_{ratio} = 1 - \left(\frac{r_i}{r_o}\right)^2$$

$$= 1 - \left(\frac{153.3}{157.3}\right)^2$$

$$= 0.0502$$

And using a loss factor of 0.017 for PVC

$$E_{LOSS} = (4\pi^2 f^2 C_A L)D_F F_{ratio}$$

$$= (4\pi^2(1x10^6)^2(3.6278x10^{-12})(300x10^{-6})(0.017)(0.0502)$$

And

$$E_{LOSS} = 36.67x10^{-6}$$

And

$$R_\varepsilon = X_L E_{LOSS}$$
$$= 1885(36.67x10^{-6})$$

$$R_\varepsilon = 0.06913$$

Finally, the effective Q of the coil at 1 MHz is

$$Q_{eff} = Q_{coil}\alpha$$
$$= \frac{X_L}{R_{coil} + R_\varepsilon}\alpha$$
$$= \frac{1885}{3.883 + 0.06913}\alpha$$
$$= 477.0(\frac{(4.0x10^6)^2 - (1.0x10^6)^2}{(4.0x10^6)^2})$$
$$= 447.0$$

11.12 WORKED EXAMPLE-2

In worked example-1, the diameter of the coil was 157.3 mm. This is quite a large diameter former. Say we decide to wind the coil on a more manageable sized former and choose 90 mm instead. Using the same pitch size, how many turns are required?

Determine the number of turns required

Using Equation 11.9, determine the length of the coil. Where

$$a = \mu\pi D^2$$
$$= \mu_0\pi(90x10^{-3})^2$$
$$= 3.1978x10^{-8}$$
$$b = -4P^2L$$
$$= -4(1.575x10^{-3})^2(300x10^{-6})$$
$$= -2.9768x10^{-9}$$
$$c = -4(0.45)P^2LD$$
$$= -4(0.45)(1.575x10^{-3})^2(300x10^{-6})(90x10^{-3})$$
$$= -1.2056x10^{-10}$$

The coil length is

$$\ell = \frac{-b + \sqrt{b^2 - 4ac}}{2a}$$

$$\ell = \frac{2.9768x10^{-9} + \sqrt{(2.9768x10^{-9})^2 - 4(3.1978x10^{-8})(1.2056x10^{-10})}}{2(3.1978x10^{-8})}$$

And

$$\ell = 0.1236$$

Using Equation 11.8 to determine the number of turns

$$N = \frac{\ell}{P}$$

$$= \frac{0.1236}{1.575x10^{-3}}$$

$$= 78.48$$

Rounding up to whole number turns, set N =78. Recalculate the coil length using Equation 11.7

$$\ell = NP$$

$$= 78(1.575x10^{-3})$$

$$= 0.12285$$

The ℓ/D ratio becomes

$$\frac{\ell}{D} = \frac{0.12285}{90x10^{-3}}$$

$$= 1.365$$

Using this ℓ/D ratio in Equation 11.5 to calculate the Nagaoka coefficient

$$K = \frac{2}{\pi}\left(\frac{\ell}{D}\right)\left[\ln(1 + \frac{\pi D}{2\ell}) + \frac{1}{2.3004 + 3.2219\left(\frac{\ell}{D}\right) + 1.7793\left(\frac{\ell}{D}\right)^2}\right]$$

$$= \frac{2}{\pi}(1.365)\left[\ln(1 + \frac{\pi 0.7326}{2}) + \frac{1}{2.3004 + 3.2219(1.365) + 1.7793(1.365)^2}\right]$$

$$= 0.7522$$

Using this value in Equation 11.6 to calculate the inductance

$$L = \frac{\mu \pi D^2 N^2}{4\ell} K$$

$$= \frac{\mu_0 \pi (90x10^{-3})^2 (78)^2}{4(0.12285)} 0.7522$$

$$= 298x10^{-6}$$

Note the effective Q of this coil is 380, a drop of 15% from the 157.3 mm diameter coil.

12

Mutual Inductance

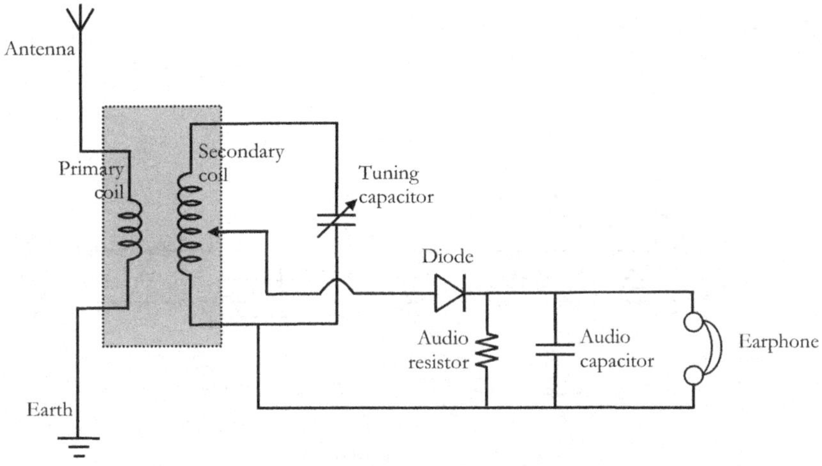

12.1 ELECTRICALLY CONNECTED COILS

Consider two coils connected in series to an alternating power source, as shown in Figure 12.1. The AC current through the coils will generate a time-varying magnetic field around each coil. Each coil will try to counteract the change in magnetic flux surrounding the coil by producing a back emf resulting in a voltage across each coil e_{11} and e_{22}.

$$e_{11} = N_1 \frac{d\phi_1}{dt}$$

And

$$e_{22} = N_2 \frac{d\phi_2}{dt}$$

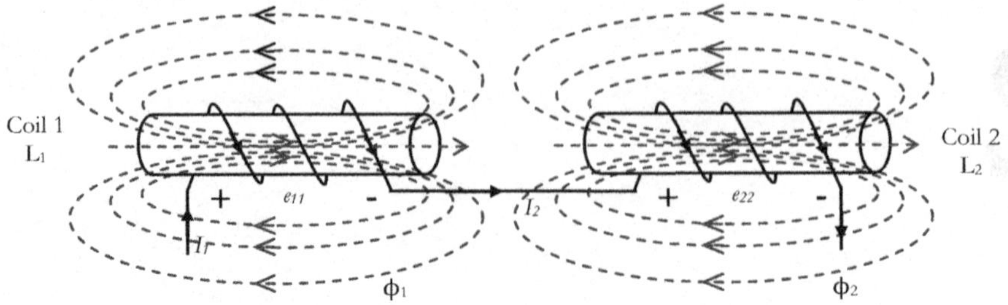

Figure 12.1
Magnetic fields surrounding series connected coils.

However, some of the magnetic flux ϕ_1 around coil 1 will intercept (cut) coil 2, as shown in Figure 12.2. The magnetic flux cutting coil 2 coming from coil 1 is given by ϕ_{21}. Coil 2 will now try to resist this change in flux coming from coil 1 by creating another back emf. This back emf will result in another voltage e_{21} across coil 2 and is proportional to the change in the magnetic flux coming from coil 1. The emf across coil 2, due to the changing magnetic field of coil 1, is

$$e_{21} = N_2 \frac{d\phi_{21}}{dt}$$

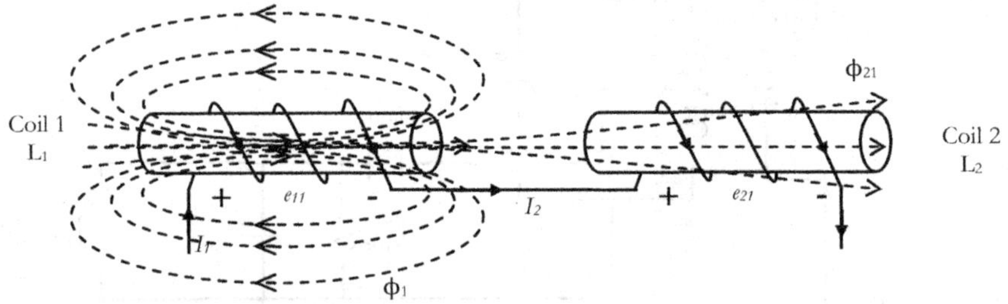

Figure 12.2
Magnetic field of coil 1 cutting coil 2.

We now have two components of voltage across coil 2; one due to the change in flux resulting from the current running through the coil itself, self-inductance, and another due to the change in magnetic flux cutting coil 2 from coil 1. The total voltage across coil 2 is the sum of both these voltage components.

$$e_2 = N_2 \frac{d\phi_2}{dt} + N_2 \frac{d\phi_{21}}{dt} \qquad 12.1$$

Similarly, for coil 1, the change in magnetic flux coming from coil 2 cutting coil 1 will result in an induced voltage, see Figure 12.3. And the induced voltage in coil 1 due to coil 2 is

$$e_{12} = N_1 \frac{d\phi_{12}}{dt}$$

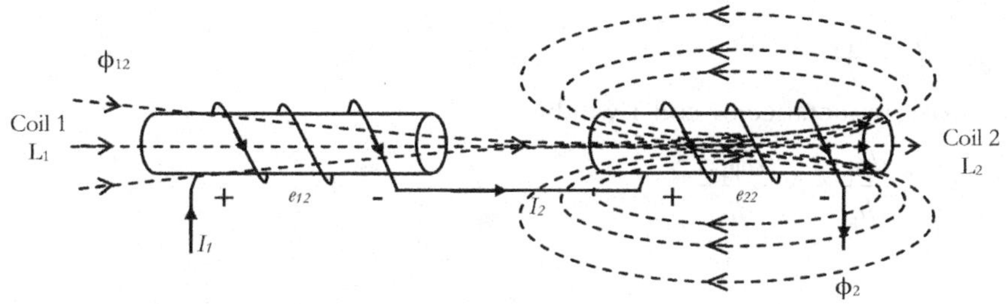

Figure 12.3
Magnetic field of coil 2 cutting coil 1.

And the total voltage across coil 1 will be

$$e_1 = N_1 \frac{d\phi_1}{dt} + N_1 \frac{d\phi_{12}}{dt} \qquad 12.2$$

Looking at the magnetic field interactions in terms of the coil's inductance values, the self-inductance of a coil is proportional to the change in flux due to the change in current that caused the flux change (Equation 5.16). That is, the self-inductance for coil 1 is

$$L_1 = N_1 \frac{d\phi_1}{di_1}$$

And for coil 2

$$L_2 = N_2 \frac{d\phi_2}{di_2}$$

Also, the change in inductance of coil 2 due to the change in the magnetic field of coil 1 is,

$$M_{21} = N_2 \frac{d\phi_{21}}{di_1}$$

And the change in inductance of coil 1 due to the change in the magnetic field of coil 2 is,

$$M_{12} = N_1 \frac{d\phi_{12}}{di_2}$$

This change in the inductance of one coil due to the change in the magnetic field of the other coil is called ***mutual inductance***.

The mutual inductance between two coils is proportional to the instantaneous change in magnetic flux linking one coil due to the instantaneous change in current through the other coil.

Mutual inductance thus changes the total inductance of a coil. The total inductance of coil 2 is now

$$L_{Total2} = N_2 \frac{d\phi_2}{di_2} + N_2 \frac{d\phi_{21}}{di_1}$$

And

$$L_{Total2} = L_2 + M_{21}$$

Similarly, the total inductance of coil 1 will be

$$L_{Total1} = N_1 \frac{d\phi_1}{di_1} + N_1 \frac{d\phi_{12}}{di_2}$$

And

$$L_{Total1} = L_1 + M_{12}$$

In this instance, the magnetic flux ϕ_1 and ϕ_2 are in phase, and the back emf will also be in phase. Therefore the mutual inductance increases the overall inductance of both coils. If the magnetic flux created by the current is 180° out of phase with the magnetic flux coming from the other coil, as shown in Figure 12.4, then the mutual inductance will be subtracted from the total value.

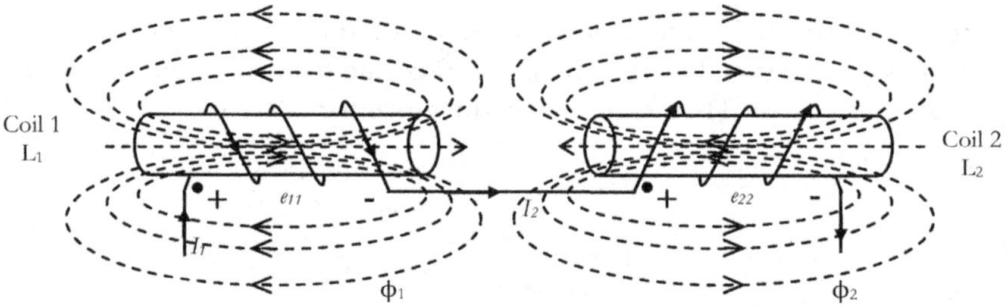

Figure 12.4
Magnetic fields surrounding series connected contra-wound coils are 180° out of phase.

Then

$$L_{Total2} = L_2 - M_{21}$$

And

$$L_{Total1} = L_1 - M_{12}$$

Coil Winding Direction

In Figure 12.4, the winding direction used in coil 1 differs from coil 2, resulting in a different direction of the magnetic field. If we view the direction of the windings facing from one end of the coil (which end does not matter due to symmetry), we can assign a winding direction to a coil. Place an arrow on the lead pointing towards you, then trace the winding back, as shown in Figure 12.5. The winding direction will either be clockwise or anti-clockwise. *The winding direction of a coil is intrinsic to the coil and does not depend on its orientation.*

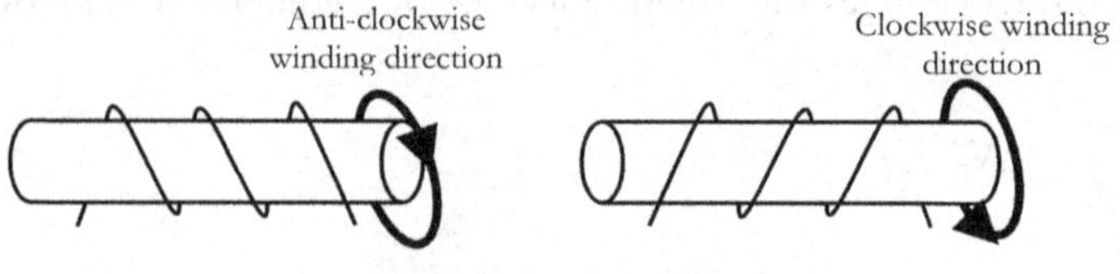

Figure 12.5
Coils will have a clockwise or anti-clockwise winding direction.

Mutual Inductance Reciprocity

The voltage across a coil can be expressed as a function of the coil's inductance and current. Using Equation 12.2, we can write

$$e_1 = N_1 \frac{d\phi_1}{dt} + N_1 \frac{d\phi_{12}}{dt}$$
$$= N_1 \frac{d\phi_1}{di_1}\frac{di_1}{dt} + N_1 \frac{d\phi_{12}}{di_2}\frac{di_2}{dt}$$

Since inductance is,

$$L = N\frac{d\phi}{di}$$

we can write

$$e_1 = L_1 \frac{di_1}{dt} + M_{12}\frac{di_2}{dt}$$

And similarly, for coil 2

$$e_2 = L_2 \frac{di_2}{dt} + M_{21}\frac{di_1}{dt}$$

Due to the conservation of energy, the sum of the total energy contained in both coils must be conserved. Mutual inductance can be thought of as coil 1 giving up some of its energy to coil 2 and coil 2 giving up some of its energy to coil 1. The total energy given to coil 2 from coil 1 may be different to the energy given to coil 1 from coil 2. However, as inductance is proportional to the rate of change (not the magnitude of the change), the mutual inductance on coil 2 due to the current in coil 1 (M_{21}), and the mutual inductance on coil 1 due to the change in current in coil 2 (M_{12}) are the same. That is,

$$M_{12} = M_{21}$$ Mutual inductance **reciprocity**

This is called the mutual inductance ***reciprocity theorem***.

Total Coil Inductance

The total inductance of a coil due to in-phase mutual inductance is,

$$\boxed{L_T = L + M}$$ Total inductance of a coil due to in-phase mutual inductance (same winding direction) 12.3

The total inductance of a coil due to 180° out of phase mutual inductance is,

$$\boxed{L_T = L - M}$$ Total inductance of a coil due to 180° out of phase mutual inductance (contra winding direction) 12.4

The total circuit inductance for the series connected coils with the same winding direction is

$$\boxed{L_T = L_1 + L_2 + 2M}$$ Total inductance of coils in series due to in-phase mutual inductance (same winding direction) 12.5

And the total circuit inductance for the series connected coils with opposite (or contra) winding directions is

$$\boxed{L_T = L_1 + L_2 - 2M}$$ Total inductance of coils in series due to 180° out of phase mutual inductance (same contra winding direction) 12.6

Figure 12.6 shows different series connected coil configurations giving additive and subtractive mutual inductance values. Figures 12.6a and 12.6b show two coils with the same winding direction, with configuration 12.6a giving additive mutual inductance and 12.6b giving subtractive mutual inductance. In Figures 12.6c and 12.6d, the coils have different winding directions (contra-wound coils), with 12.6c giving subtractive mutual inductance and 12.6d giving additive mutual inductance. Using these configurations and Equations 12.5 and 12.6, you can measure the mutual inductance between two coils. See the Chapter on measuring techniques.

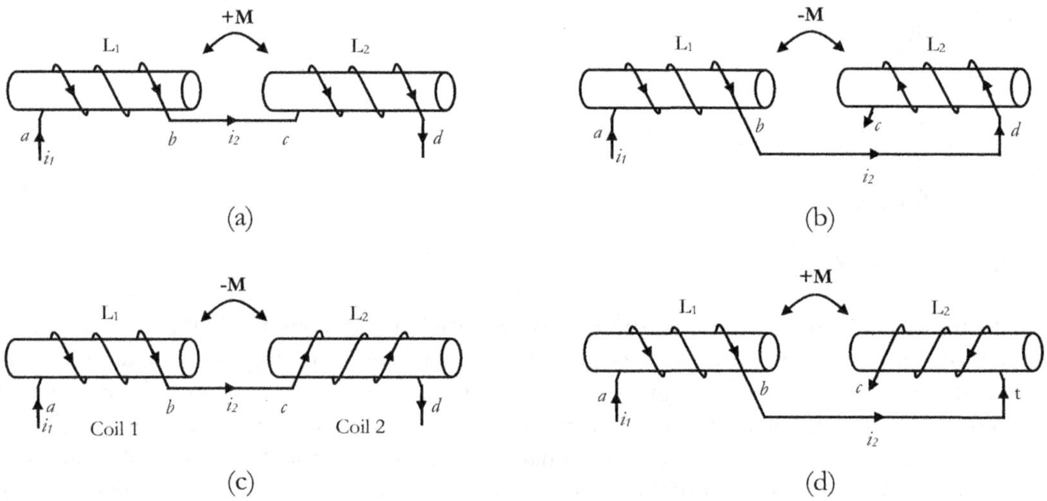

Figure 12.6
Circuit configurations giving additive and subtractive mutual inductance.

12.2 CALCULATING MUTUAL INDUCTANCE

In this section, we will derive two methods of calculating mutual inductance between coils. I have included both, as the first method is based on calculations that you might come across in other books or lectures and requires some assumptions to be true. The second method is similar but a little more complicated but does not rely on assumptions. We will compare both methods to see which gives the best results.

Using Single Magnetic Field Strength

We will first derive the common Equation given for the mutual inductance between two coils. Consider the shape of the magnetic field in and around the coils shown in Figure 12.1. All the magnetic field lines outside a coil exist inside a coil. The area outside the coil is infinite, and the area inside is finite and small. Therefore the magnetic flux density **B** is greatest inside the coil and is the overwhelming cause of the mutual inductance. For the following derivation, two assumptions are made:

1. The magnetic field strength responsible for the mutual inductance is contained inside the coil.
2. The magnetic field strength is the same throughout the cross-sectional area of the coil.

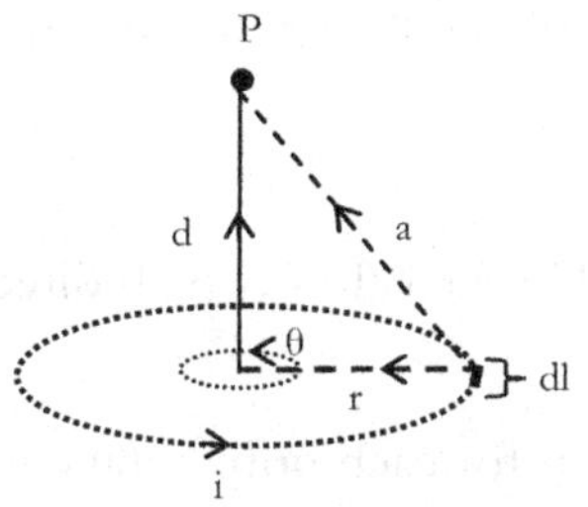

Figure 12.7
The magnetic field at point P.

Given a single loop of wire, as shown in Figure 12.7, we will use the Biot-Savart law to determine the magnetic field strength at point P at some distance d from the centre of the coil.

$$\mathbf{B} = \frac{\mu_0 i}{4\pi}\oint \frac{dl \times \hat{a}}{|a|^2}$$

where $|a|^2$ is the distance between the circumference of the loop to point P and is given by,

$$|a|^2 = r^2 + d^2 \qquad 12.7$$

And **â** is the unit vector pointing along vector **a**. Using the circumference as the origin

$$\hat{a} = \frac{-r.\hat{r} + d.\hat{d}}{\sqrt{r^2 + d^2}} \qquad 12.8$$

Substituting 12.7 and 12.8 into Biot-Savart, we get

$$B = \frac{\mu_0 i}{4\pi}\oint \frac{dl \times (-r.\hat{r} + d.\hat{d})}{(r^2 + d^2)\sqrt{r^2 + d^2}}$$

As dl is some small section of the circumference, we can express this in terms of angle θ as

$$dl = d\theta(r.\hat{\theta})$$

And we can write,

$$B = \frac{\mu_0 i}{4\pi}\oint \frac{d\theta(r.\hat{\theta}) \times (-r.\hat{r} + d.\hat{d})}{(r^2 + d^2)^{3/2}}$$

Now

$\hat{\theta} \times \hat{r} = -\hat{d}$ and

$\hat{\theta} \times \hat{d} = \hat{r}$

So

$$B = \frac{\mu_0 i}{4\pi} \oint \frac{d\theta(r^2\hat{d}) + d\theta(r.d.\hat{r})}{(r^2 + d^2)^{3/2}}$$

The unit vector $\hat{d}$ is the component of the magnetic field in the d (vertical) direction, and in the above Equation, is given by the term

$$d\theta(r^2\hat{d})$$

The component of the magnetic field in the r (horizontal) direction is given by

$$d\theta(r.d.\hat{r})$$

Note that as we move around the loop for each horizontal component of the magnetic field, there will be another horizontal component 180^0 apart from one another, and they will cancel out. Therefore the entire magnetic field in the horizontal direction is zero.

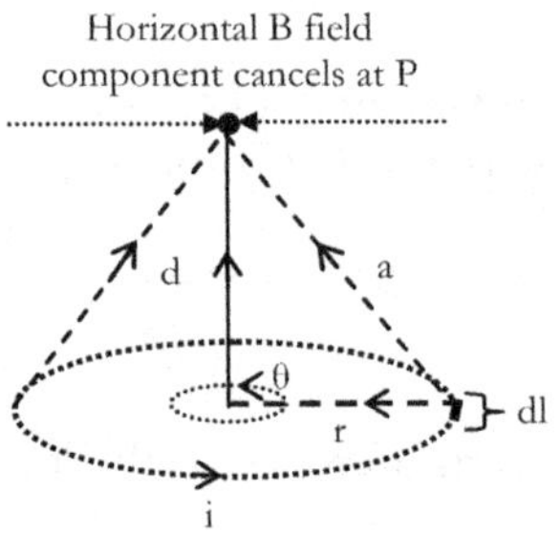

Figure 12.8
The horizontal components of the magnetic field cancel at the centre of the loop.

Now we can write

$$B = \frac{\mu_0 i}{4\pi} \oint \frac{d\theta(r^2)}{(r^2 + d^2)^{3/2}} \qquad 12.9$$

Integrating from 0 to 2π (the entire circular loop), we get

$$B = \frac{\mu_0 i(r^2)}{4\pi(r^2 + d^2)^{3/2}} \int_0^{2\pi} d\theta$$

Solving,

$$B = \frac{\mu_0 i(r^2)2\pi}{4\pi(r^2 + d^2)^{3/2}}$$

Finally, we can write the magnetic field density as

$$B = \frac{\mu_0 i(\pi r^2)}{2\pi(r^2 + d^2)^{3/2}} \qquad \text{Magnetic field density at a distance d from a loop of wire} \qquad 12.10$$

As πr^2 is the cross-sectional area of the loop, then

$$B = \frac{\mu_0 iA}{2\pi(r^2 + d^2)^{3/2}} \qquad 12.11$$

For a system of two coils, each consisting of a single loop, the magnetic flux cutting coil 2 from coil 1 can be written as

$$\phi_{21} = B_{21} \bullet A_2$$

$$= \frac{\mu_0 i_1 A_1 A_2}{2\pi(r_1^2 + d^2)^{3/2}}$$

And mutual inductance between the loops is

$$M_{21} = \frac{d\phi_{21}}{di_1} = \frac{\mu_0 A_1 A_2}{2\pi(r_1^2 + d^2)^{3/2}}$$

If we have two coils with multiple loops, where N_1 is the number of loops in coil 1, and N_2 is the number of loops in coil 2. To get the mutual inductance between the coils, we add the contributions of each of the loops in coil 1 has on the loops in coil 2. However, the distance between the loops is not constant, so we will select the distance d as being the distance between the centres of the two coils. If the two coils have different diameters, select the smallest radius coil for r. And by the reciprocity theorem, the mutual inductance between two coils becomes

$$\boxed{M = \frac{\mu_0 N_1 N_2 A_1 A_2}{2\pi(r^2 + d^2)^{3/2}}} \qquad \text{Mutual inductance between two coils} \qquad 12.12$$

Using Multiple Magnetic Field Strengths

The above method for calculating mutual inductance assumes the magnetic field is the same throughout the coil. And that only the magnetic field at the centre of the coil needs to be determined and then applied to all the loops. This may be a reasonable assumption for small-diameter coils using Ferrite cores. However, Figure 12.9 shows a more realistic depiction of the shape of the flux lines passing through the secondary. The larger the secondary is, the greater the angle of the flux lines passing through the coil's cross-sectional area. Also, the larger the primary coil is, the greater the disparity between the field strength at the centre than it is towards the circumference of the loop.

We can improve the mutual inductance calculation by using different magnetic field strengths along the radius of the primary coil and applying these to the secondary. In Figure 12.10, point P is no longer situated along the central axis of the loop but is some distance r from the circumference and is situated directly above the X-axis. Point P intercepts the X-axis at point B. Let point A be some arbitrary location on the circumference of the loop, with R being the radius of the loop. The angle θ is the angle between points B and A. If we fix points P and B, and allow A to rotate around the Z-axis (the angle theta will range from 0 to 360 degrees), we can calculate the magnetic field strength at P due to the current in the loop. Due to symmetry, the total magnetic field strength calculated at P will be the same for all points r distant from the circumference of the loop. From the diagram, we get the following relationships.

$$a_x = R\cos\theta$$

$$a_y = R\sin\theta$$

$$b_x = R - r$$

$$\mathbf{a} = (a_x - b_x)(-\hat{i}) + a_y(-\hat{j}) + d(\hat{k})$$

$$|\mathbf{a}| = \sqrt{(a_x - b_x)^2 + a_y^{\,2} + d^2}$$

$$\hat{a} = \frac{\mathbf{a}}{|\mathbf{a}|}$$

$$= \frac{(a_x - b_x)(-\hat{i}) + a_y(-\hat{j}) + d(\hat{k})}{\sqrt{(a_x - b_x)^2 + a_y^{\,2} + d^2}}$$

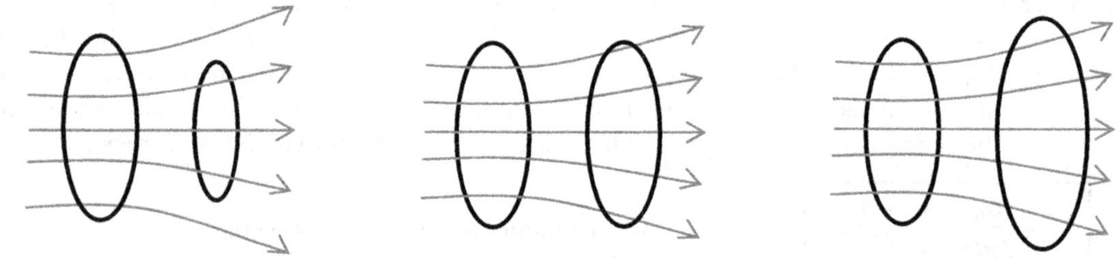

Primary diameter larger than the secondary. Primary diameter the same as the secondary. Primary diameter smaller than the secondary.

Figure 12.9

The small incremental change dl on the circumference of the loop can be expressed in cylindrical coordinates as

$$dl = R.d\theta.\hat{\theta}$$

$$\hat{\theta} = -\sin\theta\hat{i} + \cos\theta\hat{j}$$

Substituting into Biot-Savart

$$\mathbf{B} = \frac{\mu_0}{4\pi}\oint\frac{id\bar{l}\times\hat{a}}{|a|^2}$$

we get

$$\mathbf{B} = \frac{\mu_0 iR}{4\pi}\oint\frac{d\theta.\hat{\theta}\times[(a_x - b_x)(-\hat{i}) + a_y(-\hat{j}) + d(\hat{k})]}{((a_x - b_x)^2 + a_y^{\,2} + d^2)^{3/2}}$$

$$= \frac{\mu_0 iR}{4\pi}\oint\frac{d\theta[(a_x - b_x)(\hat{\theta}\times(-\hat{i})) + a_y(\hat{\theta}\times(-\hat{j})) + d(\hat{\theta}\times\hat{k})]}{((a_x - b_x)^2 + a_y^{\,2} + d^2)^{3/2}} \quad 12.13$$

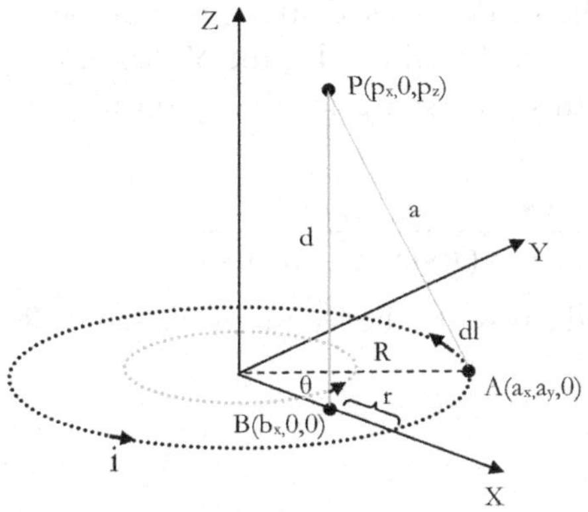

Figure 12.10
Magnetic field strength at point P.

The cross-product of $\hat{\theta}$ with each of the unit vectors is given below.

$$\hat{\theta}\times\hat{i} = -\sin\theta(\hat{i}\times\hat{i}) + \cos\theta(\hat{j}\times\hat{i})$$
$$= \cos\theta(-\hat{k})$$
$$\hat{\theta}\times\hat{j} = -\sin\theta(\hat{i}\times\hat{j}) + \cos\theta(\hat{j}\times\hat{j})$$
$$= -\sin\theta(\hat{k})$$
$$\hat{\theta}\times\hat{k} = -\sin\theta(\hat{i}\times\hat{k}) + \cos\theta(\hat{j}\times\hat{k})$$
$$= -\sin\theta(-\hat{j}) + \cos\theta(\hat{i})$$

Substituting these values into Equation 12.13, we have

$$\mathbf{B} = \frac{\mu_0 iR}{4\pi}\oint\frac{d\theta[(a_x - b_x)\cos\theta\hat{k} + a_y(\sin\theta\hat{k}) + d(\sin\theta(\hat{j}) + \cos\theta(\hat{i}))]}{((a_x - b_x)^2 + a_y{}^2 + d^2)^{3/2}}$$

$$= \frac{\mu_0 iR}{4\pi}\oint\frac{d\theta[(R\cos\theta - R + r')\cos\theta\hat{k} + R\sin\theta(\sin\theta\hat{k}) + d(\sin\theta(\hat{j}) + \cos\theta(\hat{i}))]}{((R\cos\theta - R + r)^2 + (R\sin\theta)^2 + d^2)^{3/2}}$$

$$= \frac{\mu_0 iR}{4\pi}\oint\frac{d\theta[R\cos^2\theta\hat{k} - R\cos\theta\hat{k} + r'\cos\theta\hat{k} + R\sin^2\theta\hat{k} + d\sin\theta(\hat{j}) + d\cos\theta(\hat{i})]}{((R\cos\theta - R + r)^2 + (R\sin\theta)^2 + d^2)^{3/2}}$$

$$= \frac{\mu_0 iR}{4\pi}\oint\frac{d\theta[R(\cos^2\theta + \sin^2\theta)\hat{k} - R\cos\theta\hat{k} + r'\cos\theta\hat{k} + d\sin\theta(\hat{j}) + d\cos\theta(\hat{i})]}{((R\cos\theta - R + r)^2 + (R\sin\theta)^2 + d^2)^{3/2}}$$

Finally, we arrive at,

$$\mathbf{B} = \frac{\mu_0 iR}{4\pi}\oint\frac{d\theta[(R - R\cos\theta + r\cos\theta(\hat{k}) + d\sin\theta(\hat{j}) + d\cos\theta(\hat{i})]}{((R\cos\theta - R + r)^2 + (R\sin\theta)^2 + d^2)^{3/2}} \qquad 12.14$$

This Equation will give us the magnetic field strength with the magnetic field vector split into its constituent components in the X direction ($\hat{\imath}$), the Y direction ($\hat{\jmath}$) and the Z direction ($\hat{k}$). We are only interested (on this occasion) in the vertical $\hat{k}$ direction component that is perpendicular to the loop. So,

$$\mathbf{B} = \frac{\mu_0 iR}{4\pi} \oint \frac{[(R - R\cos\theta + r\cos\theta)]d\theta}{((R\cos\theta - R + r)^2 + (R\sin\theta)^2 + d^2)^{3/2}} \qquad 12.15$$

Note when θ is zero degrees, and r is set to the length of the radius R, we get

$$\cos(0) = 1$$

$$r = R$$

and

$$\mathbf{B} = \frac{\mu_0 i}{4\pi} \oint \frac{R^2 d\theta}{(R^2 + d^2)^{3/2}}$$

which is the same as Equation 12.9.

The equation for the mutual inductance contribution of loop 1 to loop 2 at r distance from the circumference of loop1, and at a separation distance of d is

$$\frac{\phi_{21}}{di_1} = M_{21} = [\frac{\mu_0 R}{4\pi} \oint \frac{[(R - R\cos\theta + r\cos\theta)]d\theta}{((R\cos\theta - R + r)^2 + (R\sin\theta)^2 + d^2)^{3/2}}]A_2 \qquad 12.16$$

To get the total mutual inductance, we need to calculate and sum the mutual inductance for circumference distances ranging from r=0 to r=R. Numerically we can split r into some fraction of the radius R, say r=R/10. Using summation notation, we can write this as

$$M_{21} = \frac{\mu_0 R}{4\pi} \sum_{i=1}^{N_r} A_i \int \frac{[(R - R\cos\theta + r_i\cos\theta)]d\theta}{((R\cos\theta - R + r_i)^2 + (R\sin\theta)^2 + d^2)^{3/2}}$$

With the calculated magnetic field applied to some cross-sectional area A_i of loop 2. We now need to deal with the integral. In the previous section, the integral was simple and solvable. However, this integral is somewhat more difficult, and we can use numerical methods to solve it. If we divide θ into some fraction of 2π, ranging from 0 to 2π we can write,

$$M_{21} = \frac{\mu_0 R}{4\pi} \sum_{i=1}^{N_r} A_i \sum_{j=1}^{N_\theta} \frac{[(R - R\cos\theta_j + r_i\cos\theta_j)]\Delta\theta}{((R\cos\theta_j - R + r_i)^2 + (R\sin\theta_j)^2 + d^2)^{3/2}}$$

where $\Delta\theta = 2\pi/N_\theta$. If we have two coils with multiple loops, where N_1 is the number of loops in coil 1, and N_2 is the number of loops in coil 2. To get the mutual inductance between the coils, we add the contributions that each of the loops in coil 1 has on all the loops in coil 2, and we can write

$$\boxed{M = \frac{\mu_0 R}{4\pi} \sum_{m=1}^{N_1} \sum_{n=1}^{N_2} \sum_{i=1}^{N_r} A_i \sum_{j=1}^{N_\theta} \frac{[(R - R\cos\theta_j + r_i\cos\theta_j)]\Delta\theta}{((R\cos\theta_j - R + r_i)^2 + (R\sin\theta_j)^2 + d_{m,n}^2)^{3/2}}} \qquad 12.17$$

To solve, first split the circle into small slices Δθ. Set θ to zero and calculate the change in magnetic flux for the current value of θ. Next, add Δθ to θ to get the new value of θ and calculate the change in magnetic flux for this value of θ. Continue until θ reaches 360° (2π). We now have the vertical component of the magnetic field, at a distance d from the primary coil, at displacement r from the edge of the primary. That is, we have the magnetic field at point P within a loop of the secondary coil. We now apply this value to some portion of the area of loop 2 to get a small portion of the total mutual inductance. We then change the distance from the circumference r_i and redo the calculations until r_i is equal to the radius of the loop.

Figure 12.11 should help to visualise how Equation 12.17 works. In Figure 12.11, the magnetic field density B_1 calculated at P1, is on the edge of loop 1 of coil 2 and is not applied; that is, the area A_1 is zero. The field B_2 at P2 is applied to area A_2, and B_3 is applied to area A_3 etc. In this instance, we have subdivided a loop into five regions, so i ranges from 1 to 5.

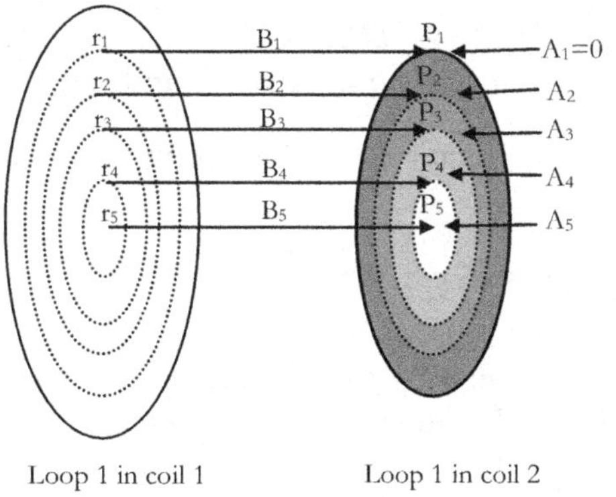

Figure 12.11
Mutual inductance between two loops of different diameters.

The sum of the magnetic fluxes on loop 1 of coil 2 is the mutual inductance between loop 1 of the primary coil and loop 1 of the secondary coil. The index n is increased by one, so n=2, which is the second loop in coil 2 and the process is repeated until all the loops in the secondary coil have been processed. Then the primary coil index is increased by 1, so m =2 and the process is repeated until all the loop combinations have been calculated and summed.

Comparison of Methods

Table 12.1 lists the coils used to compare the calculated mutual inductances using Equations 12.12 and 12.17 with measured mutual inductance values.

Coil	Measured Inductance (uH)	Diameter (mm)	Length (mm)	Number of turns	Pitch (mm)
PC2	27.26	90	8	12	0.67
SC6	302.5	75	38	62	0.61
PC3	33.3	103	8	12	0.67
SC1	332.0	103	28	48	0.58
SC7	315.0	75	35	62	0.56

Table 12.1
Coils used in mutual inductance measurements and calculations.

Table 12.2 compares the two methods for calculating mutual inductance between coils. A $\Delta\theta$ of 3.6° (j =1to100) was used in Equation 12.17, with the radius of the primary loop split into five sections (i=1to5). The distance between the coils was set at 50 mm for each test.

Primary	Secondary	Mutual inductance measured (uH)	Eq. 12.12		Eq. 12.17	
			M	% error	M	% error
PC2(m=1to12)	SC6(n=1to60)	5.98	6.20	3.68	6.01	0.50
PC3(m=1to12)	SC1(n=1to48)	9.49	12.10	27.50	10.09	6.32
SC7(m=1to62)	SC1(n=1to48)	21.55	29.10	35.03	21.52	0.14

Table 12.2
Mutual inductance measurements and calculations.

From the % error, we can see that Equation 12.17 is the more accurate of the two methods; and will be used in this book.

12.3 THE COUPLING COEFFICIENT

The proportion of the magnetic flux cutting the secondary coil from the primary coil is called the ***coupling coefficient***. This ratio is also the same for both coils and will always be between 0 and 1. It is typically denoted by the letter k.

$$k = \frac{\phi_{SP}}{\phi_P} = \frac{\phi_{PS}}{\phi_S}$$

Writing the mutual inductance due to the magnetic flux from the primary cutting the secondary, we can write

$$M_{SP} = N_S \frac{d\phi_{SP}}{di_P}$$

$$= N_S k \frac{d\phi_P}{di_P}$$

And for the flux cutting the primary from the secondary,

$$M_{PS} = N_P \frac{d\phi_{PS}}{di_S}$$

$$= N_P k \frac{d\phi_S}{di_S}$$

Multiplying M_{SP} by M_{PS}, we get

$$M_{SP} M_{PS} = (N_S k \frac{d\phi_P}{di_P})(N_P k \frac{d\phi_S}{di_S})$$

$$M^2 = k^2 (N_P \frac{d\phi_P}{di_P})(N_S \frac{d\phi_S}{di_S})$$

$$M^2 = k^2 (L_P L_S)$$

And

$$M = k\sqrt{L_P L_S}$$

If we know the self and mutual inductances, we can calculate the coupling coefficient using,

$$\boxed{k = \frac{M}{\sqrt{L_P L_S}}} \qquad \text{Coupling coefficient} \qquad 12.18$$

12.4 MAGNETICALLY COUPLED COILS

Coils not electrically connected rely on mutual inductance to propagate a signal from one section of the circuit to another. Typically one coil is connected to a power supply called the primary coil. The signal from the primary coil, by the action of magnetic induction, induces a signal in the other coil. The coil with the induced signal is called the secondary coil.

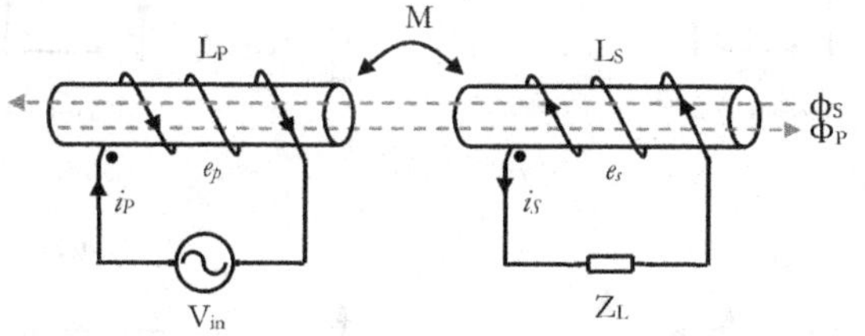

Figure 12.12
Magnetically coupled coils.

Consider two coils L_P (the primary coil) and L_S (the secondary coil), with the primary coil connected to an alternating power source, as shown in Figure 12.12. The voltage induced across the secondary coil L_S due to mutual inductance is given by Faraday's law,

$$e_{SP} = M\frac{di_P}{dt}$$

The emf across the secondary coil, due to the mutual inductance, acts like a voltage source that powers the secondary coil's circuit. Using the circuit in Figure 12.12, we will examine the voltage and current phase relationship between the primary and secondary coil. Let the current across the primary coil be given by

$$i_p = I_P \cos(\omega t)$$

The voltage drop across an inductor leads the current by 90°, see Figure 7.8, so we can write the voltage across the primary coil as

$$\begin{aligned} v_p &= V_P \cos(\omega t + \frac{\pi}{2}) \\ &= -V_P \sin(\omega t) \end{aligned}$$

Using Faraday's law, the voltage induced in the secondary coil is

$$e_{SP} = M\frac{di_P}{dt}$$

$$e_{SP} = M\frac{d(I_P \cos(\omega t))}{dt}$$
$$= \omega M(-I_P \sin(\omega t)))$$

And

$$e_{SP} = -X_M I_P \sin(\omega t)$$

Note the voltage across the secondary coil is in phase with the primary coil's voltage.

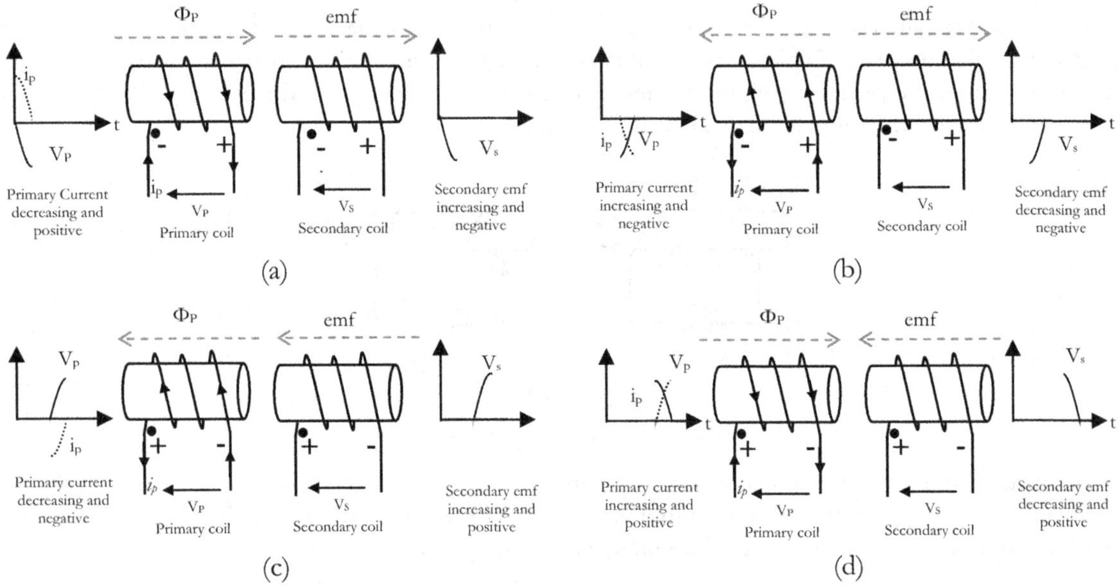

Figure 12.13
Open circuit voltage phase relationship for coils with the same winding direction.

Viewing this relationship in terms of Lentz's law, we will examine the voltage relationship with an open circuit (no load) secondary. In Figure 12.13, we have two magnetically coupled coils with the same clockwise winding direction. In Figure 12.13a, the current starts high across the primary and then drops in magnitude, and the magnetic field around the primary coil will follow suit. The magnetic field cutting the secondary coil likewise starts high and drops in magnitude. The secondary coil wants the magnetic field to be constant and will counteract this magnetic field drop by creating an emf in the same direction as the magnetic field. In 12.13b, the primary current is negative and increasing; the secondary coil will try and stop this increase by creating an emf in the opposite direction. In 12.13c, the primary current is negative and decreasing, and the secondary will generate an emf in the same direction to stop the magnetic field from decreasing. Finally, in Figure 12.13d, the primary current and magnetic field is increasing in the positive direction, and the secondary will generate an emf in the opposite direction to prevent the increase.

Figure 12.14 shows an oscilloscope image of the voltage phase relationship between the voltage across the primary coil and the open circuit voltage across the secondary coil for coils wound in the same direction.

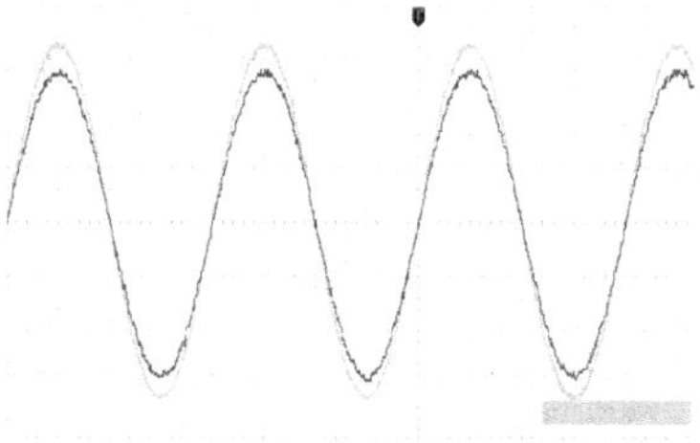

Figure 12.14
Voltage phase relationship between the primary and open-circuit secondary.

In Figure 12.15, the secondary coil is replaced with a counter-clockwise wound coil. The same current and voltage will appear across the primary as in Figure 12.13, and the emf produced in the secondary will also be in the same direction as before. However, due to the different winding direction of the secondary coil, the current will flow in a different direction, and the polarity of the voltage across the secondary with respect to the primary will change by 180°. Note the different locations of the dot.

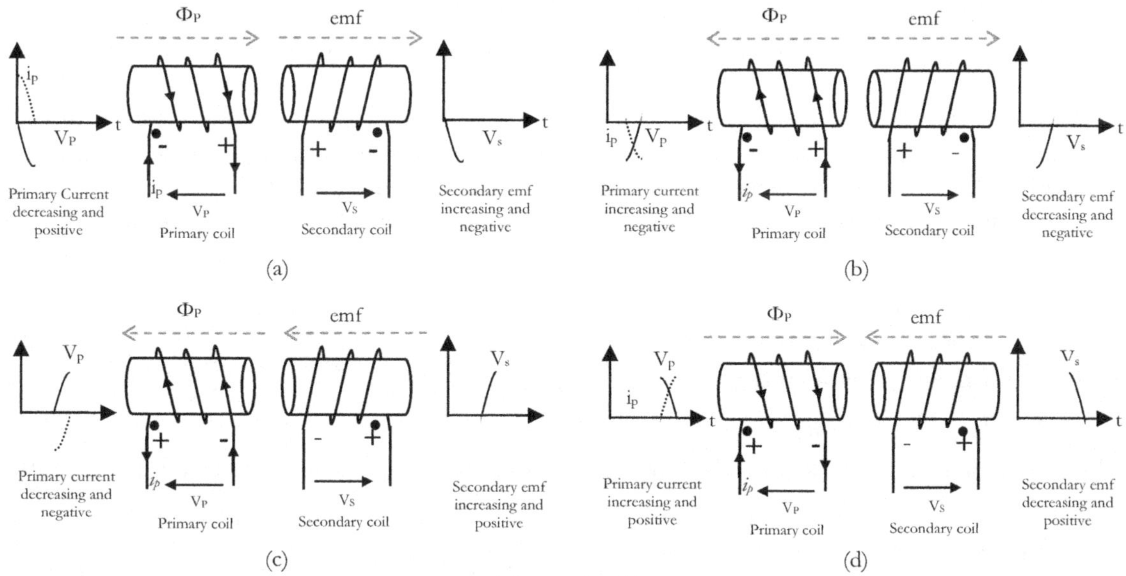

Figure 12.15
Open circuit voltage phase relationship for coils with contra-winding direction.

The Dot Convention

In each of the diagrams in Figures 12.13 and 12.15, there is a dot placed next to a coil's lead. This dot represents the voltage phase relationship between the voltage across the primary and the open circuit voltage across the secondary coil. If the primary coil voltage is measured with the positive lead of an oscilloscope probe placed at the dot with the negative lead connected to the other end of the coil, and another probe is set up in the same configuration across the secondary coil, both the primary and secondary voltages would be in phase. That is, the dot represents the position on the coil where both the primary and secondary voltages are in phase.

Furthermore, if you know the current direction flowing into and out of the coils, you can use the dot to determine whether the mutual inductance is additive or subtractive. The rule for determining if mutual inductance is additive or subtractive is:

> *If the current through both of the coils leaves the dot or both currents flow into the dot, as the currents move across the coils, then the mutual inductance is additive. Otherwise, the mutual inductance is subtractive.*

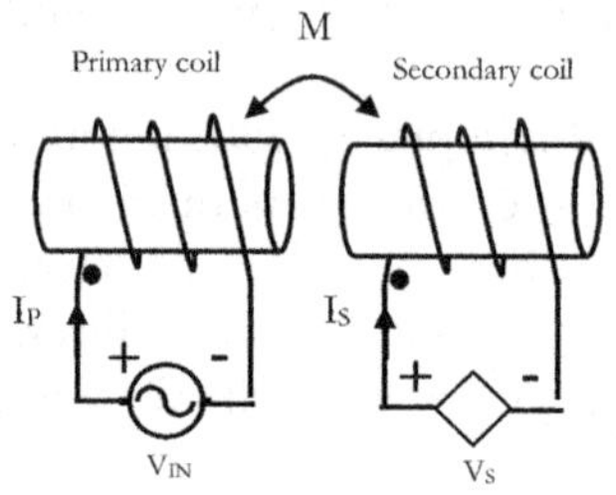

Figure 12.16
Circuit diagram of magnetically coupled coils.

For two magnetically coupled coils, as shown in Figure 12.16, both the primary and secondary currents enter the dot and then travel across the coils. Therefore the mutual inductance is treated as additive. Let us examine what it means to say that the mutual inductance is additive or subtractive in a circuit. We know that the total voltage across an ideal coil magnetically coupled with a second coil is

$$e_1 = L_1 \frac{di_1}{dt} + M \frac{di_2}{dt}$$

In phasor form, we can write this as

$$\mathbf{V}_1 = \mathbf{I}_1 jX_{L1} + \mathbf{I}_2 jX_M$$

When the current through both coils is the same, as is the case of series connected coils, we can write

$$\begin{aligned}\mathbf{V}_1 &= \mathbf{I}_1 (jX_{L1} + jX_M) \\ &= \mathbf{I}_1 (j2\pi f L_1 + j2\pi f X_M) \\ &= \mathbf{I}_1 j2\pi f (L_1 + M)\end{aligned}$$

And the total inductance is the sum of the self-inductance and mutual inductance. When the current across the second coil travels in the opposite direction with respect to the first coil, the current will be 180° out of phase, and we can write

$$\begin{aligned}\mathbf{V}_1 &= (\mathbf{I}_1 jX_{L1} - \mathbf{I}_1 jX_M) \\ &= \mathbf{I}_1 (jX_{L1} - jX_M) \\ &= \mathbf{I}_1 j2\pi f (L_1 - M)\end{aligned}$$

And the mutual inductance is subtractive. However, when the two currents have some other phase relationship (not 0° or 180°), then the total inductance, or the inductance the circuit sees, will not simply be the addition or subtraction of the mutual inductance but will have a complex phasor relationship. This is the case in air-core magnetically coupled transformers used in crystal radios. We will examine this further when dealing with air-core transformers.

13

The Crystal Radio Transformer

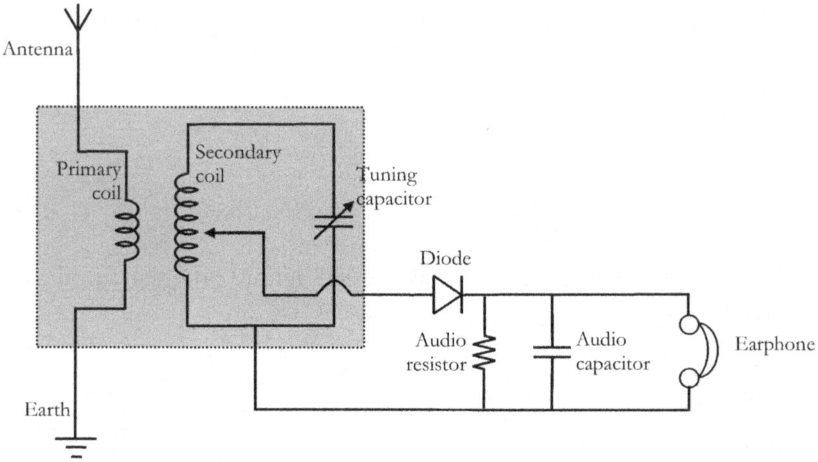

13.1 THE IDEAL TRANSFORMER

Transformers are devices that transfer power from one coil to another via mutual inductance. Figure 13.1 shows the circuit diagram for an iron-core transformer. The two coils are electrically isolated from one another, and the magnetic flux is mostly contained inside the core material, which can be made from iron or ferrite. The material and design of the core ensure a high degree of efficiency in transferring power via the magnetic field. An ***ideal transformer*** has a theoretical efficiency of 100%. Although impossible to obtain due to core and wiring loss, well-designed iron-core transformers can be 95% efficient or higher. Due to this high efficiency, iron-core transformers are typically treated as ideal.

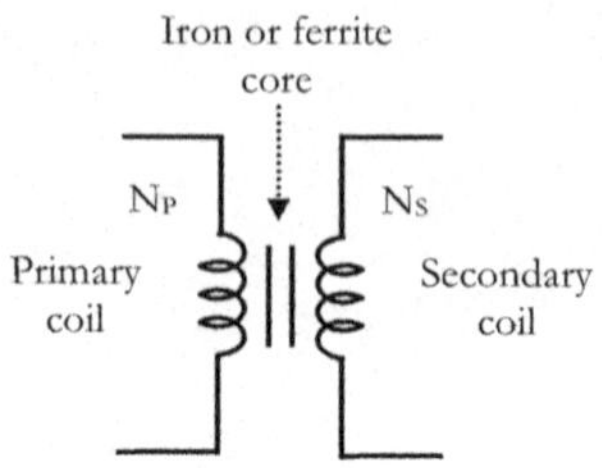

Figure 13.1
Iron-core transformer.

For an ideal transformer, the coupling coefficient is equal to 1. That is, all the magnetic field generated by the primary coil cuts the secondary coil. Say we attached the primary coil to a sinusoidal power source. The sinusoidal current flowing through the primary is given by

$$i_p = I_P \sin(\omega t)$$

Since the coupling coefficient k is 1, the magnetic flux through the primary is the same as through the secondary, and $\phi_P = \phi_S = \phi_M$. As the magnific field is in phase with the current, we can write

$$\phi_p = \Phi_P \sin(\omega t)$$

where Φ_P is the maximum magnetic flux. The voltage across the primary is given by

$$\begin{aligned} e_p &= N_P \frac{d\phi_P}{dt} \\ &= N_P \frac{d(\Phi_P \sin(\omega t))}{dt} \\ &= N_P \omega \Phi_P \sin(\omega t + 90^o) \end{aligned}$$

As we are treating the transformer as ideal, we will ignore any real-world losses, and the induced voltage across the secondary is

$$\begin{aligned} e_S &= N_S \frac{d\phi_S}{dt} \\ &= N_S \frac{d\phi_P}{dt} \\ &= N_S \frac{d(\Phi_P \sin(\omega t))}{dt} \\ &= N_S \omega \Phi_P \sin(\omega t + 90^o) \end{aligned}$$

Dividing the voltage across the primary by the voltage across the secondary, we can write

$$\frac{e_P}{e_S} = \frac{N_P \omega \Phi_P \sin(\omega t + 90^o)}{N_S \omega \Phi_P \sin(\omega t + 90^o)}$$

And

$$\frac{e_P}{e_S} = \frac{N_P}{N_S}$$

This means that the ratio of the instantaneous voltage across the primary and secondary coils is the same as the ratio of the number of turns in the primary to the number of turns in the secondary. The ratio N_P/N_S is called the ***turns ratio*** and is commonly represented by a lowercase "a", that is

$$a = \frac{N_P}{N_S}$$ Turns ratio of a transformer 13.1

Knowing the turns ratio of a transformer and the voltage across the primary, we can calculate the voltage across the secondary coil as

$$e_S = e_P \frac{N_S}{N_P} = \frac{e_P}{a}$$ Voltage across the secondary coil 13.2

A transformer with a turns ratio < 1 is called a ***step-up transformer***, as the voltage is stepped-up in relation to the primary voltage. A transformer with a turns ratio > 1 is called a ***step-down transformer*** as the voltage is decreased or stepped-down in relation to the primary voltage.

In an ideal transformer, 100% of the power is transferred to the secondary so we can write,

$$P_P = P_S$$

$$e_P i_P = e_S i_S$$

Writing as a ratio of currents,

$$\frac{i_P}{i_S} = \frac{e_S}{e_P}$$

$$\frac{i_P}{i_S} = \frac{N_S}{N_P}$$

The primary to secondary current ratio is equal to the inverse of the turns ratio. This means if the voltage across the secondary is higher than the voltage across the primary, the current flowing in the secondary reduces by the same ratio ensuring the power transfer is equal. The secondary current is given by,

$$i_S = i_P \frac{N_P}{N_S} = i_P a$$ Current across the secondary coil 13.3

Using the current-to-voltage relationship given by Equation 13.2 and Equation 13.3, we can write

$$\frac{i_P}{i_S} = \frac{e_S}{e_P}$$

$$\frac{e_P}{i_S} = \frac{e_S}{i_P}$$

$$\frac{e_P}{i_P a} = \frac{e_S}{i_S/a}$$

$$\frac{e_P}{i_P} = a^2 \frac{e_S}{i_S}$$

From Ohm's law, impedance is voltage divided by current, therefore

$$\boxed{Z_P = a^2 Z_S} \qquad \text{Primary coil impedance} \qquad 13.4$$

Given the load on the secondary coil is in parallel with the secondary coil, the voltage across the load would be the same as the voltage across the secondary coil, and the current through the load would be the same as the current through the secondary coil. That is,

$$\frac{e_S}{i_S} = \frac{V_L}{i_L}$$

And we can write

$$Z_S = Z_L$$

This tells us that the impedance of the primary coil is dependent on the impedance of the load attached to the secondary coil. This is referred to as ***reflected impedance***, and

$$\boxed{Z_P = a^2 Z_L} \qquad \text{Ideal transformer reflected impedance} \qquad 13.5$$

13.2 THE NON-RESONANT AIR-CORE TRANSFORMER

In our discussion on ground resistance, it was determined that using a single coil for crystal radios was not a good idea (due to the large antenna and ground losses). Two coils would be needed to ensure a high Q tank circuit (the tuning circuit). We need to use a transformer as part of our crystal radio design. Although iron-core transformers can efficiently transfer power, the core and wire losses will result in coils with extremely low Qs. All the radio stations in your location would bleed through, and you would have no selectivity. We need a low-loss transformer with an adjustable coupling coefficient to select the required Q for our needs.

We can make our own transformer by placing two cylindrical air coils next to each other with the ability to alter the separation distance. Adjusting the separation distance will change the mutual inductance and the coupling coefficient. A transformer constructed using two air-core coils is called an ***air-core transformer***. The circuit diagram for an air-core transformer is shown in Figure 13.2.

An air-core transformer used in a circuit is shown in Figure 13.3 with an unknown secondary load impedance Z_L. The load impedance may be resistive, capacitive or inductive. We will examine this generalised circuit before treating the secondary circuit as a resonant tank circuit. The antenna is modelled as a voltage source with some resistance and capacitance feeding into the primary coil of our air-core transformer.

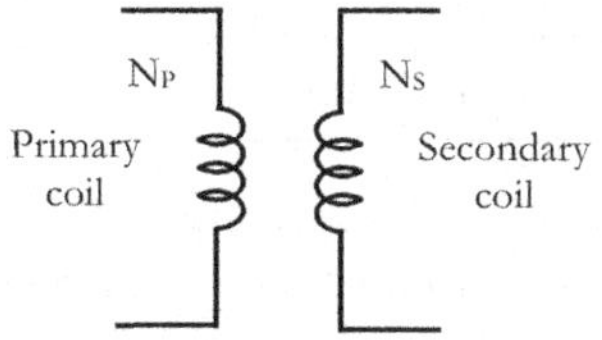

Figure 13.2
The air-core transformer.

All the energy in the secondary coil will come from the primary coil via mutual inductance. Analysing the circuit in the phasor domain, the voltage induced across the secondary can be modelled as a voltage source dependent on the primary current and mutual inductance. For a given frequency, the mutual inductance has a reactance given by

$$X_M = 2\pi fM$$

From the Chapter on mutual inductance, we know that the voltage across a magnetically coupled coil (ignoring resistance) is given by

$$\mathbf{V}_1 = \mathbf{I}_1 jX_{L1} + \mathbf{I}_2 jX_M$$

For the secondary coil voltage, this is

$$\mathbf{V}_S = \mathbf{I}_S jX_S + \mathbf{I}_P jX_M$$

And similarly, the voltage across the primary coil is

$$\mathbf{V}_P = \mathbf{I}_P jX_P + \mathbf{I}_S jX_M$$

The primary current I_P flowing through the mutual inductance produces a voltage e_{sp} across the secondary coil shown as a dependent voltage source on the secondary side (the triangle in the circuit diagram). Similarly, the secondary current I_S flowing through the mutual inductance produces a dependent voltage source e_{ps} on the primary side.

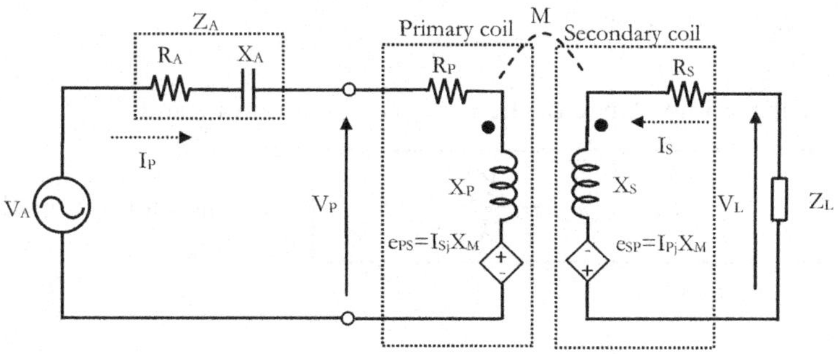

Figure 13.3
Air-core transformer circuit diagram.

Examining the primary side of the transformer using KVL, the sum of all the voltage sources and the voltage drops around the loop must be zero. For the primary side, we can write

$$\boxed{\mathbf{V_A} = \mathbf{I_P} R_A + \mathbf{I_P} R_P - \mathbf{I_P} jX_A + (\mathbf{I_P} jX_P + \mathbf{I_S} jX_M)}$$ Primary KVL 13.6

Applying KVL to the secondary side, we can write

$$\boxed{0 = \mathbf{I_S} R_S + (\mathbf{I_S} jX_S + \mathbf{I_P} jX_M) + \mathbf{I_S} \mathbf{Z_L}}$$ Secondary KVL 13.7

As the voltage source for the secondary circuit is only due to the induced voltage, we can write

$$-\mathbf{I_P} jX_M = \mathbf{I_S} R_S + \mathbf{I_S} jX_S + \mathbf{I_S} \mathbf{Z_L}$$

And solving for the secondary current Is gives

$$\boxed{\mathbf{I_S} = \frac{-\mathbf{I_P} jX_M}{R_S + jX_S + \mathbf{Z_L}}}$$ Secondary coil current 13.8

The impedance of the secondary circuit is

$$\mathbf{Z_S} = R_S + \mathrm{j}\mathrm{X_S} + \mathbf{Z_L}$$ Impedance of secondary circuit 13.9

The voltage measured across the secondary coil (including resistance) is

$$\mathbf{V_S} = \mathbf{I_S} R_S + \mathbf{I_S} jX_S + \mathbf{I_P} jX_M$$ Secondary coil voltage 13.10

and includes the induced voltage. The voltage drop across the load is

$$\mathbf{V_L} = \mathbf{I_S} \mathbf{Z_L}$$ Secondary load voltage 13.11

If we substitute the value for **Is** given by Equation 13.8 into the KVL equation of primary circuit Equation 13.6, we get

$$\mathbf{V_A} = \mathbf{I_P} R_A + \mathbf{I_P} R_P - \mathbf{I_P} jX_A + \mathbf{I_P} jX_P + \left(\frac{-\mathbf{I_P} jX_M}{R_S + jX_S + \mathbf{Z_L}} \right) jX_M$$

$$= \mathbf{I_P} R_A + \mathbf{I_P} R_P - \mathbf{I_P} jX_A + \mathbf{I_P} jX_P + \left(\frac{\mathbf{I_P} X_M^2}{R_S + jX_S + \mathbf{Z_L}} \right)$$

$$= \mathbf{I_P} \left[R_A + R_P - jX_A + jX_P + \left(\frac{X_M^2}{R_S + jX_S + \mathbf{Z_L}} \right) \right]$$

Using Ohm's law, the input impedance of the primary side is V_A/I_P, and

$$\boxed{\mathbf{Z_I} = R_A + R_P - jX_A + jX_P + \left(\frac{X_M^2}{R_S + jX_S + \mathbf{Z_L}} \right)}$$ Primary input impedance 13.12

The primary input current is given by

$$\mathbf{I_P} = \frac{\mathbf{V_A}}{\mathbf{Z_I}}$$ Primary current 13.13

And the voltage across the primary coil is

$$\mathbf{V_P} = \mathbf{I_P} \left[(R_P + jX_P) + \frac{X_M^2}{R_S + jX_S + \mathbf{Z_L}} \right]$$ Primary coil voltage 13.14

The term $X_M^2/(R_S + jX_S + \mathbf{Z}_L)$ is called the ***coupled impedance*** or reflected impedance. When the secondary impedance is capacitive ($\mathbf{Z}_L$ > jXs), the reflected impedance will look inductive (+j), and the primary coil's inductance will appear to increase. When the secondary impedance is inductive ($\mathbf{Z}_L$ < jXs), the reflected impedance will look capacitive (-j), and the primary coil inductance will appear to decrease. This reflected impedance term is only good for non-resonant circuits. Do not use this to analyse a resonant circuit. The reflected impedance becomes a bit more involved when dealing with resonant circuits, and we will cover this in the next section.

In the Chapter on mutual inductance, we saw the open-circuit voltage across the secondary coil was in phase with the voltage across the primary coil. When a load is added, a phase shift will occur, depending on the impedance of the load. Figure 13.4 shows the voltage phase shift of about -80° with respect to the primary voltage when a 5.1k resistor was added as a load to the secondary coil. From Equation 13.14, a resistive load will result in the reflected impedance looking slightly capacitive.

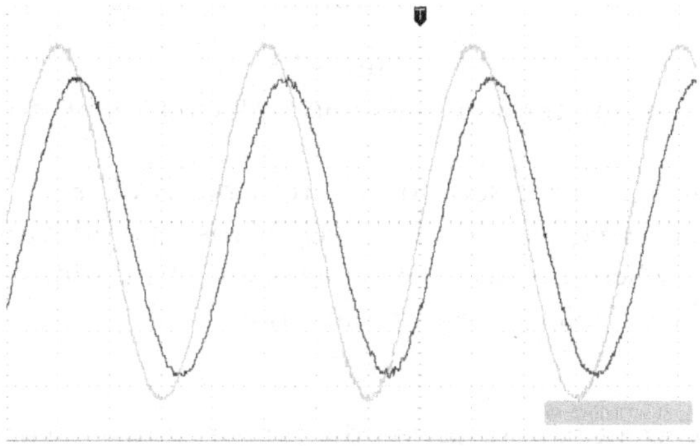

Figure 13.4
Phase shift of secondary voltage when the load is resistive.

13.3 SECONDARY RESONANT AIR-CORE TRANSFORMER

When the secondary circuit is at resonance, the secondary circuit impedance will look resistive. However, using Equation 13.8 to determine the secondary current at resonance will give incorrect results. Not only does it fail to take into account any dielectric losses, or the self-resonant frequency, it does not take into account the loss due to mutual inductance. And as you progress through the book, you will see many other losses that need to be accounted for. We could try and modify the transformer equations as we go along to account for these losses, or we could determine the Q of the coil and use the Q value in a set of equations. This is the approach I will take.

As the quality factor can be thought of as a multiplier of some initial voltage across the coil, we will determine and treat some initial induced voltage from the primary coil into the secondary and multiply this by Q to determine the voltage across the secondary coil at resonance. To determine the Q of the secondary coil, we first need to identify where the energy losses occur.

Q of an Unloaded Resonant Secondary

Figure 13.5 shows a circuit diagram with the secondary side forming a tank circuit. Apart from the capacitor required to achieve resonance, there is no other load attached to the circuit, and we will

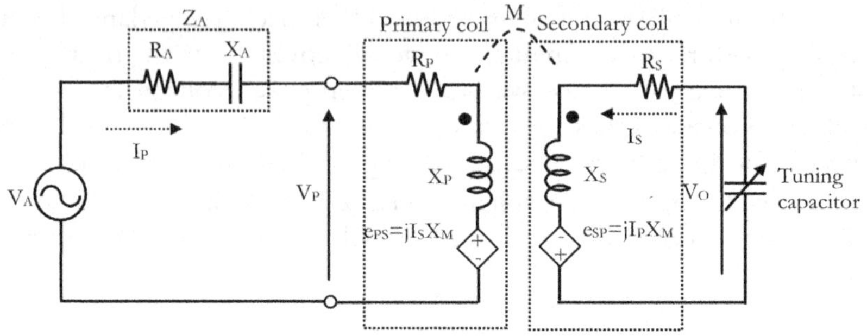

Figure 13.5
Circuit diagram for crystal radio tank circuit with no load.

refer to this as an unloaded resonant secondary. From the KVL Equation 13.6, for the primary coil, we see that the secondary coil induces a voltage across the primary coil. We will denote this induced voltage as e_{PS}, and when the secondary is resonant, the resonant secondary current $\mathbf{I_{S0}}$ will be responsible for this induced voltage, and we can write

$$e_{PS} = \mathbf{I_{S0}} jX_M$$

This voltage will cause a current to flow across the primary circuit, separate from the current due to the voltage source V_A. As the antenna impedance is given by the components Z_A in the circuit, V_A represents an ideal voltage source, and an ideal voltage source has no impedance to the flow of current. Therefore, when analysing the current due to the induced voltage e_{PS}, we treat V_A as a short, and the KVL equation for this induced voltage only is

$$\begin{aligned} 0 &= \mathbf{I_{PS}} R_A + \mathbf{I_{PS}} R_P + (\mathbf{I_{PS}} jX_P + \mathbf{I_{S0}} jX_M) - \mathbf{I_{PS}} jX_A \\ -\mathbf{I_{S0}} jX_M &= \mathbf{I_{PS}} R_A + \mathbf{I_{PS}} R_P + \mathbf{I_{PS}} jX_P - \mathbf{I_{PS}} jX_A \\ &= \mathbf{I_{PS}} (R_A + R_P + jX_P - jX_A) \end{aligned}$$

where $\mathbf{I_{PS}}$ is the current flowing through the primary circuit due to this induced voltage. Rearranging for $\mathbf{I_{PS}}$, we get

$$\mathbf{I_{PS}} = \frac{-\mathbf{I_{S0}} jX_M}{R_A + R_P + jX_P - jX_A}$$

The impedance in the denominator is the impedance of the primary circuit when it is isolated. That is when the secondary circuit has been removed, or the secondary coil is open-circuited. We will refer to this impedance as the uncoupled impedance as there is no mutual inductance term, and it is given by

$$\boxed{\mathbf{Z_{PU}} = R_A + R_P + jX_P - jX_A} \quad \text{Primary uncoupled impedance} \quad 13.15$$

As this induced current passes through the primary circuit, there will be real losses due to resistance. We will refer to this loss as the ***mutual inductance loss.*** This energy loss will reduce the Q of the resonant secondary coil. We need to incorporate this loss when calculating Q. Expressing $\mathbf{I_{PS}}$ as RMS current, we can write

$$I_{PS(RMS)} = \frac{I_{SO(PEAK)}}{\sqrt{2}} \frac{X_M}{Z_{PU}}$$

The average energy loss in the primary due to this current over one cycle (period T) is

$$E_{PS} = \left(\frac{I_{SO(PEAK)}}{\sqrt{2}} \frac{X_M}{Z_{PU}} \right)^2 (R_A + R_P)T$$

Using the definition of Q, we can determine the unloaded Q of the secondary coil of our transformer as

$$Q_{TANK} = \frac{2\pi(\frac{1}{2} I^2_{SO(PEAK)} L_S)}{E_{coil} + E_{PS}}$$

$$= \frac{2\pi(\frac{1}{2} I^2_{SO(PEAK)} L_S)}{\left(\frac{I_{SO(PEAK)}}{\sqrt{2}} \right)^2 R_{coil} T + \left(\frac{I_{SO(PEAK)}}{\sqrt{2}} \frac{X_M}{Z_{PU}} \right)^2 (R_A + R_P)T}$$

$$= \frac{2\pi f L_S}{R_{coil} + \left(\frac{X_M}{Z_{PU}} \right)^2 (R_A + R_P)}$$

If we now take into account the reduction in Q due to the self-resonant frequency, the unloaded Q of the tank circuit becomes

$$Q_{TANK} = \frac{2\pi f L_S}{R_{coil} + \left(\frac{X_M}{Z_{PU}} \right)^2 (R_A + R_P)} \alpha$$

Unloaded tank circuit Q 13.16

where R_{coil} includes all the intrinsic losses of the secondary coil, such as the AC resistance (including the proximity effect), dielectric losses and any core losses (for air coils, there are no core losses). R_A is the antenna resistance, R_P is the primary coil's AC resistance, and α is the Q self-resonant factor. The mutual inductance loss in the above term is

$$R_M = \left(\frac{X_M}{Z_{PU}} \right)^2 (R_A + R_P)$$

Mutual inductance loss 13.16b

Secondary Voltage at Resonance

We know that the initial voltage or current of a resonant circuit is boosted by the Q value. To determine the resonant voltage across the secondary coil, we need to know the initial voltage induced into the secondary coil from the primary.

Consider the circuit shown in Figure 13.6 with the secondary coil without a load attached (open-circuited). At the initial moment, when current starts to flow across the primary circuit, there will be no reflected impedance. Nothing happens instantaneously, and it takes time (although a very short amount of time) for mutual inductance to occur. So the primary input impedance is the uncoupled impedance given by Equation 13.15. The initial current across the primary coil is then

$$\boxed{\mathbf{I_{PU}} = \frac{\mathbf{V_A}}{\mathbf{Z_{PU}}}}$$ Primary uncoupled current 13.17

And the initial primary coil voltage is

$$\mathbf{V_{PU}} = \mathbf{I_{PU}}\left[(R_P + jX_P)\right]$$ Primary coil uncoupled voltage 13.18

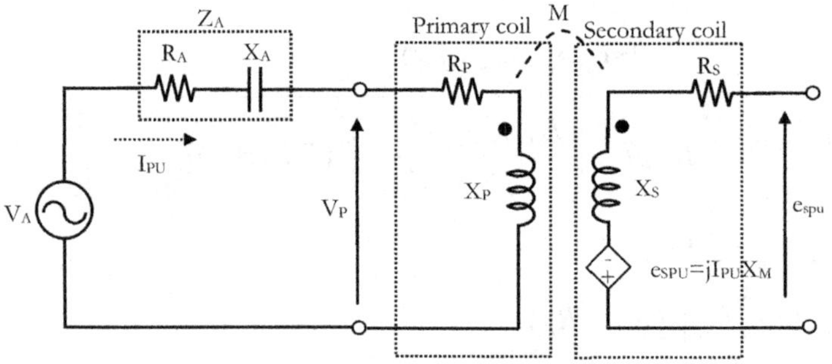

Figure 13.6
Initial voltage induced across the secondary coil.

This initial primary current will induce an emf across the secondary coil. A coil does not have to be in a closed circuit for an induced emf to occur. For an open-circuited coil, the emf will cause a current to run up and down the coil, like an antenna. This coil behaviour is confirmed and used in the chapter on measurement techniques, where the self-resonant frequency of an open-circuit coil is measured. Note that an emf must be created first before any current can flow, and this initial emf will also be induced in a circuit with a capacitor connected as the load. When dealing with circuit analysis in the phasor domain, it is assumed that the circuit is in a steady or quasi-steady state. In this instance, we are dealing with the circuit before this has occurred. See Figure 8.6, showing the build-up of voltage over time across a resonant coil. Therefore, the initial voltage induced across the secondary coil is given by

$$\boxed{\mathbf{e_{spu}} = -\mathbf{I_{PU}}\, jX_M}$$ Initial induced secondary voltage 13.19

Now after some time, at resonance, the total secondary current I_{SO} is the addition of the induced current and the circulating current, as shown in Figure 13.7. The induced current, when resonance has been achieved, will provide just enough energy to make up for the energy losses and

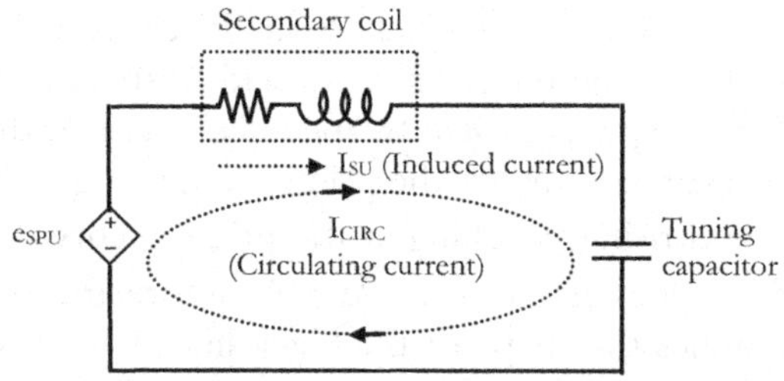

Figure 13.7
Secondary current flow.

keep the circuit resonant. That is

$$\mathbf{I_{S0}} = \mathbf{I_{SU}} + \mathbf{I_{CIRC}}$$

The voltage drop across the coil at resonance will be the initial induced voltage times the quality factor. That is

$$\boxed{\mathbf{V_{S0}} = Q_{TANK}\mathbf{e_{spu}}} \qquad \text{Secondary output voltage at resonance} \qquad 13.20$$

At resonance, the current will be in phase with the voltage (the circuit impedance looks resistive). The resonant current flows across the secondary coil, and using Ohm's law, we can write

$$\boxed{\begin{aligned}\mathbf{I_{S0}} &= \frac{\mathbf{V_{S0}}}{|\mathbf{Z_S}|} \\ &= \frac{Q_{TANK}\mathbf{e_{spu}}}{|R_S + jX_S|}\end{aligned}} \qquad \text{Secondary resonant current} \qquad 13.21$$

That is, the voltage across the coil is divided by the magnitude of the coil's impedance.

At resonance, the voltage across the secondary coil will be phase shifted with respect to the primary coil voltage by around -90° as shown by the oscilloscope image in Figure 13.8. The

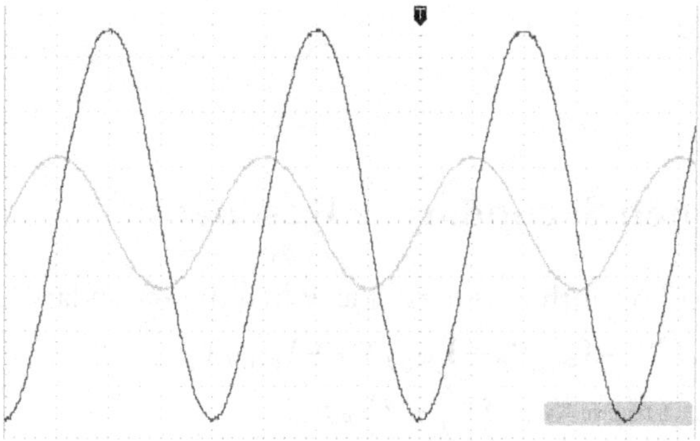

Figure 13.8
Phase shift of non-loaded resonant secondary output voltage.

primary voltage is grey, and the secondary is black. Note different scales were used for the primary and secondary voltages. To determine the phase shift with respect to the primary voltage, look at the trigger point (origin) of the graph as shown by the check mark at the top. Determine where the amplitude and shape of the secondary match the primary. In this graph, the black line as it crosses the origin is at a minimum and rising. Looking at the primary curve at +90°, the grey line is at a maximum and falling, and so does not match. At 90° before the origin, the grey curve is at a minimum and rising and matches the shape of the black line. Therefore the secondary voltage is -90° out-of-phase with the primary.

Voltage Gain Method

Another method to calculate the output voltage of the secondary coil is to multiply the primary coil's voltage by the gain of the transformer. The ratio of the secondary output load voltage to the voltage across the primary coil is the voltage gain of a transformer. That is

$$V_{GAIN} = \frac{V_O}{V_{PU}}$$

And the secondary output voltage is

$$V_O = V_{GAIN} V_{PU}$$

At resonance, the voltage gain is multiplied by the Q of the tank circuit. So we can write

$$\boxed{V_O = Q_{TANK} V_{GAIN} V_{PU}} \qquad \text{Secondary voltage at resonance using gain} \qquad 13.22$$

In a normal iron core transformer, the voltage gain is given by the turns ratio. For an air-core transformer, the voltage gain is given by the square root of the ratio of the coil's inductance values multiplied by the coupling coefficient.

$$\boxed{V_{GAIN} = k\sqrt{\frac{L_S}{L_P}}} \qquad \text{Air-core transformer gain} \qquad 13.23$$

If we know the mutual inductance, we can calculate the coupling coefficient using Equation 12.18.

$$k = \frac{M}{\sqrt{L_P L_S}}$$

Primary Voltage when Secondary is Resonant

From the KVL Equation 13.6 for the primary side, when the secondary is resonant, we can write

$$\begin{aligned} \mathbf{V_A} - \mathbf{I_{S0}} jX_M &= \mathbf{I_{PS0}} R_A + \mathbf{I_{PS0}} R_P - \mathbf{I_{PS0}} jX_A + \mathbf{I_{PS0}} jX_P \\ &= \mathbf{I_{PS0}} (R_A + R_P - jX_A + jX_P) \end{aligned}$$

And the primary input current is

$$\mathbf{I_{PS0}} = \frac{\mathbf{V_A} - \mathbf{I_{S0}} jX_M}{(R_A + R_P - jX_A + jX_P)}$$

That is

$$\mathbf{I_{PS0}} = \frac{\mathbf{V_A} - \mathbf{I_{S0}} jX_M}{\mathbf{Z_{PU}}} \qquad \text{Steady-state primary current when secondary is resonant} \qquad 13.24$$

$\mathbf{I_{PS0}}$ is the primary current at steady-state when the secondary has finally become resonant. And the voltage across the primary coil is the voltage due to the primary current and the mutual inductance voltage from the secondary. Therefore the voltage across the primary coil when the secondary is resonant is

$$\mathbf{V_{PS0}} = \mathbf{I_{PS0}}(R_P + jX_P) + \mathbf{I_{S0}} jX_M \qquad \text{Voltage across the primary coil when the secondary is resonant} \qquad 13.25$$

Reflected Impedance and Primary Resonance

There are two ways in which the primary circuit may become resonant. Firstly, if the capacitive reactance of the antenna is matched by the inductive reactance of the coil, then the uncoupled primary input impedance will become resistive, and the circuit will be resonant. When this condition occurs, the transformer is said to be double-resonant (we will also refer to this condition as uncoupled resonance). It is extremely unlikely that double-resonance will be achieved without using a tuning capacitor in the primary circuit. We will discuss the use of a tuning capacitor and double-resonance in a later section.

Even if we design the primary circuit to be resonant at a particular frequency using a coil only, chances are it will be close to resonance but not quite at resonance. When the primary circuit is not uncoupled resonant, the reflected impedance will have a reactive component (either capacitive or inductive). This reflected reactive component might be such that the primary circuit may become coupled resonant.

The crystal radio will react differently depending on the primary resonance type. When the primary is uncoupled resonant, the reflected impedance looks resistive as the primary and secondary currents are in phase. Both the uncoupled and coupled primary impedance will be resonant at the same frequency regardless of separation distance. The initial induced voltage across the secondary coil will be at a maximum (for a given separation distance). However, as the uncoupled primary impedance is at a minimum, the mutual inductance losses will be at a maximum, resulting in a relatively low Q.

When a crystal radio is not uncoupled resonant, the initial secondary induced voltage will be less than the maximum, but the uncoupled impedance will be higher, resulting in less mutual inductance losses, and the Q of the secondary will be higher. It is the combination of the Q and the initial induced voltage that ultimately determines the resonant voltage across the secondary coil. With a maximum occurring for a non-double-resonant transformer when the primary becomes coupled resonant.

Performing circuit analysis for a non-double-resonant transformer, we can express the KVL equation for the primary circuit when the secondary is resonant in terms of antenna voltage. We can write

$$\mathbf{V_A} = \mathbf{I_{PS0}} R_A + \mathbf{I_{PS0}} R_P - \mathbf{I_{PS0}} jX_A + \mathbf{I_{PS0}} jX_P + \mathbf{I_{S0}} jX_M$$

Dividing the voltage $\mathbf{V_A}$ by the primary current $\mathbf{I_{PSO}}$ we get the input impedance of the primary circuit when the secondary is resonant. That is

$$\boxed{\mathbf{Z_I} = R_A + R_P - jX_A + jX_P + \frac{\mathbf{I_{S0}} jX_M}{\mathbf{I_{PS0}}}}$$ Primary input impedance when the secondary is resonant 13.26

The reflected impedance is given by the term

$$\frac{\mathbf{I_{S0}} jX_M}{\mathbf{I_{PS0}}}$$

The reactive component of the reflected impedance always opposes the reactance of the primary circuit's uncoupled impedance. When the uncoupled primary impedance looks inductive, the reflected impedance looks capacitive (subtractive), and when the uncoupled primary impedance looks capacitive, the reflected impedance looks inductive (additive). The reflected impedance can make the primary circuit resonant. This occurs when

$$-jX_A + jX_P + imag\left\{\frac{\mathbf{I_{S0}} jX_M}{\mathbf{I_{PS0}}}\right\} = 0$$

And the primary circuit looks resistive with an input impedance given by the resistive terms only. That is

$$\mathbf{Z_I} = R_A + R_P + R_R$$ Primary input impedance when the primary is resonant

where R_R is the real (resistive) component of the reflected impedance. Note that this impedance is seen by the antenna, and the current drawn from the antenna (the line current) is given by

$$\mathbf{I_A} = \frac{\mathbf{V_A}}{\mathbf{Z_I}}$$ Primary line current when the primary is resonant

The line current $\mathbf{I_A}$ represents the energy required to maintain the secondary resonance, which is the energy lost in the system.

For AC circuits, maximum power transfer occurs when the load impedance is the complex conjugate of the source impedance; that is, it becomes resistive. In our crystal radio transformer, maximum power transfer occurs when both the primary and the secondary circuits are resonant at the same frequency (both are in-phase and look resistive).

The energy to maintain the system at resonance (the line current) is provided just at the right time to the secondary (like pushing someone on a swing at the right moment). Note that the voltage induced across the secondary at this steady state stage is not the same as the initial induced voltage. That voltage is due to the primary coil's uncoupled impedance.

The closer the uncoupled impedance is to resonance for a given frequency, the higher the initial induced voltage is across the secondary for a given distance. The initial induced voltage across the secondary depends on both the separation distance and mutual inductance. When the separation distance is small, the initial induced voltage is high, but the Q of the secondary will be low due to large mutual inductance losses. When the distance is greater, a smaller voltage is induced, but the Q will be higher. At some distance, due to the combination of the initial induced voltage and Q, a peak secondary resonant voltage will occur. This maximum also coincides when the primary coupled input impedance becomes resonant. Whether the primary becomes resonant or not depends on how far the uncoupled impedance is from resonance and the reactive component of the reflected impedance.

Capacitance Required for Secondary Resonance

When both the primary and secondary circuits have settled into the steady-state condition, the steady-state primary current $\mathbf{I_{PSO}}$ induces a voltage across the secondary. From the KVL Equation 13.7 for the secondary circuit, we can write

$$-j\mathbf{I_{PSO}}X_M = \mathbf{I_{SO}}R_S + \mathbf{I_{SO}}jX_S - \mathbf{I_{SO}}jX_C$$

Dividing through by the resonant current to get the impedance we have,

$$\frac{-j\mathbf{I_{PSO}}X_M}{\mathbf{I_{S0}}} = R_S + jX_S - jX_C$$

For the secondary to be resonant, the magnitude of the tuning capacitor's reactance must equal the inductive reactance it sees. The reactance the capacitor sees depends on the phase relationship between the voltage and the current. Therefore the capacitor sees the inductive reactance of the coil and the reactive mutual inductance component in the induced voltage. So

$$jX_C = jX_S + \frac{j\mathbf{I_{PSO}}X_M}{\mathbf{I_{S0}}}$$

The magnitude of the capacitive reactance for resonance becomes

$$\boxed{X_C = X_S + imag\left\{\frac{\mathbf{I_{PSO}}X_M}{\mathbf{I_{S0}}}\right\}}$$ Capacitive reactance for secondary resonance 13.27

The capacitance required for resonance depends on the impedance of the primary circuit, including the antenna length, the primary current and the circulating current in the secondary. It is also dependent on the Q of the secondary circuit and the mutual inductance between the coils. As the mutual inductance depends on the separation distance, you will need to retune the capacitor when you change the distance between the coils. Also, if you increase or decrease the Q by selecting a different tap for audio power, you will need to retune again. The amount of retuning will range from almost nothing to a noticeable difference in the rotation of the tuning capacitor. The higher the Q, the less noticeable this change will be. If we take into account the axial self-capacitance of the secondary coil, the capacitance value required becomes

$$\boxed{C_S = \frac{1}{2\pi f X_C} - C_{LS(axial)}}$$ Capacitance at resonance 13.28

Assuming the resonant frequency is much lower than the secondary coil's self-resonant frequency.

13.4 EFFECT OF SEPARATION ON RESONANT AIR-CORE TRANSFORMERS

Having developed equations to determine the Q and the resonant voltage, we will now examine the accuracy of these equations for an unloaded resonant crystal radio tank circuit. We will look at the effect separation distance between the primary and secondary coils has on voltage and Q for different input capacitance values (representing different antenna lengths). Using coil PC3 (33 uH) as the primary and coil SC1 (331 uH) as the secondary (see appendix 2 for coil specifications), a 1-volt peak-to-peak 1 MHz sine wave was injected into the primary coil with a capacitor in series. The output resistance of the signal generator is 50 Ω. The secondary coil's Q and output voltage were measured. Correction factors were applied to the measured Q and voltage values representing the estimated losses due to the measurement equipment as described in the Chapter on measurement techniques. The input capacitance values chosen were; 0 pF (no series capacitance), 100, 470, 751, 1000 and 2,200 pF, representing a range of antenna capacitance values. The predicted and measured values were plotted as graphs shown in Figure 13.9. The lines represent the calculated values, and the symbols are the measured values.

The first thing to note is how well the predicted and measured values of both Q and resonant secondary coil voltage coincide, thus increasing confidence that the model reflects reality. Examining the output voltages (in black), we see that for each graph (except 13.9b) that the voltage increases with distance to a maximum and then trails off. This voltage maximum occurs when the primary circuit becomes resonant (Equation 13.26 becomes resistive). Also, note the higher the primary input reactance is (either inductive or capacitive), the closer the two coils need to be to achieve primary circuit resonance. As the coils move closer together, the reflected impedance increases until the reflected impedance reactive component is of the same magnitude as the primary impedance reactance. It does not matter whether the primary uncoupled impedance looks inductive or capacitive; resonance can still be achieved. From this, we can make the following observation.

The reflected impedance seen by the primary circuit at secondary resonance always acts to reduce the primary circuits' reactance. If this reduction in primary circuit reactance is such that the reactance goes to zero, then primary coupled circuit resonance is achieved.

Looking at graph 13.9b, the shape of the voltage curve is different than the others, and the maximum voltage occurs at the minimum distance. This is because the uncoupled reactance of the primary circuit is quite large, much larger than the other graphs, and the reflected impedance never becomes great enough to make the circuit look resonant. For graph 13.9b, a 100 pF capacitor was used as the simulated antenna capacitance, which equates to an antenna of about 11 metres; far too short. Resonance could be achieved, however, if the primary coil inductance was sufficiently increased.

Looking at the curves for the Q, we see that Q increases rapidly as the separation distance between the coils increases and then tends to flatten out. This is because Q is also a function of mutual inductance. The closer the coils, the greater the mutual inductance losses are, and as the mutual inductance decreases, the losses likewise decrease. As the distance increases, the Q tends towards the isolated effective Q of the coil. However, the initial induced voltage in the secondary coil also decreases with distance, and the increase in Q is insufficient to prevent the resonant secondary voltage from decreasing. The Q in Figure 13.9b, however, remains relatively constant and high. This is because the uncoupled primary impedance is very high, and the change in impedance due to the mutual inductance is insufficient to make much of a difference in the mutual inductance losses.

The smallest capacitor used in the experiment to achieve primary resonance was 470 pF (40-metre antenna), and the highest was 2,200 pF (60-metre antenna). However, this was only for a single frequency of 1 MHz. Figure 13.10 graphs the voltages and Qs for 530 and 1,700 kHz, the bottom and top of the AM band, respectively. For the 40-metre antenna, Figure 13.10a, resonance

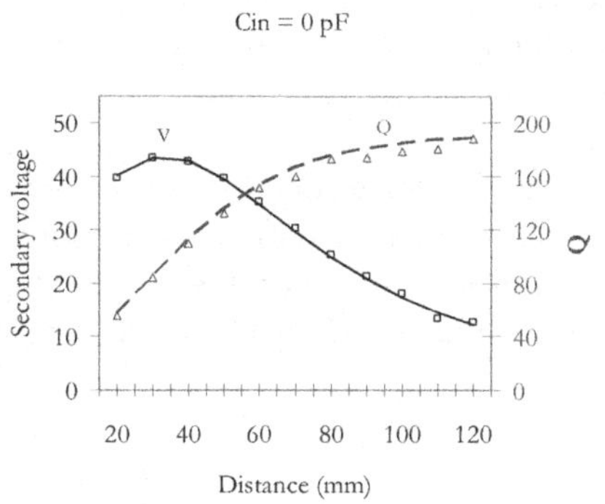

(a)) Primary input reactance = +j209

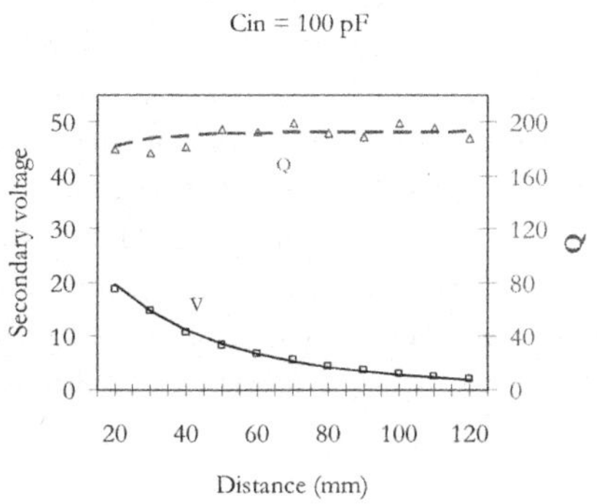

(b) Primary input reactance = -j1,328

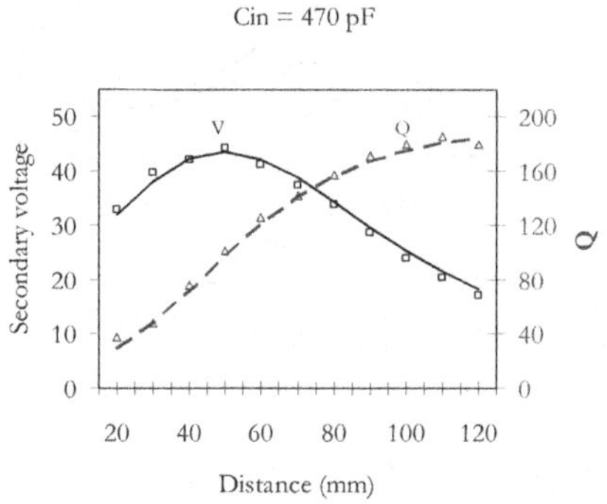

(c) Primary input reactance = -j129

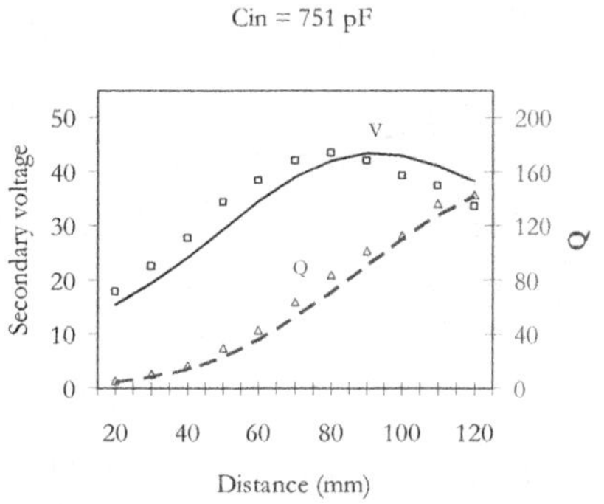

(d) Primary input reactance = -j3

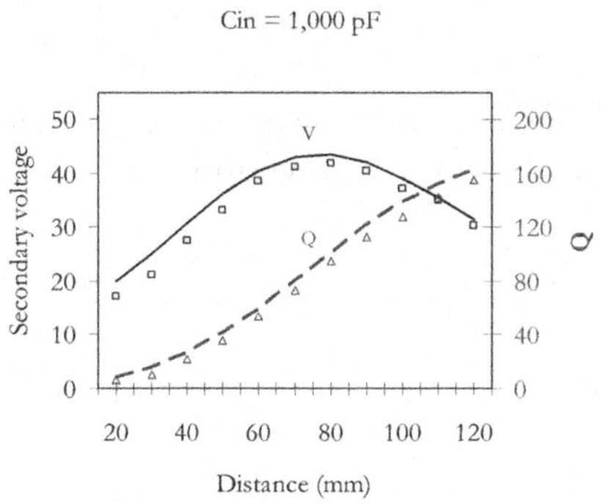

(e) Primary input reactance = +j50

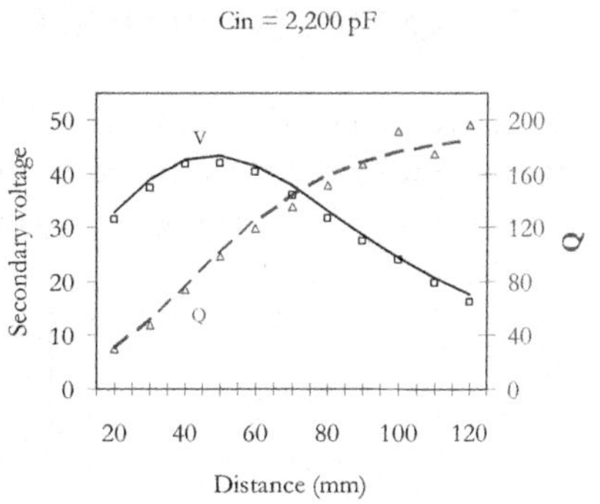

(f) Primary input reactance = +j137

Figure 13.9
Distance versus coil voltage and Q with different input capacitance.

was not achieved at 530 kHz. For the 60-metre antenna, both frequencies became resonant. The best approach when using a single coil and the antenna capacitance to attain resonance is to match the inductance of the coil to the lower frequency and then use taps for higher frequencies. The closer we make the primary circuit uncoupled resonant, the less we need to rely on the reflected impedance for resonance. You should be able to attain resonance by adjusting the separation distance and selecting the correct tap. We will explore this further in later Chapters.

You may wish to use a wiper-arm to change the inductance to ensure a closer match for resonance; however, I am not a fan of wiper-arms (as previously discussed), and exact resonance can be achieved by using a tuning capacitor.

We can consider the distance at which the maximum voltage occurs as the minimum separation distance of the coils. Moving the coils closer will result in a reduction of both Q and voltage, so there is nothing to gain. Moving the coils further apart, however, results in a higher Q increasing the selectivity of the crystal radio with some loss of voltage. This is a trade-off you may wish to make if there is interference due to other stations.

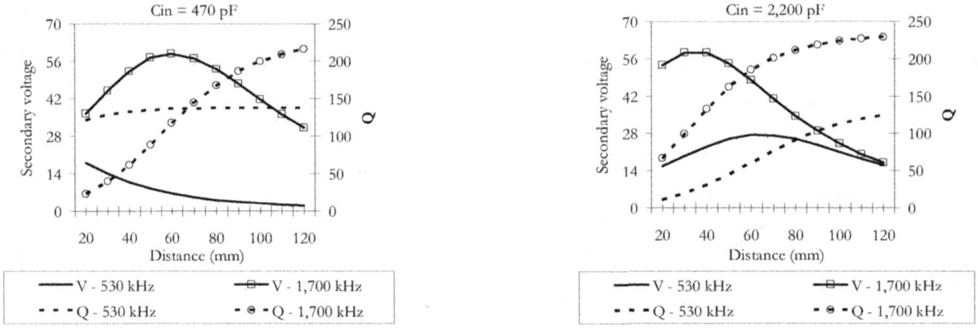

Figure 13.10
Distance versus voltage and Q for different frequencies and input capacitance.

13.5 AUTOTRANSFORMERS

So far, we have discussed transformers comprising two separate coils with no direct electrical connection. A transformer can be built using a single coil with taps, as shown in Figure 13.11. This type of transformer is called an autotransformer.

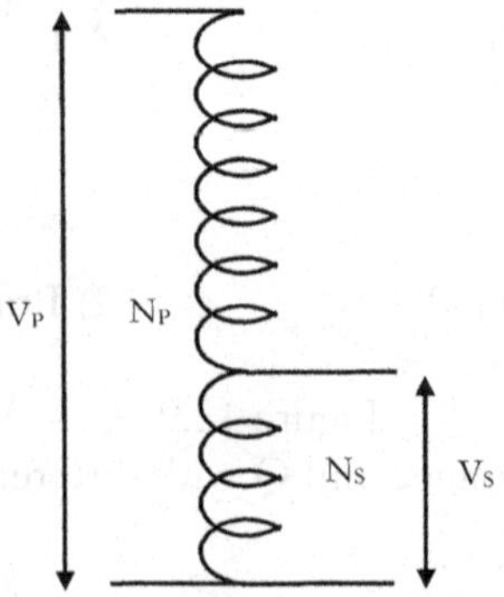

Figure 13.11
The autotransformer.

In a crystal radio, the secondary coil is tapped to provide power to the demodulator and audio circuitry and can be treated as an autotransformer. The voltage across the secondary tank circuit at resonance can be treated as a voltage source, and we can model the secondary side as an ideal voltage source in parallel with an autotransformer, as shown in Figure 13.12. This autotransformer is a step-down transformer, and the voltage across the load will be less than the voltage across the entire coil.

As typical with transformers, the section connected to the power supply is called the primary, and the section connected to the load is called the secondary. However, as this autotransformer is itself part of a transformer, this designation might be a little confusing, so we will continue to refer to the secondary coil as the secondary coil consisting of tapped and untapped sections, with the tapped section connected to the load.

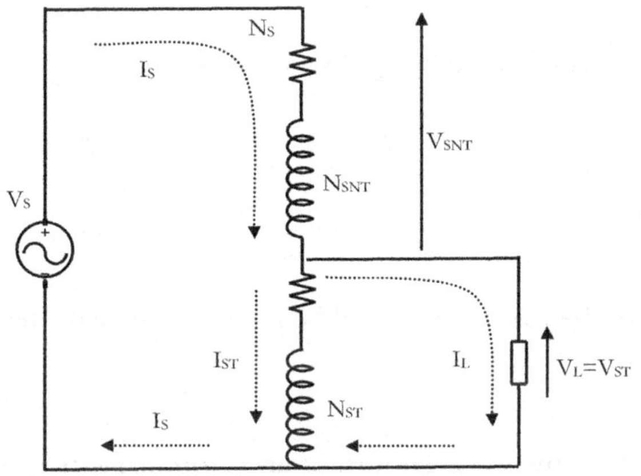

Figure 13.12
Modelling the secondary tank circuit as an autotransformer.

Consider Figure 13.12. The secondary coil of N_S loops of wire is subdivided into two sections; a tapped section of N_{ST} turns with a load attached and an unloaded section of N_{SNT} windings. These two coil sections can be treated as two separate coils with self-inductance L and mutual inductance M. The total inductance of the untapped coil given by

$$L_{T(SNT)} = L_{SNT} + M$$

And the total inductance of the tapped coil section N_{ST}, given by

$$L_{T(ST)} = L_{ST} + M$$

The total inductance of these two coils adds up to the self-inductance of the secondary coil N_S.

$$\begin{aligned} L_S &= L_{T(SNT)} + L_{T(ST)} \\ &= (L_{SNT} + M) + (L_{ST} + M) \\ &= L_{SNT} + L_{ST} + 2M \end{aligned}$$

If we know the total self-inductance of the secondary coil N_S and the self-inductance of the two coil sections, either by measurement or calculation, we can calculate the mutual inductance between the two coil sections.

$$M = \frac{L_S - L_{SNT} - L_{ST}}{2}$$

This mutual inductance is simply part of the inherent mutual inductance between windings in a single coil and is built into the inductance formula for coils. So no matter where the tap is placed, the total inductance of both coil sections will add up to the self-inductance of the single coil N_S. For example, the self-inductance of the first tap on coil SC1 (331 uH) was measured to be 28 uH, and the untapped section was measured to be 221 uH. The mutual inductance between the sections is, therefore

$$M = \frac{331 - 221 - 28}{2}$$
$$= 41uH$$

And the inductance of the tapped and untapped sections is

$$L_{T(ST)} = 28 + 41$$
$$= 69uH$$
$$L_{T(SNT)} = 221 + 41$$
$$= 262uH$$

And the total inductance of the two sections will be the inductance of the coil.

$$L_S = 262 + 69$$
$$= 331uH$$

The volts per turn is the same for each turn in an autotransformer, so the ratio of the tapped coil voltage to the voltage across the entire secondary coil is given by the turns ratio, that is

$$\frac{V_{ST}}{V_S} = \frac{N_{ST}}{N_S}$$

And the voltage across the tapped section of the coil is

$$V_{ST} = \frac{N_{ST}}{N_S} V_S$$

The load is in parallel with the tapped coil, so the load voltage is the same as the tapped coil voltage, that is

$$\boxed{V_L = V_{ST} = \frac{N_{ST}}{N_S} V_S}$$ Voltage across the autotransformer load 13.29

Similarly, the voltage across the untapped coil N_{SNT} is

$$V_{SNT} = \frac{N_{SNT}}{N_S} V_S$$ Voltage across the untapped autotransformer 13.30

Examining the current flow through the circuit in Figure 13.12, the current I_S from the power supply travels across windings N_{SNT} until it reaches the tap. At the tap, the current splits, with some of the current travelling across the load I_L and some across the tapped coil I_{ST}. At the common rail, the branch currents combine and flow back to the power supply. That is

$$I_S = I_{ST} + I_L$$

The line current I_S flowing across the untapped section of the coil N_{SNT} and can be calculated using Ohm's law.

$$I_S = \frac{V_{SNT}}{Z_{SNT}}$$ Line current across the secondary 13.31

Using the current divider rule, we can calculate the current across the load and the tapped coil. The current across the tapped section of the coil is

$$\boxed{I_{ST} = \frac{Z_L}{Z_L + Z_{ST}} I_S}$$ Current across the tapped section 13.32

And using the current divider rule for the load, the current across the load is

$$\boxed{I_L = \frac{Z_{ST}}{Z_L + Z_{ST}} I_S}$$ Current across the load 13.33

The impedances of the untapped section is

$$\boxed{Z_{SNT} = \frac{N_{SNT}}{N_S} Z_S}$$ Impedance of the untapped section 13.34

The impedances of the tapped section is

$$\boxed{Z_{ST} = \frac{N_{ST}}{N_S} Z_S}$$ Impedance of the tapped section 13.35

13.6 Q OF THE LOADED TANK CIRCUIT

In section 13.3, we derived a formula for the Q of an unloaded resonant tank circuit where all the circulating current passes through the coil. With a loaded tank circuit, the circulating current flows across the untapped section of the coil and then splits. Some current flows through the load and some through the tapped section of the coil. In Figure 13.12, we have a load attached to a tap. If we expressed the power dropped across each section in terms of the circulating current, we could derive a formula for the Q.

The power loss across the untapped section of the coil is

$$P_{SNT} = I_S^2 R_{SNT}$$

where R_{SNT} is the resistance component of the untapped impedance given by Equation 13.34. Note that the coil resistance values include the skin and proximity effect losses.

For the tapped section, the ratio of the untapped to the tapped voltage is given by

$$\frac{V_{ST}}{V_{SNT}} = \frac{N_{ST}}{N_{SNT}}$$

and

$$V_{ST} = \frac{N_{ST}}{N_{SNT}} V_{SNT}$$

$$I_{ST} Z_{ST} = \frac{N_{ST}}{N_{SNT}} I_S Z_{SNT}$$

$$I_{ST} = I_S \frac{N_{ST}}{N_{SNT}} \frac{Z_{SNT}}{Z_{ST}}$$

The power loss across the tapped section of the coil is

$$P_{ST} = I_{ST}^2 R_{ST}$$

$$= \left(I_S [\frac{N_{ST}}{N_{SNT}} \frac{Z_{SNT}}{Z_{ST}}] \right)^2 R_{ST}$$

The load is in parallel with the tapped section, so the voltage across the load is the same as the voltage across the tapped section so that we can write

$$V_L = V_{ST}$$

$$I_L Z_L = I_{ST} Z_{ST}$$

$$I_L = \left(I_S \frac{N_{ST}}{N_{SNT}} \frac{Z_{SNT}}{Z_{ST}} \right) \frac{Z_{ST}}{Z_L}$$

$$= \left(I_S \frac{N_{ST}}{N_{SNT}} \frac{Z_{SNT}}{Z_L} \right)$$

And the power dropped across the load is

$$P_L = I_L^2 R_L$$

$$= \left(I_S \frac{N_{ST}}{N_{SNT}} \frac{Z_{SNT}}{Z_L} \right)^2 R_L$$

Finally, we need to take into account the power loss due to mutual inductance. The power loss from the secondary coil due to the current induced in the primary is given by Equation 13.16b as an equivalent resistance value R_M, and is

$$R_M = \left(\frac{X_M}{Z_{PU}} \right)^2 (R_A + R_P)$$

In the unloaded tank circuit, the circulating current flowed through the entire coil, and so the current flowed through this entire equivalent resistance value. With the load attached, the current is different for the untapped and tapped sections of the coil, and we must apportion these currents to R_M. Treating R_M as a resistance value distributed along the length of the coil, the portion of this resistance in the untapped section is

$$R_{M(NT)} = \frac{N_{SNT}}{N_S} R_M$$

And the power loss is

$$P_{M\,(NT)} = I_S^2 \frac{N_{SNT}}{N_S} R_M$$

For the tapped section, the resistance portion is

$$R_{M(T)} = \frac{N_{ST}}{N_S} R_M$$

And the power loss is

$$P_{M(T)} = I_{ST}^2 \frac{N_{ST}}{N_S} R_M$$

$$= \left(I_S [\frac{N_{ST}}{N_{SNT}} \frac{Z_{SNT}}{Z_{ST}}] \right)^2 \frac{N_{ST}}{N_S} R_M$$

The total mutual inductance power loss is, therefore

$$P_M = P_{M(NT)} + P_{M(T)}$$

$$= I_S^2 \frac{N_{SNT}}{N_S} R_M + \left(I_S [\frac{N_{ST}}{N_{SNT}} \frac{Z_{SNT}}{Z_{ST}}] \right)^2 \frac{N_{ST}}{N_S} R_M$$

$$= I_S^2 R_M \left(\frac{N_{SNT}}{N_S} + \frac{N_{ST}}{N_S} \left(\frac{N_{ST}}{N_{SNT}} \frac{Z_{SNT}}{Z_{ST}} \right)^2 \right)$$

Expressing these power losses in terms of resistance values, we have

$$R_{PSNT} = R_{SNT}$$ Loss in the non-tapped section of the coil

$$R_{PST} = \left([\frac{N_{ST}}{N_{SNT}} \frac{Z_{SNT}}{Z_{ST}}] \right)^2 R_{ST}$$ Loss in the tapped section of the coil

$$R_{PL} == \left(\frac{N_{ST}}{N_{SNT}} \frac{Z_{SNT}}{Z_L} \right)^2 R_L$$ Loss across the load

$$R_{PM} = R_M \left(\frac{N_{SNT}}{N_S} + \frac{N_{ST}}{N_S} \left(\frac{N_{ST}}{N_{SNT}} \frac{Z_{SNT}}{Z_{ST}} \right)^2 \right)$$ Mutual inductance loss

Inserting these losses into the equation for Q, we have

$$\boxed{Q_{CIRCUIT} = \frac{X_S}{R_{PSNT} + R_{PST} + R_{PL} + R_{PM}} \alpha}$$

This analysis assumes that the same current is passing through the load during the entire cycle.

Due to the action of the diode and the earphone in the crystal radio, this is not the case. In a later Chapter, we will modify these values to arrive at a Q value for the operating crystal radio.

13.7 DOUBLE-RESONANT AIR-TRANSFORMER

When both the primary and secondary circuits use a tuning capacitor to achieve resonance, we say the transformer is double-resonant. For a tuning capacitor to be effectively used in the primary circuit, the antenna capacitance must be at least twice the maximum capacitance value of the tuning capacitor. The tuning capacitor is in series with the antenna impedance, and the upper capacitance value will be limited by the smallest capacitance in series. This will be covered further in the Chapter on double-resonant crystal radios. Figure 13.13 shows the circuit diagram for an unloaded double resonant air-core transformer.

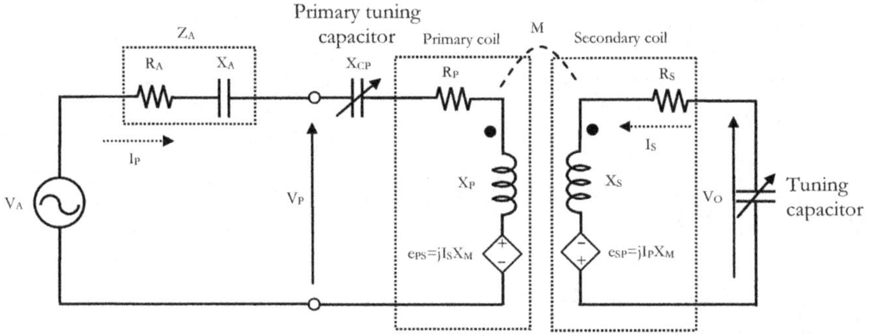

Figure 13.13
Double-resonant air-core transformer circuit diagram.

Capacitance for Resonance

When the primary circuit is resonant, the input impedance will be resistive, and the primary uncoupled current $\mathbf{I_{PU}}$ will be in phase with the primary voltage with a zero phase angle. That is

$$\mathbf{I_{PU}} = I_{PU}\angle 0^\circ$$

And the secondary resonant voltage will be

$$\mathbf{V_{SO}} = -Q_{TANK} I_{PU}\angle 0^\circ\, jX_M$$

With the secondary resonant circulating current being

$$\mathbf{I_{SO}} = \frac{-Q_{TANK} I_{PU}\angle 0^\circ\, jX_M}{|R_S + jX_S|\angle 0^\circ}$$

$$\mathbf{I_{SO}} = \frac{Q_{TANK} I_{PU} X_M}{|R_S + jX_S|}\angle -90^\circ$$

The induced voltage across the primary coil due to the secondary coil will be

$$\mathbf{e_{PS}} = \left[\frac{Q_{TANK} I_{PU} X_M}{|R_S + jX_S|} \angle -90^\circ \right] X_M \angle 90^\circ$$

$$= \left[\frac{Q_{TANK} I_{PU} X_M}{|R_S + jX_S|} \right] X_M \angle 0^\circ$$

The phase angle is zero, so it is in phase with the primary source voltage. At steady-state, the primary current will be

$$\mathbf{I_{PS0}} = \frac{V_A \angle 0^\circ - e_{PS} \angle 0^\circ}{(R_A + R_P) \angle 0^\circ} = I_{PS0} \angle 0^\circ$$

and is resistive. Subsequently, the reflected impedance is purely resistive, and the capacitance required for primary resonance will not change. Adding the reactance of the primary tuning capacitor to the primary coupled input impedance, we get

$$\mathbf{Z_I} = R_A + R_P - jX_A - jX_{CP} + jX_P + \frac{\mathbf{I_{S0}} X_M}{\mathbf{I_{PS0}}}$$

And when both the primary and secondary are resonant, the primary input impedance becomes

$$\boxed{\mathbf{Z_I} = R_A + R_P + \frac{I_{S0} X_M}{I_{PS0}}}$$ Primary input impedance at double-resonance is purely resistive 13.36

And the primary capacitance value for resonance is

$$\boxed{X_{CP} = -X_A + X_P}$$ Primary capacitive reactance value at double-resonance 13.37

At the steady-state condition, the KVL for the secondary circuit is

$$I_{PSO} X_M \angle -90^\circ = I_{SO} \angle -90^\circ (R_S + jX_S - jX_C)$$

And

$$R_S + jX_S - jX_C = \frac{I_{PSO} X_M \angle -90^\circ}{I_{SO} \angle -90^\circ} = \frac{I_{PSO} X_M}{I_{SO}}$$

Therefore the induced voltage is also resistive. So the tuning capacitance value for secondary resonance is

$$\boxed{X_C = X_S}$$ Secondary capacitive reactance value at double-resonance 13.38

We will refer to the condition when both Equation 13.37 and Equation 13.38 are true as double-resonance. And we will call this resonant condition uncoupled-resonance. That is, mutual inductance plays no part in making the circuits look resonant. However, this condition may not easily be achieved.

When adjusting both the primary and secondary tuning capacitors until we hear a station, the chances that both circuits are double-resonant is remote. Both the primary and secondary tuning capacitors will see reactance in the mutual inductance and may, in fact, be coupled-resonant. We adjust the primary tuning capacitor until a maximum volume is reached, and then we retune the secondary for maximum volume. It is possible that the difference in volume between double-resonance and coupled-resonance is not noticeable, and we may still be slightly coupled-resonant.

We should therefore assume that we are at coupled-resonance, and when we change the distance between the two coils, we should retune the circuits.

Double Resonance Over-coupling

If the coupling between the two coils is high, this double resonance can produce strange results when attempting to tune into a station. When the primary and secondary coils are brought close together, the reactance due to the coupling may be high enough that the primary becomes resonant at a different station than expected. This produces a strange result in the secondary tuning. A station at a higher frequency may now become secondary resonant when the secondary tuning capacitor is tuned to a lower frequency. That is, the station appears at the wrong location. There are tunable stations on either side of a middle frequency that can appear to be at the wrong frequency. There is a double hump that occurs, as shown in Figure 13.14. When this peculiar phenomenon occurs, the coils are said to be over-coupled. When the distance is increased, the two humps move closer to the middle frequency and at a distance called the critical-coupling, the humps merge. We, therefore, want to keep the distance between the two coils at, or greater than, the critical coupling distance. As a rule, keep the separation distance to at least one secondary coil diameter to prevent over-coupling.

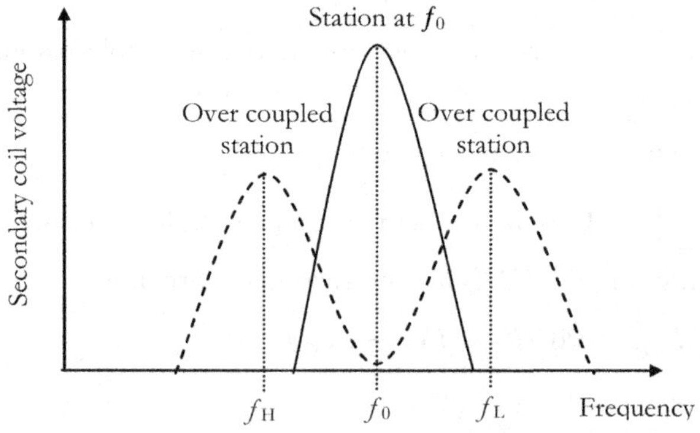

Figure 13.14
Stations appearing at the wrong locations due to over-coupling.

14

Diodes

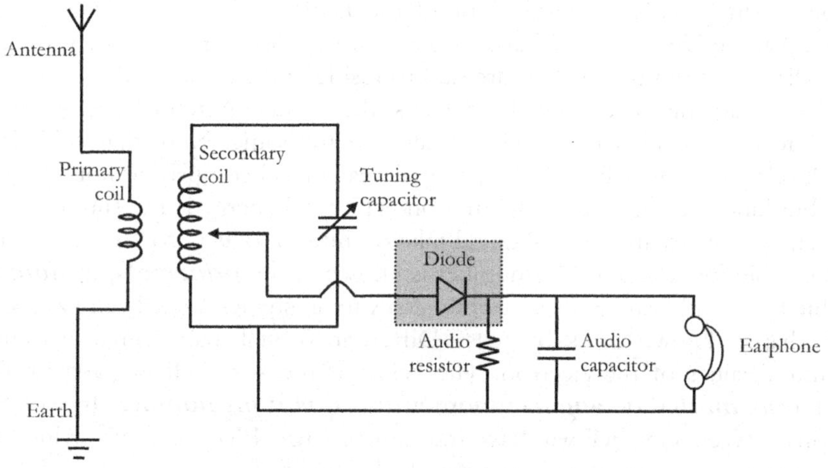

14.1 OVERVIEW

Diodes are devices that present a high resistance to the flow of current in one direction and a low resistance to current flow in the opposite direction. Figure 14.1 shows an AC voltage signal fed into a diode. The resistance to the current across the diode during the positive transition of the signal is very small, allowing the signal to pass. The resistance to the current, when the voltage is reversed, is very high, effectively blocking the current from passing. The result is that half the signal is removed, allowing the positive portion to pass.

To understand how diodes work, we need to investigate energy bands and how different types of materials, when joined, create energy potential differences at their junctions.

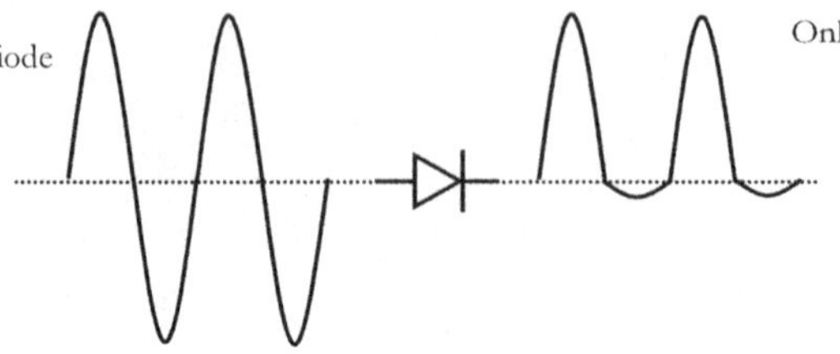

Figure 14.1
AC signal through a diode.

14.2 ELECTRON ORBITALS AND ENERGY

Consider a single isolated atom. Around the nucleus, there are a number of locations where electrons can reside called orbitals. Orbitals are grouped into shells. Orbitals having the same average distance from the nucleus are said to reside in the same shell. As each orbital in a shell has the same average distance from the nucleus, the average potential energy (as measured from the nucleus) for each orbital in a shell will also be the same. Note that orbitals can have different energy levels within a shell, with the energy in each level comprising not only the average potential energy but kinetic energy due to its motion. The total energy is the sum of both the potential and kinetic energy. Shells are quantised and take on integer values from n=1,2,... up to the outermost shell allowable for an atom. This number is known as the ***principal quantum number***.

Orbitals not only vary in distance but also vary in shape. A shell can have a number of different orbital shapes known as sub-shells. Different orbital paths give rise to different angular momentum values of the electron. The value of the sub-shell is given by ℓ=0,1,2...n-1 and is called the ***azimuthal*** or ***angular momentum quantum number***. In isolated atoms, electrons occupying a given sub-shell will have the same energy. Electrons occupying higher sub-shells will have slightly more energy than lower sub-shell orbitals.

The orientation of the sub-shells can also differ. One can think of these as orbitals with different orbital plains, and the electrons orbit around the nucleus at different angles. The value associated with the orbital orientation is called the ***magnetic quantum number***. And the number describing the orientation ranges from m_ℓ=-ℓ, 0, +ℓ. For instance, a sub-shell at ℓ=2 will have the following allowable magnetic quantum numbers -2, -1, 0, 1, 2.

Finally, an electron has an intrinsic spin and is described by the ***spin quantum number*** "s". An electron can take on one of two values; either spin-up (½) or spin-down (-½), with each orbital comprising a maximum of two electrons, one with spin-up and one with spin-down.

In summary, shells are comprised of sub-shells, and sub-shells are comprised of orbitals, with the first three quantum numbers describing a unique orbital. While it is convenient to think of electrons as solid little lumps, one can also consider them to be vibrations in a field. And the formation of orbitals as standing waves around an atom.

Now let's consider a hypothetical atom with an electron sitting on the nucleus, as shown in Figure 14.2a. If we move the electron away from the nucleus, we perform work on the electron by overcoming the electrical attraction between the positively charged nucleus and the negatively charged electron. If we now move the electron so that it is placed in a valid orbital, then the electron will gain potential energy. If we move the electron further away to another valid orbital in a higher shell, say the valence (outermost) shell, then the electron has a higher potential energy than before. The further away an orbital is from the nucleus (on average), the greater the potential energy of the occupying electron. Note that sometimes the zero potential position is set at an infinite distance from the nucleus, and sometimes the zero potential is set at the end of the valence band. Regardless of where you decide zero potential is, the potential energy is always measured from some reference point, the same way voltages are measured in a circuit with reference to the common rail.

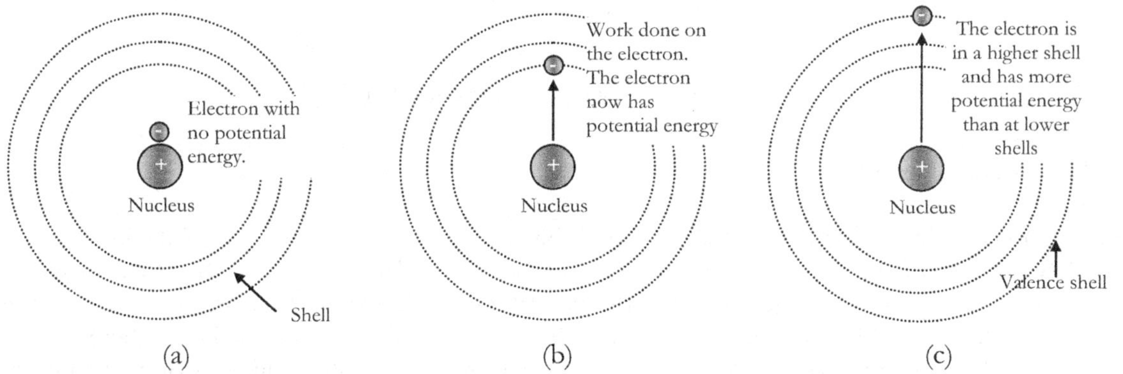

Figure 14.2
Potential energy of an electron within an isolated atom.

Orbitals in different sub-shells represent different energy states, and for an electron to move into a higher state, it must receive the correct amount of energy, or it will not move. Similarly, for an electron to move down to a lower energy orbital, it must release an exact amount of energy to occupy a lower state. The energy required to move between orbitals is ***quantised*** and specific, and any random amount of energy will not do.

14.3 ENERGY BANDS

In the isolated atom discussed above, the energy an electron has in a given sub-shell is very specific. However, when a collection of atoms combine to form a solid (as shown in Figure 14.3), the interaction between the atoms causes the allowable energy levels within a sub-shell to be smeared over a range rather than a specific value. Changing the energy an electron has due to its relationship to other electrons in the solid. As the valence electrons intimately interact with other atoms, this effect is greatest in the valence shell. The range of allowable energy levels for valence electrons is called the valence band. The difference between the energy states is very small. However, there are a lot of them, in the order of 10^{20} states in the valence band. Therefore the range of energies an electron can have within the valence band of a solid is considered continuous for modelling purposes.

Inside a material lattice, there is another electron energy band, higher than the valence band, called the ***conduction band***. Electrons occupying the conduction band are not bound to specific atoms and can move freely within the lattice. Under the influence of an electric field, these electrons will be accelerated, causing a current to flow. We will only be interested in the valence and conduction bands as these two bands determine whether a material is a conductor, semiconductor or insulator.

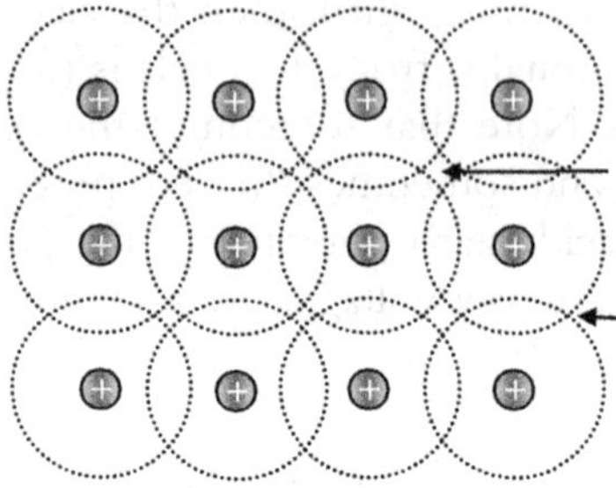

Figure 14.3
Material lattice with a conduction band.

The Fermi Level

As described above, the first three quantum numbers, n, ℓ and m_ℓ, describe the orbital in which an electron occupies, and the last number, "s" describes the electron's spin. Due to a quantum mechanical effect known as the ***Pauli Exclusion Principle***, electrons within an atom cannot have the same set of quantum numbers. As the first three numbers define an orbital and, therefore, an energy level, a maximum of only two electrons can occupy any orbital; one electron with spin-up and one with spin-down.

Consider a block of covalently bonded material chilled down to absolute zero, 0 Kelvin. The atoms and electrons have no kinetic energy, but due to the exclusion principle, all the electrons cannot reside in a single energy state but will reside at their lowest "possible" potential energy states.

If we now heat the material, some electrons will gain sufficient thermal energy (kinetic energy) to move to a higher orbital. As we continue to heat the material, more and more electrons will gain sufficient energy to move to a higher orbital state. The thermal energy gained by an electron to occupy a valid state is statistical in nature, and the distribution curve is given by an equation called the ***Fermi-Dirac distribution*** f(E),

$$f(E)=\frac{1}{1+e^{(E-E_F)/kT}} \qquad \text{Fermi-Dirac distribution} \qquad 14.1$$

where E is the energy of a valid available state, k is Boltzmann's constant, and T is the temperature in Kelvin. The variable E_F is called the ***Fermi level***, and it represents the maximum energy an electron in an atom can have when the material is at 0 Kelvin (absolute zero). Above absolute zero, the probability that an electron will possess the Fermi level energy is 50%. The Fermi level energy value differs depending on the material, and if two different materials are joined, a new Fermi level will be established.

Consider Figure 14.4. At 0 Kelvin, the probability that an electron will have an energy level at or below the Fermi level is 1. As the temperature increases, the shape of the distribution changes. However, the Fermi level probability has changed to 0.5. The probability of an electron occupying energy states below the Fermi level is high, and the probability of an electron in a higher orbital is low. As we increase the temperature further to 300 K, the distribution flattens out even further, with an increased probability that electrons will occupy available higher energy orbitals. However, the Fermi level probability always remains at 0.5 regardless of the temperature.

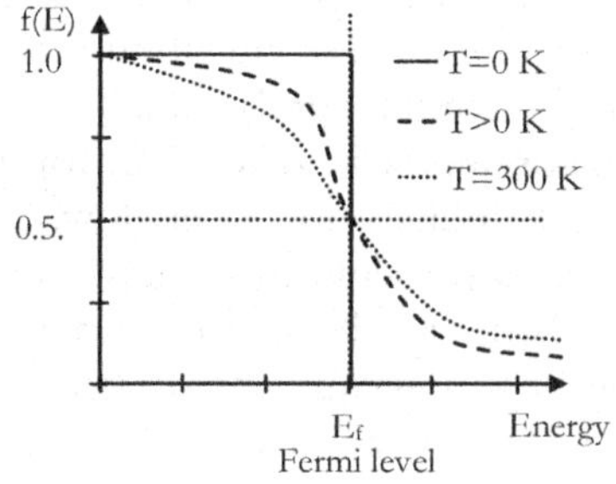

Figure 14.4
Fermi-Dirac distribution.

We can represent band energies, the Fermi-Dirac distribution curve and the Fermi level in a diagrammatical form known as a ***band diagram***, as shown in Figure 14.5. For a metal, the valence and conduction bands overlap. For a semiconductor (materials that are not good conductors but are not insulators), the valence and conduction bands are spread apart with a gap between the two bands. For an insulator, there is a large energy gap between the valence and conduction bands. The energy difference between the top of one band and the bottom of the next band is called a ***band gap***. A band gap is a region of energy that electrons cannot have. When we refer to the band gap, we are talking about the band gap between the valence and conduction bands, as these are the energy levels we are interested in.

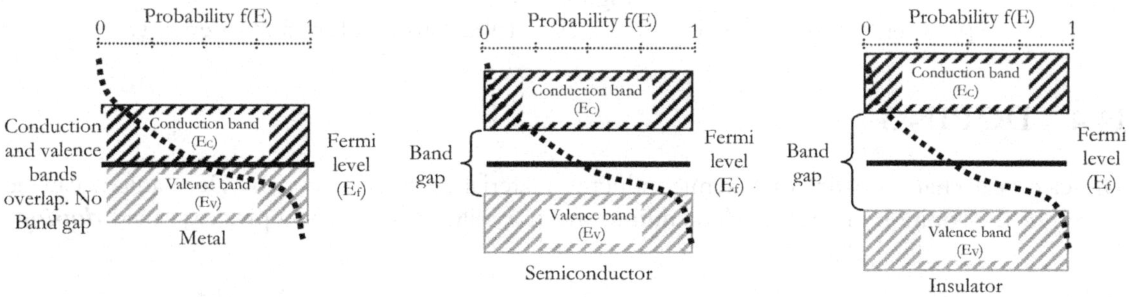

Figure 14.5
Energy band diagram of metal, semiconductor and insulator.

For metals, the Fermi level resides between the overlapping bands, and there is no band gap. For semiconductors and insulators, the Fermi level resides inside the band gap. This does not mean an electron can exist in the band gap. The distribution of elections in any band is given by the ***density of states*** function, which gives the density of allowed electron energy states for a given energy level within a band. To determine the concentration distribution of electrons at a given energy level, the density of states g(E) is multiplied by the Fermi-Dirac distribution f(E).

$$D_N = g(E)f(E)$$ Electron density distribution within a band 14.2

From this, we get the number of charges per unit volume (the number of charges per cubic centimetre). Each band has its own density of states distribution function, and for a band gap, it is zero, so the distribution of electrons in a band gap is also zero, see Figure 14.6.

Looking at the Fermi-Dirac distribution for metal in Figure 14.5, we can see that the probability of finding an electron in the conduction band is high. The probability that an electron is in the conduction band for a semiconductor is lower than for a metal. And the probability that an electron is in the conduction band in an insulator is very low.

The band gap between the levels represents the relative amount of energy required for an electron to jump from the valence band to the conduction band. In metals, the valence and conduction bands overlap, resulting in a large number of electrons in the conduction band. For semiconductors, more energy is required for an electron to jump from the valence band to the conduction band. For insulators, the gap is very large, requiring the electron to gain more energy before a jump can occur, resulting in even fewer electrons in the conduction band.

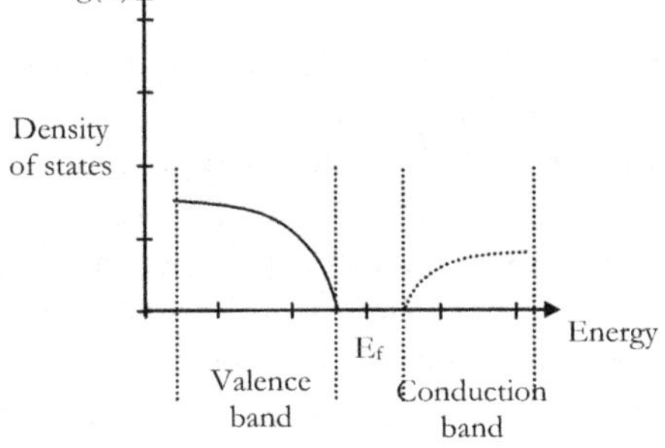

Figure 14.6
Typical semiconductor density of states distributions as a function of energy.

14.4 DOPING

The electrical characteristics of a semiconductor material such as Silicon or Germanium can be altered by the addition of another element into the crystalline structure in a process called ***doping***.

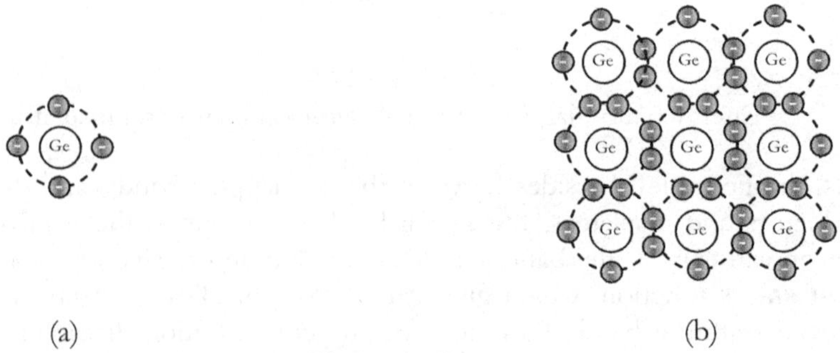

Figure 14.7
Pure Germanium crystal.

Consider a Germanium atom, as shown in Figure 14.7a, with four electrons in its valance shell.

Germanium will combine with other Germanium atoms to form a stable crystal lattice by sharing its four outer electrons with other atoms in a covalent bond, Figure 14.7b. In this configuration, very few electrons will reside in the conduction band.

Now consider the element Arsenic. It has five electrons in its valence shell, Figure 14.8a. If we add a little Arsenic to the Germanium crystal, four of the valence electrons of the Arsenic will be shared with the Germanium forming a strong covalent bond. However, there will now be an extra electron in the crystalline structure that is not covalently bonded, Figure 14.8b. This electron is free to move and can easily enter the conduction band. Note that the material is still overall electrically neutral as there are just as many electrons as there are protons. This material is called N-type material, N for negative, and electrons are said to be the majority current carriers.

If instead of doping Germanium with Arsenic, say we doped it with Indium having three valence electrons, as shown in Figure 14.9a. The three valance electrons will covalently bond with

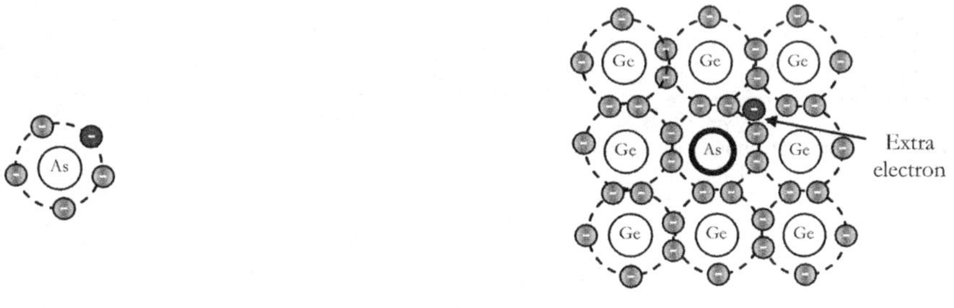

(a) (b)

Figure 14.8
Germanium crystal doped with Arsenic (N-type material).

the Germanium, however, not all the available covalent bonding places are full, and any passing electron will be attracted to this position, Figure 14.9b. The absence of an electron required for a stable bond can be considered a ***hole*** in the bonding where any free electrons can favourably reside. This is called P-type material (P for positive), and the holes are said to be the majority current carriers, even though they don't actually move. However, once again, this material is overall electrically neutral. Devices made with two different majority carriers are called ***bipolar-junction*** devices.

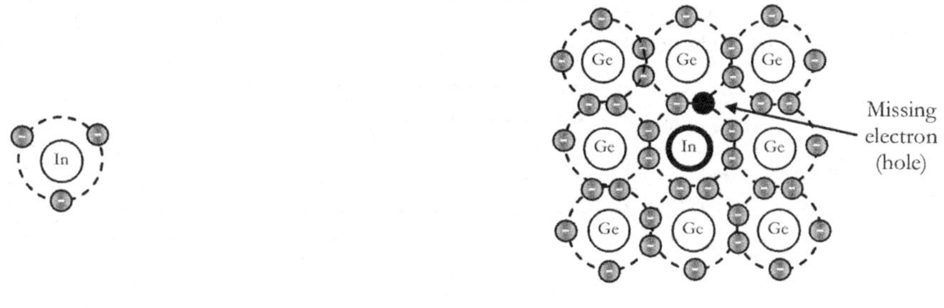

(a) (b)

Figure 14.9
Germanium crystal doped with Indium (P-type material).

Doping has the effect of changing the Fermi level. For N-type material, the Fermi level moves towards the conduction band, indicating there will be more electrons in the conduction band than before the Germanium was doped. For P-type material, the Fermi level moves towards the valence band, indicating fewer electrons in the conduction band but with more holes in the valence band. See Figure 14.10. Doping has very little effect on the valence and conduction band energies, however, and they essentially remain unchanged.

The band levels for undoped materials are called ***intrinsic band levels***, and the Fermi level is called the ***intrinsic Fermi level***. When different materials are joined (say N type and P type material), a new Fermi level for the entire material will be established.

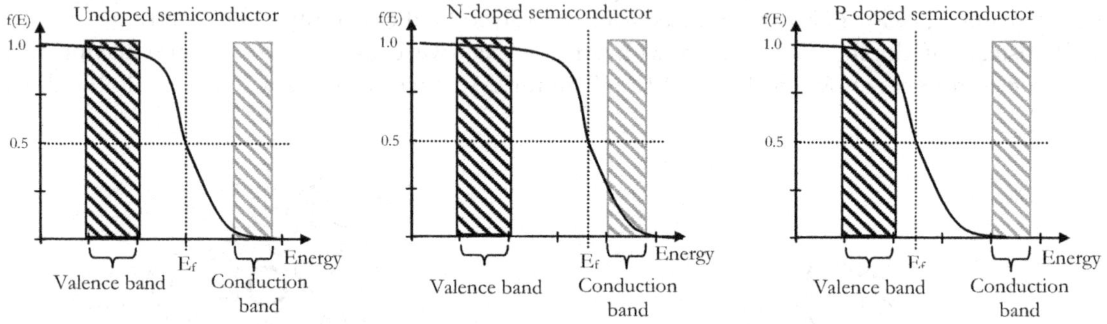

Figure 14.10
Shifting the Fermi level by doping.

14.5 THE PN JUNCTION

Consider we have some P and N material, and we bring them together. Before any time has elapsed, the initial energy band diagram at this instant in time is shown in Figure 14.11. Note that the conduction and valence bands are the same for each material, as doping does not affect these band energies very much. Due to thermal energy, there will be some electrons in the N-type

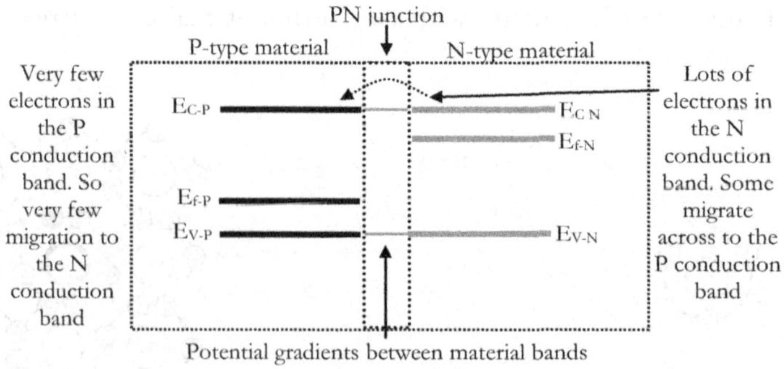

Figure 14.11
PN junction at initial contact.

material conduction band that will migrate across to the P-type material conduction band. The number of electrons in the P side conduction band is vanishingly small (see Fermi-Dirac distribution in Figure 14.10), so very few electrons will migrate from the P side to the N side. Due to the P material's probability distribution, many migrating electrons from the N side will lose energy and end up in the P material's valence band. These migrating electrons are attracted to the holes in the P material's valence band, resulting in a region near the material's junction with excess electrons. Both the P and N materials were initially electrically neutral; now, the P side junction region will look negative, and the region near the junction on the N side, due to the loss of electrons, will look positive. See Figure 14.12. Together these regions are called the depletion region as they are depleted of their majority charge carriers, as shown in Figure 14.12b.

This process continues until the Fermi level energy becomes the same across the entire material. The relative positions of the conduction and valence bands to the Fermi level for each of the separate materials remain the same, however. If we now redraw Figure 14.11, aligning both materials to the new Fermi level, the conduction and valence bands in the P material shift with respect to the bands in the N material.

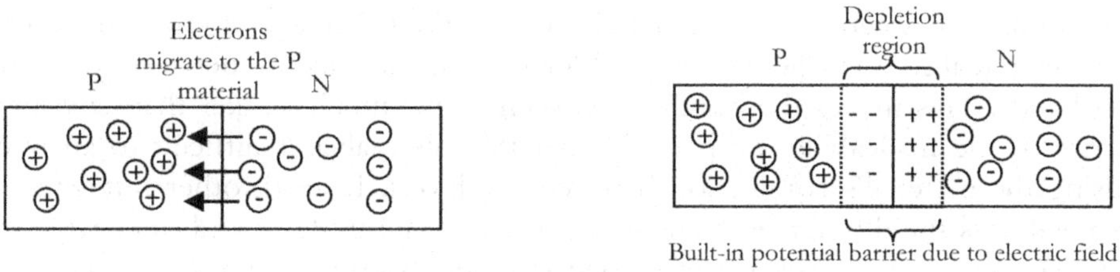

Figure 14.12
P and N-type material bonded together.

Looking at Figure 14.13, the start of the conduction band for the P material is at a higher energy level than for the N material. Where the two materials join, the energy levels form a gradient due to the migration of electrons. This results in a phenomenon known as band bending across the PN junction, where a voltage gradient forms between the bands. The Fermi level has equalised, resulting in a potential difference V_{bi} between the two conduction band energy levels. This potential difference is known as the ***built-in voltage barrier*** or ***built-in potential barrier***.

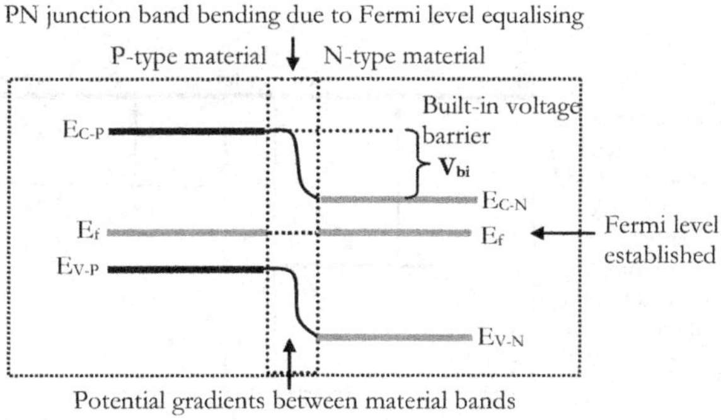

Figure 14.13
PN junction at thermal equilibrium.

This potential barrier must be overcome in order for significant amounts of current to flow from the N material to the P material. For Silicon PN diodes, this barrier potential is typically around 0.6 to 1.0 volts. Note that the built-in potential barrier cannot be directly measured by a multimeter. The potential exists at the junction between the two materials, and there is no net electric field outside the junction to cause current to flow. That is, it is not a voltage source. It should be thought of as an energy barrier to overcome.

In the P material, the start of the conduction band energy level is higher than for the N material. However, as very few electrons exist in the P material's conduction band, very few electrons will migrate across; and this extremely small leakage current can generally be ignored.

14.6 SCHOTTKY DIODES

There is another family of diodes where a junction is formed between metal and semiconductor material. Typically N-type material is used, and we shall examine the MN-type diode. In metal, the valence and conduction bands overlap, and the Fermi level sits inside these overlapping energy bands. The bonding between atoms in a metal is due to the collection of electrons residing in this overlapping area (called metallic bonding), blurring the distinction between the valence and conduction bands. This makes the bonding between atoms much weaker than covalent bonding and gives rise to the malleability property of metals. This makes it difficult to analyse the MN junction using the material's conduction band energy levels. Instead, other reference points are used. For metal, it is the difference in energy between the Fermi level and the energy required to remove an electron from the surface of a solid into the external vacuum. This energy value (in volts) is given by the ***work function* Φ**, and

$$\Phi_M = E_{VAC} - E_F \qquad \text{Work function (volts)} \qquad 14.3$$

where E_{VAC} is the energy an electron will have once freed from the lattice and is in the vacuum, and E_F is the Fermi level.

For a semiconductor, the energy is referenced from the start of the conduction band to the vacuum energy and is called the ***semiconductor electron affinity***, and

$$\chi = E_{VAC} - E_C \qquad \text{Semiconductor electron affinity (volts)} \qquad 14.4$$

where E_C is the energy level at the bottom of the conduction band.

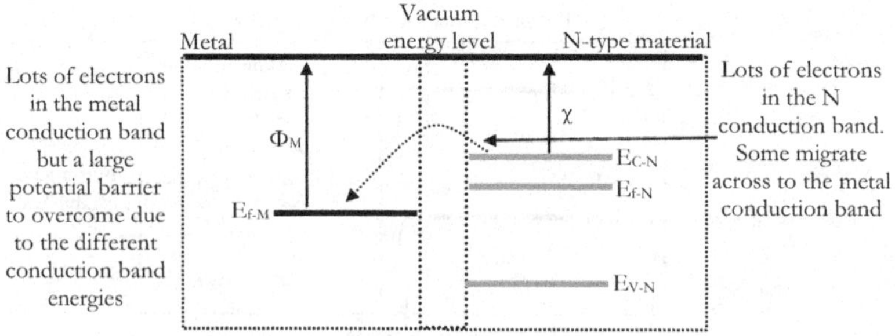

Figure 14.14
MN junction at initial contact.

Figure 14.14 shows the energy levels at initial contact between a metal and N-type material, before band bending has occurred. The Fermi level (and most of the conduction band) of the metal is at a lower level than the bottom of the semiconductor's conduction band and forms a barrier (known as the Schottky barrier) to the movement of electrons from the metal to the N material. In PN-type diodes, there is no difference between the intrinsic conduction band energies of each of the materials, so this potential does not exist.

On the N side, the electrons in the conduction band are at a higher energy state and can easily migrate across to the metal. The migration of electrons continues until the Fermi level of the material is established at equilibrium, as shown in Figure 14.15. A positively charged depletion region exists on the N side. The majority carriers in the metal are also electrons, and due to the vast numbers of free electrons, no depletion region exists on the metal side. Materials using the same type of majority carriers are called ***unipolar junction*** devices.

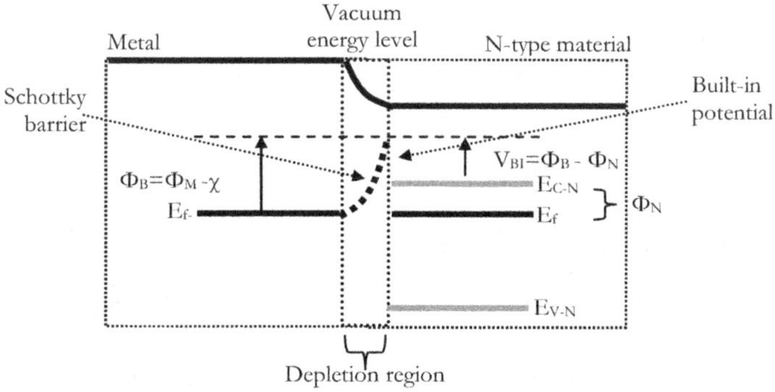

Figure 14.15
MN junction at thermal equilibrium.

There is, however, a dense but narrow region of negative charge build-up on the metal side due to the electrostatic attraction to the positive ions on the N side, see Figure 14.16. With the Schottky barrier preventing electrons from moving to the N side. This negative region on the metal side forms a potential barrier to the movement of electrons from the N side (like charges repel). We now have two potential differences; the Schottky barrier and the built-in voltage barrier.

Just like the PN diode, this built-in voltage barrier must be overcome to allow the flow of current from the N material to the metal. The built-in voltage of a Schottky diode is smaller than the built-in voltage of a PN junction diode and ranges from around 0.3 volts to 0.5 volts. Germanium point contact diodes are a form of Schottky diodes, with a small area of contact between the metal (cat's whisker) and the semiconductor material.

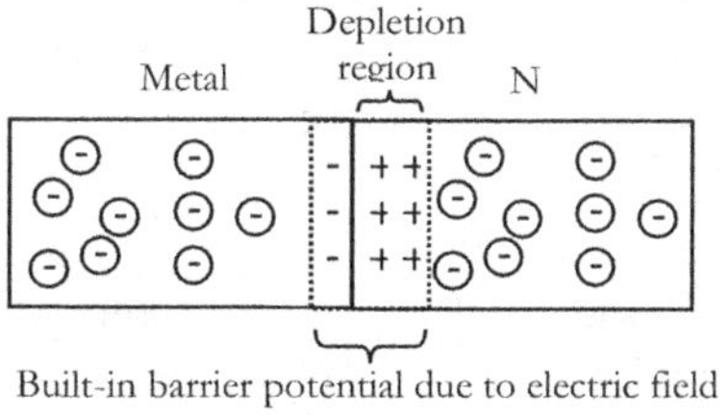

Figure 14.16
Metal and N-type material bonded together.

14.7 BIASING THE DIODE

Applying an external DC voltage across a diode is called biasing the diode. The diode may be biased in three ways; unbiased, forward biased or reverse biased. In the following example, we will use a PN junction diode, but the same description applies to Schottky diodes.

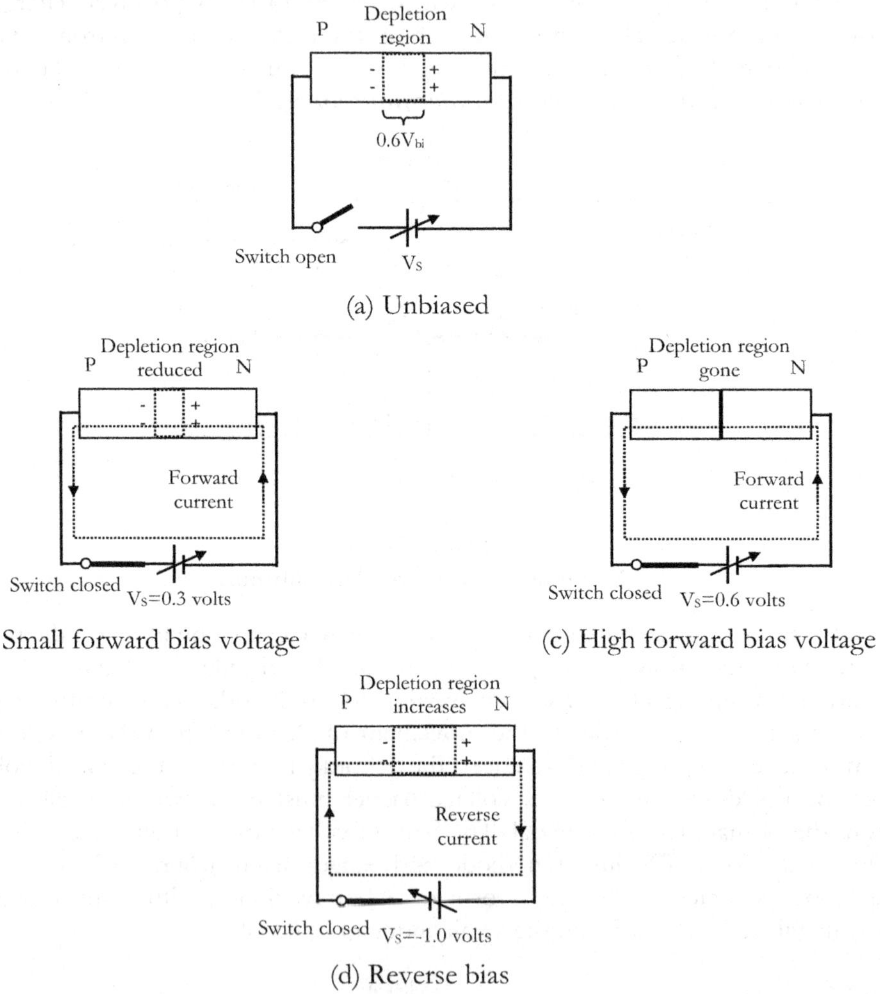

Figure 14.17
Diode biasing.

In Figure 14.17a, we have a PN diode with a built-in potential barrier of 0.6 volts, connected to a variable voltage source in series with a switch. When the switch is open, no external voltage is applied across the diode, and the diode is said to be ***unbiased***.

In Figure 14.17b, the voltage source is set to 0.3 volts. The positive side of the battery is connected to the P material, and the negative side is connected to the N material. The switch is closed, and 0.3 volts are applied across the diode. This applied voltage is in opposition to the built-in voltage barrier and reduces the barrier potential. When the built-in potential barrier is decreased by an applied voltage, the diode is said to be ***forward biased***.

This reduction in the barrier potential allows some current (electron current flow) to flow across the diode from the N side to the P side and through the circuit. You may think that no current will flow as we have not overcome the barrier potential. However, there will be electrons with enough thermal energy to overcome this reduced barrier. And the greater the reduction in the barrier, the more current will flow. If you were to measure the voltage across the diode, you would see a positive voltage drop. The current flowing across the internal diode resistance, from Ohm's law, is a voltage drop.

If the applied voltage is increased to 0.6 volts (Figure 14.17c), the barrier potential disappears, and we are only left with some material and connection resistance, and a lot of current will flow through the circuit. In real-world circuits, a resistor is added in series to prevent excessive current that could damage the diode. Notice that as the forward bias voltage increases, the depletion width decreases.

If the voltage source is reversed, as shown in Figure 14.17d, the applied voltage increases the barrier potential. The P material conduction band has only a few states occupied, and some electrons will flow across the junction resulting in a very small current. This tiny current is called the ***reverse current*** and is typically in the order of micro or even nanoamps. When the applied voltage causes the potential barrier to increase, the diode is said to be ***reverse biased.*** When the diode is reversed biased, the depletion region expands.

The following is the circuit symbol for a diode. The thick black line is the ***cathode*** of the diode (the side connected to the negative potential), indicating the side where electrons enter the component when forward biased. The diode lead connected to the positive potential is called the ***anode***. For Schottky MN and PN silicon diodes, the N material side is the cathode.

Diode

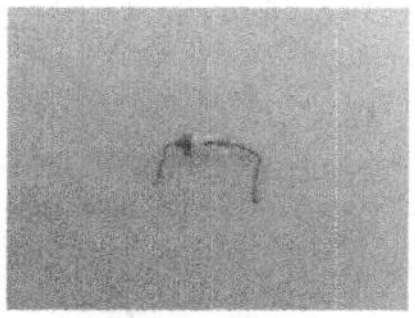

The circuit diagram for a Schottky diode is

Schottky diode

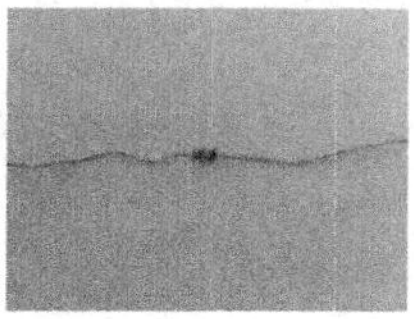

14.8 THE I-V CURVE

If we graph the DC current versus the voltage across an ideal diode, we get a curve resembling that shown in Figure 14.18 called the ***I-V curve***. At the knee of the curve, significant current begins to flow. The voltage around this area is called the ***turn-on voltage***. Schottky diodes have a typical turn-on voltage of around 0.2 to 0.4 volts, and Germanium point contact diodes with about 0.3 to 0.5 volts. This term, however, is a little misleading. As we can see from the graph, current flows across the diode well below the so-called turn-on voltage. Also, the current at which this voltage occurs is not defined, and some multimeters will show the voltage across the diode at 0.5 mA of current and others at 1 mA, giving different turn-on voltages for the same diode. Also, note that the barrier potential is not the same as the turn-on voltage. The barrier potential is typically larger than the turn-on voltage.

When the diode is forward biased such that the barrier potential approaches zero, a lot of current will flow in the order of tens or hundreds of milliamps for small signal diodes and many amps for power diodes. In electronic circuits dealing with tens of milliamps or more, the small current before the knee is ignored, and the diode is treated something like a switch. However, the voltages in crystal radios can range anywhere from well below to well above the knee, so we cannot ignore voltages below the so-called turn-on voltage. In fact, we cannot ignore the small current when the diode is reverse biased.

Notice from the I-V curve that when a diode is reversed biased, the small amount of reverse current is quite constant for a large voltage range. Also, note the difference in scale for the reverse current and forward current. There is a point, however, at which the reverse bias voltage is so large that the barrier potential becomes inconsequential and a large amount of reverse current flows. This voltage is known as the ***breakdown voltage***, and the current at this voltage is called the ***avalanche current***. These breakdown voltages typically occur in the tens or even hundreds of volts. Therefore this region does not concern us, as the voltages in a crystal radio are very much smaller than this.

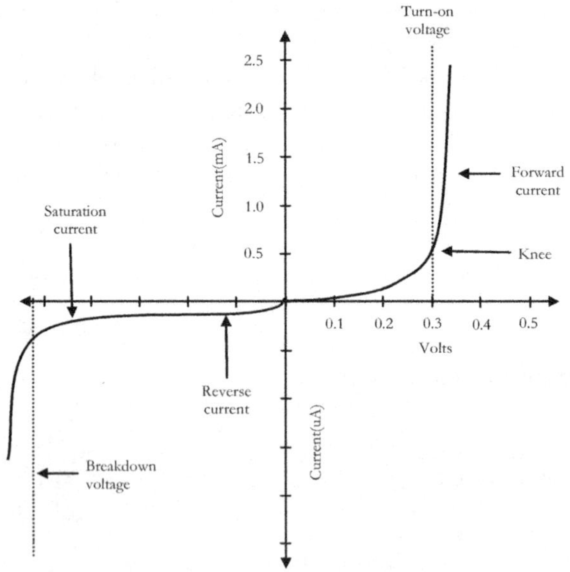

Figure 14.18
Ideal diode I-V curve.

For an idealised diode, the current flowing through a diode is given by the following equation

$$I_D = I_S(e^{\frac{qV_D}{kTn}} - 1) \qquad \text{Ideal diode current equation} \qquad 14.5$$

where I_S is the saturation current, V_D is the voltage dropped across the diode, **n** is the ideality factor, q is the charge of an electron, k is Boltzmann's constant, and T is the absolute temperature in kelvin. The term q/kT is approximately 25.85mV at 300K (room temperature) and is written as

V_T. Substituting V_T in Equation 14.5, we get

$$I_D = I_S(e^{\frac{V_D}{nV_T}} - 1)$$ Ideal diode current equation 14.6

The ideality factor is typically between one and two, and sometimes higher, and represents how close a diode fits this ideal diode equation. A value of one means the diode closely matches the ideal curve. The ideality factor can change with voltage, and sometimes multiple ideality factors are given depending on the voltage range of interest. It is common for diodes not to fit this equation very well at all. In these circumstances, we can fit a power or a polynomial equation to the I-V curve for better results.

When the voltage across the diode is negative (reversed biased), the exponential term is less than one. As the negative voltage increases, this term becomes smaller and smaller. When the reverse bias voltage is large, this term tends to zero, and the reverse current flowing across the diode is simply the negative of saturation current, that is $-I_S$. The saturation current can be considered as the reverse bias current flow when a large negative voltage is placed across the diode but below the breakdown voltage. For small signal Germanium and Schottky diodes, the saturation current is typically measured at around minus ten or twenty volts (see measurement techniques Chapter). For a Germanium 1N34A diode, the saturation current is about 30 uA at minus 10 volts but can vary from diode to diode. The saturation current is also temperature dependent. On many specification sheets, diode measurements are made at 25° Celsius. As a rule of thumb, the saturation current doubles for every 10° C increase in temperature.

14.9 DC AND DYNAMIC RESISTANCE

From Ohm's law, a component's resistance (or impedance) is calculated by dividing the voltage drop across a component by the current flowing through the component. For a resistor, this ratio is constant. However, looking at the I-V curve in Figure 14.18, we see the ratio of voltage to current for a diode is not constant, and the diode's resistance changes depending on the voltage and current. For a constant DC voltage, the resistance at a given voltage is simply

$$R_D = \frac{V_D}{I_D}$$ DC resistance of a diode 14.7

where V_D is the voltage drop across the diode, and I_D is the current passing through the diode at the given voltage. This is called the ***diode's DC resistance.***

When a small time-varying signal, such as a sine wave, is applied across the diode, as in Figure 14.19, the voltage and current relationship varies depending on the instantaneous value of the signal at any given time. When a diode is DC biased so that the operating point is above the potential barrier, the I-V curve looks relatively constant, and the resistance is also fairly constant and appears Ohmic (line a resistor). If a small AC signal is passed across the diode, the average resistance is given by the change in voltage divided by the change in current. The average resistance due to a varying voltage signal is called the ***diode's dynamic resistance*** and is designated by a lowercase "r" and is given by

$$r_D = \frac{\Delta V_D}{\Delta I_D}$$ Dynamic AC resistance of a diode 14.8

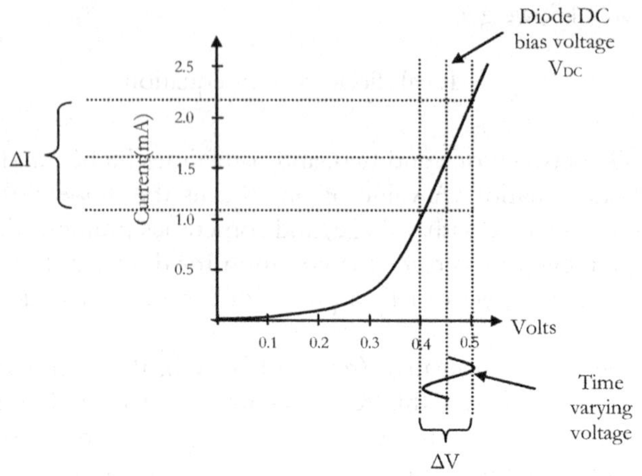

Figure 14.19
Dynamic resistance.

In a crystal radio, the voltage induced across a coil's tap is applied across the diode. Before any charge is placed on the earphone, the signal will appear across the diode, as shown in Figure 14.20a, and the voltage across the diode will range from some negative peak value to some positive peak value. The instantaneous resistance changes dramatically. Using the dynamic

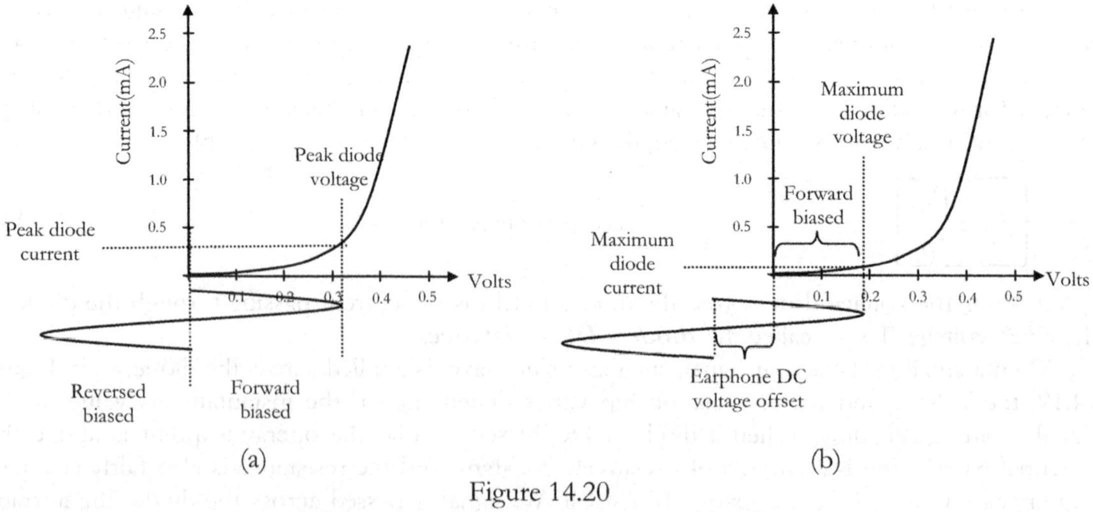

Figure 14.20
AC signal across the diode in a crystal radio.

resistance will give an average resistance. We will divide the crystal radio analysis into two sections; one when the diode is forward biased and one when the diode is reversed biased. We will use the average forward and average reverse voltage of the AC signal to get an average resistance (dynamic resistance) in each bias mode. As the diode is not forward DC biased, the voltage range will be from 0 to the average sine wave voltage. Therefore the dynamic resistance given by 14.8 will be the same as the DC resistance given by 14.7.

Eventually, there will be some DC voltage build-up across the earphone (discussed in the next Chapter), resulting in the diode being negatively DC biased. This results in the diode being forward biased for a fraction of the total signal period, and the peak voltage (and average) across the diode will be reduced, as shown in Figure 14.20b.

14.10 DIODE CAPACITANCE

Examining the depletion region of a diode, either in reverse or forward bias modes, we see that there is always a separation of charge. This separation of charge acts similar to the charge stored on the plates of a capacitor, and as such, the diode has capacitance. The charge separation around the depletion region is called the ***junction capacitance***, or the ***depletion layer capacitance***, see Figure 14.21.

Also shown in Figure 14.21 is another charge distribution that is highest near the depletion region and falls away to the edges of the diode. This charge distribution is due to minority carriers and is called the ***diffusion capacitance***. We will examine both types of capacitance, starting with the junction capacitance.

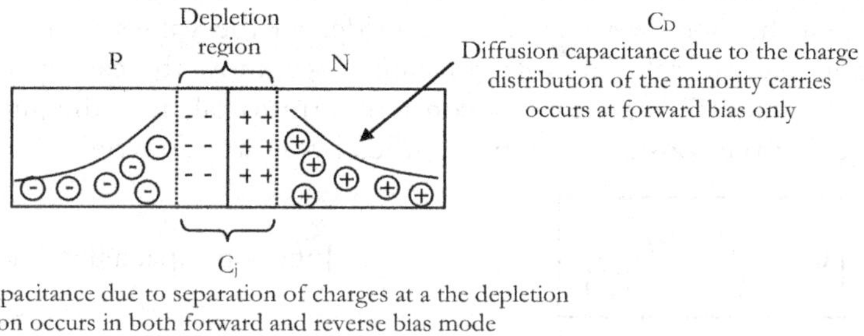

Figure 14.21
Diode junction and diffusion capacitance.

Junction Capacitance

The value of the junction capacitance is inversely proportional to the width of the depletion region; the narrower the depletion region width, the greater the capacitance. Figure 14.22 plots the junction capacitance as a function of diode voltage. In reverse bias mode, as the negative voltage increases, the depletion region width increases, causing the junction capacitance to decrease. In forward bias mode, the depletion region width decreases, causing the junction capacitance to increase.

Junction capacitance is often stated in a diode's specification sheet at a reverse bias of -1 volt at a frequency of 1 MHz. Typically, junction capacitance values range from a fraction of a pico-farad to many pico-farads. When the diode's forward voltage drop approaches the built-in potential, the junction capacitance becomes very large and can be in the order of nano-farads. However, the capacitance flattens out and never entirely disappears, as shown in Figure 14.22. The junction capacitance for a Schottky diode is given by the following equation

$$C_j = \sqrt{\frac{qD_N \varepsilon A^2}{2}} \sqrt{\frac{1}{(V_{bi} - V_D)}} \quad 14.9$$

where q is the charge of an electron, D_N is the doping density, ε is the material permittivity, and A is the surface area of contact at the junction. This equation is only valid for forward voltages below the built-in voltage of the diode.

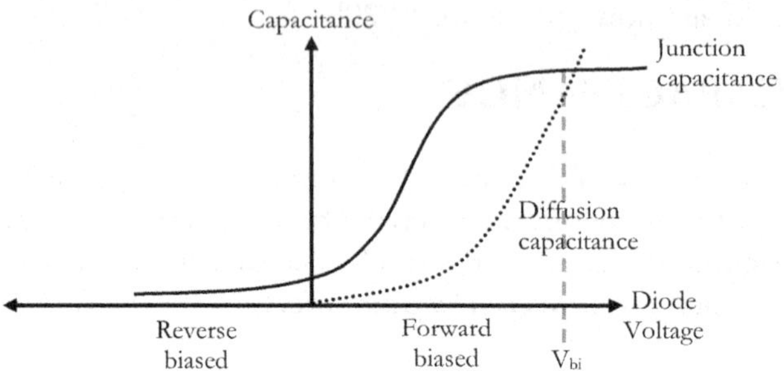

Figure 14.22
Diode capacitance versus diode voltage.

All the values in Equation 14.9 are constant for a given diode, with the diode applied voltage V_D being the only variable. If we consider all the values under the first square root term as a constant, we can calculate this constant if we know the capacitance for a given voltage and the built-in voltage. We can use this constant term to calculate the junction capacitance for any diode voltage. Let the constant term be signified by β_D we can write

$$C_j = \sqrt[N]{\frac{\beta_D}{(V_{bi} - V_D)}}$$ Junction capacitance of a diode 14.10

where V_D is the voltage drop across the diode, V_{bi} is the built-in potential, N is the root index (typically 2) and β_D is the accumulation of all the constant values giving a single value. Equation 14.10 is only valid for diode drops less than the built-in potential. At or above the built-in potential, you can set the voltage difference in the denominator to some small arbitrary value, say a nano-volt (10^{-9}) and use this constant value at or above the built-in voltage.

The root index N is generally 2, that is we take the square root. However, not all diodes follow the square root law and the root index can be different for all voltage levels, or can change for a range of voltage levels. However, the deviation from the square root is typically small and only has a noticeable impact on our model for large reverse bias values at or below a reverse bias of one volt or more. For modelling purposes set the index to 2.

If we know or are given the junction capacitance at some voltage, we can calculate β_D, the intrinsic properties of a diode, by rearranging 14.10 to give

$$\beta_D = C_j^N (V_{bi} - V_D)$$ Intrinsic properties of a diode 14.11

This value of β_D can then be used in Equation 14.10 to calculate the junction capacitance for any given reverse or forward voltage value less than the built-in voltage.

For example, the typical junction capacitance of a BAT85 Schottky diode is given in the specification sheet as 10.0 pF at a reverse voltage of 1 volt. The built-in potential voltage is rarely given in a specification sheet. If the built-in voltage is not given, select a value of 0.4 volts for a

Schottky and 0.5 volts for a point contact Germanium as a guesstimate. The built-in potential is not given, so we will use a value of 0.4 volts. Using Equation 14.11, we get

$$\beta_D = C_j^2(V_{bi} - V_D)$$
$$= (10.0x10^{-12})^2(0.4 + 1.0)$$

And

$$\beta_D = 1.4x10^{-22}$$

Using this value of β_D in Equation 14.10, we can calculate the junction capacitance for a given applied voltage

$$C_j = \sqrt{\frac{\beta_D}{(V_{bi} - V_D)}}$$
$$= \sqrt{\frac{1.4x10^{-22}}{(0.4 - V_D)}}$$

At a forward diode voltage drop of 0.1 volts (V_D=0.1), the junction capacitance is

$$C_j = \sqrt{\frac{1.4x10^{-22}}{0.3}}$$
$$= 21.6x10^{-12}$$

The junction capacitance is proportional to the surface contact area between the semiconductor materials at the junction. Typically the surface contact area between the metal and the semiconductor of a Schottky diode is smaller than for PN diodes, resulting in lower junction capacitance values. For Germanium point contact diodes, the contact surface area is even smaller than for a typical Schottky and results in very low junction capacitance values.

In the Chapter on measurement techniques, a technique is described for measuring both the junction capacitance and the built-in potential of a diode. Note that these are very tricky measurements and are susceptible to noise, and small differences in measurement values can make a big difference to the calculated result. I took several measurements per diode and used the average value. I do not suggest performing these measurements as a useful way of selecting diodes. I performed these measurements only to prove the model, and I have included them for completeness. Later we shall develop rules for selecting candidate diodes. Simply buy the diodes and test them in your crystal radio to see how they perform.

Diffusion Capacitance

As shown in Figures 14.21 and 14.22, there is another type of diode capacitance due to the minority carrier distribution across the diode. This type of capacitance occurs at forward bias when the minority carrier charges distribute themselves across the bulk of both materials. This change in charge distribution due to a change in voltage gives rise to capacitance called the ***diffusion capacitance*** and is expressed as

$$C_D = \frac{dQ_D}{dV} \qquad 14.12$$

where dQ_D is the change in charge distribution, and dV is the change in voltage.

From the N side, the electrons migrate across the diode and the depletion region, leaving behind a distribution of positive minority carriers (holes). These holes are created when a free electron leaves the atom, and the holes are destroyed when an electron recombines. The creation and destruction of holes continue while the diode is forward biased. On the P side, the electrons are considered the minority carriers (for a PN diode) and combine with the holes in the P material to form a negative minority carrier (electron) distribution, see Figure 14.21. The average time for a minority carrier to exist before recombination is called the ***average transit time*** τ_t, or mean carrier lifetime. The amount of charge in transit required for recombination is given by

$$Q_D = I_D \tau_t \tag{14.13}$$

where I_D is the diode current. Substituting Equation 14.13 into 14.12, the diffusion capacitance can be written as

$$C_D = \frac{dI_D}{dV} \tau_t \tag{14.14}$$

Using the ideal-diode Equation 14.6 for the current I_D and taking the derivative with respect to the voltage, we get

$$\begin{aligned} C_D &= \frac{d\left[I_S \left(e^{\frac{V_D}{nV_T}} - 1 \right) \right]}{dV} \tau_t \\ &= \frac{1}{nV_T} \left[I_S e^{\frac{V_D}{nV_T}} \right] \tau_t \end{aligned}$$

And the diffusion capacitance is approximately

$$C_D = \frac{\tau_t I_D}{nV_T}$$

However, often the ideal diode equation does not fit very well for real-world diodes, especially at higher voltages. As this capacitance will change as the signal varies, we will deal with average values and calculate the diffusion capacitance as the change in current divided by the change in voltage (the inverse of the dynamic resistance) times the average transit time. That is from Equation 14.14, we get

$$\boxed{C_D = \frac{\Delta I_D}{\Delta V_D} \tau_t} \qquad \text{Diffusion capacitance of a diode} \tag{14.15}$$

If we know the value of the diffusion capacitance for a given voltage, we can calculate the average transit time τ_t by

$$\boxed{\tau_t = C_D \frac{\Delta V_D}{\Delta I_D}} \qquad \text{Average transit time} \tag{14.16}$$

We can then use the value of τ_t in Equation 14.15 to calculate the diffusion capacitance for a given forward voltage. Small signal Schottky diodes have a typical average transit time of around 50-200 nano-seconds.

For metal junction MN diodes, there will be some charge distribution in the semiconductor material. However, the average transit time will be much smaller for Schottky diodes than for PN junction diodes; therefore, Schottky diodes will have much smaller diffusion capacitance values.

14.11 REVERSE RECOVERY TIME

When a diode changes from forward to reverse bias, the depletion region is reformed. The time taken to reform the depletion region is called the ***Reverse recovery time*** and is an important factor in determining a diode's response to high-frequency signals. The shorter the recovery time, the better the diode responds to high frequencies. If the period of the input signal is too quick to allow the depletion region to reform, the input signal will not be faithfully reproduced, and the output signal will look distorted. The depletion region in a Schottky diode is smaller than for a PN, and so the time taken to reform the depletion region is less than the time taken for a PN diode; this makes the Schottky (and Germanium point contact) diodes suitable for high-frequency applications.

14.12 DIODE CIRCUIT EQUIVALENT MODEL

In forward bias mode, the diode will have both junction and diffusion capacitance in parallel with the dynamic resistance. The diode will also have some small series resistance due to the lead wires connecting with the semiconductor material, as shown in Figure 14.23a. In reverse bias mode, the diffusion capacitance does not exist, so our reverse bias model only has the junction capacitance, as shown in Figure 14.23b.

(a) Forward bias diode model (b) Reverse bias diode model

Figure 14.23
Diode circuit equivalent models.

14.13 DIODE SPECIFICATION SHEETS

Table 14.1 lists the most common diode parameters given in a diode's specification sheet. Even though the turn-on voltage is the most often "discussed" parameter, this value is seldom given. As we have seen, this is a rather nebulous term with no exact definition. Instead, specification sheets give one or more values of forward current for specified forward voltages. Also, the values are given at some specific temperature, typically 25 °C.

Quite often, a diode's maximum values are given in a specification sheet. Table 14.2 lists the typical maximum values specified. Exceeding these values will most likely result in damage to the diode.

Diode electrical characteristics				
Parameter	Explanation	Symbol	Example value	Typical test condition
Forward voltage drop	The DC voltage drop across a diode at a given forward current.	V_F	1 V	Forward current of 5 mA
Reverse leakage current	The DC current at a given reverse bias voltage	I_R	30 uA 500 uA	10 volts reverse bias 50 volts reverse bias
Junction capacitance	The capacitance across the junction. Typically at a reverse voltage of -1 volts.	Cj	4 pF	-1 volts, 1 MHz frequency
Reverse recovery time	The time it takes for the diode to reform the depletion region after being reverse biased. This will determine the maximum useful frequency for the diode. The shorter the time, the higher the frequency.	t_{rr}	8 nS	

Table 14.1
Typical electrical diode characteristics given in a specification sheet.

Diode absolute maximum ratings				
Parameter	Explanation	Symbol	Example value	Typical test condition
Peak reverse voltage (Breakdown voltage)	The reverse voltage at which the avalanche current will occur.	PIV	65 volts	
Peak forward surge current (Non-repetitive)	The maximum current for a brief period as might occur when first applying power.	I_{FSM}	500 mA	1 second
Peak forward surge current (Repetitive)	The maximum repetitive peak voltage that will occur for an AC signal.	I_{FSR}	300 mA	
Forward continuous current	Maximum forward DC current before damage occurs.	I_F	200 mA	
Max operating junction temperature	Maximum temperature at which the diode will operate.	T_J	+75 °C	

Table 14.2
Typical maximum electrical diode ratings given in a specification sheet.

14.14 DIODE PARAMETERS FOR THE MODEL

To incorporate a diode in our crystal radio model, we need the following

1. A forward bias I-V current equation as a function of diode voltage.
2. A reverse bias I-V current equation as a function of diode voltage.
3. The junction capacitance at a given voltage, typically at -1 volts.
4. The built-in voltage of the diode.
5. The average transit time to calculate the diffusion capacitance.

The ideal diode current Equation 14.6 rarely fits a real-world diode. So we need to create an I-V curve and apply a best-fit equation to the data (see measurement techniques Chapter). From experience, a power equation or a polynomial typically will fit the data best. You will need to create two equations; one for the forward bias and one for the reverse bias. Luckily spreadsheet computer programs are good at determining best-fit curves. Enter your I-V data into the spreadsheet and try different function types to see which fits best.

In this book, we will use three different diodes for testing and analysis; a 1N34A, a BAT85 and an X-Schottky. The X-Schottky is an unknown Schottky diode I purchased online. It was sold as a 1N34A. It is definitely not a 1N34A; however, as luck would have it, is a good diode for crystal radios. Unfortunately, it is unbranded and, therefore, unknown. The equations used to determine the current for each diode are given in Tables 14.3, 14.4 and 14.5. All the forward current equations (mA) are given by power equations, and all the reverse currents (μA) are given by quadratic equations.

Next, the junction capacitance is required. Usually, this is given in the diode's specification sheet. However, for the X-Schottky, I determined the junction capacitance using the method outlined in the measurement techniques Chapter. The junction capacitance was determined to be 4.7 pF at -1 volts. The junction capacitance for the 1N34A is given in the specification sheet as 1 pF at -1 volts. Specifications sheets for the BAT85 give junction capacitances ranging from 6 to 10 pF at -1 volts. I measured the junction capacitance to be 5.8 pF at -1 volts, so 6 pF was chosen.

To determine the junction capacitance as a function of diode voltage, we also require the barrier potential (built-in voltage). Can we measure the turn-on voltage and use this as the built-in voltage? Let us examine whether this is a valid thing to do. We know when the applied voltage across the diode is equal to the built-in voltage; the depletion region all but disappears. Therefore the resistance across the diode is due to the semiconductor material and some wiring resistance, and is fairly constant over a small voltage range. There should be a section of the I-V curve that is linear above the knee. If we draw a straight line from the end of the curve, the voltage where the I-V curve deviates from the straight line will give us a reasonable estimate of the built-in voltage. The built-in voltage will be the voltage just above where the two graphs deviate.

Consider the forward bias I-V curve for the X-Schottky given in Figure 14.24. The voltage where the I-V curve and the straight line start to coincide is about 0.35 volts. The built-in voltage was also measured using the technique described in the measurement techniques Chapter and was found to be 0.37 volts. The turn-on voltage was measured to be 0.177 volts and is well below the knee and therefore can not be used as the built-in voltage.

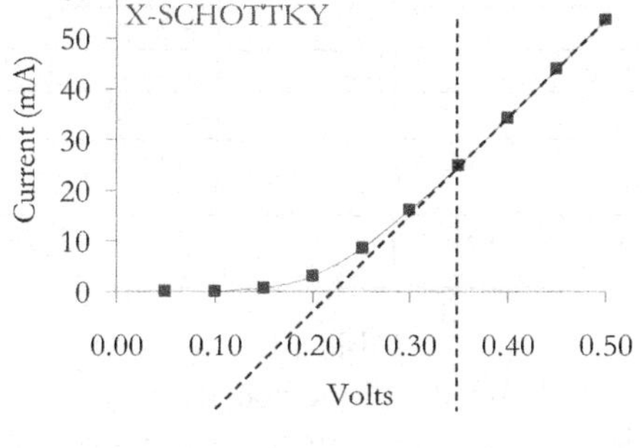

Figure 14.24
Forward bias I-V curve for the X-Schottky.

Applying the I-V curve method to determine the built-in voltage for the BAT85 and the

1N34A, we get a built-in voltage for the BAT85 of 0.4 volts and 0.55 volts for the 1N34A, as shown in Figure 14.25. The measured turn-on voltages were 0.235 for the BAT85 and 0.351 for the 1N34A. Both turn-on voltages are well below the knee in the non-linear section of the curve.

The multimeter used to measure the turn-on voltages applies 3.25 volts to the diode with a 2 kΩ resistor in series. Older multimeters may use lower voltages. The lower the applied voltage, the lower the displayed turn-on voltage will be.

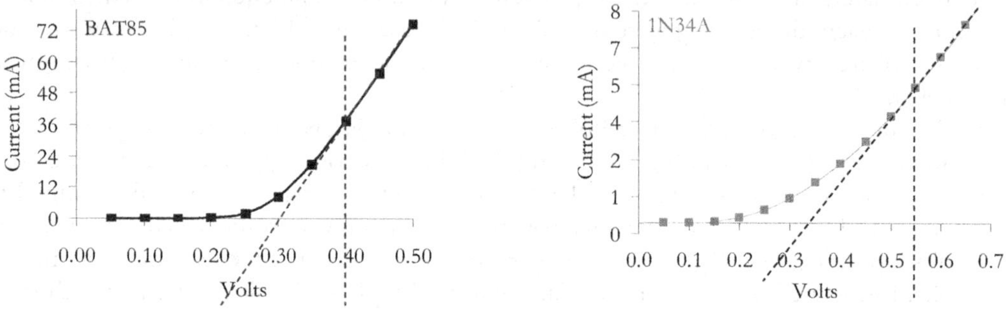

Figure 14.25
Forward bias I-V curve for the BAT85 and 1N34A.

An empirical equation that estimates the build-in voltage from the measured turn-on voltage is given by equation 14.17.

$$V_{BI} = (V_{ON})^{0.6} \qquad \text{Estimation of barrier potential} \qquad 14.17$$

Applying Equation 14.17 to the measured turn-on voltages, we get built-in voltages of 0.35 for the X-Schottky, 0.42 for the BAT85 and 0.53 volts for the 1N34A. These are very close to the values derived using the graphical method. To determine what voltage your multimeter applies, set the multimeter to the diode function and measure the voltage using another multimeter. If your multimeter applies a different voltage (10 % or more) to the diode, make some I-V curves and derive your own equation.

Parameter	Symbol	Value	Test condition
Built-in potential barrier	V_{BI}	0.37	
Junction capacitance	Cj	4.71 pF	-1 volt
Parasitic capacitance	Cp	1.5 pF	(due to wiring)
Average transit time	τ_t	100 nS	
Forward current equation (mA)	$I_D = (1942)V_F^{4.105}$		
Reverse current equation (µA)	$I_R = (-0.2234)V_R^2 - 0.2135V_R - 2.2398$		

Table 14.3
X-Schottky diode specifications.

Lastly, we need to determine the average transit time to calculate the diffusion capacitance. You can use the technique described in the measurement techniques Chapter (a very fiddly and challenging measurement), or you can use my measurements as a guestimate; 60 nS for a Germanium point contact diode and about 125 nS for a Schottky diode. The greater the average transit time, the greater the diffusion capacitance.

We shall examine the role diodes play in crystal radios and their effect on performance in subsequent chapters.

Parameter	Symbol	Value	Test condition
Built-in potential barrier	V_{BI}	0.4	
Junction capacitance	Cj	6.0 pF	-1 volt
Parasitic capacitance	C_P	1.5 pF	(due to wiring)
Average transit time	τ_t	120 nS	
Forward current equation (mA)	$I_D = (2947)V_F^{5.2778}$		
Reverse current equation (μA)	$I_R = (0.0004)V_R^2 + 0.1243V_R - 0.0973$		

Table 14.4
BAT85 diode specifications.

Parameter	Symbol	Value	Test condition
Built-in potential barrier	V_{BI}	0.55	
Junction capacitance	Cj	1.0 pF	-1 volt
Parasitic capacitance	C_P	1.5 pF	(due to wiring)
Average transit time	τ_t	60 nS	
Forward current equation (mA)	$I_D = 27.262V_D^{3.125}$		
Reverse current equation (μA)	$I_D = 0.0981V_D^2 + 1.0019V_D - 0.4430$		

Table 14.5
1N34A diode specifications.

15

Demodulation

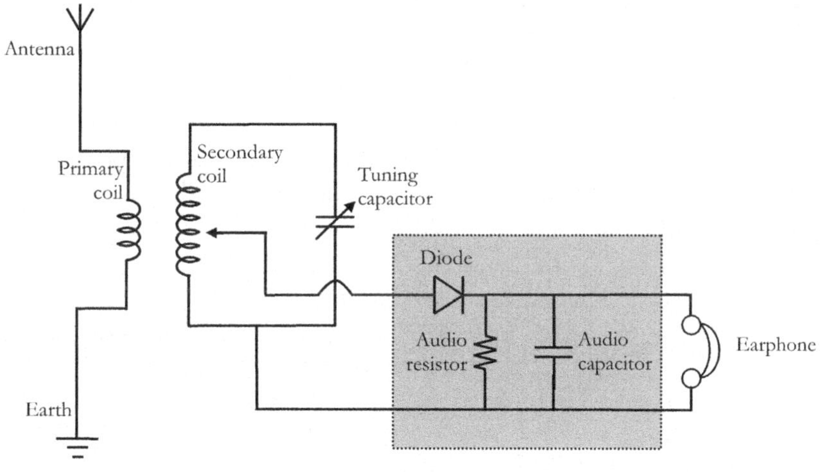

15.1 MODULATION

Commercial AM Radio frequencies range from 530 to 1,700 kHz; however, detectable audio signals range from 20 to 20 kHz. How is it then possible for us to hear these comparatively low frequencies in high-frequency radio waves?

At a radio station, they will have an electronic device called an oscillator. An oscillator produces a constant pure sine wave at a given frequency. Consider Figure 15.1 representing a radio station broadcasting at 1 MHz. The oscillator will be set to produce a 1 MHz sine wave. Next, the station will have an audio input device that converts sound waves into electrical signals (a transducer), such as a microphone. This audio signal may then go through some audio amplification depending on the source of the signal. In this example, we will set the audio input to a 1 kHz tone for simplicity. In reality, music and voice will consist of a complicated waveform containing many frequencies.

Both the RF signal and audio signal are fed into a device called a ***modulator*** and undergo the process of ***modulation***. The modulator combines the carrier wave and audio signal in a non-linear fashion to produce an output waveform containing both the carrier frequency and the encoded audio signal. This modulated signal may then be amplified and fed into the transmitting antenna.

The amplitude of the carrier wave is modified by both the frequency and amplitude of the audio signal resulting in a waveform where the peaks follow the audio signal; that is, the carrier is amplitude modulated. The audio frequencies are not present in the broadcast wave, but the audio signal is encoded in the envelope of the broadcast signal. Notice that the audio signal is encoded in both the positive and negative transitions of the amplitude, as shown in Figure 15.1.

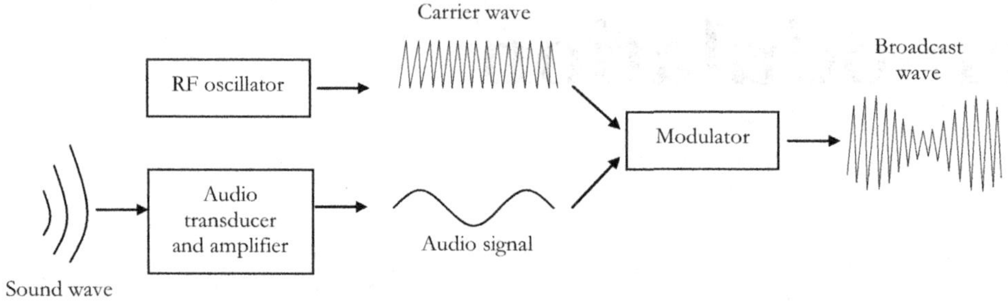

Figure 15.1
Amplitude modulation.

When the carrier wave is modulated with the audio signal, the peak of the audio signal adds with the carrier wave's amplitude, and the troughs of the audio signal subtract from the carrier wave's amplitude, as shown in Figure 15.2.

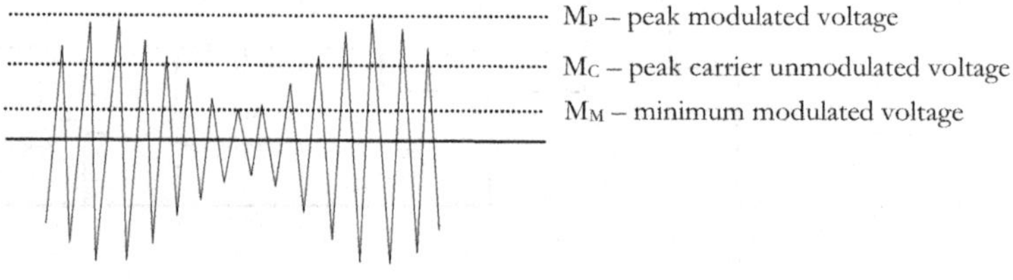

Figure 15.2
Modulation index.

The ratio of the modulated maximum peak amplitude to the modulated minimum amplitude is called the modulation index and is given by

$$m = \frac{M_P - M_M}{M_P + M_M} \qquad \text{AM modulation index} \qquad 15.1$$

where M_P is the peak value of the modulated signal, and M_M is the minimum value of the modulated signal. The peak modulated voltage is given by

$$M_P = M_C(1+m) \qquad \text{Peak modulated voltage}$$

For an unmodulated carrier wave, the amplitude will be constant with the amplitude given by M_C.

Looking at the broadcast signal in the frequency domain, shown in Figure 15.3, the signal contains the original carrier wave frequency and frequencies above and below the carrier wave corresponding to the audio frequencies. That is, the modulated signal is composed of

$$f_m = f_c \pm f_a$$

where f_c is the carrier wave frequency and f_a is the audio frequency. The collection of frequencies higher than the carrier wave is known as the upper side-band, and the frequencies below the carrier frequency are called the lower side-band. In this simple example, the 1 MHz frequency is the carrier wave and the upper side-band is at 1.001 MHz, and the lower side-band is at 0.999 MHz. The spread of the upper and lower side-bands from the carrier is the reason why there must be at least a gap of 9 or 10 kHz between each station, depending on the country.

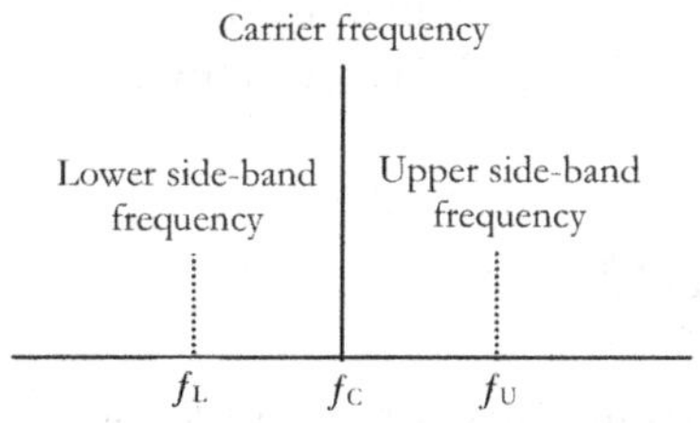

Figure 15.3
Radio wave in the frequency domain.

15.2 DETECTION

If we trace a smooth line across the envelope of the modulated carrier wave, the outline will represent the audio waveform, in this case, a single 1 kHz sine wave. However, as we can see from Figure 15.4, there are two versions of the input signal, both 180° out of phase with one another. If both envelope signals are fed directly into an earphone, the positive signal will cancel the negative signal, and we will hear nothing. We must therefore remove one of the two envelopes; it does not matter which one, either the positive or negative.

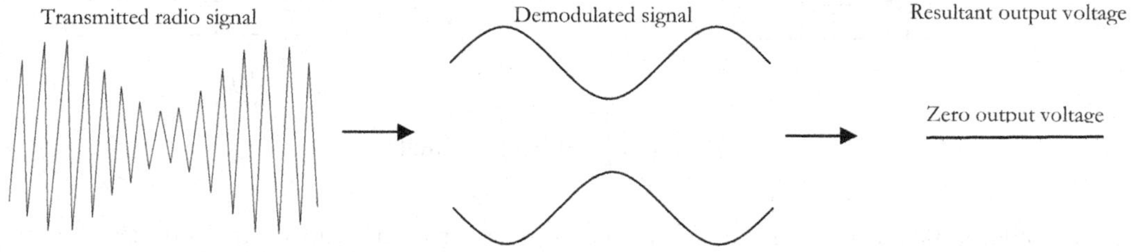

Figure 15.4
Positive and negative signals cancel without detection.

Selecting an envelope can easily be accomplished by passing the radio wave through a diode, as shown in Figure 15.5. A diode used in this fashion is often referred to as a ***detector***. From the early days of radio, the term ***detection*** simply described the ability of a circuit to detect the existence of a radio signal. In modern times, the terms detection and demodulation are often used interchangeably.

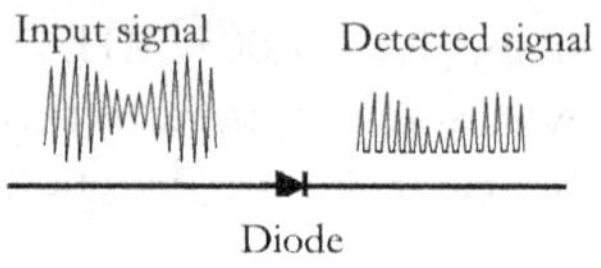

Figure 15.5
Signal detection.

15.3 DEMODULATION

The process of extracting the original signal from an RF wave is called ***demodulation***. Now that we only have the positive going waveform, we can demodulate the signal and extract the original audio signal. This is accomplished by passing the detected radio wave into a demodulator. For an AM crystal radio, demodulation is achieved by passing the detected waveform into a low-pass filter, as shown in diagram 15.6.

In Figure 15.6a, the input voltage is rising, causing the capacitor C1 to charge up. The value of C1 is chosen so that the capacitive reactance is small at RF frequencies, most of the current will flow onto the capacitor plates, and very little will flow across the resistor. The capacitor continues to charge until the maximum voltage has been reached. Then, as the input voltage starts to decrease, the capacitor begins to discharge through the resistor, see Figure 15.6b. This assumes we have an ideal diode that prevents current from flowing back through the diode.

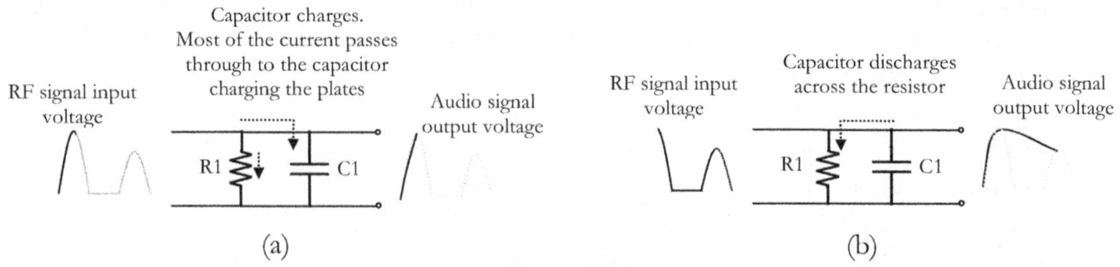

Figure 15.6
Extracting the audio signal.

As the charge flows from the capacitor, the voltage across the capacitor decreases. The rate of discharge is proportional to both the value of the capacitor and the resistor. The higher the resistance value, the longer it takes the capacitor to discharge. Selecting the right value for the capacitor and resistor will result in an output waveform (the voltage across the capacitor) that closely matches the original audio signal, as shown in Figure 15.7. This type of demodulator is known as an edge-detector as it tries to follow the edges (the peak voltage values) of the waveform.

In the case of an AM radio station, this demodulated signal will be the sound broadcasted from the station. You can now listen to the station by piping this signal out to an audio transducer, such as a piezoelectric earphone or an audio amplifier and speaker.

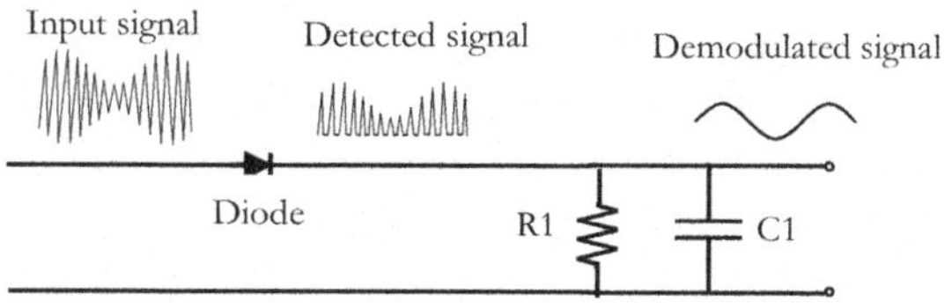

Figure 15.7
AM signal demodulation.

15.4 AUDIO EDGE DETECTOR DESIGN

Consider a capacitor discharging across a resistor, as shown in Figure 15.6b. The rate at which a capacitor discharges across a circuit with resistance is shown in Figure 15.8. Initially, the rate of

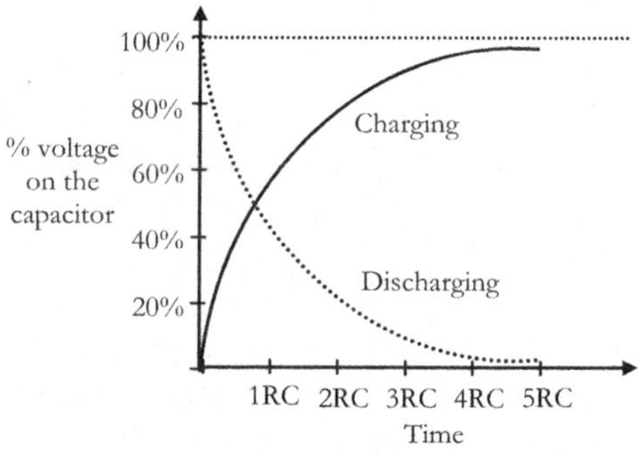

Figure 15.8
Charge and discharge response of an RC circuit.

discharge is high and then tappers off. That rate of discharge is related to the capacitance value multiplied by the resistance, and is called the time constant τ

$$\boxed{\tau = RC} \qquad \text{Time constant} \qquad 15.2$$

The voltage across the capacitor at a given time is given by the equation

$$\boxed{V_C = V_I e^{\frac{-t}{RC}}} \qquad \text{Voltage across a discharging capacitor} \qquad 15.3$$

where V_I is the initial voltage across the capacitor, RC is the time constant, and t is the time in seconds. After one time constant, 63% of the charge will have been removed, and the voltage will decrease by 63%. After five time constants, very little voltage will be left on the capacitor, as such, five time constants is considered the time required to fully discharge a capacitor.

Note the curve for charging a capacitor using a DC power source is the inverse of the discharge curve, and while charging, the voltage across the capacitor is given by

$$V_C = V_S(1-e^{-t/RC})$$

Voltage across a charging capacitor (DC) 15.4

where V_S is the DC power supply voltage. This equation is only valid if the power source is DC. For AC circuits, normal circuit analysis applies, and Ohm's law is used.

Now consider the demodulator circuit shown in Figure 15.7. When the diode is forward biased, the capacitor will charge up. As the RF signal voltage drops below the charge on the capacitor, the capacitor discharges. In Figure 15.9, a small section of the detected RF signal is shown. The capacitor charges and discharges for each RF cycle. The depth of the discharge is related to the RC time constant and the RF frequency. If the RC time constant is close to the RF period, then the voltage frequency across the capacitor will closely match the RF frequency, and no audio will be detected. As the RC time constant increases, the envelope will start to approach high audio frequencies, and very little of the lower audio frequencies will be heard. Increasing the RC time constant further will bring out more of the lower audio frequencies and less of the higher frequencies. Thus the RC time constant determines how the audio sounds.

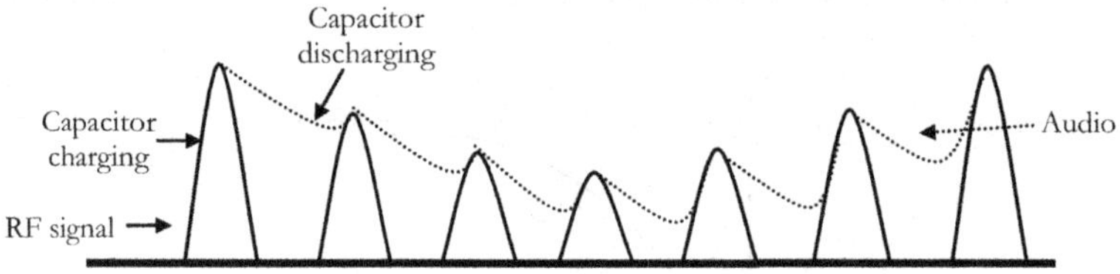

Figure 15.9
Audio capacitor charging and discharging.

The voltage drop in amplitude across the RC circuit during an RF cycle is related to the frequency of the original audio signal and the modulation index. The RC value required to provide reasonable demodulation of a particular audio frequency is given by

$$RC = \frac{\sqrt{1-m^2}}{f_A m}$$

Selecting a time constant 15.5

where f_A is the audio frequency of interest, and m is the modulation index. This does not mean that other frequencies will not be demodulated; it simply means that for the given frequency, the demodulation should be close to optimal for a simple edge detector.

The typical human ear is very responsive to a 1 kHz signal. Say the modulation index of the station is 0.8. What would be the time constant required to provide optimal demodulation at 1 kHz?

$$RC = \frac{\sqrt{1-0.8^2}}{10^3(0.8)}$$
$$= 0.00075$$

That is 0.75 mS. Say the capacitor value is 2.2 nH. What resistor should we use?

$$RC = 0.00075$$

And

$$R = 0.75x10^{-3} / 2.2x10^{-9}$$
$$= 340{,}909$$

As we shall see in the next Chapter, the piezoelectric earphone has a large capacitance value of around 15 nF. The earphone is in parallel to the edge detector circuit. So quite often, the piezoelectric earphone is used as the capacitor in the edge detector circuit. In this case, the resistor required would be 50 kΩ. You could use a standard 51 or 47 k resistor, with the 47 kΩ being the most commonly used value used in crystal radios.

Ultimately the resistance value will be determined by the audio transducer frequency range, personal preference (more bass or treble) and the effect on the coil's Q. A low resistance value will result in more current flowing across the resistor, which will result in a small reduction in Q compared to a higher resistance value. This effect is often grossly exaggerated as we shall see later when discussing Q losses. In fact, using a 10 k resistor will result in a reduction of Q of less than 1% compared to a 150 k resistor. Using a low-value resistor will result in a better treble response, and using a high-value resistor will result in a better bass response. However, using a very high resistance can result in a reduction in audio amplitude and possible distortion in a piezoelectric earphone, especially for strong stations.

Table 15.1 lists resistance values for different optimal audio frequencies and audio capacitor values for a typical 0.8 modulation index value.

Optimal audio frequency	Resistance value using a 2.2 nF audio capacitor only	Resistance value using the piezoelectric earphone (15 nF)
500 Hz	682 K	100 K
1,000 Hz	341 K	50 K
5,000 Hz	68 K	10 K

Table 15.1
Audio resistor values from optimal audio frequency (0.8 modulation index).

15.5 EAR FREQUENCY RESPONSE

The range of audio frequencies detected by the human ear is between 20 to 20,000 Hz. The actual range that any individual can hear varies quite a lot, and is largely dependent on age. The older you are, the narrower the frequency response, and the higher the threshold of sound pressure required to detect a particular frequency. Graph 15.10 shows the typical frequency response for a young person, 18 to 25, with good hearing. The horizontal axis shows frequency, and the vertical axis shows the sound pressure level (SPL) in decibels. Each contour represents the sound pressure level required for a person to perceive each frequency at the same volume. Each contour is referenced to a 1 kHz tone for a given decibel level and is called a ***Phon***. The "0 Phon" contour shows the sound pressure level required for each frequency to sound as loud as a 1 kHz signal at 0 dB. The "40 Phone" contour is when the reference signal is at 40 decibels, etc.

The sound pressure level is measured relative to 20 micro Pascals of pressure (P_0). So a sound pressure level of 20 micro Pascals is zero decibels. A Pascal is the pressure that one Newton of force applied over an area of one metre exerts on an object. So 20 uP is a very tiny amount of pressure indeed. Atmospheric pressure at sea level, at 20 °C, is about 101,353 Pascals (14.7psi). P_0

is considered to be the threshold pressure required to detect a 1 kHz tone. The formula to calculate the sound pressure level in decibels is

$$SPL = 20\log_{10}\left(\frac{P}{P_0}\right) \qquad \text{Sound pressure level in decibels} \qquad 15.6$$

where P is the sound pressure level in Pascals.

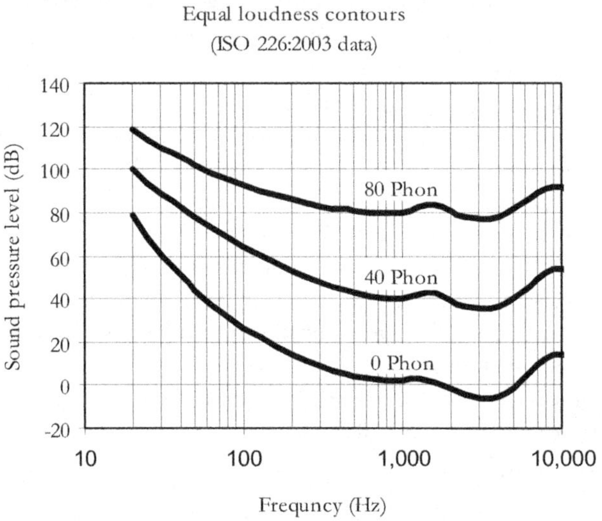

Figure 15.10
Ear response to SPL as a function of frequency.

From the graph, we can see that the human ear is most sensitive to frequencies around 4 kHz, and is not very sensitive to 100 Hz signals. With 100 Hz signals requiring much greater pressures to be detected by the human ear. Between 500 to 6,000 Hz, the SPL is reasonably flat. Table 15.2 relates SPL in decibels to the sound levels you may have experienced.

dB (SPL)	Description of the sound source
0	A mosquito at 2-metres distance.
20	Whisper at 1-metre distance.
40	Normal conversation at a 1-metre distance.
60	Loud conversation at a 1-metre distance.
80	Busy city street
100	Jackhammer at 1-metre.
120	Jet takeoff at 100 metres.

Table 15.2
SPL related to experienced sound-level situations.

The scale is logarithmic, so an increase of 10 dB translates to a 3.2-fold increase in pressure, and an increase of 20 dB translates to a 10-fold increase in sound pressure. Going from a whisper to the sound of a jet taking off would require a 100,000-fold increase in pressure. Prolonged exposure to 90 dB or more can result in permanent hearing damage.

The human ear's response to sound level changes is somewhat subjective, but a rule of thumb is a 10 dB increase in sound pressure levels equates to being twice as loud. A 1 dB increase is not noticeable. A 2 dB change in SPL is just noticeable, and a 3 dB change is easily noticeable.

Given that we are most sensitive to audio signals in the 500 to 6,000 Hz range, and given we are not using any audio amplification, the best bet in hearing a signal would be to design our edge detector to be optimal somewhere in this range.

16

The Piezoelectric Earphone

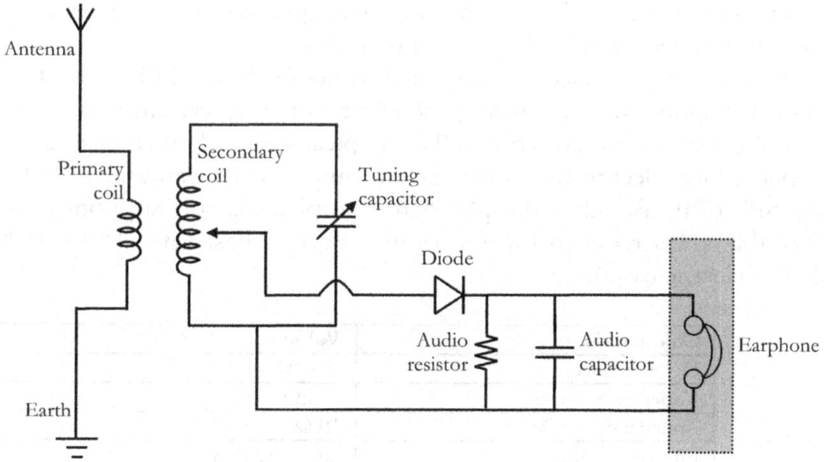

16.1 CONSTRUCTION

Piezoelectric materials produce an electric charge when twisted, squeezed, or deformed in some manner. Conversely, a voltage applied across the surface of piezoelectric materials will cause it to deform. If this voltage is AC, the crystal will vibrate at the applied frequency; and if this voltage waveform represents an audio signal, the vibrating crystal will reproduce the sound. The ubiquitous piezoelectric earphone has been the audio transducer of choice for generations of crystal radio enthusiasts. Its sensitivity to extremely small input signals and low cost made it an obvious choice.

Piezoelectric earphones are made by sandwiching piezoelectric material, barium titanate or quartz, between two thin metal plates of aluminium or brass. This assembly is glued into an earphone enclosure with the outer edges of the plates pressed between the two sides of the earphone enclosure, thus keeping the outer edges rigid while allowing the rest of the assembly to vibrate. This acts like a vibrating diaphragm, similar to how a speaker works. This vibrating diaphragm moves the air molecules producing sound pressure waves. The power of the sound wave is very small indeed and must be funnelled directly into the ear for it to be heard clearly. In a piezoelectric earphone, a small hole above the crystal channels the pressure sound wave along a narrow tube that extends into the ear canal. The sound eventually emerges very close to the eardrum.

16.2 SENSITIVITY

The sensitivity of a speaker, or earphone, is a measure of its ability to transform electrical signals into sound waves. A signal is fed into the device, and the sound pressure is measured some distance away. For speakers, a one-watt, one-kHz sound wave is applied to the speaker with a sound presser level gauge positioned 1 metre away. The sound pressure level, SPL, is given in decibels, and the sensitivity value is specified in decibels of sound pressure level per watt, dB SPL/watt. Typical speaker values are above 90 dB SPL/W.

Earphones are tested using a human head analogue with ears designed to replicate the structure of the human ear. Sound pressure gauges are located at the position of the eardrum. A one-milliwatt, one-kHz signal is applied and the sound pressure is measured at the eardrum. Typical values for earphones are 90 dB SPL/mW and greater.

The sensitivity of piezoelectric earphones is about 70 dB SPL/mW. However, piezoelectric earphones can respond to much lower power levels than typical earphones. A sensitive 20 Ω in-ear earphone will generate a SPL of around 0 dB (typical limit of human hearing) when a 400 nW one-kHz is applied. Piezoelectric earphones have an impedance of around 10 kΩ at one-kHz, and will generate a SPL of 0 dB with a 100 pW signal. So piezoelectric earphones will respond to signals 1/4000th of the power required for normal in-ear earphones. Table 16.1 lists the characteristics of a typical piezoelectric earphone.

Characteristic	Value
DC resistance	> 20M
Impedance at 1 kHz	10 kΩ
Impedance at 1 MHz	10 Ω
Frequency range	200 to 8,000 Hz
Sensitivity at 1 kHz	70 dB SPL/mW
Capacitance	15 nF

Table 16.1
Typical piezoelectric earphone characteristics.

16.3 CAPACITANCE

Sandwiching non-conductive material between two metal plates creates a capacitor, and the piezoelectric earphone has a typical capacitance of around 15 nF. As the earphone is connected in parallel with the edge detector, Figure 16.1, this capacitance is added to the existing capacitance.

As the earphone capacitance is much higher than that used in the edge detector, we can remove the edge detector capacitor and use the earphone's capacitance to provide the demodulation.

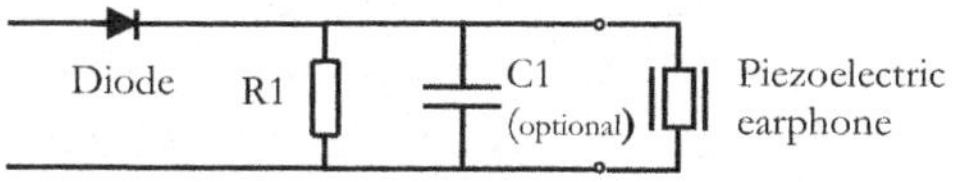

Figure 16.1
Edge detector with earphone.

Figure 16.2 shows the effect of removing a demodulation capacitor and using the earphone alone for demodulation. A 1-MHz carrier wave was modulated with a 1-kHz sine wave, with a modulation index of 0.8. This was fed into the circuit, as shown in Figure 16.1. There is not a noticeable difference between the two demodulated waveforms, demonstrating that a piezoelectric earphone can be used alone for demodulation.

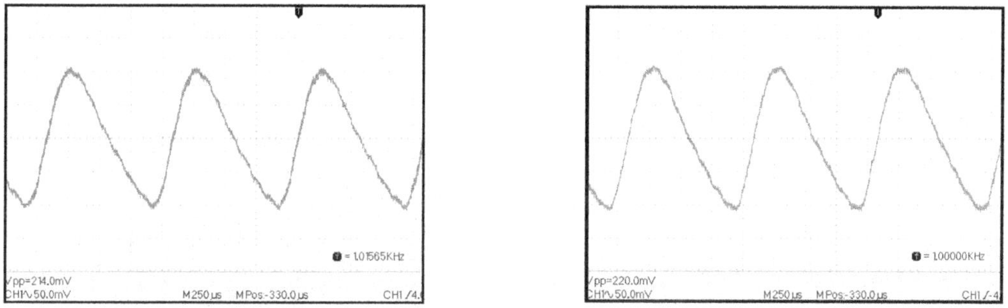

(a) Edge detector with 1 nF capacitor. (b) Edge detector with earphone only.

Figure 16.2
1 kHz demodulated waveform.

16.4 ENERGY LOSS

Piezoelectric earphones convert electrical energy into sound energy by causing the piezoelectric material to vibrate or distort. The total input energy is equal to the stored energy plus the mechanical energy of vibration, expressed mathematically as

$$E_I = E_S + E_M \qquad \text{Total energy stored in a piezoelectric earphone} \qquad 16.1$$

where E_I is the total input electrical energy, E_S is the stored energy in the form of an electric field between the plates (capacitance), and E_M is the energy used to vibrate the crystal (mechanical energy). The stored energy can be modelled as a capacitor, with the mechanical energy loss as some resistor value, as shown in Figure 16.3a. How the piezoelectric earphone responds to external circuitry is somewhat complicated. It responds differently for a purely resistive load than it does for a load with inductive or capacitive reactance. It will also respond differently to different levels of voltage and signal types.

DC Loss Using a Load Resistor

During the positive transitions of the RF signal, a charge is placed across the earphone. We need to examine how the earphone reacts during the discharge phase and how it converts the electrical energy into mechanical energy. If we place a charge across the earphone and measure the discharge RC time constant τ using an Oscilloscope, assuming the capacitance is constant, we can determine the equivalent resistance by manipulating Equation 15.2, and the equivalent earphone mechanical resistance is given by

$$\boxed{R = \frac{\tau}{C_E}} \qquad \text{Equivalent earphone resistance} \qquad 16.2$$

A reduction in the time constant represents a decrease in the equivalent resistance and an increase in the electrical energy converted into mechanical energy. In Figure 16.3b, a purely resistive load has been added in parallel to the earphone. Say we place a charge across the earphone in circuit 16.3b

(a) Earphone circuit model. (b) Earphone with load resistance attached.

Figure 16.3
Equivalent circuit model of a piezoelectric earphone.

and measure the time constant τ_M. If we take the ratio of the measured time constant τ_M to the calculated time constant τ_L, any difference between the two time constants represents a loss of energy not accounted for by the load resistor R_L. If we multiply R_L by this ratio, we get an apparent load resistance R_M. That is

$$\frac{R_M}{R_L} = \frac{\dfrac{\tau_M}{C_E}}{\dfrac{\tau_L}{C_E}}$$

$$= \frac{\tau_M}{\tau_L}$$

$$R_M = R_L \frac{\tau_M}{\tau_L}$$

This ratio represents the proportion of the earphone's stored energy used as electrical energy E_S. Say the ratio was 0.25; this means the load resistance appears to be only 25% of its actual value. Or expressed another way, the resistance has decreased by 0.75 (75%). That is, 75% of the energy was converted to mechanical energy leaving only 25% to be used as electrical energy. We will refer to this ratio as the DC earphone ***available electrical energy factor*** κ_{DC} (kappa DC). And is given by

$$\kappa_{DC} = \frac{R_M}{R_L} = \frac{\tau_M}{\tau_L}$$ Available electrical energy factor 16.3

κ_{DC} will always reduce the apparent value of the load resistance to represent a virtual increase in current flow, thus reducing the time constant. And the apparent resistance is given by

$$R_M = \kappa_{DC} R_L$$ Apparent load resistor 16.4

The value of the apparent resistance will be used to adjust the discharge time constant of the earphone to calculate the amount of charge retained (and voltage) on the earphone during the discharge portion of a cycle. We will see how this is used in calculating the Q of the crystal radio later.

Subtracting 1 from the available energy factor gives us the proportion of the stored electrical energy converted to mechanical energy. We will refer to this as the DC earphone ***mechanical energy factor***, ξ_{DC} (Xi DC).

$$\xi_{DC} = 1 - \kappa_{DC}$$ Mechanical energy factor 16.5

By placing different load resistors in parallel to the earphone, we can measure the discharge time constant and determine κ_{DC} for various loads and voltages. Table 16.2 shows the results of various load resistors connected in parallel to the earphone. The measurements were repeated for various earphone voltages.

DC voltage	**R_L=20 k τ_L=300 uS**		**R_L=47 k τ_L=705 uS**		**R_L=100 k τ_L=1,500 uS**		**R_L=1,000 k τ_L=15,000 uS**	
	τ_M (uS)	κ_{DC}	τ_M (uS)	κ_{DC}	τ_M (uS)	κ_{DC}	τ_M (uS)	κ_{DC}
0.1	185	0.6167	340	0.4823	800	0.5333	9000	0.6000
0.5	185	0.6167	320	0.4539	710	0.4733	7400	0.4933
1.0	185	0.6167	320	0.4539	690	0.4600	7400	0.4933
2.5	170	0.5667	295	0.4184	660	0.4400	7200	0.4800

Table 16.2
Available electrical energy factor for various loads and voltages.

Notice the change in apparent resistance is also a function of voltage. The DC voltage appearing across the earphone is related to the strength of the station and the diode impedance and can range from a few millivolts up to a volt or more. Higher DC voltages are possible for extremely powerful or nearby stations.

Choosing 0.1 volts DC as a base and graphing each of the DC earphone available electrical energy factors as a function of load resistance, we get the graph shown in Figure 16.4. The graph is split into three line segments with different equations. We can use these functions to determine κ_{DC} for any resistance at 0.1 volts DC. We will call this value κ_R, where R is the load resistance in kΩ. Note that for a large enough R, $\kappa_R > 1$. The proportion of available electrical energy can never be greater than 1 (100%), so always set the maximum κ_R to 1. The equations for κ_R are

$$\kappa_R = -0.005R + 0.7163 \quad \text{for R <= 47 k} \qquad 16.6a$$

$$\kappa_R = 0.001R + 0.4371 \quad \text{for R <= 100 k} \qquad 16.6b$$

$$\kappa_R = 0.00007R + 0.5259 \quad \text{for R > 100 k} \qquad 16.6c$$

where R is resistance in kΩ. We have equations as a function of resistance for a single voltage value. However, we want a κ_{DC} for any voltage, so we need to modify the value of κ_R for voltage. We will multiply κ_R with a voltage modification factor κ_{volt} to determine κ_{DC} as a function of resistance and voltage.

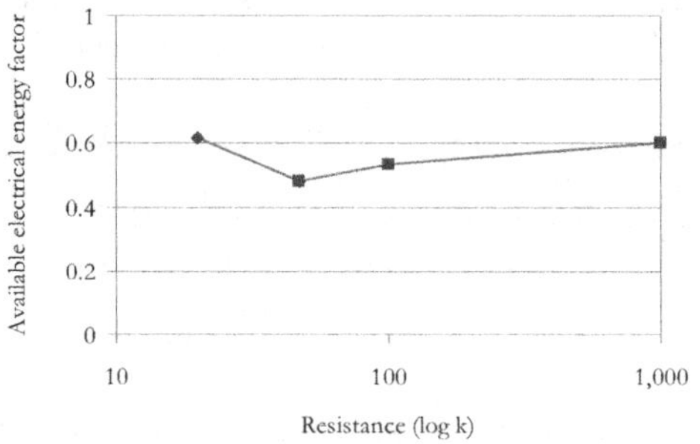

Figure 16.4
DC earphone available electrical energy factor as a function of load resistance at 0.1 volts.

In Table 16.2, the last column is for a resistance of 1 MΩ. At these high resistance values, the earphone will not demodulate audio signals very well. We can take the change between time constants for different voltages as the effect of voltage on the earphone. As κ_R will be κ_{DC} at 0.1 V, the κ_{volt} values will be relative to the time constants at 0.1 volts. Table 16.3 list the change in time constant with respect to voltage.

DC voltage	R_L=20 k	R_L=47 k	R_L=100 k	R_L=1,000 k
	κ_{volt}	κ_{volt}	κ_{volt}	κ_{volt}
0.1	1.00	1.00	1.00	1.00
0.5	1.00	0.94	0.89	0.82
1.0	1.00	0.94	0.86	0.82
2.5	0.92	0.87	0.83	0.80
5.0	0.79	0.88	0.85	0.87

Table 16.3
Earphone response due to voltage.

The graphs of κ_{volt} as a function of voltage for a load resistance of 1 MΩ are shown in Figure 16.5. Applying straight lines between the values, we arrive at a series of equations as a function of voltage that can be used to provide a value of κ_{volt}.

$$\kappa_{volt} = -0.4444V + 1.0444 \qquad \text{for V <= 0.5} \qquad 16.7a$$

$$\kappa_{volt} = -0.012V + 0.8308 \qquad \text{for 0.5 < V <= 2.5} \qquad 16.7b$$

$$\kappa_{volt} = 0.0267V + 0.7333 \qquad \text{for V <= >2.5} \qquad 16.7c$$

The Available electrical energy factor can now be calculated as a function of resistance and voltage as

$$\boxed{\kappa_{DC} = \kappa_R \kappa_{volt}} \qquad \text{Available electrical energy factor} \qquad 16.8$$

The lower the κ_{DC} value, the lower the Q, as it has the effect of lowering the parallel resistance to the earphone resulting in a greater current flow and loss. Stated another way, as κ_{DC} gets smaller, the earphone mechanical energy factor ξ_{DC} rises. Note that this reduction in the operating Q of a crystal radio is a good thing. It means the energy is finding its way into the earphone and producing sound.

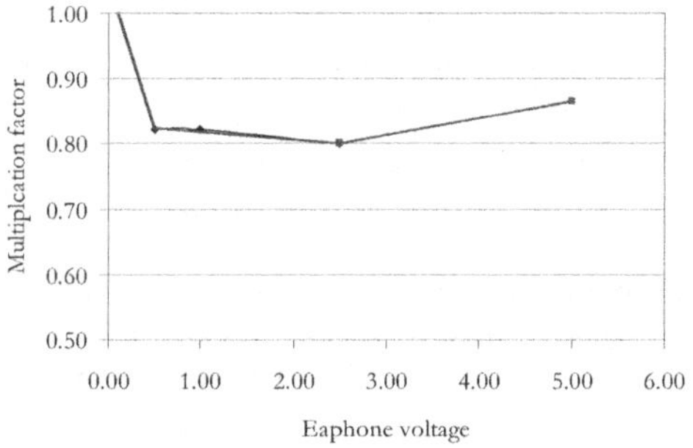

Figure 16.5
DC adjustment electrical energy factor as a function of voltage.

AC Loss

As well as mechanical loss due to a DC voltage across the earphone, there is also mechanical loss due to AC voltage. The piezoelectric crystal will respond even to high-frequency voltages. If we measure the peak-to-peak voltage across the earphone (without a load resistor) while applying an AC voltage of 0.1 volts peak-to-peak at different frequencies; and then calculate the expected voltage using circuit analysis, we can express the ratio of the measured to calculated voltage as an earphone ***AC available electrical energy factor***, κ_{AC} as a function of frequency. Figure 16.6 shows the graph of κ_{AC} as a function of frequency. Applying a log best fit gives us the following equation.

$$\boxed{\kappa_{AC} = -0.4217\ln(f) + 3.5274} \qquad \text{AC available electrical energy factor} \qquad 16.9$$

where the frequency f is in kHz.

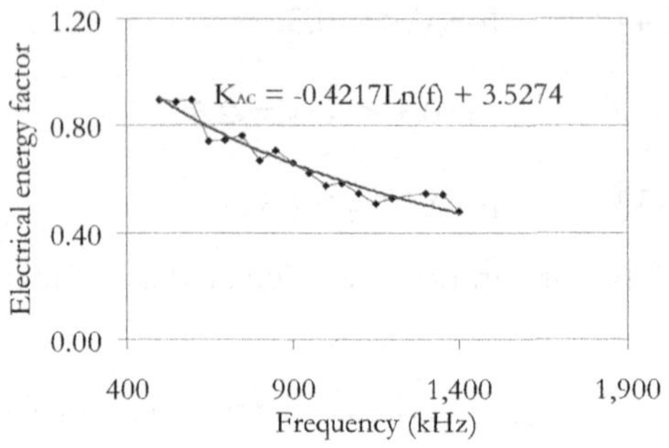

Figure 16.6
AC earphone available electrical energy as a function of frequency.

Total Loss

We can now determine the available electrical energy factor as a function of resistance, voltage and frequency. If we now multiply the DC earphone available electrical energy factor by the AC earphone available electrical energy factor, we arrive at the ***total available electrical energy factor***, κ

$$\kappa = \kappa_{AC}\kappa_{DC} \qquad \text{Total available electrical energy factor} \qquad 16.10$$

And the total earphone mechanical energy factor is

$$\xi = 1 - \kappa \qquad \text{Total mechanical energy factor} \qquad 16.11$$

No Load Adjustment

When an audio load resistor is not placed in the circuit, the earphone discharges across the diode and the coil's tapped section. The load is now complex and has resistance and capacitive and inductive reactance. The earphone responds differently to this complex load than it does when a pure resistor is placed across the earphone. In reverse bias, the diode's resistance is typically much higher than a load resistor of, say 47 KΩ. This results in a change in the time constant that is largely related to the change due to voltage, that is, κ_{volt}, and we can use Equation 16.7 to calculate this value. However, there is still a component due to the magnitude of the resistance, κ_R. From Table 16.3, we will set κ_R to the values at the 0.5 volts row as an approximation and refer to this value as κ_{RD}. And we have

$$\kappa_{RD} = 1.00 \qquad \text{for R <= 20 k} \qquad 16.12a$$

$$\kappa_{RD} = 0.94 \qquad \text{for R <= 47 k} \qquad 16.12b$$

$$\kappa_{RD} = 0.89 \qquad \text{for R <= 100 k} \qquad 16.12c$$

$$\kappa_{RD} = 0.82 \qquad \text{for R > 100 k} \qquad 16.12d$$

The equation for the available electrical energy factor 16.8 is still valid, but to distinguish the different values of κ_R, we will refer to it as κ_{DCD}, and

$$\boxed{\kappa_{DCD} = \kappa_{RD}\kappa_{volt}} \qquad \text{Available electrical energy factor for diode only} \qquad 16.8b$$

16.5 DETERMINING THE EARPHONE DC VOLTAGE

The DC voltage placed across the earphone greatly affects the performance of the crystal radio. The greater the DC voltage across the earphone, with respect to the peak tap voltage, the greater the operating Q of the crystal radio. The charge placed across the earphone is due to the RF signal, and the following analysis is for RF frequencies and not audio frequencies. During a complete RF cycle, charge is added and removed from the earphone. Over a number of cycles, if the charge added is greater than the charge removed, then the voltage across the earphone increases until the charge added equals the charge removed, leaving the earphone with a DC voltage.

For a steady unmodulated signal, a steady DC voltage will be placed across the earphone. If the signal is modulated, as from a radio station, this DC voltage will rise and fall in line with the modulated signal as the earphone charges and discharges. A modulated signal with different peaks and troughs will cause the diode impedance to change and, thus, the Q of the radio to vary. The following analysis will be for an unmodulated signal, and a modification to the model will be added to simulate modulation.

Charging Phase

Let us assume there is no charge present on the earphone. During the first positive transition of the voltage from the tap, the diode will be forward biased and current will begin to flow, placing a charge on the earphone. The amount of charge, and therefore the voltage across the earphone, is determined by the diode's forward impedance, Z_{DF}, the load resistor and capacitance, the length of time the diode is conducting, and the peak voltage level across the tap. Figure 16.7a shows the circuit diagram for a demodulator when the diode is forward biased.

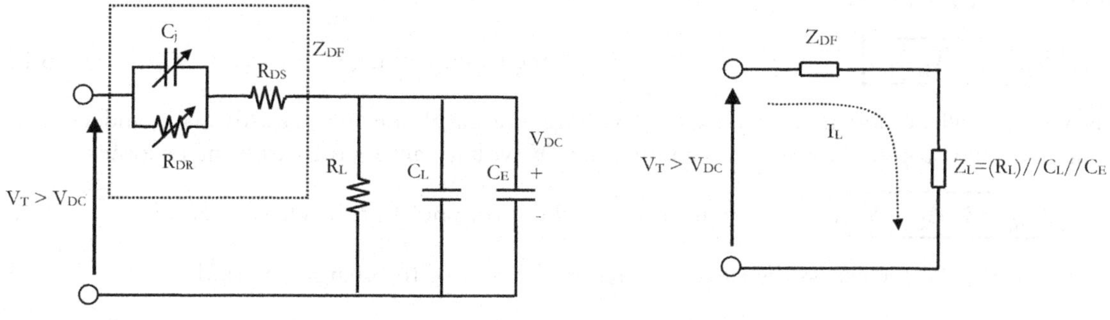

Figure 16.7
Diode forward biased – earphone charging.

The piezoelectric earphone has a very large capacitance due to its construction and will be treated as a capacitor. This is in parallel to the optional load capacitor C_L, and the total load capacitance is just the addition of both the earphone capacitance and the value of the load capacitor. The parallel combination of the piezoelectric earphone and any load capacitor will just be referred to as the earphone or capacitor and will be treated as a single component. The typical capacitance value of the piezoelectric earphone is around 15 nF and, at a frequency of 1 MHz, presents an impedance of 10 Ω.

The forward diode impedance Z_{DF} is in series with the load impedance Z_L, shown in Figure 16.7b, and the load impedance is the parallel combination of the load resistor and the capacitors (earphone). For the diode to be forward biased, the peak tap voltage must be greater than any DC voltage (V_{DC}) across the earphone. Initially, the earphone will not be charged, and V_{DC} will be zero. The peak forward voltage applied to the demodulator will be the difference between the DC voltage on the earphone and the peak amplitude of the signal. That is,

$$\mathbf{V_{F(peak)}} = \mathbf{V}_{\mathbf{T}(peak)} - \mathbf{V_{DC}} \quad \text{Peak forward voltage across the demodulator} \quad 16.13$$

where $V_{T(peak)}$ is the peak tap voltage, and V_{DC} is the DC voltage stored on the earphone. This forward voltage is dropped across the diode and the earphone. That is,

$$\begin{aligned} \mathbf{V_{F(peak)}} &= \mathbf{V_{DF}} + \mathbf{V_L} \\ &= \mathbf{I_L Z_{DF}} + \mathbf{I_L Z_L} \end{aligned} \quad 16.14$$

where V_{DF} is the voltage drop across the diode, and V_L is the voltage drop across the earphone. Given a tap voltage V_T and the load impedance Z_L we need to determine the forward diode impedance Z_{DF} and the series current I_L. However, the diode impedance is dynamic and depends on the current. Typically in circuit analysis, the voltage across the diode is simply set to the turn-on voltage and the circuit is analysed. However, because we are dealing with very small currents and voltages that have large impacts on the performance of the crystal radio, we must be somewhat more accurate. We can apply an iterative approach by initially setting the voltage across the diode to be half the peak forward applied voltage and calculating the diode impedance for this given voltage. We can then calculate the series current and solve the equation. We can then adjust the current, up or down, until the voltage drops minus the applied voltage is some arbitrary small value. See Appendix 3 for calculating diode voltage.

As all the load components are in parallel with the earphone, the voltage across the load and, therefore, the earphone is

$$\boxed{\mathbf{V}_{TOPUP} = \mathbf{I_L Z_L}} \quad \text{Earphone top-up voltage} \quad 16.15$$

This voltage will be added to any existing voltage the earphone may already have and can be considered as the top-up voltage. And so the total DC voltage across the earphone becomes

$$\boxed{\mathbf{V_{DC}} = \mathbf{I_L Z_L} + \mathbf{V_0}} \quad \text{DC voltage across the earphone after the charge period} \quad 16.16$$

where V_0 is the voltage across the earphone before the start of the charging period.

Discharging Phase

During the reverse bias period, the earphone will discharge across any load resistor and through the diode. During the discharge phase, some or all of the charge is removed from the earphone.

As Q is determined for a complete cycle, we must know how long the charging and discharging phases last and what fraction of a cycle they occur for. The time the charging phase lasts is denoted by T_{ON}. We can express the charging phase as a fraction of a cycle L_{ON} (as shown in Figure 16.8) and

$$L_{ON} = \frac{\theta_2 - \theta_1}{2\pi}$$ Fraction of the cycle the diode is forward biased 16.17

where

$$\theta_1 = \sin^{-1}(V_{DC} / V_T)$$
$$\theta_2 = \pi - \theta_1$$ 16.18

V_{DC} is the DC voltage across the earphone, and V_T is the peak signal tap voltage. And the charging time is simply this fraction multiplied by the signal's period, that is

$$T_{ON} = L_{ON} T$$ Time the diode is forward biased 16.19

And the discharge fraction and time are given by

$$L_{OFF} = 1 - L_{ON}$$ Fraction of the cycle the diode is reversed biased 16.20

and

$$T_{OFF} = L_{OFF} T$$ Time the diode is reversed biased 16.21

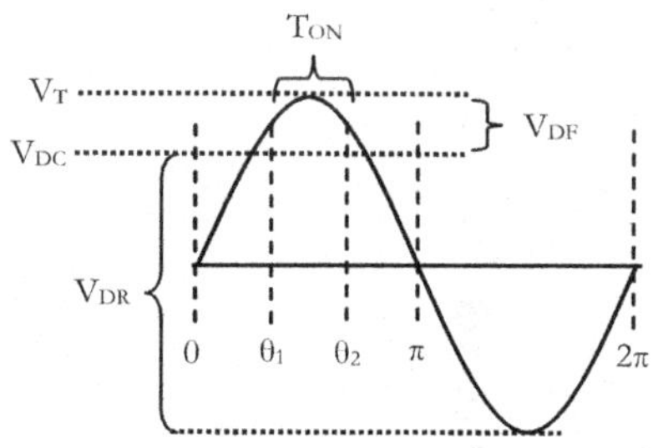

Figure 16.8
Fraction of the cycle period the diode forward biased.

During the charging phase, the tap voltage eventually drops below the voltage on the capacitor, the diode will become reversed biased, and the capacitor will begin to discharge. The capacitor will discharge across any load resistor and through the diode. The amount of charge removed from the capacitor during the discharge period depends on the RC time constant formed between the capacitor and the total ***apparent resistance*** along the discharge path. This includes any mechanical energy loss and electrical energy loss. This decrease in voltage or voltage drop across the earphone after the discharge period is given by

$$V_{DROP} = V_{DC}(1 - e^{\frac{-(T-T_{ON})}{R_R C_E}})$$ Decrease in earphone voltage over discharge time 16.22

where V_{DC} is the voltage across the earphone at the start of the discharge period, and R_R is the magnitude of the apparent resistance across the discharge path.

The reverse bias earphone DC voltage discharge path is shown in Figure 16.9. The peak reverse voltage applied across the diode will be the addition of any DC voltage on the earphone and the negative peak tap voltage, as shown in Figure 16.9. As we are dealing with DC voltages, and as the earphone discharges, the voltage changes, so the reverse bias diode impedance will change. We will use the average voltage for the reverse bias voltage and calculate the diode impedance using this average voltage value.

$$V_{DR} = -0.707V_T + V_{DC}$$ Average reverse bias diode voltage 16.23

The reverse diode resistance at this average reverse bias voltage is

$$R_{DR} = abs(\frac{V_{DR}}{I_{DR}})$$ Average reverse bias resistance 16.24

And the average reverse bias diode impedance is

$$Z_{DR} = (R_{DR} // X_{Cj}) + R_{DS}$$ Average reverse bias impedance 16.25

where X_{Cj} is the reactance of the diode's junction capacitance at the given frequency. And the total impedance of this branch is

$$Z_R = Z_{DR} + Z_{ST}$$ Total impedance of diode branch 16.26

When determining the apparent resistance of the discharge path, we must take into account the mechanical energy loss across the earphone. First, consider the current flowing through the load resistor R_L shown in diagram 16.9a. The apparent value of the load

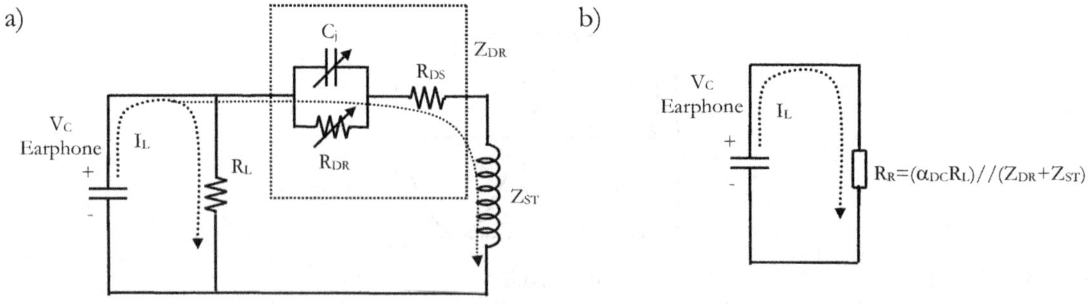

Figure 16.9
Diode reverse biased – earphone discharging.

resistor will be modified by the earphone's available electrical energy factor κ_{DC}, and

$$R'_L = \kappa_{DC} R_L$$ 16.27

Next, the current flowing through the diode will be determined by the diode's impedance in series with the coil's tap impedance with the total impedance of this branch given by Z_R. The resistance of the diode is not directly in parallel with the earphone and does not have the same effect as the load resistor. As charge is removed from the earphone, the reverse bias impedance changes, and typically the diode's reverse bias resistance will be much greater than the load resistance. Therefore, the load resistor will be the dominant factor, and the final value of the discharge path apparent resistance becomes

$$R_R = |(\kappa_{DC} R_L)//(Z_{DR} + Z_{ST})| \quad \text{Discharge resistance with a load resistor} \qquad 16.28$$

If a load resistor is not used, the earphone discharges across the diode, and the earphone discharge time is now greatly dependent on the diode's reverse bias impedance. Much more so than when a load resistor is used. As the reverse bias voltage increases the junction capacitance decreases. However, the rate at which the junction capacitance decreases does not always follow the square root term as given by equation 14.10 and can drop at a different rate. The rate at which the reverse bias junction capacitance changes, can have a large impact on the charge placed across the earphone, and therefore the Q and performance of the radio. This is particularly true when the reverse bias resistance is high and the reverse bias junction capacitance is small.

To model this behaviour, I allow for two root index values. One for voltages above -1 volt and another for voltages below -1 volt. I have included this configuration in the model for completeness. However, when we design and build our crystal radios, a load resistor for demodulation will always be included. Reverse bias voltages greater than one volt will only occur with extremely high-Q coils for close and powerful radio stations. And even when this condition occurs, the discharge path is mainly through the demodulation resistor and the radio will not be sensitive to these small changes in the diode's reverse bias impedance. For the diode only case, the discharge resistance now becomes the reverse bias impedance of the diode, and the apparent discharge resistance becomes

$$R_R = |\kappa_{DCD}(Z_{DR} + Z_{ST})| \quad \text{Discharge resistance across diode} \qquad 16.29$$

We are now in a position to calculate this reduction in voltage (V_{DROP}) across the earphone during the discharge phase using Equation 16.22. The voltage across the earphone after discharging will be the voltage across the earphone after the charge period V_{DC} minus the voltage dropped during the discharge period V_{DROP}. That is,

$$\mathbf{V}_0 = V_{DC} - \mathbf{V}_{DROP} \quad \text{DC voltage across the earphone after the discharge period} \qquad 16.30$$

V_0 will be the voltage across the earphone at the beginning of the next charge cycle to be used in Equation 16.13. If not all the charge is bled off from the capacitor during the discharge phase, then during the next charge phase, more charge will be placed on the earphone. The increase in charge now placed across the earphone will be less than the increase in charge in the previous cycle as the diode will now have some reverse bias applied due to the earphone voltage. While the increase, or top-up charge V_{TOPUP} is greater than the charge loss V_{DROP}, the voltage across the earphone will continue to grow for the number of cycles a modulated waveform takes to reach a maximum, that is ¼(f_c/f_m) cycles. Or when the top-up charge equals the discharge charge. At this stage, a DC voltage will be placed across the earphone with some ripple as shown in Figure 16.10 and the process of calculating the DC voltage is complete.

This DC voltage on the capacitor reduces the time the diode is in forward biased mode per cycle, as the voltage across the tap must be greater than the DC voltage on the earphone for the diode to be forward biased. The higher the relative DC voltage on the earphone, the less time the diode is conducting per cycle. So the higher the DC voltage is in relation to the tap voltage, the less energy will be taken from the coil per cycle. As such, a higher relative DC voltage results in a higher crystal radio Q value, see Figure 16.11.

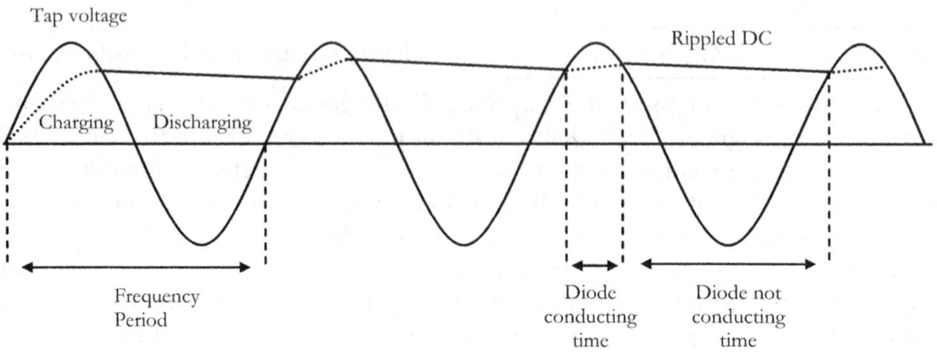

Figure 16.10
Earphone charging and discharging over many cycles.

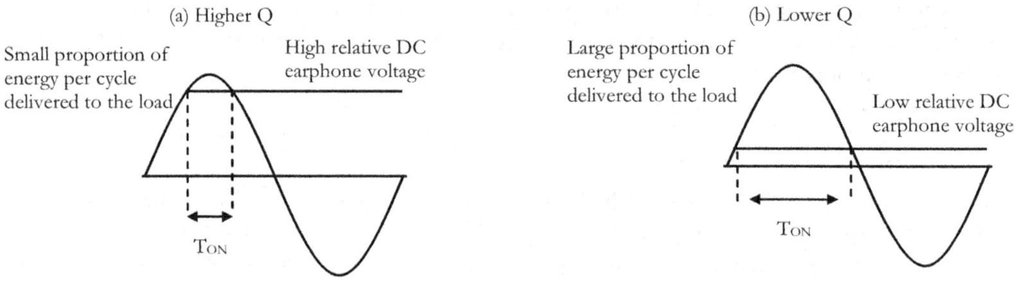

Figure 16.11
The relationship between Q and earphone DC voltage.

16.6 AUDIO VOLTAGE AMPLITUDE

The audio signal is recreated by the edge detector as the earphone (capacitor) charges and discharges. For audio demodulation, we deal with audio frequencies. As the audio signal will contain many frequencies, we shall select a single frequency and calculate the voltage drop for that frequency. Most, if not all, signals will contain a 1 kHz frequency, and we will choose this frequency for analysis.

The maximum DC voltage will be placed on the earphone when the modulated signal is at a peak. The DC voltage will then discharge, following the edge of the modulated signal, and for a 1 kHz signal, the minimum voltage will occur at half the 1 kHz period, that is 0.5 milliseconds. As the impedance of the diode will change as the earphone discharges, use the RMS value of the reverse bias voltage as given by Equation 16.23, that is

$$V_{DR} = -0.707(V_T + V_{DC})$$

Use this voltage to determine the diode's impedance at 1 kHz. At audio frequencies, the diode's capacitive reactance will be in the order of one thousand times greater than for RF and so we are only interested in the diode's resistance value. Calculate the available electrical energy factor, and the audio apparent discharge path resistance R_R becomes

$$R_R = |(\kappa_{DC} R_L)//(R_{DR} + R_{ST})| \qquad \text{Audio resistance with a load resistor} \qquad 16.31$$

When no load resistor is used, then

$$R_R = |\kappa_{DCD}(R_{DR} + R_{ST})| \qquad \text{Audio resistance with no load resistor} \qquad 16.31a$$

The maximum voltage drop is defined by the modulation index m. So the voltage discharged during the 0.5 milliseconds period, for the 1 kHz audio signal will be

$$V_{AUDIO} = mV_{DC}(1 - e^{\frac{-(0.5x10^{-3})}{R_R C_E}}) \qquad \text{1 kHz audio signal amplitude} \qquad 16.32$$

where m is the modulation index (typically 0.8). The audio voltage V_{AUDIO}, will be used to calculate the sound pressure level of the piezoelectric earphone.

16.7 MODIFYING THE PIEZOELECTRIC EARPHONE

There is a significant problem with the piezoelectric earphones sold at the current time. One side of the piezoelectric crystal is not soldered to the lead wire, and the connection is made by gluing the lead onto the crystal. There is a notch in the earphone case that should apply some pressure to keep the lead in place, but this does not often work, and the lead wire eventually detaches from the crystal, causing the earphone to fail.

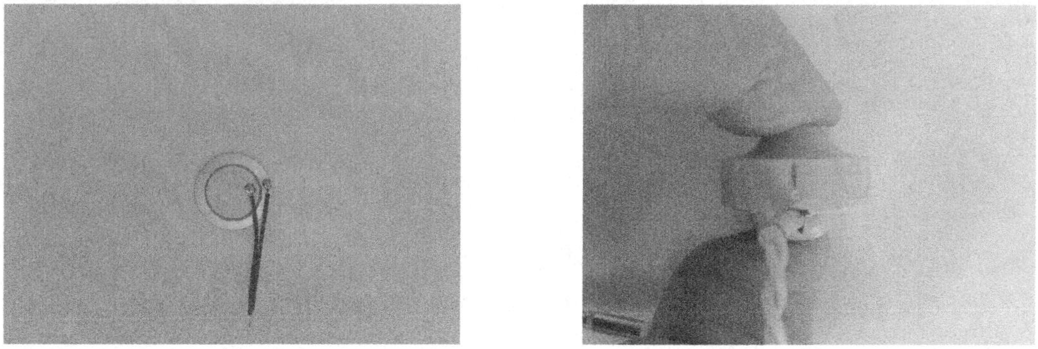

(a) (b)

Figure 16.12
The piezoelectric earphone.

You can fix a broken earphone by replacing the crystal with a 20 mm piezoelectric disk, Figure 16.12a. These can be purchased quite cheaply online, with wires already soldered onto the disk. They are typically sold in various pack sizes. Get more than one, just in case you get a bad crystal.

The earphone needs to be opened up to remove the old disk. There is a join around the earphone, as shown in Figure 16.12b. Using a pencil, run a line across the join near where the lead enters the case. This will allow you to realign the two halves of the case when you glue them back together. Hold the case between your thumb and pointing finger, and cut into the join using a sharp utility knife. Do not press down too hard. Make small cuts along a small section of the case. The knife should not dig into the case but should cut in a smooth action. If the knife jams, do not force it, as this is the way you will cut yourself. Take your time and keep your fingers out of the way. After cutting a millimetre or so into the case, use a small flat-head screwdriver to try and pry the case apart; see Figures 16.13a and 16.13b. Press down hard on the screwdriver and try and get the head into the seam. If it does not snap apart, use a slightly larger screwdriver. If it refuses to crack, keep cutting with the utility knife, making the cut longer and deeper along the seam. Some cases crack open quite easily, while others can be very difficult. Don't worry if the edge looks somewhat messy due to all the cutting and gouging. You can smooth this area a bit using sandpaper, and if you want, you can cover the seam using some coloured adhesive tape.

a)

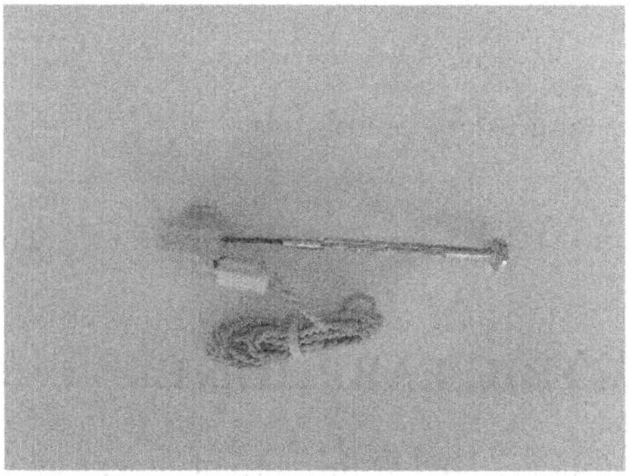

b)

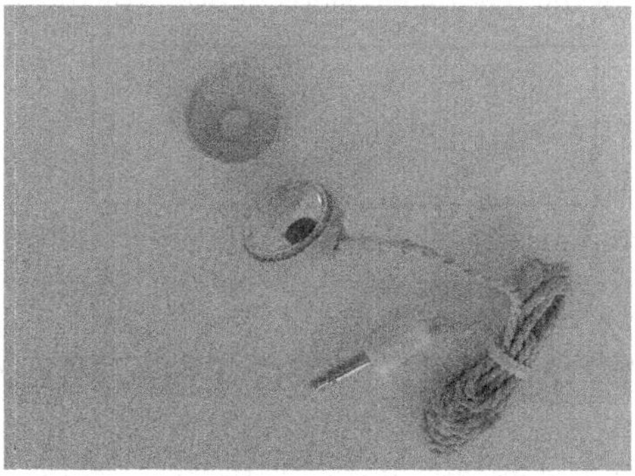

c)

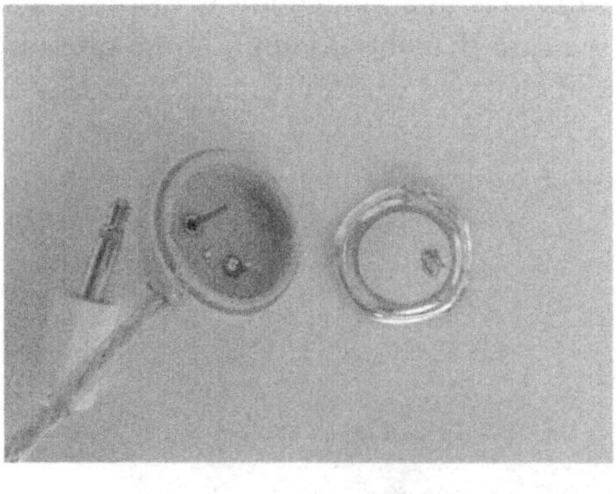

d)

Figure 16.13
Replacing the old piezoelectric crystal.

When you have finally cracked open the case, use the utility knife to remove the old piezoelectric crystal. The outer edge of the crystal rests on a lip running around a sound cavity. Push the blade of the utility knife between the crystal and the lip and pry upwards along the length of the lip. When detached, use a pair of wire cutters to cut the lead wires going to the crystal. See Figure 16.13c.

Solder the new crystal in place; see Figure 16.13d. You may have to shorten the leads from the new crystal to about 10 mm. Gently push the lead wires into the cavity making sure the exposed wires are not touching. Run a bead of epoxy glue around the lip and clamp the crystal in place. Use some soft material over the crystal for protection from the clamp (I use a square of balsa wood cut to size). See Figure 16.14a. Set it aside, and while the glue dries, continue with the modifications.

There is a notch in the top part of the case that is supposed to hold the glued lead wire to the crystal. Some are larger than others. Scrape this notch flat using the utility knife or a file; we don't want it anymore.

Now cut an o-ring from some spare thin cardboard, like from a cereal box. The outside diameter should be the same size as the crystal, 20 mm. The width of the o-ring only needs to be about 2 mm or so. It is used to help keep the edge of the crystal fixed so that the crystal can vibrate, similar to how a regular speaker is fixed and the speaker material vibrates to produce sound waves. Push this o-ring into the top of the case. See Figure 16.14b.

When the glue has dried (about half an hour or so), remove the clamp and protective material. Run a bead of glue around the joining seam, align the two halves of the case along the pencil mark made earlier and clamp the two halves together. See Figure 16.14c.

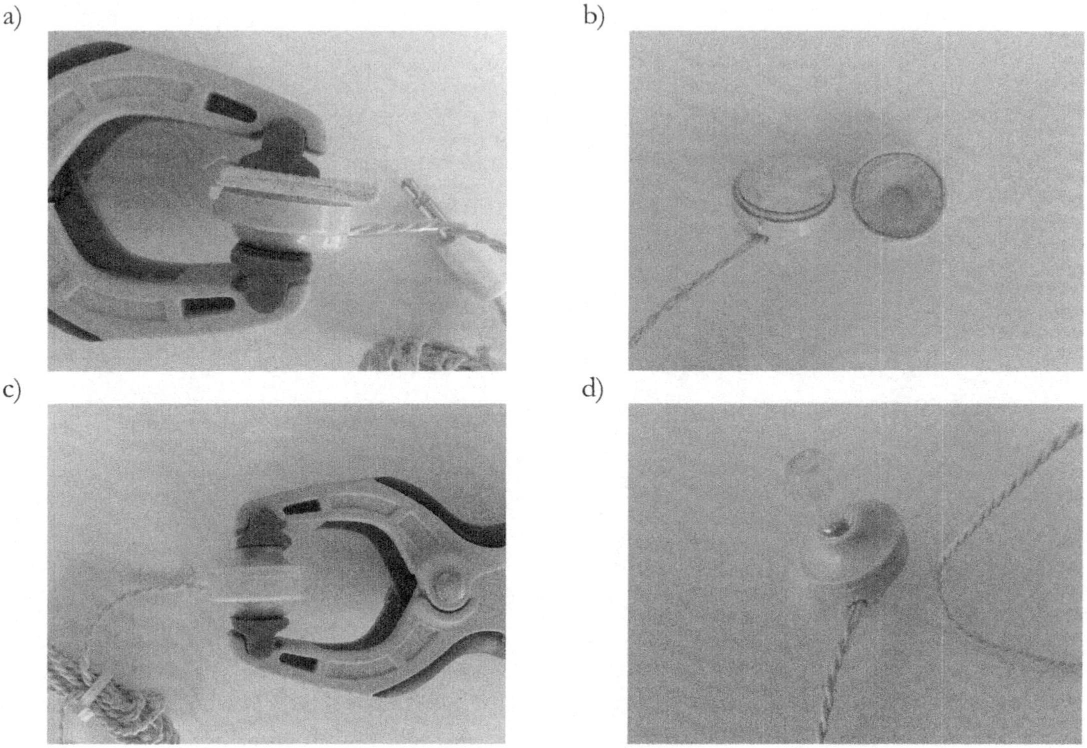

Figure 16.14
Gluing the new crystal and case.

When the glue has dried, you can clean up the join using the utility knife, and if you wish, cover the join with a strip of self-adhesive material or tape. This is only for cosmetic purposes and does not affect the sound quality. Screw in the inner ear tube and test it out. See Figure 16.14d. I generally find the modified earphone has a better bass response than the original, which is more to my taste.

Try the earphone out in a pocket radio to check if it is working. Make sure you turn the volume down first. If the earphone is not working, then most likely the wires have come apart from the crystal. If the volume is low and there is very little bass response, it is most likely that the crystal is not seated and has come loose. Or you forgot to insert the o-ring. You can open up the case and check. Fix any problem and re-glue the crystal and case. If the earphone still does not work, then you may have a bad crystal. Try another one.

17

The Complete Model

17.1 OVERVIEW

In this Chapter, we will complete the model and calculate the operating Q of a crystal radio. In our discussion regarding the crystal radio transformer (Chapter 13), we derived an equation for determining the Q of the secondary coil when attached to a resistive load. From the Chapters on demodulation and the piezoelectric earphone, we know the load is complex containing a diode that permits more current to flow in one direction than the other. Therefore to analyse the Q of a crystal radio, we will split the analysis into two main parts; one where the diode is forward biased and one where the diode is reversed biased.

The demodulator (the load) is connected to a tap of the secondary coil, and the voltage across the tap is dependent on the Q of the secondary coil with the load attached. The impedance of the diode, and therefore the secondary Q, is dependent on the applied voltage across the tap. But we do not know this voltage until the loaded Q has been calculated. To solve this problem, we could calculate the secondary coil's unloaded Q and the resonant voltage, as described in Chapter 13. We could use this voltage to determine the initial voltage across the coil's first loop, as described in the section on autotransformers, and calculate the Q with the load attached to the first loop. The first loop loaded Q value could then be used to approximate the initial voltage when the load is attached to the second loop. We then recalculated the Q for the load attached to the second loop, and repeat this process for each loop of the coil.

When determining Q, power loss is calculated with reference to the peak current. Due to the DC voltage on the earphone, we must transform the load impedance such that when the power loss is calculated, it is in relation to the peak current. This needs to be done so that the Q calculation is independent of current; otherwise, we would have actual current values in the equation.

When discussing the loss across the load (the demodulator), we will derive equivalent resistance values R_{ELEC} and R_{MECH}. For a given circuit current, these represent the equivalent resistance values required to give the total energy loss due to the earphone's action on the circuit. Normal circuit analysis is performed on the circuit using actual circuit values to derive the proportion of current passing across lossy resistance values. This lossy resistance value is due to the current flowing across the diode's resistance. The earphone is almost a short at RF frequencies, and any load resistor can be ignored when calculating the RF current. Only after the circuit has been analysed to determine the current branches and current proportions are R_{ELEC} and R_{MECH} applied (replacing the diode's resistance value) to give the loss of energy and to determine the Q.

17.2 POWER LOSS DURING FORWARD BIAS

First, we will consider the power losses when the diode is forward-biased. After calculating the unloaded Q of the secondary, use Equation 13.29 to determine the tap voltage and then calculate the DC earphone voltage as described in Chapter 16. Note that a number of RF cycles are required to place a stable DC voltage across the earphone, and it is after this steady DC voltage has been placed across the earphone that we then calculate Q for a single RF cycle.

Consider Figure 17.1, showing the equivalent circuit diagram for a crystal radio when the diode is forward-biased. After the DC voltage has been calculated, the peak voltage dropped across the demodulator (the load) is given by

$$\mathbf{V_{L(ON)}} = \mathbf{V_{T(peak)}} - \mathbf{V_{DC}} \qquad \text{Voltage across the load}$$

Use $V_{L(ON)}$ to calculate the forward voltage drop across the diode (Appendix 3) and the diode's forward bias resistance and capacitance (junction and diffusion), as described in Chapter 14. We now have all the voltages and impedances to perform the forward bias analysis.

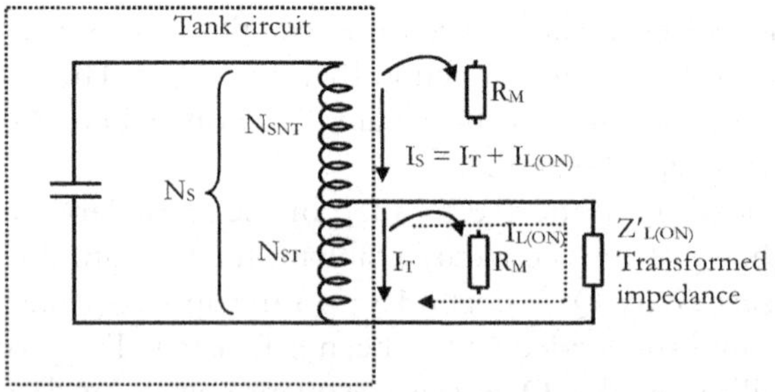

Figure 17.1
Circuit current when the diode is forward biased.

Coil Losses

During forward bias, the peak current across the load is

$$\mathbf{I_{L(ON)}} = \frac{\mathbf{V_{T(peak)}} - \mathbf{V_{DC}}}{\mathbf{Z_{L(ON)}}} \qquad \text{Load current} \qquad 17.1$$

where $V_{T(peak)}$ is the peak tap voltage, V_{DC} is the DC voltage on the earphone, and $Z_{L(ON)}$ is the total impedance of the load at forward bias. The total load impedance is the forward bias diode impedance Z_{DF} in series with the earphone in parallel with the load resistor. That is

$$\mathbf{Z}_{\mathbf{L(ON)}} = \mathbf{Z}_{\mathbf{DF}} + \mathbf{Z}_{\mathbf{E}} \\ = \mathbf{Z}_{\mathbf{DF}} + \mathbf{R}_{\mathbf{L}}//\mathbf{X}_{\mathbf{C}} \qquad 17.2$$

Using Ohm's law, we can find the equivalent impedance that will give us the load current in terms of the peak tap current as if the DC voltage on the earphone didn't exist. This allows us to express losses in terms of peak current when determining Q. The equivalent transformed load impedance is

$$\mathbf{Z'}_{\mathbf{L(ON)}} = \frac{\mathbf{V}_{\mathbf{T(peak)}}}{\mathbf{I}_{\mathbf{L(ON)}}} = R'_L - jX'_L$$

Equivalent transformed load impedance 17.3

Figure 17.1 shows the circuit diagram during forward bias, with the demodulator circuit shown as a single equivalent transformed load impedance Z'_L. The mutual inductance loss is shown as an impedance R_M. Furthermore, the load impedance reduces the overall impedance of the coil and will change the peak current value passing through the coil (from the tuning capacitor). The change in the peak current can be expressed as a ratio of the non-loaded coil impedance $Z_{S(OFF)}$ (just the coil impedance) divided by the loaded coil impedance $Z_{S(ON)}$.

$$I_\Delta = \left| \frac{\mathbf{Z}_{\mathbf{S}}}{\mathbf{Z}_{\mathbf{S(ON)}}} \right| \qquad 17.4$$

The coil impedance with the load at forward bias $Z_{S(ON)}$ is the non-loaded section of the coil impedance in series with the loaded coil section. That is

$$\mathbf{Z}_{\mathbf{S(ON)}} = \mathbf{Z}_{\mathbf{SNT}} + \mathbf{Z}_{\mathbf{ST}}//\mathbf{Z'}_{\mathbf{L}}$$

Coil impedance when forward biased 17.5

There are three branch currents, as shown in Figure 17.1, so we will write three loss terms. The first loss term is the loss through the non-loaded section of the coil N_{SNT}. Taking into account the change in current, the loss due to the coil resistance and mutual inductance, this loss term becomes

$$\boxed{P_{SNT(ON)} = I_\Delta^2 (R_{SNT} + \frac{N_{SNT}}{N_S} R_M)}$$

Power loss across the non-loaded section of the coil 17.6

The next loss term to consider is the loss through the loaded section of the coil. Note that the current is divided between the load and the tapped section of the coil, so we will use the current divider rule to get the tap coil current.

$$I_{\Delta ST(ON)} = \left| \frac{\mathbf{Z'}_{\mathbf{L(ON)}}}{\mathbf{Z'}_{\mathbf{L(ON)}} + \mathbf{Z}_{\mathbf{ST}}} \right|$$

Proportion of current through the tapped coil section 17.7

Note that $I_{\Delta ST(ON)}$ is not the actual current but the proportion of the peak current flowing across the tap section of the coil using the transformed load impedance. Taking into account the change in current, the loss due to coil resistance and mutual inductance, the loss across the loaded section of the coil becomes

$$P_{ST(ON)} = (I_{\Delta} I_{\Delta ST(ON)})^2 (R_{ST} + \frac{N_{ST}}{N_S} R_M)$$ Power loss across the loaded section of the coil 17.8

Load Losses

Now we will consider the loss across the demodulator (the load). Figure 17.2 represents the untransformed load circuit with the input impedance $Z_{L(ON)}$ given by

$$\mathbf{Z_{L(ON)}} = \mathbf{Z_{DF}} + \mathbf{R_L}//\mathbf{X_C}$$ 17.9

The impedance of the diode Z_{DF} is

$$\mathbf{Z_{DF}} = (R_{DF} // X_{Cj}) + R_{DS}$$ Forward bias diode impedance 17.10

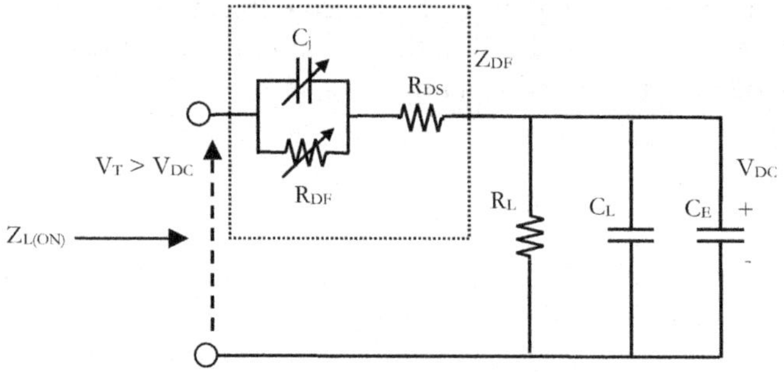

Figure 17.2
Load when diode forward biased.

The value of the load resistance R_L is orders of magnitude higher than the earphone reactance (10 Ω) at RF frequencies, and virtually all the current will flow onto the earphone and an insignificant amount across the load resistor. So we can safely ignore R_L, and the equivalent circuit is shown in Figure 17.3a. From Figure 17.3a, we can see that the loss across the demodulator during forward bias is the loss due to the diode's forward bias resistance R_{ZDF}. The reactance of the diode capacitance and the earphone capacitance do not represent losses.

From our discussion on piezoelectric earphones in Chapter 16, we represented the electrical energy loss due to the presence of the earphone as a transformation of the resistance in parallel with the earphone. During forward bias, the resistance values across the earphone are the diode resistance and the tap section of the coil resistance. This resistance is transformed using the total remaining electrical energy κ. As power loss is given by $P=I^2R$, the power loss resistance R_{ELEC} is given by

$$R_{ELEC} = real\{\kappa^2 (\mathbf{Z_{DF}} + \mathbf{Z_{ST}})\}$$ Electrical energy loss resistance 17.11

The resistor R_L is in parallel to the earphone, and the peak voltage across the earphone after charging is given by $V_{DC} + V_{TOPUP}$ so calculate κ using this voltage and the load resistance R_L for the frequency of interest. R_{ELEC} represents the electrical loss of the current passing across the diode's forward bias resistance and the influence of the piezoelectric earphone on the loss. If a load resistor is not present, then **κ** is calculated using the real (resistive) part of the diode's forward bias impedance R_{ZDF}.

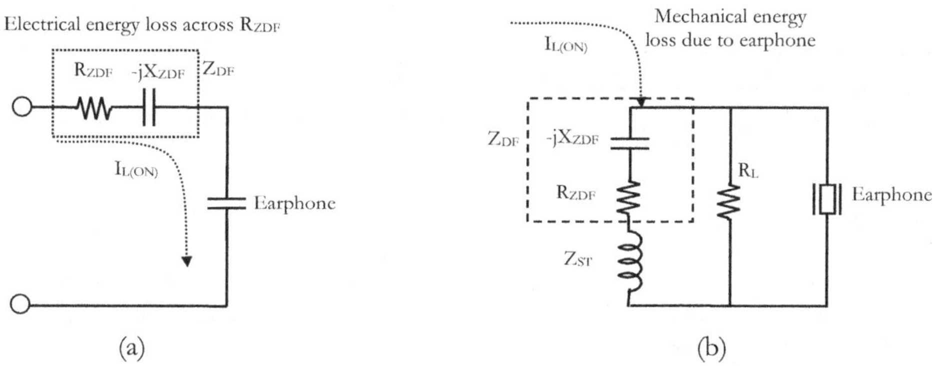

Figure 17.3
Electrical and mechanical losses across the load.

Having taken care of the electrical loss, we must now take into account the earphone's mechanical loss. The mechanical loss factor is given by (1-κ). The mechanical energy loss is determined by what the earphone "sees" as the load in parallel with the earphone. The equivalent circuit representing the mechanical loss of the earphone is shown in Figure 17.3b. The major influence on the mechanical energy loss is due to the load resistance R_L with the diode impedance in parallel. Therefore the mechanical energy loss resistance during forward bias is given by

$$\boxed{R_{MECH} = real\{(\mathbf{Z_{DF}} + \mathbf{Z_{ST}})//(1-\kappa)^2 R_L\}}$$ Mechanical energy loss resistance 17.12

Note that the mechanical loss resistance is only the resistive (real) part of the phasor.

If a load resistor is not present, then **κ** is calculated using the real (resistive) part of the diode's forward bias impedance R_{ZDF}, and R_{MECH} becomes

$$\boxed{R_{MECH} = real\{(1-\kappa)^2(\mathbf{Z_{DF}} + \mathbf{Z_{ST}})\}}$$ Mechanical energy loss resistance when a load resistor is not used 17.13

The proportion of the current flowing into the load from the coil, using the current divider rule, is

$$I_{\Delta L(ON)} = \left|\frac{\mathbf{Z_{ST}}}{\mathbf{Z'_{L(ON)}} + \mathbf{Z_{ST}}}\right|$$ Proportion of current through the load 17.14

This tap current splits when passing through the diode. Some pass through the diode's forward bias resistance, and some across the forward bias capacitance. Using the current divider rule, we get the proportion of current passing through the diode's resistance as

$$I_{\Delta DF} = \left|\frac{\mathbf{X_{C_{jF}}}}{\mathbf{R_{DF}} + \mathbf{X_{C_{jF}}}}\right|$$ Proportion of current through diode resistance 17.15

The equivalent electrical resistance R_{ELEC} and mechanical resistance R_{MECH} need to be adjusted so that they represent components of the transformed impedance Z'_L. This is done by multiplying the resistive components by the ratio

$$\beta_L = \left|\frac{\mathbf{Z'_L}}{\mathbf{Z_L}}\right| \qquad 17.16$$

Finally, the forward bias loss across the load is

$$\boxed{P_{L(ON)} = (I_\Delta I_{\Delta L(ON)} I_{\Delta DF})^2 \beta_L (R_{ELECT} + R_{MECH})} \qquad \text{Load energy loss} \qquad 17.17$$

Total Forward Bias Loss

The total loss during the forward bias phase is the sum of the losses times the fraction of the period the diode is in forward bias L_{ON}. Calculate L_{ON} as described in Chapter 16. The final forward bias loss then becomes

$$\boxed{P_{(ON)} = L_{ON}(P_{SNT(ON)} + P_{ST(ON)} + P_{L(ON)})} \qquad \text{Total forward bias loss} \qquad 17.18$$

17.3 POWER LOSS DURING REVERSE BIAS

Figure 17.4 shows the current path when the diode is in reverse bias. As in the forward bias mode

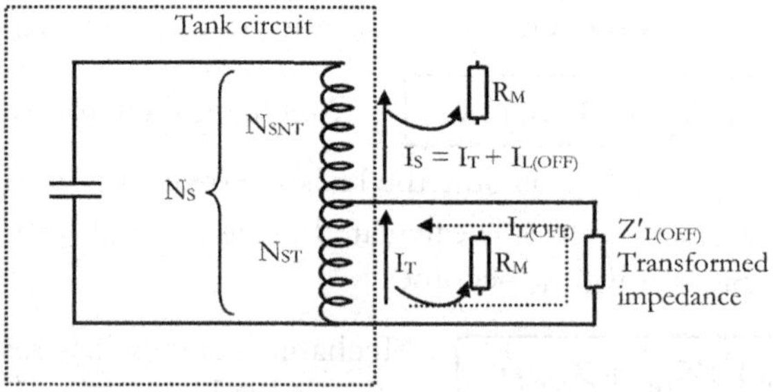

Figure 17.4
Diode reverse biased.

there are three current paths, and we will derive losses for each branch. During the reverse bias phase, at the start of the discharge period, the earphone voltage is $V_{DC} + V_{TOPUP}$ and at the end of the period is $V_{DC} - V_{DROP}$, with V_{DC} being the average. The voltage dropped across the load is

$$\mathbf{V_{L(OFF)}} = -(\mathbf{V_{T(peak)}} + \mathbf{V_{DC}}) \qquad \text{Voltage across the load}$$

The average voltage across the diode during reverse bias is

$$\mathbf{V_{D(OFF)}} = -(0.707 * \mathbf{V_{T(peak)}} + \mathbf{V_{DC}}) \qquad \text{Voltage across the diode}$$

Use this voltage to calculate the diode's reverse bias resistance and capacitance.

Coil Losses

The loss through the non-loaded section of the coil is the loss due to the coil resistance of this section and the mutual inductance. We write this loss term as

$$P_{SNT(OFF)} = (R_{SNT} + \frac{N_{SNT}}{N_S} R_M) \quad \text{Reverse bias power loss across the non-loaded section of the coil} \quad 17.19$$

Next, we will consider the loss through the loaded section of the coil. The peak reverse bias current across the load is

$$\mathbf{I}_{\mathbf{L(OFF)}} = \frac{\mathbf{V}_{\mathbf{T(peak)}} + \mathbf{V}_{\mathbf{DC}}}{\mathbf{Z}_{\mathbf{L(OFF)}}} \quad \text{Reverse bias peak load current} \quad 17.20$$

The load impedance Z_L is the reverse bias diode impedance Z_{DR} in series with audio impedance Z_E.

$$\begin{aligned}\mathbf{Z}_{\mathbf{L(OFF)}} &= \mathbf{Z}_{\mathbf{DR}} + \mathbf{Z}_{\mathbf{E}} \\ &= \mathbf{Z}_{\mathbf{DR}} + \mathbf{R}_{\mathbf{L}} // \mathbf{X}_{\mathbf{C}}\end{aligned} \quad 17.21$$

The equivalent transformed load impedance is

$$\begin{aligned}\mathbf{Z}'_{\mathbf{L(OFF)}} &= \frac{\mathbf{V}_{\mathbf{T(peak)}}}{\mathbf{I}_{\mathbf{L(OFF)}}} \\ &= R'_L - jX'_{\mathrm{L}}\end{aligned} \quad \text{Reverse bias equivalent transformed load impedance} \quad 17.22$$

The current is divided between the load and coil, so we will use the current divider rule to get the tap coil current ratio

$$I_{\Delta ST(OFF)} = \left| \frac{\mathbf{Z}'_{\mathbf{L(OFF)}}}{\mathbf{Z}'_{\mathbf{L(OFF)}} + \mathbf{Z}_{\mathbf{ST}}} \right| \quad \text{Proportion of current through tapped coil section} \quad 17.23$$

And the loss due to the coil resistance and mutual inductance for the tapped section of the coil is

$$P_{ST(OFF)} = I^2_{\Delta ST(OFF)} (R_{ST} + \frac{N_{ST}}{N_S} R_M) \quad \text{Reverse bias power loss across the loaded section of the coil} \quad 17.24$$

Load Losses

Figure 17.4 shows the electrical and mechanical losses during reverse bias. If a load resistor is used, calculate the electrical energy loss **κ** using the DC voltage across the earphone V_{DC} and the load resistance value R_L. If a load resistor is not present, then **κ** is calculated using the real (resistive) part of the diode's reverse bias impedance R_{ZDR}. The resistance representing the electrical loss of the circuit is

$$R_{ELEC} = real\{\kappa^2 (\mathbf{Z}_{\mathbf{DR}} + \mathbf{Z}_{\mathbf{ST}})\} \quad \text{Reverse bias electrical loss resistance} \quad 17.25$$

The resistance representing the mechanical loss across the earphone is given by the resistive value of the reverse bias diode's impedance in parallel with the load resistor and is

$$R_{MECH} = real\{(\mathbf{Z_{DR}} + \mathbf{Z_{ST}})//(1-\kappa)^2 R_L\}$$ Mechanical loss resistance 17.26

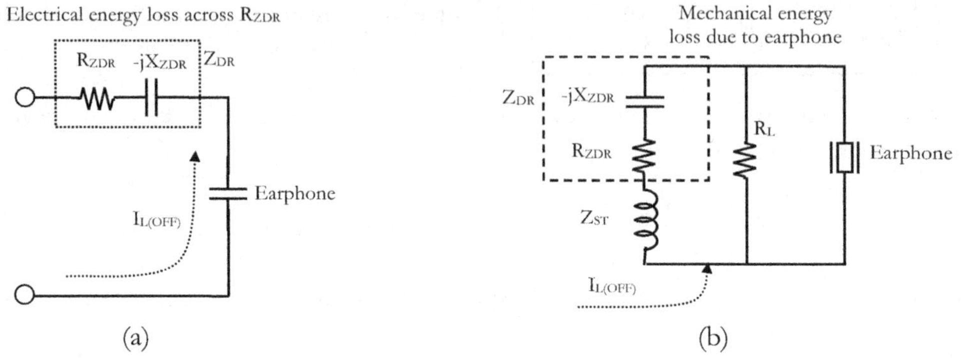

Figure 17.5
Reverse bias electrical and mechanical losses across the load.

If a load resistor is not present, then the mechanical loss becomes

$$R_{MECH} = real\{(1-\kappa)^2(\mathbf{Z_{DR}} + \mathbf{Z_{ST}})\}$$ Mechanical loss resistance no load resistor 17.27

The proportion of the current from the coil passing through the load is

$$I_{\Delta L(OFF)} = \left| \frac{\mathbf{Z_{ST}}}{\mathbf{Z'_{L(OFF)}} + \mathbf{Z_{ST}}} \right|$$ Proportion of current through the load 17.28

This current splits when passing through the diode. Some pass through the diode's reverse bias resistance, and some across the reverse bias capacitance. Using the current divider rule, we get the proportion of current passing through the diode's resistance as

$$I_{\Delta DR} = \left| \frac{X_{C_{jR}}}{R_{DR} + X_{C_{jR}}} \right|$$ Proportion of current through diode resistance 17.29

As in the forward bias phase, the equivalent electrical resistance R_{ELEC} and mechanical resistance R_{MECH} need to be adjusted so that they represent components of the transformed impedance Z'_L. Finally, the reverse bias load loss becomes

$$P_{L(OFF)} = (I_{\Delta L(OFF)} I_{\Delta DR})^2 \beta_L (R_{ELECT} + R_{MECH})$$ Reverse bias load energy loss 17.30

Total Reverse Bias Loss

The total loss during the reverse bias phase is the sum of the losses times the fraction of the period L_{OFF} the diode is in reverse bias mode and is

$$P_{(OFF)} = L_{OFF}(P_{SNT(OFF)} + P_{ST(OFF)} + P_{L(OFF)})$$ Total reverse bias loss 17.31

17.4 THE Q OF A CRYSTAL RADIO

To calculate the Q of a crystal radio, we combine both the forward bias and reverse bias losses, and the Q becomes

$$Q_{CR} = \frac{X_S}{P_{L(OFF)} + P_{L(ON)}} \alpha \qquad \text{Circuit Q of a crystal radio} \qquad 17.32$$

Unmodulated Signal Test

A series of measurements were made using the crystal radio setup given in Table 17.1 to determine the accuracy of this model. Comparisons between the measured and calculated values of Q and tap voltage are shown in Figures 17.6 to 17.8. The calculated values are shown as lines and the measured values as symbols. The demodulator circuit was attached to taps 11, 24, 36 and 48 of the secondary coil, and the Q and tap voltage were measured. The power supply used for the test has an output impedance of 50 Ω. A 2,200 pF capacitor was placed in series with the power supply to simulate antenna capacitance. An unmodulated 0.1 volt peak-to-peak 1MHz sine wave was injected into the primary coil. The distance between the primary and secondary coils was set to 50 mm for each measurement. Measurements were taken with and without a 47 kΩ load resistor for the X-Schottky, 1N34A and BAT85 diodes.

Characteristic	Value
Primary coil	PC3 (33.3 uH)
Secondary coil	SC1 (332 uH)
Separation distance	50 mm
Carrier frequency	1 MHz
Signal	Unmodulated
Measurements at windings	11, 24, 36, 48

Table 17.1
Crystal radio configuration used for Q and tap measurements.

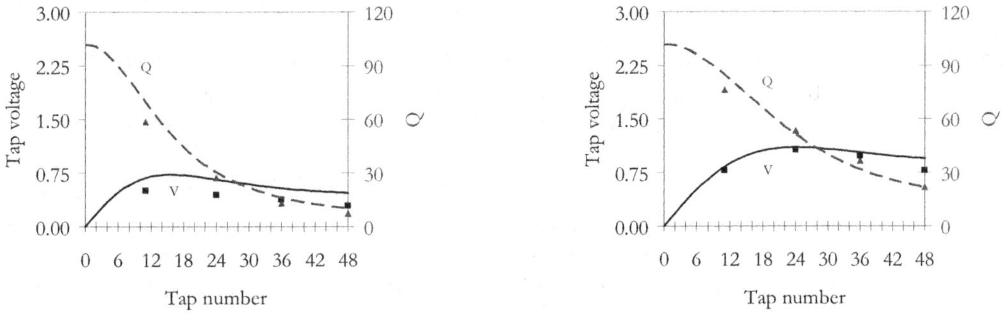

(a) 47 k load resistor.

(b) No load resistor.

Figure 17.6
X-Schottky tap voltage and Q.

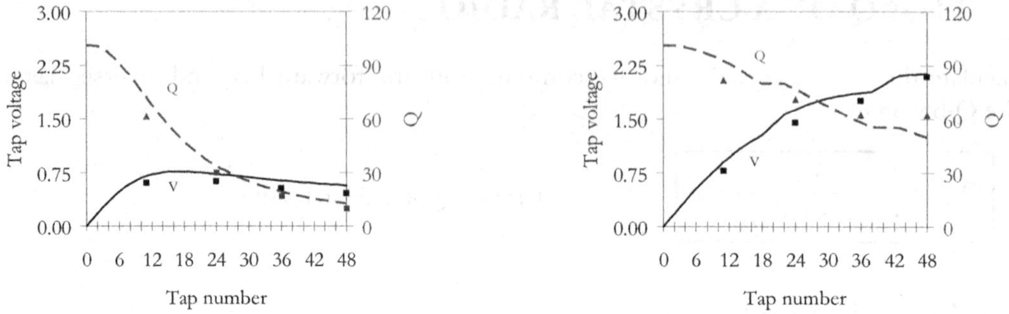

(a) 47 k load resistor.

(b) No load resistor.

Figure 17.7
1N34A tap voltage and Q.

Notice how the model predicted the overall shapes of the curves very well. And that it predicted the different shapes in the Q and tap voltages between the crystal radio using a load resistor and for the crystal radios discharging across the diode (no load resistor).

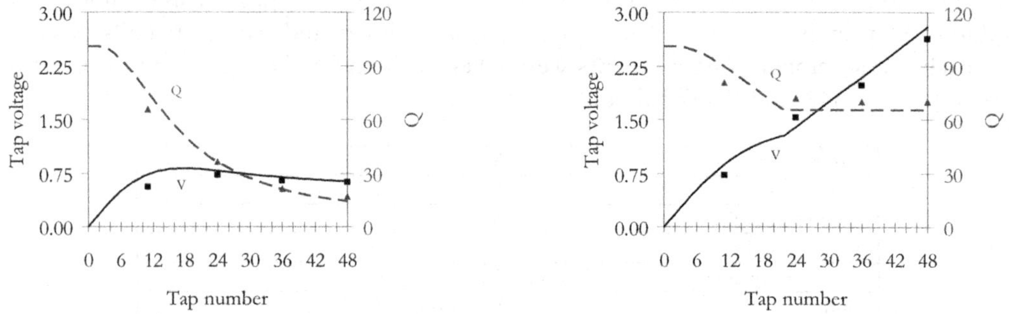

(a) 47 k load resistor.

(b) No load resistor.

Figure 17.8
BAT85 tap voltage and Q.

17.5 THE DEMODULATED AUDIO SIGNAL

When a carrier wave is modulated there will be just as many RF cycles with amplitudes above the carrier wave peak as there are below the carrier wave peak. Therefore the average cycle amplitude will be the carrier amplitude. This means the model does not need to be changed to take into account modulation.

To test the model, a 0.1 volt peak-to-peak 1 MHz carrier was modulated using a 1 kHz signal with an 80% modulation index. This signal was fed into the crystal radio setup as described above. For each tap position, the demodulated 1 kHz peak-to-peak voltage across the earphone was measured. Figures 17.9 to 17.11 show the measured values as squares and the calculated values as solid lines.

Clearly, the model is a good predictor of the piezoelectric earphone audio voltages. The audio response voltages and operating Qs are combined in single graphs for easy comparison, as shown in Figures 17.12 and 17.13. Notice that when a 47 kΩ resistor is not used for demodulation, the audio voltage decreases. As the reverse bias diode resistance will be much greater than 47 kΩ, the optimal audio frequency being demodulated will not be at 1 kHz, but will be at a lower frequency.

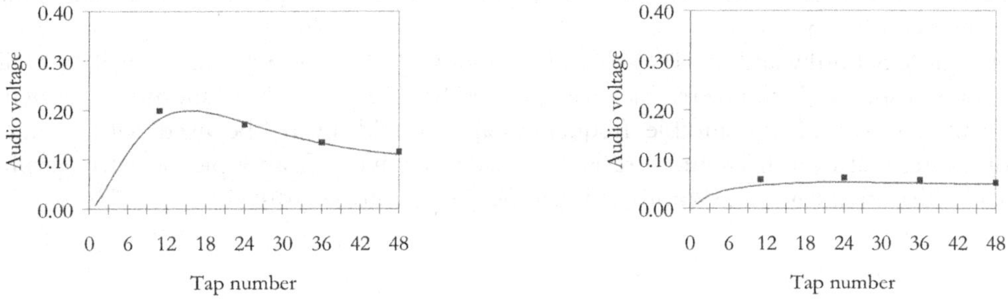

(a) 47 k load resistor. (b) No load resistor.

Figure 17.9
Earphone audio voltage as a function of tap number for X-Schottky.

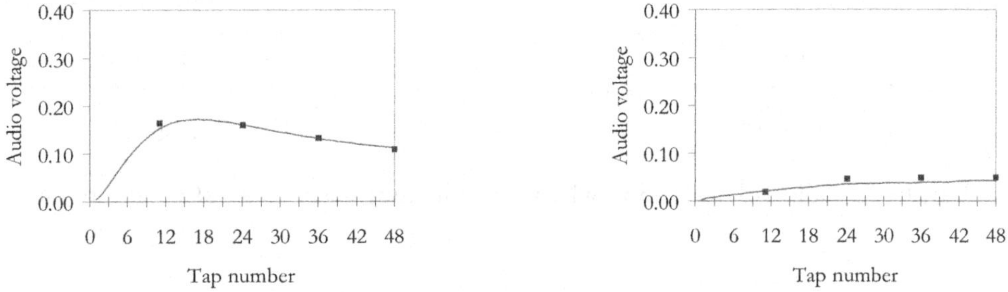

(a) 47 k load resistor. (b) No load resistor.

Figure 17.10
Earphone audio voltage as a function of tap number for 1N34A.

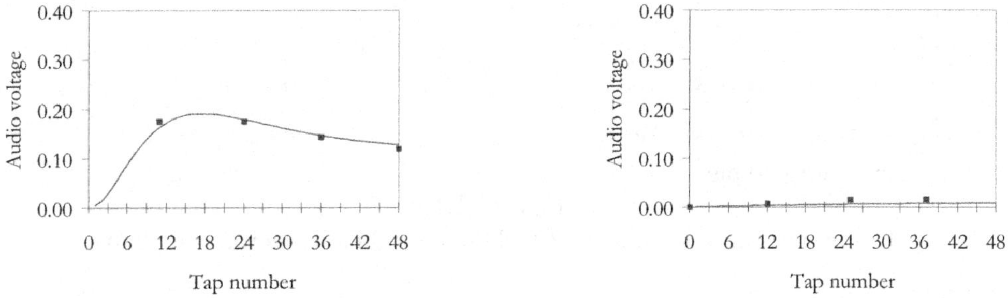

(a) 47 k load resistor (b) No load resistor.

Figure 17.11
Earphone audio voltage as a function of tap number for BAT85.

How well a diode demodulates depends on the value of the reverse bias resistance. For high-power stations, the reverse bias voltage will be high and the reverse bias resistance will likewise be high, and the audio signal will not demodulate optimally at 1 kHz. This may result in some high frequency audio distortion. For low-power stations, the reverse bias voltage will not be as great, and the reverse bias resistance will be lower. Therefore, for low-power stations the audio signal will not be as distorted.

Both the X-Schottky and the 1N34A will demodulate the audio reasonably well, but will have a greater bass response. The reverse bias for the BAT85 is in megaohms and does not demodulate the signal very well at any audible frequency, and should always be used with an audio load resistor. Note, that even if an audio resistor is used, the audio from a piezoelectric earphone can sound a bit distorted for loud stations as it can be driven into overdrive.

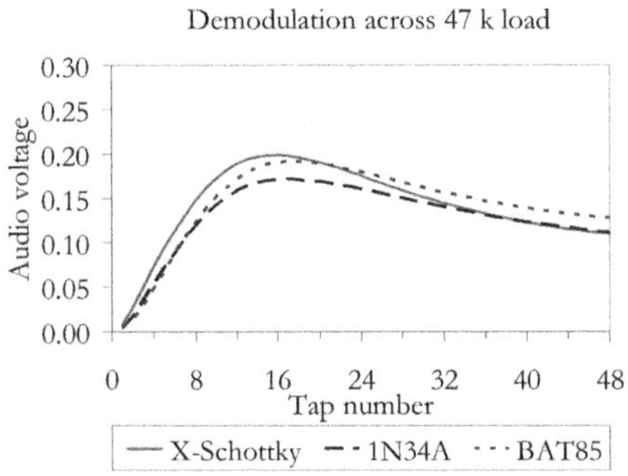

(a) Audio voltage.

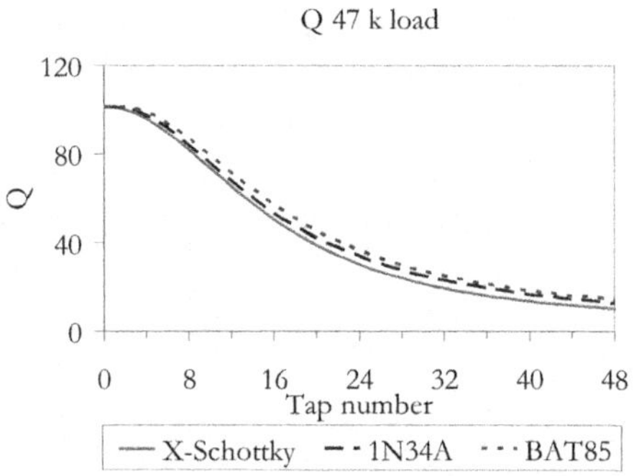

(b) Q.

Figure 17.12
Earphone audio voltage and Q when demodulating across a 47 kΩ audio resistor.

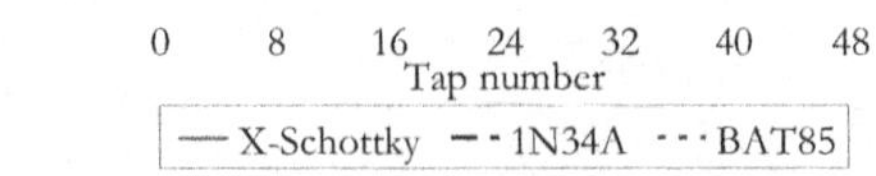

(a) Audio voltage.

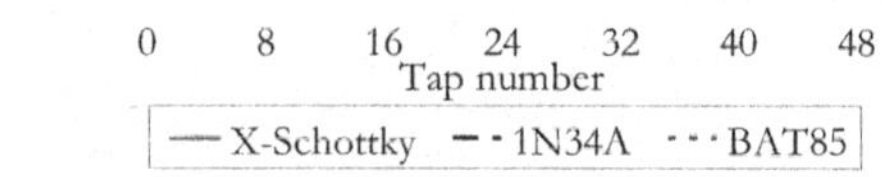

(b) Q.

Figure 17.13
Earphone audio voltage and Q when demodulating across the diode.

From Figure 17.12b and 17.13b we can see that when the signal is being demodulated, the Q drops quite significantly. When the signal is not being demodulated, as for the BAT85 curve in Figure 17.13, the Q remains high. This means that energy is not being converted to sound, and much less energy is removed from the coil.

In a crystal radio, the operating Q value is largely irrelevant. One simply selects the coil separation distance and tap position to provide the loudest signal without other station interference. A small operating Q (provided we have built a high Q coil and we are not losing energy elsewhere due to a poor design) means we are converting electrical energy into sound, which is precisely what we want. A high-Q coil, however, provides us with more options to sacrifice some volume for the ability to separate channels.

17.6 EARPHONE VOLUME

Recall from section 15.5 that the ear responds to sound pressure levels (SPL) and is logarithmic in nature. With the SPL value given by

$$L_P = 20\log_{10}(\frac{p}{p_0})$$

where P_0 is considered to be the threshold pressure needed to detect a 1 kHz tone. The minimum voltage across the earphone I can detect at 1 kHz is 0.001 volts peak-to-peak. Assuming the pressure level developed by the movement of the piezoelectric crystal is directly proportional to the voltage across the earphone, we can replace the pressure values of P and P_0 with voltage values. Therefore, for a piezoelectric earphone, the SPL is given by

$$\boxed{L_P = 20\log_{10}(\frac{V_E}{0.001})} \quad \text{Piezoelectric earphone SPL} \quad 17.33$$

where V_E is the peak-to-peak voltage across the earphone. Equation 17.33 is graphed in Figure 17.14 with earphone voltages ranging from 0.001 (just audible) to 1 volt (very loud). Notice a small increase in voltage at lower voltage levels results in a greater increase in volume. The same increase in voltage at higher voltage levels results in a much smaller increase in SPL and volume. As a rule of thumb, an increase in the SPL of 2 dB and below is not noticeable, and an increase of 3 dB is just noticeable. A station with a SPL of 53 dB will sound just as loud with a SPL of 51 dB. An increase of 10 dB will sound twice as loud. The ear's response to sound pressure levels is important when selecting tap positions for the secondary coil.

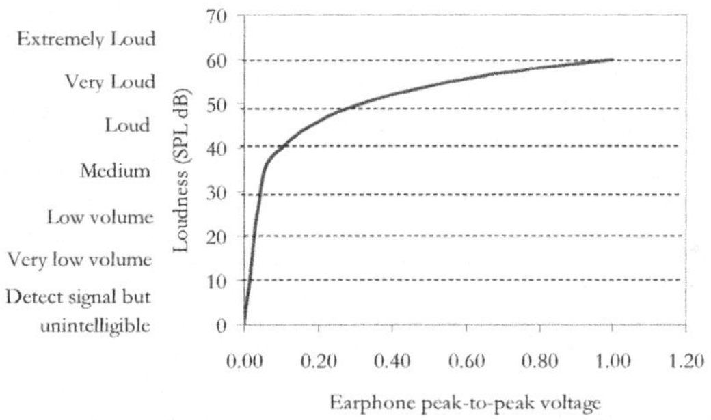

Figure 17.14
SPL as a function of earphone voltage.

Figure 17.15 shows the audio voltages for each of the diodes demodulating across the 47 kΩ expressed as volumes in SPL dB. From Figure 17.12a we would expect a large difference in volume going from tap 11 to tap 48. However, due to the logarithmic nature of the human ear, this is not the case. The greatest voltage difference is for the X-Schottky with the voltage ranging from 0.18 volts at tap 11 to 0.11 volts at tap 48. The voltage at tap 11 is over 60% of the voltage at tap 48. However, the SPL at tap 11 is 45 dB and at tap 48 is 41 dB. A noticeable difference, but not 60% louder.

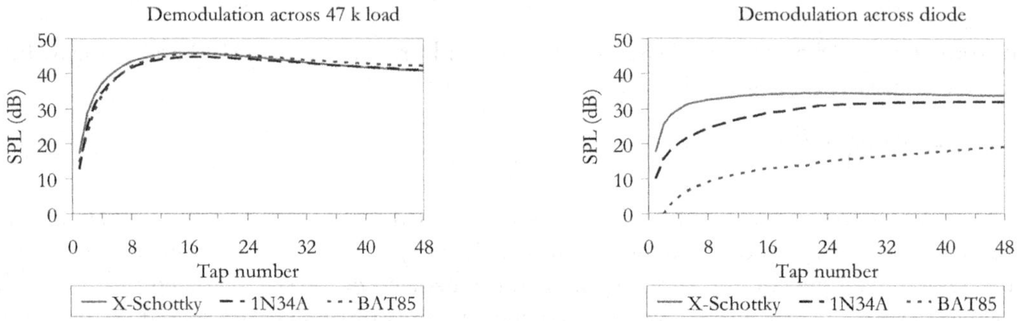

(a) (b)

Figure 17.15
Earphone volume for each diode, shown as SPL dB with an antenna voltage of 0.1 Vpp.

17.7 DIODE SELECTION

Consider the combined I-V curves for the diodes specified in Chapter 14, given in Figure 17.16. From the forward bias I-V curve, we would expect the 1N34A to be a poor performer. However as shown in figures 17.12 and 17.13, although the volume is lower than for the X-Schottky and the BAT85, it is not hugely different. And the Q is slightly better than the X-Schottky. This is because the earphone DC voltage is accumulated over many cycles, and even a small charge can build up over many cycles provided the reverse bias resistance is high. And the reverse bias resistance for the 1N34A is higher than the reverse bias resistance of the X-Schottky.

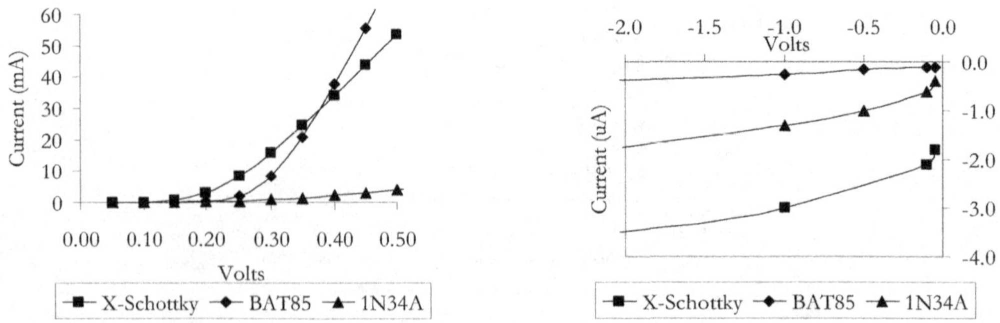

(a) Forward bias I-V (b) Reverse bias I-V

Figure 17.16
Diode I-V curves.

So although the forward bias impedance of the diode is important, the reverse bias impedance is a very important factor when selecting diodes. This is particularly true when considering low-power stations. The forward voltage across the diode provided by a low power station may be well below 0.1 volts. And the difference between the forward bias currents for the diodes may now be not as great as that above the turn-on voltage. This means the reverse bias impedance becomes the dominant factor.

Consider Figure 17.17 which shows the audio volume (SPL dB) and the operating Q of the crystal radio with a very low antenna voltage of 0.005 volts peak-to-peak. At this antenna voltage, not enough charge is placed on the earphone to produce an audio signal for the X-Schottky. As very little charge is placed on the earphone, the Q will be low. A diode that performs well with medium or high signal levels may not perform as well at very low signal levels.

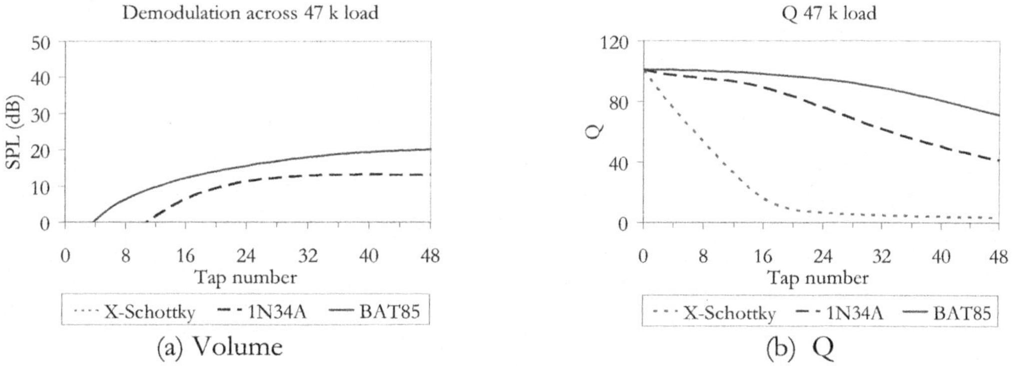

Figure 17.17
Volume and Q with an antenna voltage of 0.005 Vpp.

As a rule of thumb, *select diodes with low forward bias impedances and high reverse bias impedances. With a greater weight given to the reverse bias impedance.* Note, that diode characteristics can be different for the same make and model of diode. When selecting diodes, select a few different models of diodes and simply try them out in the crystal radio to see which performs better.

18

Crystal Radio Design

18.1 OVERVIEW

In this Chapter, we will go through the steps required to determine the capacitance and inductance values of the components when designing a crystal radio, starting with the primary coil inductance value needed for a given antenna to achieve resonance for selected frequencies. We will then determine the range of the tuning capacitor required to span the AM frequency band. Next, we will determine the secondary coil's inductance required for use with the given tuning capacitor. And finally, we will determine the tap positions on the secondary coil required to provide power to the demodulator.

18.2 DETERMINE THE PRIMARY COIL INDUCTANCE

We want the primary circuit to be uncoupled resonant to provide maximum power transfer from the primary to the secondary. To achieve uncoupled resonance, we need to know the antenna's capacitance for a given frequency. And the inductive reactance of the primary coil must match the antenna's capacitive reactance. That is

$$X_P = X_A$$

Exact uncoupled resonance cannot be achieved for every possible frequency within the AM band if only a coil is used. Using each winding as a tap, even a swiper arm, will not result in exact resonance for all frequencies. We can achieve resonance for all frequencies by using a tuning capacitor in the primary circuit, known as double-resonance. We will cover double-resonance in a later Chapter. In this Chapter, we will concentrate on using tapped primary coils and examine how they work.

For a tapped primary coil, the number of frequencies that can be uncoupled resonant equals the number of taps plus one (for the entire coil itself). For frequencies that are not uncoupled resonant (and that is quite likely to be all the stations), we must rely on gaining maximum power transfer by making the coil coupled resonant. That is

$$X_P = X_A - \left| imag\left\{ \frac{\mathbf{I}_{\mathbf{S0}} jX_M}{\mathbf{I}_{\mathbf{PS0}}} \right\} \right|$$

Note that we are achieving electrical resonance and not antenna resonance. These are two quite distinct properties and should not be confused with one another. Antenna resonance is achieved when a radio wave's maximum standing wave voltage is induced across the antenna. Electrical resonance occurs when some of the stored energy is released in phase with the RF signal (the same way resonance is achieved when pushing a swing at the right moment). A short antenna will only receive a portion of the available RF energy, and making it electrically resonant just means you are maximising the available current. You have in no way turned a short antenna into a resonant antenna, as I have often seen stated. A long antenna approaching antenna resonance will have a much larger standing wave voltage and a greater amount of energy available for you to use. A longer antenna approaching resonance, and made electrically resonant, will always outperform a short antenna made electrically resonant.

Determining the Primary Coil Tap Positions

Using taps to change the inductance value of the primary coil allows us to split the AM broadcast frequencies into bands. Each tap will be resonant at some frequency. First, determine the capacitance value of your antenna for a given frequency using the equations given in Chapter 10 for antenna and transmission line capacitance. Then use Equation 8.3 to determine the inductance required for that frequency. That is

$$L_{TAP} = \frac{1}{4\pi^2 f^2 C_{ANT}} \tag{18.1}$$

where C_{ANT} is the capacitance of the antenna at frequency f.

Table 18.1 gives the antenna capacitance for our standard 25-metre antenna for a selection of frequencies. Also, the capacitance value is given when a standard 10-metre transmission line is added. Alongside each antenna capacitance value is the calculated inductance required for resonance.

Frequency (kHz)	Antenna only			+Transmission line		
	Capacitance (pF)	Inductance required for resonance (uH)	Resistance (Ω)	Capacitance (pF)	Inductance required for resonance (uH)	Resistance (Ω)
530	227	397	11.0	1,521	59.3	12.5
550	228	367	11.6	1,523	55.0	13.1
750	234	192	18.3	1,555	29.0	20.0
1,000	244	104	28.5	1,612	15.7	30.4
1,250	259	63	40.4	1,696	9.6	42.5
1,700	307	29	64.1	1,951	4.5	66.4

Table 18.1
Capacitance values of the standard 25-metre crystal radio antenna and transmission line.

Once the inductance values have been determined, use the techniques in Chapter 11 for cylindrical coils to determine the geometry of your coil for a given inductance, former size, wire type and pitch. For each inductance value, you will have calculated the number of turns, as shown in Table 18.2 for our example. The largest inductance value will be given for the lowest frequency of interest, and the number of turns for this inductance will be the total number of turns in the primary coil N_{LP}. We also have the number of turns for inductance values required at the higher frequencies that we will refer to as N_{LT}. The loop number (starting from the left-hand side of the coil) for each tap in the primary coil can then be calculated using

$$W_{TAP} = N_{LP} - N_{LT}$$ Winding number for taps 18.2

Typically the antenna will be connected to the right-hand end lead of the coil with the earth attached to the tap number, but they can be swapped around as it does not matter. Each tap number for this primary coil geometry was calculated and is shown in Table 18.2.

For example, say the lowest frequency of interest is 550 kHz. From Table 18.1, we require an inductance of 367 uH for resonance. Using a 90 mm PVC former and 0.5 mm wire with 0.6 mm pitch, 56 turns are required to give an inductance of 367 uH. This will be the total number of turns in our primary coil, that is N_{LP} = 56. Note that the turn numbers have been rounded to whole numbers.

The next frequency of interest is at 750 kHz with a required inductance of 192 uH, requiring 39 turns. So the first tap to give 192 uH will be at winding number

$$\begin{aligned} W_{TAP} &= N_{LP} - N_{LT} \\ &= 56 - 39 \\ &= 17 \end{aligned}$$

Frequency (kHz)	**Antenna only**			**+Transmission line**		
	Inductance required for resonance (uH)	Turns required	Tap at turn #	Inductance required for resonance (uH)	Turns required	Tap at turn #
550	367	56	0	55.0	19	0
750	192	39	17	29.0	13	6
1,000	104	26	30	15.7	10	9
1,250	63	20	36	9.6	7	12
1,700	29	12	44	4.5	4	15

Table 18.2
Primary coil taps.

Choosing the antenna and transmission line inductance values for our example antenna, the inductance values were recalculated using integer number of turns. The primary coil specifications are given in Table 18.3. Also, the last tap at turn number 15 was dropped. The difference in inductance value for the taps at winding numbers 12 and 15 is only 5.1 uH. There will be enough coupling impedance at winding number 12 to make the 1,700 kHz frequency coupled resonant at some distance. Also, the small number of turns for the last tap will result in a very small mutual inductance and a small induced voltage unless the separation distance is very small.

Description	Value		
Coil name	PCEX-1		
Coil inductance	60.4 uH		
Coil diameter	90.5 mm		
Coil length	11.4 mm		
Number of windings	19		
Tap winding #	6	L =32 uH	ℓ=7.8 mm
	9	L=20 uH	ℓ=6.0 mm
	12	L=11 uH	ℓ=4.2 mm
Pitch	0.6 mm		
Wire	0.5 mm enamelled copper wire (core D=0.5 mm, OD=0.6 mm)		
Former material	PVC		
Former diameter	90 mm		
Former wall thickness	2 mm		
Q_{eff} (at 1MHz)	118		

Table 18.3
90 mm tapped 60.4 uH example primary coil.

Primary Coil Taps Effect on a Crystal Radio

To examine how these taps work, let's first examine the antenna-only case. For example, say we have three stations; a high power station at 550 kHz that induces a 0.1 Vpp across the antenna, and two low power stations at 1,100 kHz and 1,250 kHz that induce 0.02 Vpp across the antenna. Say we tuned the secondary coil to the station at 1,100 kHz, but the earth was at tap 0 with a primary coil inductance of 367 uH. The primary coil will be resonant at 550 kHz and will induce a voltage across the secondary coil, as shown in Figure 18.1a, with the induced voltage dependent on the separation distance. The station at 1,250 kHz will also induce a voltage across the secondary, but it will be very much smaller as it is far from primary resonant. The tuned station, at 1,100 kHz, will have its induced voltage across the secondary coil amplified by the Q of the secondary coil. However, the station at 550 kHz has an induced voltage larger than the tuned station's voltage, and we will hear both stations.

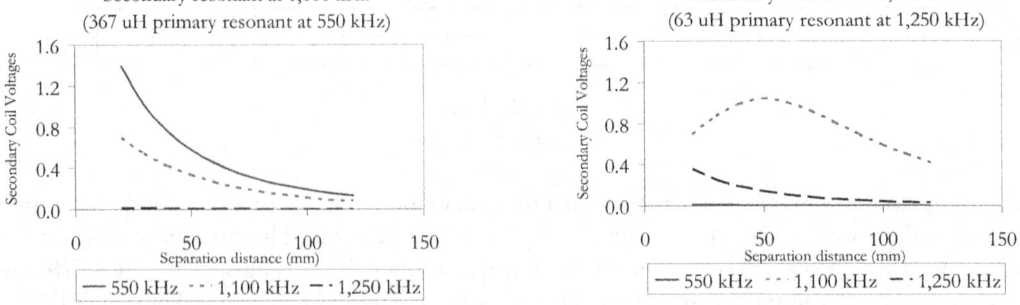

(a) Primary inductance 367 uH
Secondary tuned to 1,100 kHz.

(b) Primary inductance 63 uH
Secondary tuned to 1,100 kHz.

Figure 18.1
Interference voltage due to other stations – Antenna only.

Say we now connect the earth wire to the third tap so that the primary coil inductance is 63 uH, as shown in Figure 18.1b. At this inductance, the station at 1,250 kHz will be primary resonant and induce a higher voltage than before across the secondary. The station at 550 kHz will no longer be close to resonance, and the induced voltage across the secondary will be much lower. The voltage for the tuned station at 1,100 kHz is now closer to uncoupled resonance, and the induced voltage across the secondary will also be higher than before. This induced voltage will be boosted by the Q of the secondary, resulting in a high resonant secondary voltage, with a peak resonant voltage occurring when the primary becomes coupled resonant at about a 60 mm separation distance. There will now be little or no interference from the other stations.

Say we now add the standard transmission line. The maximum primary coil inductance required is now only 55 uH due to the large increase in transmission line capacitance. Figure 18.2a shows the secondary voltages for each frequency when the primary coil inductance is set to 55 uH. Note that the induced voltage for the 550 kHz station is still relatively high and may result in some bleed-through. When the 9.6 uH primary tap is selected, the voltage difference between the tuned station at 1,100 kHz, with respect to the other frequencies, is quite large, resulting in good selectivity.

For the antenna-only or when the transmission line is included, in both instances, the station at 1,100 kHz becomes coupled primary resonant but at different separation distances. However, the voltage across the secondary for the tuned station is slightly larger when the 9.6 uH tap is selected. This is due to two factors. One, the coupled primary resonance is achieved with a smaller separation distance when the 9.6 uH tap is selected, resulting in a larger induced voltage across the secondary. Two, the mutual inductance is lower due to the smaller primary inductance value. This results in lower mutual inductance losses and a higher Q value for the secondary.

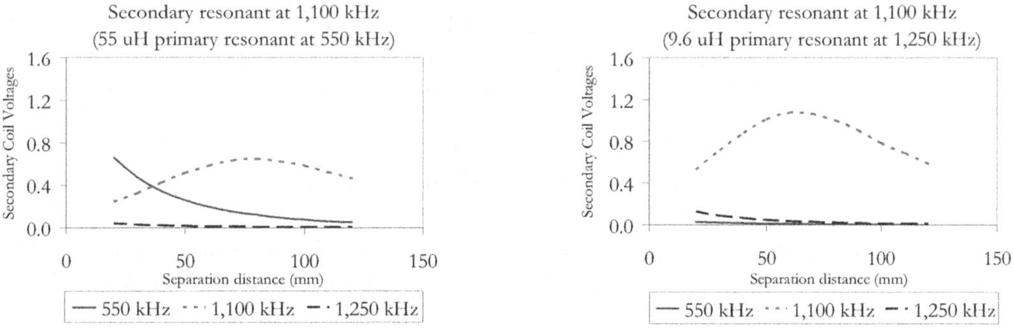

(a) Primary inductance 55 uH
Secondary tuned to 1,100 kHz.

(b) Primary inductance 9.6 uH
Secondary tuned to 1,100 kHz.

Figure 18.2
Interference voltage due to other stations – +Transmission line.

Primary Coil Tap Adjustment to the Model

The induced emf on the primary due to the magnetic field of the secondary is induced across all the loops of the primary coil, including the unconnected loops. Current will flow through the connected loops in the circuit but not the unconnected loops. The primary can be treated as an autotransformer with the voltage across the taps given by

$$V_{TAP} = \frac{N_P}{N_{TOTAL}} EMF$$

This results in a reduction of the mutual inductance loss, R_M. This reduction is equal to the proportion of the loops in the active portion of the coil divided by the total number of loops in the coil. The loss due to mutual inductance R_M now becomes

$$R_M = \left(\frac{N_P}{N_{TOTAL}} \frac{X_M}{Z_{PU}} \right)^2 (R_A + R_P) \qquad \text{Mutual inductance loss for a tapped primary} \qquad 18.3$$

where N_P is the number of turns for the given tapped inductance, and N_{TOTAL} is the total number of turns for the primary coil. R_P is the resistance, Z_{PU} is the input impedance of the primary coil across the tapped section, and R_A is the antenna resistance.

18.3 DETERMINE THE REQUIRED TUNING CAPACITOR

The medium wave broadcast band frequencies are dependent on the region in which you live. The International Telecommunications Union (ITU) has divided the world into three regions. Region 1 includes Europe, Africa and the Middle East. Region 2 includes North and South America and Greenland. Region 3 includes Australia and Asia. Regions 1 & 3 have a slightly different AM broadcast spectrum with different station spacing than Region 2, as shown in Table 18.4.

Region	Lowest carrier wave signal (kHz)	Highest carrier wave signal (kHz)	Station Spacing (kHz)	Lowest modulated signal (kHz)	Highest modulated signal (kHz)
1 & 3	531	1602	9	526.5	1606.5
2	530	1600	10	525	1605

Table 18.4
Medium wave broadcast band frequencies.

In 1988 the ITU extended the upper-frequency range of the AM broadcast band, as shown in Table 18.5. For design purposes, we will use the lowest carrier wave frequency of 530 kHz and the highest modulated signal of 1,701 kHz so that we can cover any region.

Region	Lowest carrier wave signal (kHz)	Highest carrier wave signal (kHz)	Station Spacing (kHz)	Lowest modulated signal (kHz)	Highest modulated signal (kHz)
1 & 3	1611	1701	9	1606.5	1705.5
2	1610	1700	10	1605	1705

Table 18.5
Extended medium wave broadcast band frequencies.

Having determined the frequency range of interest, we next need to determine the capacitance range required to span the frequency range. The maximum capacitance of the tuning capacitor will give us the lowest resonant frequency. Rearranging Equation 8.2 to give the lowest resonant frequency, we get

$$f_{low}^{2} = \frac{1}{4\pi^2 L C_{\max}}$$

And for the highest resonant frequency

$$f_{high}^{2} = \frac{1}{4\pi^2 L C_{\min}}$$

Taking the ratio of the highest frequency to the lowest frequency, the inductance variable L and the constant values are eliminated. If the minimum capacitance of our tuning capacitor is less than C_{min} then we can tune to a higher frequency than required, and similarly, if the maximum capacitance of the tuning capacitor is greater than C_{max} we can tune to a lower frequency than f_{low}. So we can write the ratio as an inequality, meaning that the capacitance range ratio must be equal to or higher than the square of the frequency ratio to satisfy our needs. That is

$$\boxed{\frac{f_{high}^{2}}{f_{low}^{2}} \le \frac{C_{\max}}{C_{\min}}} \qquad 18.4$$

There is an infinite combination of maximum and minimum capacitance values that can satisfy this inequality. Say we have a variable capacitor on hand, and we either measured or looked up the specifications and found it ranges from 7 pF to 140 pF. We need to determine whether this capacitor can tune over the required frequency range. The frequency ratio for our selected band is

$$\frac{f_{high}^{2}}{f_{low}^{2}} = \left(\frac{1{,}701x10^3\,Hz}{530x10^3\,Hz}\right)^2$$
$$= 10.3$$

C_{min} and C_{max} must take into account coil and any other circuit parasitic capacitance. From our analysis of coil self-capacitance, we know that coil self-capacitance is a function of frequency. The closer we get to the self-resonant frequency, the closer we get to the self-capacitance value. For frequencies far below self-resonance, the self-capacitance of the coil becomes negligible. In effect, the self-capacitance acts to reduce the apparent self-inductance of a coil at higher frequencies and does not act like your typical circuit capacitance. For this reason, the coil's self-capacitance is added to the minimum capacitance C_{min} and not the maximum capacitance. Typical coil self-resonant capacitances are around 2 or 3pF.

Circuit stray (or parasitic) capacitance due to circuit wiring, on the other hand, will act like normal circuit capacitors. The parallel stray capacitance will act like a capacitor in parallel with your tuning capacitor and will make a difference in the range of tunable frequencies. The maximum and minimum capacitance values are thus given by

$$\boxed{C_{\min} = C_{T\min} + C_{stray} + C_{coil}} \qquad 18.5a$$

$$\boxed{C_{\max} = C_{T\max} + C_{stray}} \qquad 18.5b$$

where C_{Tmin} is the minimum tuner capacitance, including trim capacitance, and C_{Tmax} is the maximum capacitance of the tuner, including trim capacitance. C_{stray} is wiring and circuit capacitance, and C_{coil} is the self-capacitance of the coil.

When selecting a tuning capacitor, keep in mind that the values specified are "nominal", and the actual values may differ by one or two pF. This is not a big deal when tuning to low frequencies, but it does make a difference at high frequencies. There will also be a screw used to adjust the capacitance range by a small amount called the ***trim capacitance screw***. Typically this will range from 2 to 10 pf. Adjust this screw to the lowest setting. Usually, the stated nominal minimum and maximum values are specified with the trim set to the minimum value. For our analysis, we will use 2 pF for stray capacitance and 3 pF for the coil capacitance. And for our given tuning capacitor, we get

$$C_{\min} = 7 + 2 + 3$$
$$= 12\,pF$$
$$C_{\max} = 140 + 2$$
$$= 142\,pF$$

And the capacitance ratio becomes

$$\frac{C_{\max}}{C_{\min}} = \frac{142x10^{-12}}{12x10^{-12}}$$
$$= 11.8$$

The frequency ratio of 10.3 and is less than 11.8, so we are just good. Note that if $C_{T\min}$ is slightly higher by 1 or 2 pF, then we could not use this capacitor, at least not for this range of frequencies, so it is just on the borderline. This is cutting it a bit fine, so I will try a more traditional 7-360 pF tuning capacitor (note that these are hard to come by). Performing the same analysis, the capacitance ratio becomes

$$\frac{C_{\max}}{C_{\min}} = \frac{362x10^{-12}}{12x10^{-12}}$$
$$= 30.2$$

The capacitance ratio is much greater than the frequency ratio, so this tuning capacitor will have no problems tuning over the range required. Note that if you chose 1,600 kHz as your highest frequency, then the frequency ratio would be 9.11, and the 140 pF tuning capacitor would be suitable. The upper value of tuning capacitors used for the AM broadcast band can range from 140 pF to 360 pF. Upper values of less than 140 pF will not be suitable. Values greater than 360 pF can be used; however, less of the full turning circle of the capacitor will be used to span the band, and very small adjustments to tuning are required making tuning more difficult.

Note that the inductance seen by the capacitor is the coil's self-inductance and a component of the mutual inductance as given by Equation 13.27. That is

$$X_C = X_S + imag\left\{\frac{\mathbf{I}_{\mathbf{PSO}} X_M}{\mathbf{I}_{\mathbf{S0}}}\right\} \qquad \text{Capacitive reactance required for resonance}$$

However, the mutual inductance component seen by the capacitor will only be a small fraction of the mutual inductance value X_M. This component will be very small compared to the typical inductance of the secondary coil and can be ignored for this analysis.

18.4 DETERMINE THE SECONDARY COIL INDUCTANCE

Having chosen the tuning capacitor, we now need to determine the secondary coil's inductance. As the capacitance ratio of our variable capacitor is 30.2 and is greater than the frequency ratio of 10.3, we have some spare capacitance so to speak. We can determine the required inductance value by setting the minimum tuner capacitance to tune to the high frequency, which means we can tune to frequencies below the lowest carrier frequency of interest. Or we can set the maximum tuner capacitance to tune to the lowest frequency and allow tuning to frequencies above the highest frequency of interest. These two conditions will set the upper and lower inductance values. Ultimately we can choose an inductance value anywhere between and including these upper and low limits.

The secondary coil inductance range

The lowest inductance we can use for our tuning capacitor is determined by setting the maximum capacitance to tune to the lowest frequency. That is

$$L_{\min} = \frac{1}{4\pi^2 f_{low}^2 C_{\max}}$$ Minimum inductance value 18.6

For the chosen capacitor, this would be

$$\begin{aligned} L_{\min} &= \frac{1}{4\pi^2 (530x10^3)^2 362x10^{-12}} \\ &= 249uH \end{aligned}$$

The maximum possible inductance is determined by setting the minimum capacitance to tune to the highest frequency of interest. That is

$$L_{\max} = \frac{1}{4\pi^2 f_{high}^2 C_{\min}}$$ Maximum inductance value 18.7

Again for our example, the maximum inductance would be

$$\begin{aligned} L_{\max} &= \frac{1}{4\pi^2 (1{,}701x10^3)^2 12x10^{-12}} \\ &= 730uH \end{aligned}$$

Choosing the Secondary Coil Inductance Value

We now have a wide range of inductance values to choose from, anywhere from 249 uH to 730 uH. Let us first examine the upper and lower limits of the inductance values. If we chose to use the 730 uH inductor, from Equation 18.7, the minimum capacitance sets the maximum inductance for the highest frequency. Now by rearranging Equation 18.4 to find the capacitance that will tune to the lowest frequency, we get

$$C_{fhigh} = C_{min}$$
$$C_{flow} = \left(\frac{f_{high}^{2}}{f_{low}^{2}} C_{min} \right)$$

Set the minimum capacitance value to resonance at the highest frequency 18.8

Using the values calculated already for our frequency range and the chosen tuning capacitor, we have

$$C_{fhigh} = 12pF$$
$$C_{flow} = 10.3(12pF)$$
$$= 123.6pF$$

Therefore to tune over the entire range of frequencies using the 730 uH inductor would require a capacitance range from 12 pF to 123.6 pF. We are, therefore, only using a portion of the entire capacitance range of our tuning capacitor (12 to 362 pF). We can calculate this portion using

$$C_{RANGE} = \frac{C_{flow} - C_{fhigh}}{C_{max} - C_{min}}$$ 18.9

So for our tuning capacitor, we get

$$C_{RANGE} = \frac{123.6pf - 12.0pf}{362pf - 12pf}$$
$$= 0.32$$

We are only using 32% of the total capacitance range.

If we decided to use the lower limit inductor of 249 uH, the lowest capacitance value to tune to the highest frequency is given by

$$C_{flow} = C_{max}$$
$$C_{fhigh} = C_{max} / \frac{f_{high}^{2}}{f_{low}^{2}}$$

Set the maximum capacitance value to resonance at the lowest frequency 18.10

And for our example, we would get

$$C_{flow} = 362pF$$
$$C_{fhigh} = \frac{362pf}{10.3}$$
$$= 35.1pF$$

And CRANGE would be

$$C_{RANGE} = \frac{362.0pf - 35.1pf}{362.0pf - 12.0pf}$$
$$= 0.93$$

Now we are using 93% of the capacitance range. Using a greater portion of the tuning capacitor range makes it physically easier to tune into stations as we are using more of the capacitor's total rotation. Using the 730 uH inductor with only 32% of the rotation will make it very difficult to tune, especially for stations close in frequency. This is one consideration when determining the secondary coil inductance to use. Other considerations are Q, self-resonant frequency, wire length required to wind the coil and the physical size of the coil.

Consider the effective Q as a function of inductance of a tightly wound coil using 0.5 mm enamel copper wire wound on a 90 mm PVC former with a 2 mm wall thickness and a pitch of 0.6 mm for different frequencies, as shown in Figure 18.3a. A coil using the same wire and former with a 1.2 mm pitch is shown in Figure 18.3b.

In Figure 18.3a, we can see that the maximum Q for the 1,700 kHz curve occurs at around 500 uH and then falls off. Prior to 500 uH, all coils at all frequencies have less Q. This will set our upper limit of inductance to 500 uH for this coil.

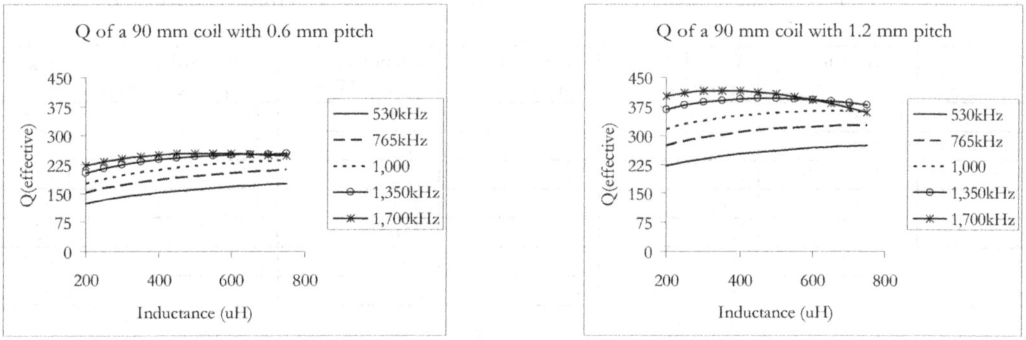

(a) (b)

Figure 18.3
Coil effective Q as a function of inductance.

In Table 18.6, the specifications for various inductance values for the tightly wound coil are given. As the inductance value increases, the Q increases; however, the useful capacitance range decreases, the self-resonant frequency decreases, and the length of wire required to wind the coil increases. At 450 uH, the tuning capacitance useful range is only about 50%.

L	ℓ/d ratio	Wire length (metres)	Number of turns	Effective Q(1 MHz)	Self-resonant frequency (MHz)	C_{RANGE}
249	0.3047	13	46	187	5.5	94%
300	0.3442	15	52	198	5.1	79%
350	0.3801	16	57	206	4.8	67%
400	0.4164	18	62	212	4.5	58%
450	0.4502	19	68	219	4.3	52%
500	0.4832	20	72	223	4.1	47%
730	0.6285	27	94	236	3.5	33%

Table 18.6
90 mm diameter coil specifications for different inductance values – 0.6 mm pitch.

Ultimately the final value chosen for the secondary coil inductance depends on what stations you have at higher frequencies and how much C_{RANGE} you can deal with, and also practical considerations such as wire length and former size. Also, note the ℓ/d ratio is not that important for closely wound coils as this ratio relates to the minimum length of wire required to wind a coil, and this mainly affects the skin effect and the coil's AC resistance. The Q of closely spaced coils, however, is overwhelmed by the proximity effect negating much of this benefit. In coils with larger pitch values, the proximity effect is greatly reduced, and the ℓ/d ratio becomes a bit more relevant.

In Figure 18.3b, the pitch of the coils was increased to 1.2 mm (twice the wire thickness). Notice the substantial increase in Q due to the reduction in the proximity effect. The maximum effective Q for a 1,700 kHz signal occurs at an inductance of around 350 uH. Table 18.7 gives the coil specifications for different inductance values using 1.2 mm pitch. Also, note the difference in wire length required to wind the coils. A 350 uH coil with a 0.6 mm pitch requires 16 metres of wire, and for a 1.2 mm pitch coil, 21 metres are required. You need to make sure you have enough wire on a spool to wind the coil. You don't want to run out of wire while you are winding.

L	ℓ/d ratio	Wire length (metres)	Number of turns	Effective Q(1 MHz)	Self-resonant frequency (MHz)	C_{RANGE}
249	0.7843	17	59	329	5.9	94%
300	0.9005	19	68	339	5.4	79%
350	1.0120	21	76	345	4.9	67%
400	1.1217	24	84	351	4.6	58%
450	1.2301	26	92	356	4.3	52%
500	1.3374	28	100	359	4.0	47%
730	1.8183	39	136	362	3.2	33%

Table 18.7
90 mm diameter coil specifications for different inductance values – 1.2 mm pitch.

In these examples, the former diameter was kept constant at 90 mm, and the ℓ/d ratio was allowed to vary. If the ℓ/d ratio was set to 0.407 (optimum wire length) for a 350 uH coil with a 1.2 mm pitch, the effective Q would increase to 384 (up from 345, an 11% increase), and the coil diameter would increase to 138 mm. This is quite a large former diameter. In a later Chapter, we will see how to achieve extremely high Qs without resorting to ridiculously sized formers by using Litz wire. Setting the ℓ/d ratio to 0.407 for the tightly wound coil would not result in an increase in Q, and the coil diameter would reduce to 87 mm.

Description	Value
Coil name	SCEX-1
Coil inductance	306 uH
Coil diameter	90.6 mm
Coil length	31.2 mm
Number of windings	52
Pitch	0.6 mm
Wire	0.5mm enamelled copper wire (core D=0.5mm, OD=0.6mm)
Former material	PVC
Former diameter	90mm
Former wall thickness	2mm
Q_{eff} (at 1MHz)	199

Table 18.8
90 mm tapped 306 uH example secondary coil.

For the tightly wound coil of 0.6 mm pitch, the 300 uH inductor was chosen. We have a higher Q than the 243 uH inductor, and the C_{RANGE} is still good at 79%. Going to higher inductance values gives us a small boost in Q with an increase in the wire length and the number of turns and a reduction in C_{RANGE}. The inductance value was recalculated using a whole number of turns, and the specification for our example secondary coil is given in Table 18.8.

18.5 DETERMINING OPTIMAL SECONDARY TAP POSITIONS

The purpose of a crystal radio is to provide maximum station volume with the least amount of station interference. Ultimately we want to know the configuration of our crystal radio to provide this maximum volume. Firstly we know the secondary resonant coil voltage is a function of separation distance. Suppose we calculate the earphone voltage for each secondary coil winding at a range of separation distances. In that case, we will determine the distance and tap position that will give us this maximum volume. The configuration for the maximum voltage will likely result in a low operating Q for the crystal radio. We may wish to sacrifice some volume for selectivity by increasing the Q. The operating Q of a crystal radio can be increased by selecting lower-end tap values on the secondary coil and increasing the separation distance between the primary and secondary coils. As stated previously, the operating Q of a crystal radio is largely irrelevant. Select the loudest volume you can get, with enough selectivity to remove station interference. What this Q value is, does not matter.

Given our selected example primary and secondary coils above, we can state the components for our example crystal radio as given in Table 18.9.

Description	Value
Crystal radio name	CREX-1
Antenna	Standard 25-metre antenna + standard trans line
Primary coil	PCEX-1
Secondary coil	SCEX-1
Diode	BAT85
Audio resistor	47 kΩ
Audio capacitor	none
Earphone	piezoelectric

Table 18.9
Example crystal radio specifications.

To determine the maximum earphone voltage start with a low-frequency value of 550 kHz and a beginning separation distance of 20 mm. Set the antenna input voltage to some value (say 0.1Vpp) and the antenna resistance and capacitance for the given frequency. Calculate the earphone voltage for each secondary coil winding and select the maximum earphone voltage. Increase the separation distance by 10 mm and redo the calculations. Finish with the separation distance at 120 mm. Next, select a higher carrier wave frequency and repeat the calculations. Finish when the frequency is the maximum frequency of interest.

Of course, this could not be done manually, and we will use the CrystalRadioDesign program to do the calculations. The results of the earphone volume, as a function of separation distance for the given frequencies, are plotted in Figure 18.4 for antenna voltages of 0.1 Vpp and 0.02 Vpp, representing loud and quiet stations.

From the graphs, select a separation distance so that the difference between the sound pressure levels is about 2 dB or less. We have chosen a 50 mm separation distance as this provides a reasonable amount of Q (at 1 MHz) to separate stations, without a noticeable decrease in volume.

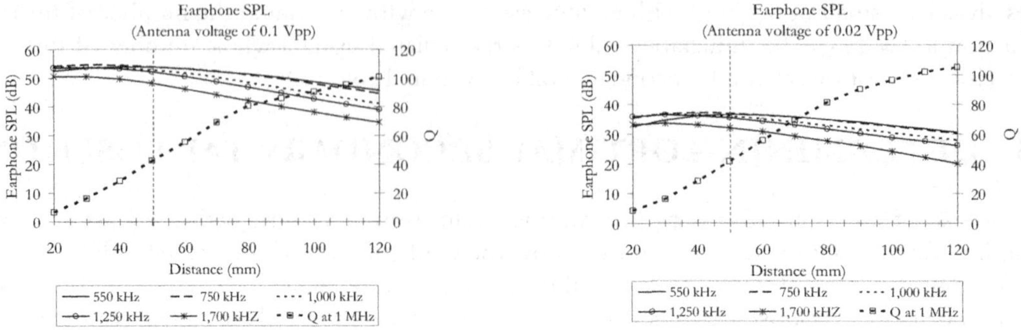

Figure 18.4
Maximum earphone SPL as a function of separation distance.

Table 18.10 lists the difference in sound pressure levels from the maximum SPL to the SPL at a distance of 50 mm. All except one value at 1,700 kHz for 0.1 V antenna voltage are smaller than 2 dB.

Frequency (kHz)	Difference in SPL from max (dB)	
	0.1 V	0.02V
550	0.90	0.00
750	0.77	0.25
1,000	0.99	0.54
1,250	1.52	0.99
1,700	2.39	1.47

Table 18.10
Difference in SPL at 50 mm separation distance from the maximum value.

We will now plot the SPL (at a distance of 50 mm) as a function of winding number to determine the tap positions, as shown in Figure 18.5. The tap positions are chosen so that for any given frequency there is a tap less that 2 dB down from the maximum volume. Therefore these tap positions will cover all available frequencies to give the optimal sound volume.

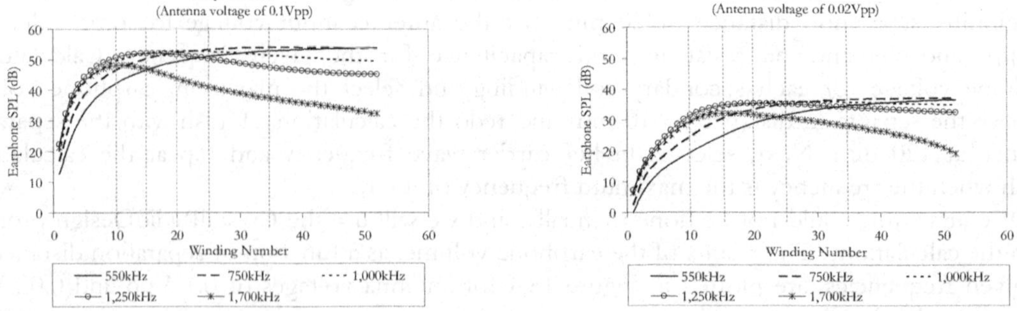

(a) (b)

Figure 18.5
Earphone SPL as a function of winding number at 50 mm separation distance.

Note that lower-frequency stations tend to require a higher tap position than higher-frequency stations. And low power stations also tend to require higher tap positions. As a rule of thumb

place your first tap at around 20% of the number of coil windings, then +10%, +10%, +10% +20%.

For the above example, 20% of 52 turns is about 10 turns, and 10% is about 5 turns. So the taps would be at 10, 15, 20, 25 and 35.

18.6 OTHER LOSSES TO CONSIDER

When constructing a crystal radio, there will be other energy losses due to the surrounding material. These losses are unavoidable if we wish to construct a working crystal set. However, we can keep these losses to a minimum.

Construction Wiring Loss

When wiring up components of the crystal set, such as the variable capacitor across the secondary coil, or wiring up the demodulator, always use the same wire gauge or thicker. If you wound a coil using 0.63 mm copper wire, do not use 0.5mm copper wire to hook up components; doing so will increase the AC losses above what is expected. This is especially true if you wind a coil using Litz wire and then use normal copper wire. The loss can be quite dramatic. If you use Litz wire to wind a coil, use the same Litz wire throughout the entire construction.

Wiring Proximity

When connecting components, try and keep any two wires as far apart for as long as possible. This will reduce any capacitance between wires and reduce the amount of magnetic induction. When connecting the diode, keep the wires of the diode straight rather than bending them. Bending the diode wires will increase any parasitic capacitance across the diode and may affect its performance.

Apart from tap wires coming from the primary and secondary coils, do not run any other wires under the coils. For instance, do not run the antenna and earth wires under the coils; feed them into the set away from the coils and then connect them to the primary coil.

Coil Mounting

When mounting the coils onto the crystal set, the surrounding material will affect the Q of the coil. Typically the chassis is made from wood, although you may use other material if you wish. Mounting the secondary coil directly onto the base, as some crystal radio designs do, results in a loss of energy as some of this energy is absorbed by the material. For wood, this results in a Q reduction of around 5%. The further you move the secondary coil away from the base, the smaller this loss becomes, and it drops off radically in a non-linear fashion. Moving the coil half a coil diameter away from the base drastically reduces this loss. So mount your coils at least one half to one full coil diameter off the base.

There is some loss due to the mounting support structures on the sides of the coil, but this seems to be much less than the base loss. So the mounting supports should be large enough to provide stable and strong support but not excessively large.

Coil Coupling Slide

Loose coupler crystal sets require the separation between the primary and secondary coils to be adjustable. Some radio designs keep the primary coil position fixed and allow the secondary coil to be moved. I opt for keeping the secondary coil fixed and allow the primary coil position to be adjusted as it is generally somewhat smaller than the secondary coil, and it does not have the demodulator attached.

A simple sliding mechanism can be made using a single long length of former material, with the secondary coil wound at one end of the former and the primary coil wound on a slightly wider diameter former that fits over the secondary coil former. This former should be long enough to accommodate the primary and secondary coils and the maximum separation distance you require. There will be some slight dielectric losses, but these appear to be quite small, as discussed in the section on dielectric losses and are in the order of about 1% at high frequencies for these coils. You could also cut out sections of the secondary former to reduce the overall amount of dielectric material to reduce these losses.

Alternatively, a sliding mechanism can be made using a long slim piece of styrene. This effectively removes any dielectric losses due to the sliding mechanism, and this is the slider we will use in this book.

Switch Losses

All switches have some small amount of capacitance across the terminals, some more than others. For poorly designed crystal sets that require a switch to change the tunable frequency range, a switch can result in a significant loss of Q and should be avoided at all costs. As we have seen, there is no need to do this if you follow the design principles given above.

Rotary switches used to select different inductance values on the primary coil are not much of a problem as the primary coil has very low Q already due to the ground and antenna losses, and the currents flowing through the wires will often be orders of magnitude lower than in the secondary resulting in low magnetic induction losses. Using rotary switches to select the tap for the demodulation on the secondary coil has already been covered, and the loss was found to be insignificant.

Summary of Crystal Radio Losses

Table 18.11 summarises the major losses in a working crystal radio.

Loss cause	Description	Actions to reduce the loss
AC resistance	Increase in the resistive loss of a wire due to the skin effect	Use thicker wire, less wire or Litz wire
Proximity effect	Loss due to the magnetic field impinging on adjacent loops causing the distribution of current to change in the wire	Increase the separation between loops (the pitch) or use Litz wire.

Loss cause	Description	Actions to reduce the loss
Self-resonance	The capacitance due to the transmission line effect and charge separation between the ends of the coil loops. Reduces the useful upper effective frequency of the coil. .	Do not make the coils too long or too short. Keep ℓ/d ratio between 0.2 and 0.6 for tightly wound coils and 0.3 to 1.0 for larger pitch coils. Do not allow the self-resonant frequency to drop much below 3.4 MHz.
Mutual inductance	Loss due to the secondary coil's magnetic field inducing a current across the primary coil. This loss can be quite high and can reduce the Q by 30% or more.	Increase the separation distance between the primary and secondary coil. Use fewer loops in the primary coil. A deeper ground rod will also reduce this loss.
Demodulation loss	Loss due to diode resistance and mechanical earphone losses when converting electrical energy into sound. The earphone mechanical loss is a good loss. Meaning the audio signal is producing sound waves for you to hear and should account for the greatest reduction in Q.	Try different diodes. As a rule of thumb, low turn-on voltage with high reverse resistance. Use a higher value load resistor or none at all. Connect the demodulator to a lower tap on the secondary coil.
Antenna losses	Loss due to ground resistance, antenna radiation resistance and AC wire resistance.	Increase the vertical length of the antenna to increase induced voltage. Keep the entire length of the antenna above the ground. Use thicker wire for the antenna. Reduce ground loss using a deeper and/or wider ground rod.
Construction losses	Loss due to placement of components on a chassis and use of switches and wiring.	Follow the suggestions given above.

Table 18.11
Crystal radio energy loss summary.

19

Building a Crystal Radio

19.1 OVERVIEW

In this Chapter, we will build our first crystal radio using 0.63 mm (22 gauge) enamelled copper wire having an outside diameter of 0.68 mm. Each coil and tap is rounded to the nearest full turn. Ultimately, the inductance and Q will depend on your ability to wind coils precisely. Even tiny deviations in the diameter of the wire or coating will affect the inductance value. You are doing well if you can wind a coil within ±5% of the stated value. The more coils you wind, the better you get at it.

Figure 19.1 shows the completed crystal radio we will build. The rotary switches are optional. However, they make selecting taps far more convenient than alligator clips. Some basic tools and electronic equipment are required and are listed in Table 19.1. The parts required to build the radio are given in Table 19.2.

Figure 19.1
The completed crystal radio.

Cut and drill all the formers and chassis parts before winding the coils. It is easier to pre-assemble and check that the pieces fit and are working without the coils and wire leads. You may be keen to wind your first coil but resist the urge until all the parts are made and tested. You may also wish to read this entire Chapter before building the radio. Use all the diagrams and pictures as guides to help in the construction.

Tools required
Wire cutter.
Sandpaper or small file
Phillips screwdriver
Wood saw
Electric drill and 2, 2.5, 3, 6, 7 and 10 mm drill bits
Soldering iron and solder
Multimeter
Masking tape and glue
Hot glue gun and glue
Pencil

Table 19.1
Tools required.

Material	Qty
0.63 mm (22 Gauge) enamelled copper wire	25 metres
1.25 mm (18 Gauge) enamelled copper wire	0.1 metres
Small signal Schottky (BAT85) or Germanium point contact diode (1N34A)	1
47 kΩ resistor	1
Piezoelectric earphone	1
Tuning capacitor (360 pF)	1
Round 33 mm knob for turning capacitor	1
Mono earphone socket	1
* Optional Rotary switch five pole	2
* Optional chicken head knobs for rotary switches	2
Screw-type terminal block (5 positions)	1
Quick fix terminal block(2 positions)	4
90 mm diameter PVC	1 metre
Foam board for slide rail	10 x 240 mm
300 x 240 mm 7 mm thick plywood	1
160 x 100 mm 3 mm thick plywood	1
180 mm length of 65 x 12 mm pine	1
270 mm length of 20 x 12 mm pine	1
Popsicle sticks	2
3 x 12 mm (4Gx12 mm) wood screws	9
3.5 x 20 mm (6Gx20 mm) wood screws	8
3.5 x 25 mm (6Gx25 mm) wood screws	4
3.5 x 30mm (6Gx30 mm) wood screws	2
3 x 15 mm nylon screws	2
3 mm nylon nuts	2

Table 19.2
Crystal radio parts list.

19.2 MAKING THE SLIDE RAIL

Cut the slide rail first, as it will be used when making the primary coil's slider sleeve. Cut a 10 x 235 mm strip of 7 mm thick foam board using a utility knife. Note if you intend to build the double-resonant crystal radio described in a later Chapter, then cut a 290 mm length of foam board as the distance between the coils needs to be increased to prevent over-coupling. Use a ruler as a straight edge while cutting. Foam board is lightweight and easy to cut while being reasonably strong. Sand the edges to make it smooth. Drill two 3 mm holes at each end, 8 mm in from the ends through the 10 mm surface. See Figure 19.2.

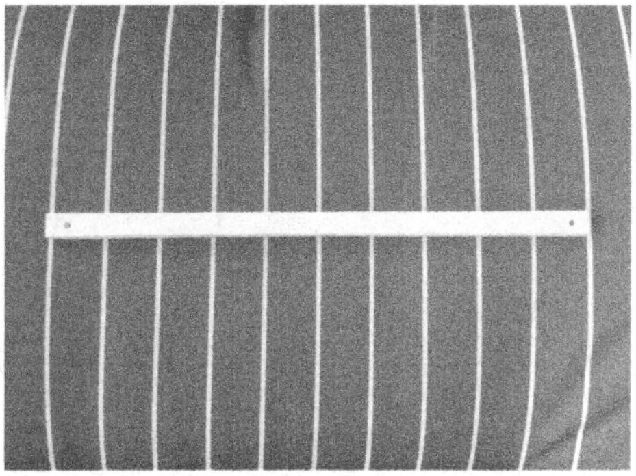

Figure 19.2
The slide rail.

19.3 MAKING THE FORMERS

For this crystal radio, three sections of PVC piping will be cut; a 20 mm length used to connect the slider rail to the left-hand side stand, a free-moving primary coil former and a fixed secondary coil former attached to the right-hand stand. All three are cut from the same diameter pipe. Each section needs to be aligned and drilled. To help in the alignment of these sections, draw registration lines along the length of the pipe. Using an angled piece of metal (or plastic), draw four lines of at least 140 mm in length and 90° apart; see Figures 19.3 and 19.4. Use a protractor to help with the angles.

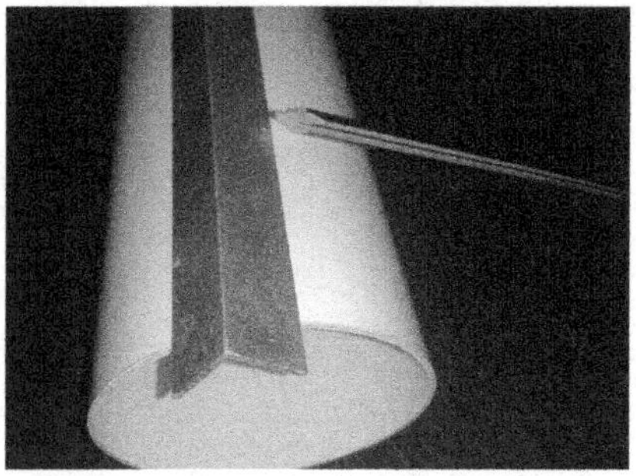

Figure 19.3
Use an angled piece of metal as a straight edge.

Now cut the pipe into three sections. Mark off a 20 mm length and cut this section of pipe. This section will be attached to the left side stand. Then cut an 80 mm length to be used as the secondary coil former and finally a 35 mm length for the primary coil former. For each pipe, arbitrarily choose one of the drawn lines as the front line and draw a cross next to this line. Identifying the front line will be used to orientate the sections as we assemble and drill. Figure 19.4

shows the lines drawn on the pipe before it is cut into sections.

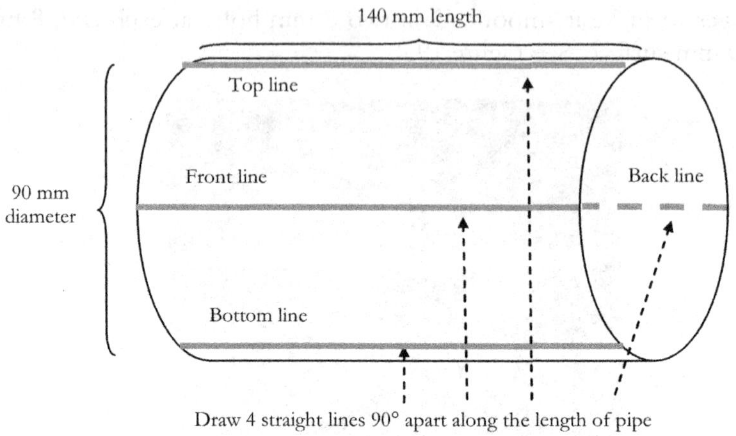

Figure 19.4
Draw registration lines on a length of PVC piping.

Making the Fixed Slide Rail Former

This length of pipe is used to align and fix the left-hand side of the slide rail. Figure 19.5 shows the dimensions and the holes required.

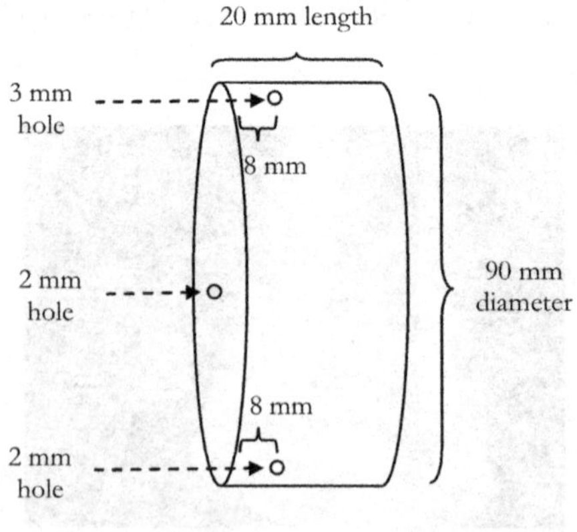

Figure 19.5
The fixed slide rail former.

Construction:

1. Cut a 20 mm length of 90 mm PVC pipe. Sand and smooth any rough edges.

2. Drill a 3 mm hole on the top line 8 mm from the left-hand end.

3. Drill 2 mm holes on the back and the bottom line 8 mm from the left-hand end.

Making the Primary Coil Former

Figure 19.6 shows the dimensions of the primary coil former.

Construction:

1. Cut a 35 mm length of 90 mm PVC pipe. Sand and smooth any rough edges.

2. Along the front line, drill four one-mm holes as specified by Figure 19.6. These holes are used to feed the lead wires of the primary coil.

3. Cut out a rectangular section of cardboard, 35 x 35 mm, for a slide rail sleeve. From inside the former hot glue one end of the sleeve about 10 mm left of the top line. I use hot glue as it forms a nice bond between the cardboard and the PVC. Place the slide rail inside the primary coil so that it is aligned with the top line, and bend the cardboard over the slide rail to make a sleeve. Ensure the rail slides easily through the sleeve, so don't make it too tight. When satisfied with the fit hot glue the other end and remove the slide rail. Generally, cardboard is not a recommended material to use in crystal radios as it may absorb water. Use coated cardboard and seal the edges using glue if you wish. I have not found the performance of the crystal radio to be diminished by using this small section of cardboard.

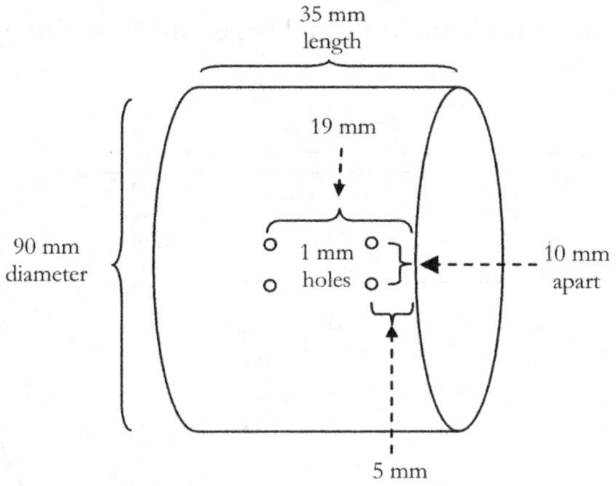

Figure 19.6
The primary coil former.

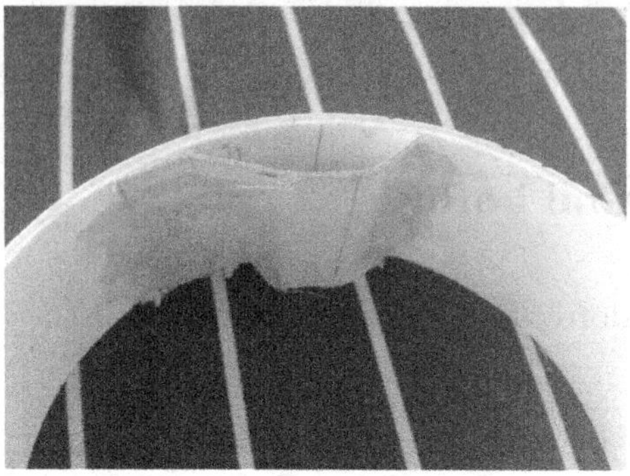

Figure 19.7
The primary coil former with sleeve.

Making the Secondary Coil Former

Figure 19.8 shows the dimensions of the secondary coil former.

Construction:

1. Cut a 65 mm length of 90 mm PVC pipe. Sand and smooth any rough edges.

2. Along the front line, drill four one-mm holes as specified by Figure 19.8. These holes are used to feed the lead wires of the secondary coil.

3. Drill a 3 mm hole on the top line 8 mm from the right-hand end.

4. Drill 2.5 mm holes at the back and bottom line, 6 mm from the right-hand end.

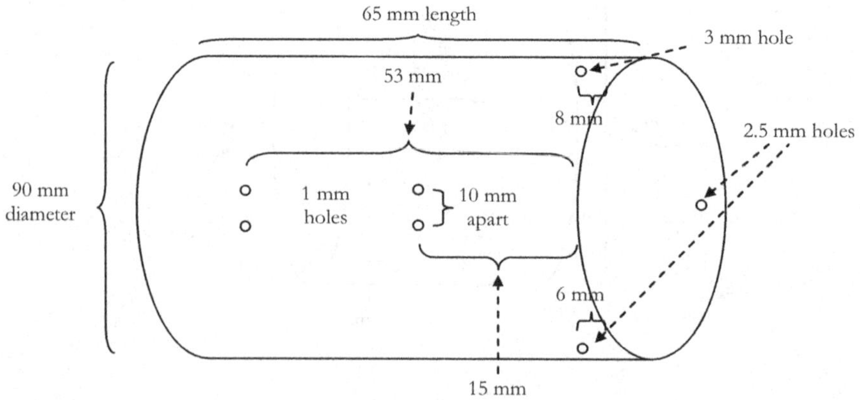

Figure 19.8
The secondary coil former.

19.4 MAKING THE CHASSIS

Using Figure 19.9 as your guide, cut a 300 x 240 mm rectangle from 7 mm thick plywood as the base for the crystal radio.

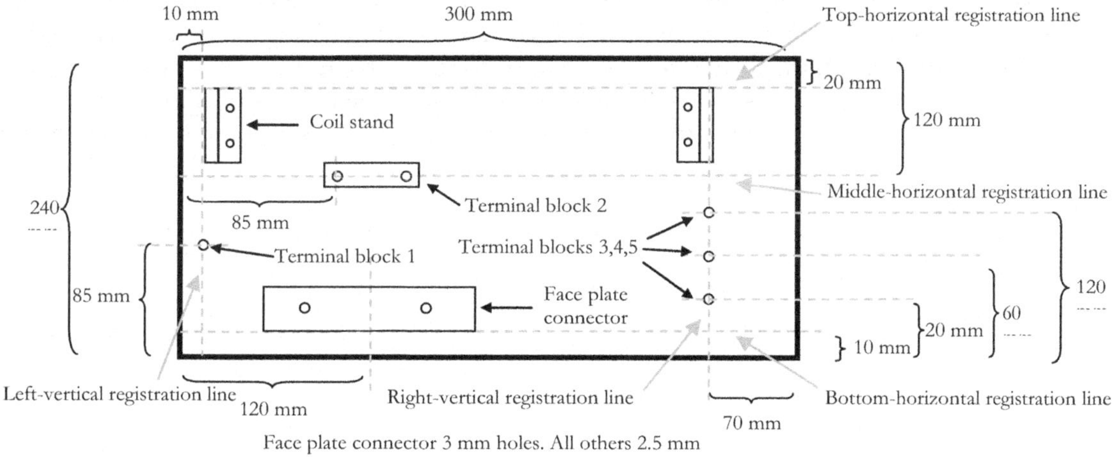

Figure 19.9
The base.

Now orient the base as shown in Figure 19.9 and draw the registration lines. Draw the top-horizontal registration line across the base 20 mm down from the top. Draw the middle-horizontal registration line across the base 120 mm down from the top. Draw the bottom-horizontal registration line across the base 10 mm up from the bottom. Draw the left-vertical registration line 10 mm in from the left and the right-vertical registration line 70 mm in from the right side.

Assemble the Coil Stands

Use Figure 19.10 as a guide. Cut two 90 mm lengths from 65 x 12 mm pine as the coil stands. Cut two 65 mm lengths from 20 x 12 mm pine to use as the coil stand connectors.

1. Drill two 3 mm holes 20 mm in from both ends of the coil stand connectors.

2. Drill two 3 mm holes 10 mm in from both ends, 6 mm up from the bottom of the coil stand.

3. Place the coil stand and the base connected on a flat surface. Bring together and align. Clamp the two together, and using the 3 mm holes in the coil stand as a template, drill a 3 mm hole about 6 mm in depth into the base connector.

4. Remove the clamps and affix the two parts together using 3.5 x 25 mm wood screws.

5. Do the same for the other coil stand and base connector.

6. Using a Popsicle stick, drill two 2.5 mm holes 20 mm from both ends.

7. Using a 3.0 x 12 mm wood screw, attach the slide rail former through the back hole to the Popsicle stick. Using your thumb to hold the Popsicle stick flat against the side of the coil stand, place the assembly on top of the left coil stand so that the bottom hole in the slide rail former touches the top of the stand. Using a pencil, mark the hole position and drill a 2 mm hole. Attach the former to the standing using a 3.0 x 12 mm screw. Now press the Popsicle flat against the stand and mark the position of the hole to attach it to the stand. Drill a 2 mm hole and attach the stick to the stand using a 3.0 x 12 mm screw. See Figure 19.11a.

8. Attach the secondary coil to the right-hand coil standing using the same procedure. See Figure 19.11b.

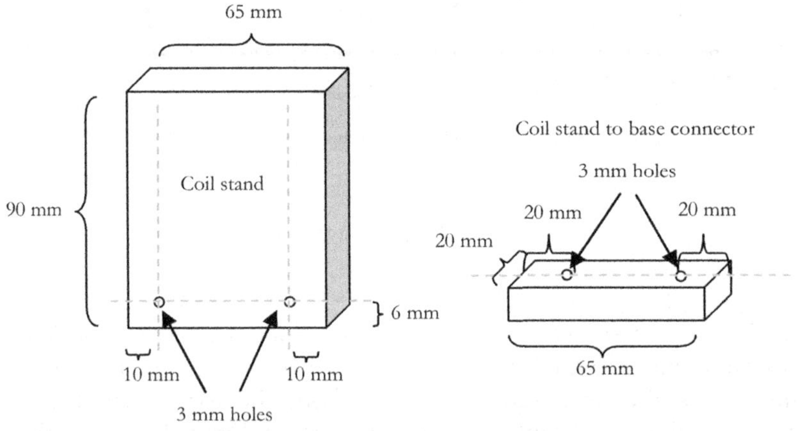

Figure 19.10
Coil stand.

(a) (b)

Figure 19.11
Left and right coil stands with formers attached.

Attaching the Coil Stands to the Base

To attach the coil train to the base, first, assemble the coil train and use the registration marks on the base to align the train and then drill the base attachment holes. Using a 3 mm nylon screw and nut, attach the slide rail to the left-hand coil former from the top. Next, slide the primary coil onto the slide rail through the slide rail sleeve. Finally, attach the slide rail to the right-hand side coil (the secondary coil). See Figure 19.12.

Place the coil train on the base and align the left side coil stand to the top and left registration lines. Mark the positions of the base connector holes. Remove the train and drill two 3 mm holes through the base. Place the train back in the base and attach the left side coil stand to the base using two 3.5 x 20 mm wood screws. Adjust the train so the primary coil moves easily across the rail and the right-hand coil stand is aligned with the top registration line. Once satisfied, mark the two base connector holes of the right-hand coil stand; remove the train and drill the holes through the base. Note that if the 290 mm slide rail were cut, then the right-hand coil stand would be closer to the right-hand edge of the base.

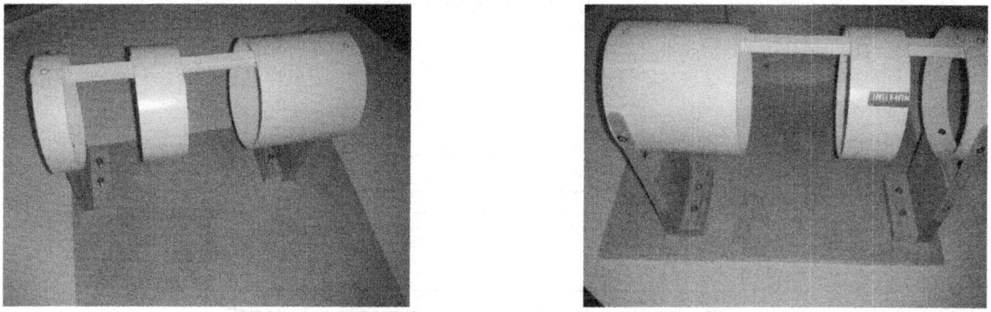

(a) (b)

Figure 19.12
Attaching the coil train to the base.

Making the Faceplate

Cut a 160 x 100 mm rectangle from 3mm thick plywood for the faceplate as shown in Figure 19.13. Drill the holes at the positions given in Figure 19.13. Cut a 140 mm length from 20 x 12 mm pine for the faceplate base connector, as shown in Figure 19.14. Drill two 3mm holes through the 12 mm side of the faceplate connector. Position the faceplate connector along the bottom horizontal registration line on the base, with the centre of the faceplate connector 120 mm in from the left-hand side. Using the left-hand hole on the connector, drill a 3 mm hole through the base and then screw the connector down using a 3.5 x 30 mm wood screw. Make sure the connector is still aligned, and then drill through the right-hand side hole, through to the base, and screw it down. See Figure 19.15a.

Next, place the faceplate flat against the connector, aligning the faceplate's centre with the connector's centre. Drill a 2 mm hole through the existing faceplate hole into the connector and attach using a 3 x 12 mm screw. Then do the same for the left and right-hand side faceplate holes. See Figure 19.15b and Figure 19.16.

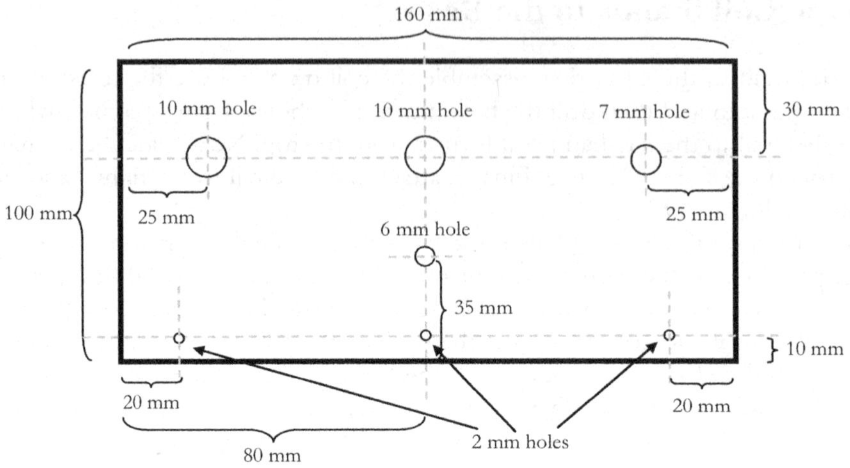

Figure 19.13
Three-ply faceplate.

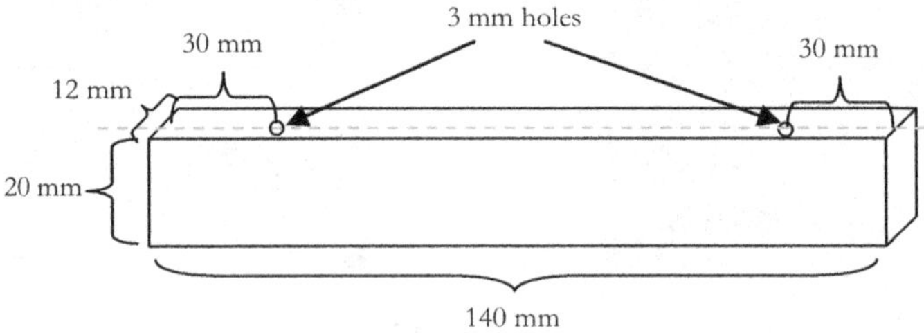

Figure 19.14
The faceplate base connector.

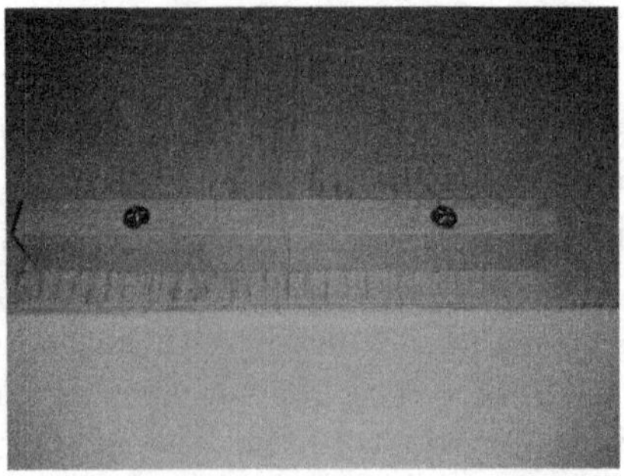

(a) Faceplate base connector.

(b)Faceplate.

Figure 19.15
The faceplate base connector and faceplate.

Adding the Terminal Blocks

Using Figures 19.9 and 19.16 as a guide, screw the terminal blocks onto the base. Drill 2.5 mm holes for the terminal blocks and affix them to the base, as shown in Figure 19.16, using 3.5 x 20 mm screws. The chassis is complete, and we can now wind the coils.

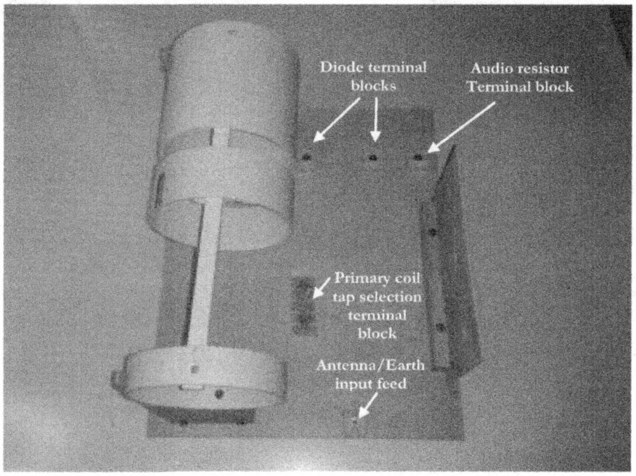

Figure 19.16
Completed chassis.

19.5 WINDING THE COILS

Coil Winding Technique

The following is a general technique that can be used for winding any cylindrical coil. Use this technique to wind the primary and secondary coils. The key to winding decent coils is keeping the wire taught with some strain as you wind the loops.

First, make a wire dispenser. Pass a rod through the spool of wire to act as an axle around which the spool can rotate. It does not matter what you use for an axle; it could be a wooden dowel rod or even a pen, just so long as the spool can rotate as you pull the wire. You could even place the spool of wire in a toilet roll dispenser. I use a wooden dowel rod as an axle with the ends attached to a pair of "helping hands", see Figure 19.17a. Next, place a large book in front of your dispenser and pull some wire across it. Then place two or more heavy books on top of the wire. This provides tension to the wire; see Figure 19.17b.

As we start winding the coil, we need to keep the windings in place and prevent them from spreading out, or unwinding as we go. We will create a backstop by cutting a 20 mm length of the 90 mm PVC pipe, cut open to create a clip. Place the backstop next to the lead holes on the left at the first loop position. Fix the backstop into place by using a couple of strips of masking tape. Or, instead of using a PVC clip, you could wind several layers of masking tape next to the starting holes to serve as a backstop. Just make sure the masking tape forms a straight edge. Next, cut a length of masking tape to cover and hold the windings while we form taps. See Figure 19.18a.

To start the coil winding, feed some wire through the two starting holes in your former, pulling out enough wire to use as a lead. Using the backstop as a straight edge, begin winding the loops; see Figure 19.18b. Do not try and wrap the wire around the former, but rotate the former so the top of the former moves towards you.

(a) Creating a wire dispenser.

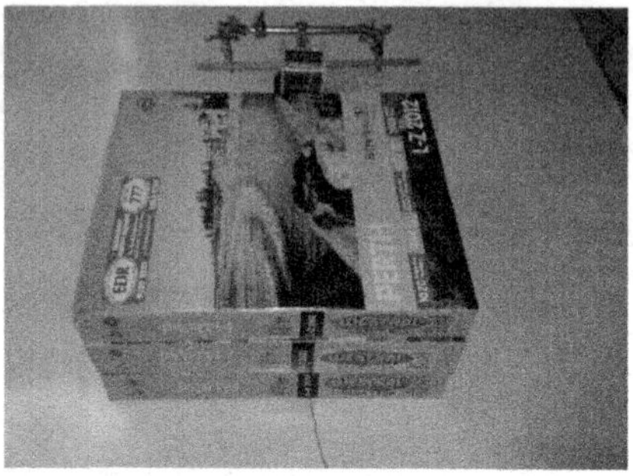

b) Applying pressure to keep the wire taut.

Figure 19.17
Setting up a wire dispenser.

Keep the wire taught as you wind. Make sure your first turn is pressed against the backstop to form a straight circular loop. Use your left-hand thumb to push the wire down and your right thumb to manoeuvre the wire to keep it straight and on course. If a loop starts to deviate from the path, stop winding and use your thumb or a finger to push it back in place and then continue winding, all the while keeping the wire taut.

When you have wound enough loops for your first tap, spread the masking tape across the existing loops to keep them in place. Using a small piece of dowel, about 5 mm in diameter, wrap the wire around the dowel from the bottom over the top and back out through the bottom. Then turn the dowel and wire a full 360°. While keeping the dowel in place, push the wire and dowel against the former and continue winding the loop, see Figure 19.18c. Once the tap loop has been formed, remove the dowel rod and continue winding. Continue in this fashion and use the masking tape to keep existing loops in place when required.

When the end of the coil has been reached, cut the wire to include the lead length and then loop the wire through the holes drilled at the end. The masking tape should be across the total length of the coil at this stage to keep the loops in place. Place some glue around the start and ending holes to keep the wire from moving. Then run some glue around the last loop between the wire and the former. Let the glue set for twenty minutes or so, and then remove the masking tape and backstop. Run some more glue around the first loop and the former, and leave it to dry for an hour.

When the glue has dried, remove about 10 mm of the enamel coating from both end leads. Ensure you remove the enamel from all sides of the wire's circumference. Then remove enough enamel from the taps so that a wire can be attached. If you have a multimeter, you can check that you have removed the enamel by taking a resistance reading; the resistance should be very low, 1 Ω or less across the leads and even less across the taps. The actual value depends on the wire and the length of wire used for the coil.

To aid in the orientation of the coils, draw a small dot next to the first loop near the left-hand lead. The left lead will be considered the first tap, and the right-hand lead will be the last tap. Figure 19.19 shows the orientation and tap numbering for the secondary coil.

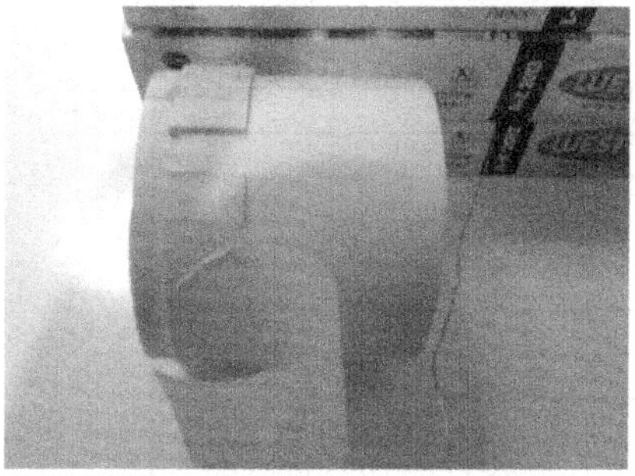

(a) Affix a backstop to the former.

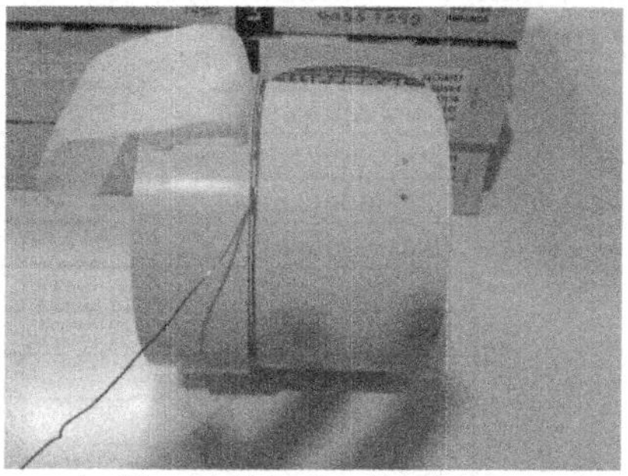

(b) Start winding using the backstop to keep the loop straight.

(c) Using a small dowel rod to wind a tap.

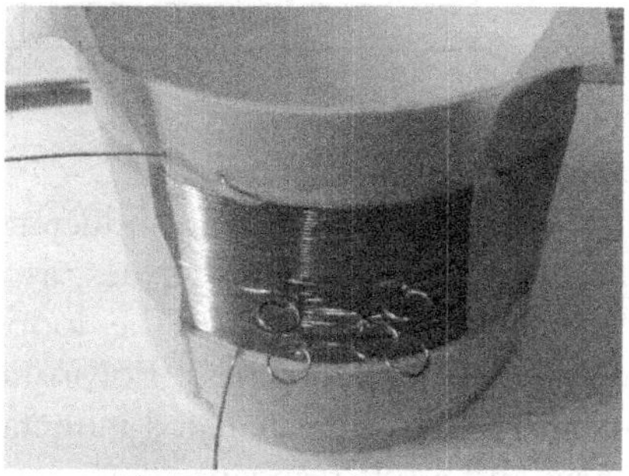

(d) The completed coil.

Figure 19.18
Winding a coil.

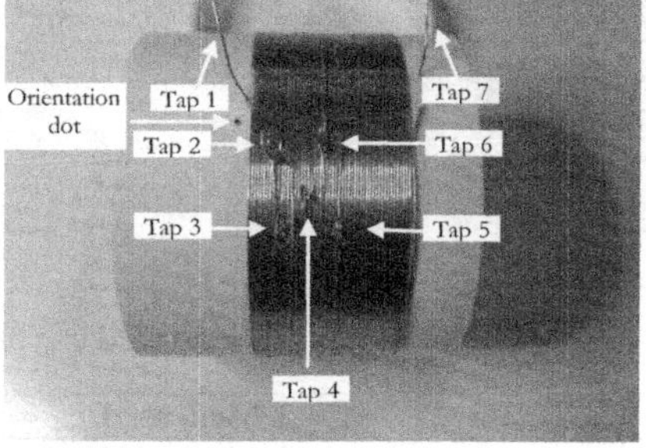

Figure 19.19
Coil orientation and tap numbers.

Winding the Secondary Coil

The secondary coil is tightly wound using 54 turns of 0.63 mm enamelled copper wire, with taps at loops 10, 15, 20, 25 and 35. The specification for the secondary coil is given in Table 19.3. The secondary coil was wound using the coil winding technique described above and is shown in Figure 19.20.

Description	Calculated values
Coil name	SC-BUILD-1
Coil inductance	305.96 uH
Coil diameter	90.68 mm
Coil length	36.72 mm
Number of windings	54
Taps at loop number	10, 15, 20, 25, 35
Pitch	0.68 mm
Wire	enamelled copper wire (core D = 0.63 mm, OD = 0.68 mm)
Former material	PVC
Former diameter	90mm
Former wall thickness	2mm

Table 19.3
The secondary coil specifications.

After winding the coil, the inductance, Q and self-resonant frequency were measured and compared with the calculated values, as shown in Table 19.4. These values are quite acceptable, given the slight deviations in the loops due to winding taps, measurement errors, and tiny variations in the wire thickness. For instance, increasing the pitch to 0.69 mm (that's just 1/100th of a millimetre) gives a calculated inductance of 304 uH (a 0.64% difference), a Q of 194 (a 4.9% difference) and a self-resonant frequency of 5.2 MHz (a 0.58% difference). So we see that a very tiny deviation in the actual diameter of the wire (including the enamel coating) or pitch value makes a difference in the eventual inductance and effective Q values for tightly wound coils.

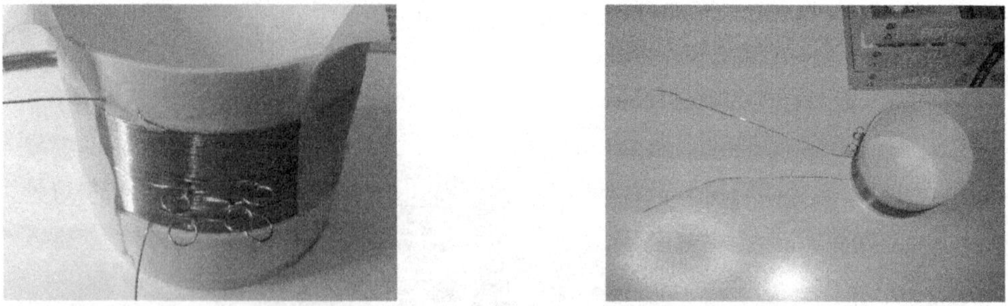

Figure 19.20
The completed secondary coil.

	Calculated	Measured	% difference
Inductance	305.96 uH	309.6	1.18
Effective Q (1 MHz)	185	193	4.14
Self-resonant frequency	5.17 MHZ	5.30 MHz	2.45

Table 19.4
Difference between calculated and measured values.

Winding the Primary Coil

The primary coil is tightly wound using 19 turns of 0.63 mm enamelled copper wire, with taps at loops 6, 9 and 12. The specification for the secondary coil is given in Table 19.5. Wind the coil using the above technique. Figure 19.21 shows the completed primary coil.

Description	Calculated values
Coil name	PC-BUILD-1
Coil inductance	58.5 uH
Coil diameter	90.68 mm
Coil length	12.92 mm
Number of windings	19
Taps at loop number	6, 9, 12
Pitch	0.68 mm
Wire	enamelled copper wire (core D = 0.63 mm, OD = 0.68 mm)
Former material	PVC
Former diameter	90 mm
Former wall thickness	2 mm

Table 19.5
The primary coil specifications.

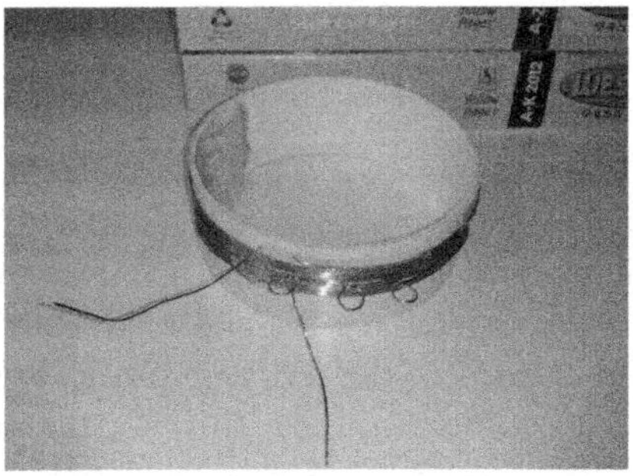

Figure 19.21
The completed primary coil former with sleeve.

Table 19.6 shows the difference between the calculated and the measured inductance values at different tap positions.

Tap	Calculated inductance (uH)	Measured inductance (uH)	% difference	# turns	Measured coil length (mm)
1-5	58.5	58.7	0.3	19	12.9
2-5	31.0	31.0	0.0	13	8.8
3-5	19.8	19.7	0.5	10	6.8
4-5	10.7	10.7	0.0	7	4.8

Table 19.6
Difference between calculated and measured values for the primary coil.

19.6 WIRING THE CRYSTAL RADIO

Wiring the Primary Coil

Follow the circuit diagram and picture in Figure 19.22 to wire the primary coil. The primary coil needs to move freely along the guide rail, so use flexible hook-up wire to connect the leads and taps from the primary coil to the primary coil terminal block. Use flexible wire with a total wire diameter equal to or greater than the diameter of the enamelled copper wire; in this case, 0.63mm. The wire length must be such that the primary coil can be moved across the entire length of the rail.

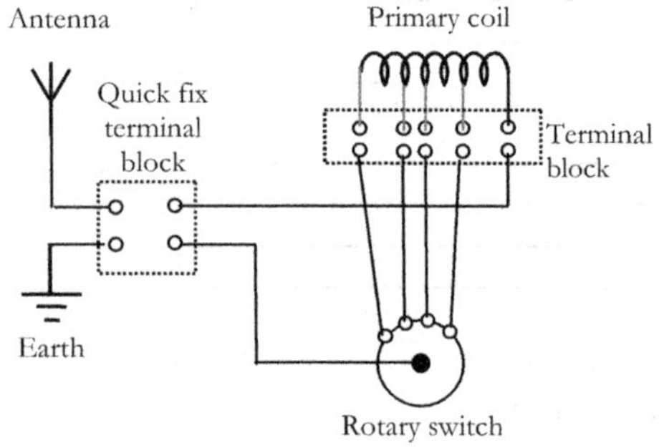

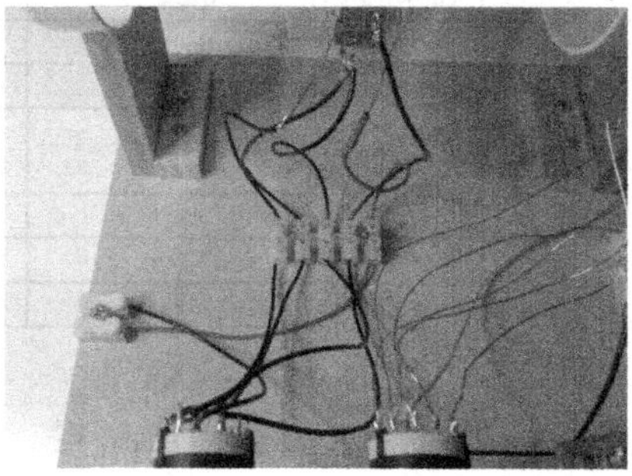

Figure 19.22
Primary coil wiring.

Wiring steps:

1. Cut three 130 mm lengths of flexible wire for the taps and two of 100 mm for the leads. Strip the insulation from the ends and solder it onto the primary coil's taps and leads. Connect the other ends to the terminal block, as shown in the diagram. You can cover the exposed solder joints with heat shrink tubing if you wish.

2. Cut a 140 mm length of flexible wire and attach one end to the antenna quick-fix terminal block and the other end to the right-hand primary coil's terminal block.

3. Cut four lengths of 140 mm flexible wire and solder one end of each wire to a terminal on the rotary switch. Switch positions 1, 2, 3 and 4. Connect the other ends to the terminal block at the appropriate positions. That is 1-1, 2-2 etc.

4. Cut another 140 mm length of flexible wire. Solder one end to the centre pin of the rotary switch, and connect the other end to the earth terminal block.

Wiring the Secondary Coil

Follow the circuit diagram in Figure 19.22 and the pictures in Figure 19.23 to wire the secondary coil.

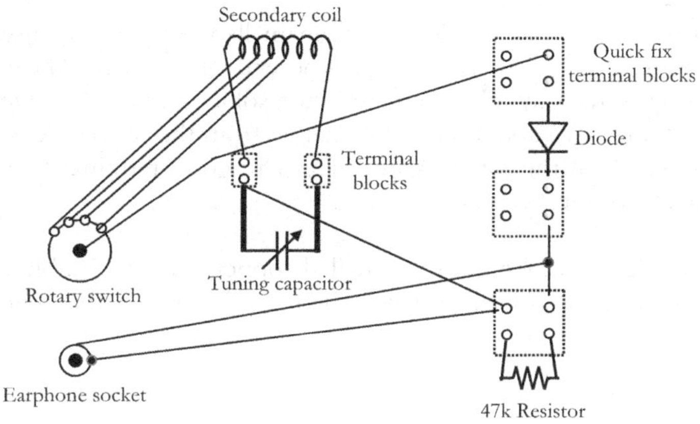

Figure 19.23
Secondary coil wiring diagram.

Wiring the tuning capacitor:

1. Wire the tuning capacitor to the secondary coil. Cut two 50 mm lengths of thick (1.25mm) enamelled copper wire and removed about 10 mm of enamel from both ends. Solder the copper wire to the terminals of the tuning capacitor, as shown in Figure 19.24a. These rigid leads keep the wiring apart, reducing lead wire capacitance and inductive losses.

2. Run the lead wires from the secondary coil to the tuning capacitor leads, cut off any excess, and remove the enamel. You can solder the coil's lead wires directly to the capacitor leads or use a terminal block. The terminal block makes the removal of the tuning capacitor or secondary coil somewhat easier if required, so this is what I use. Note that the common (or earth) will be taken from the left side of the secondary coil lead.

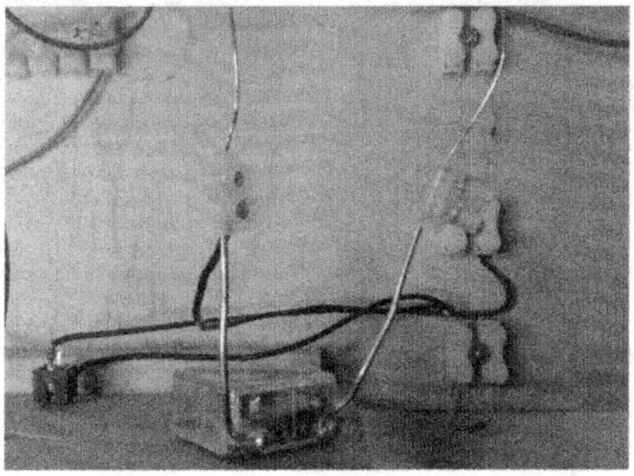

(a)

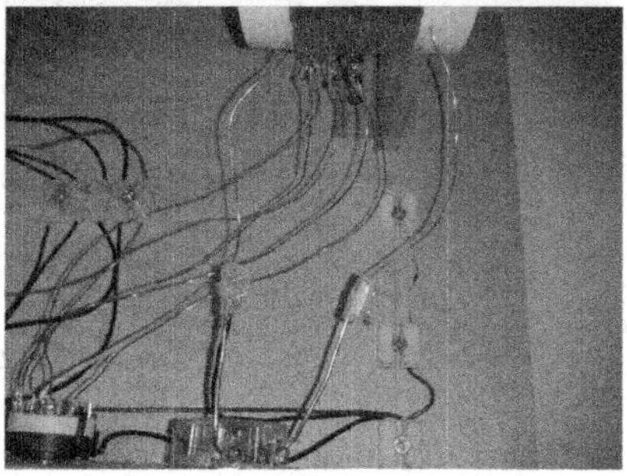

(b)

Figure 19.24
Secondary coil wiring.

Wiring the rotary switch and taps:

1. Cut a 160 mm length of the 0.63 mm enamelled copper wire used to wind the coil. Remove about 10 mm of enamel from one end of the wire. Make a small hook to fit around the tap of the secondary coil, and then solder it into place. Do this for each of the five taps. Then remove about 5 mm of enamel from the other side of the wire and solder onto the terminals of the rotary switch. Tap 1 to go to terminal 1 of the switch, tap 2 to go to terminal 2 of the switch etc.

2. Cut a 180 mm length of 0.63 mm enamelled copper wire (or flexible wire). Solder one end to the centre terminal of the rotary switch, and connect the other end to the diode terminal block.

Wiring the earphone socket:

1. Cut two 150 mm lengths of flexible wire. Connect both wires to the earphone socket, as shown in Figure 19.25a. One is the earth (or common) wire, and the other is the positive wire.

2. Cut a small length of flexible wire. Twist one end together with the positive earphone socket wire and push them into the right-hand side of the resistor terminal block. Connect the other free end to the diode terminal block.

3. Cut a 130 mm length of flexible wire. Twist one end together with the earphone socket earth wire and insert them both into the left-hand side of the resistor terminal block. Insert the other end of the wire into the left-hand side tuning capacitor terminal block.

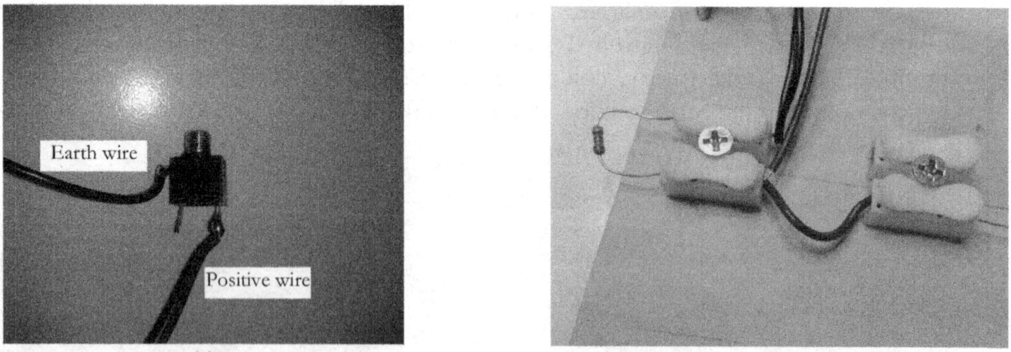

(a) (b)

Figure 19.25
Wiring the earphone socket.

Now attach the rotary switches, tuning capacitor and earphone socket to the face place and add knobs. Insert the diode between the diode terminal blocks and place the 47 kΩ resistor across the resistor terminal block. Insert the earth and antenna wires into the antenna terminal block. Finally, insert the earphone into the earphone socket, and we are ready to go. The completed radio should look like the radio shown in Figure 19.1.

19.7 USING THE CRYSTAL RADIO

Attach the antenna and earth, and plug in the earphone. Set the earth switch to tap 1 and the audio take-off to tap 6 of the secondary coil. Remember, the left-hand lead wire is considered to be tap 1. Move the primary coil up against the secondary. Starting from the lowest frequency (the tuning capacitor turned fully anti-clockwise), turn the tuning capacitor clockwise until you hear a station. If nothing is heard, select tap 2 as the earth and try again. When you hear a station, move the primary coil away from the secondary until the station volume peaks. The further away the primary coil is from the secondary, the higher the Q, but the volume may (but not always) be lower.

If you need to keep the primary coil up against the secondary to hear your local radio stations, the power of the radio stations may be quite low, or you are quite a distance from the transmission, or your primary coil is not close to being resonant at that frequency. There are three things you can do to improve the situation. You can improve your antenna by making it higher and longer, or you can improve your earth by driving the earth rod deeper into the ground. Lastly, you can improve the Q of your crystal radio by trying another diode, or you can improve the secondary coil's effective Q by winding a better coil. We will wind a better secondary coil in the next Chapter.

If you hear more than one station at a time, try moving the primary further away, or change the primary tap (earth) position to a higher tap value, or both. Choosing a lower tap position on the secondary coil also increases the Q and may also increase the volume. Lower-frequency stations tend to have a higher volume at higher secondary tap positions, and higher-frequency stations tend to have a higher volume at lower secondary tap positions.

For instance, the optimal setup for a 585 kHz station might be tap 1 for the primary, tap 5 or 6 for the secondary at a 40 mm distance between the coils. For a station at 1080 kHz, the optimal setup might be tap 3 for the primary and tap 4 or 5 for the secondary at a 40 mm distance. Also, if two loud stations that are reasonably close in frequency are bleeding through, then you may change the primary to a higher tap position and the secondary to a lower tap position to separate the stations.

You will need to experiment for a while to get the settings right. I like to find a good overall separation distance between the primary and secondary and leave it alone. I then adjust the primary and secondary tap positions depending on the station I am listening to.

If you don't hear any of your local radio stations, check your connections and the wiring of the crystal radio. Perhaps you have not adequately removed the enamel from the copper wire, or you have made a mistake in the wiring. Check the diode with a multimeter or replace the diode with a known good diode. If you are using a modern piezoelectric earphone, it is very likely defective. Try plugging your earphone into a transistor radio. Make sure the volume is turned way down low first, and then increase the volume a bit. If you do not hear the station, then your earphone is defective. See the Chapter on the piezoelectric earphone for instructions on replacing the piezoelectric crystal.

19.8 THE CRYSTAL RADIO PERFORMANCE WITH AND WITHOUT TAP SWITCHES

The frequency range covered by the crystal radio was measured to be 472 kHz to 2,550 kHz using primary taps 1-5. Even though, ultimately, the operating Q of a crystal radio is mainly irrelevant, we will measure and calculate the operating Q of the radio to compare it with our model. In this way, we can be confident that the losses occur where stated, and that our understanding of the operation of a crystal radio and the effect of the components on the operation are valid.

We will measure the Q with and without the use of switches and tap leads to see what impact they have. Figure 19.26a shows the crystal radio without the rotary switches, with alligator clip leads used to select tap positions on the primary and secondary coils. Figure 19.26b shows the tap lead wires connected to the rotary switches.

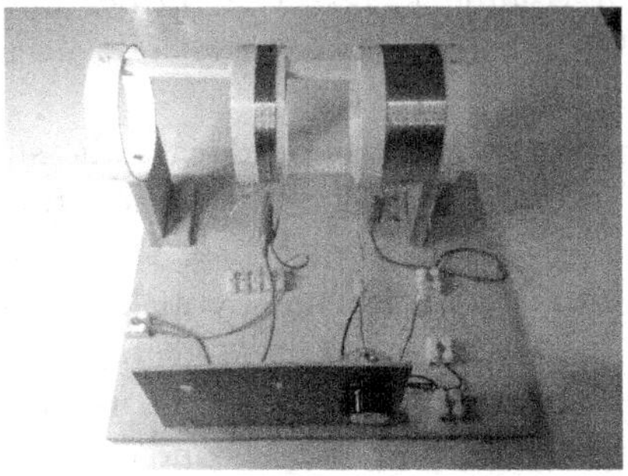

(a)

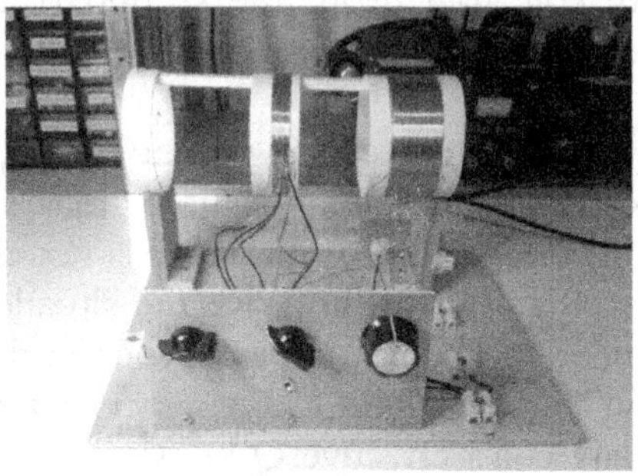

(b)

Figure 19.25
Crystal radio used with alligator clip leads instead of switches.

The Q of the crystal radio was measured by injecting an unmodulated 0.1 Vpp one-MHz signal from a function generator with 50 Ω output impedance. A 2.2 nF capacitor was connected in series to simulate the antenna capacitance. Different primary tap configurations were used, and the distance between the primary and secondary coil was kept constant at 70 mm. The demodulator was connected to tap 3 on the secondary (winding number 15). The measured Q values were divided by 0.93 to account for the measuring equipment loss, as discussed in the measurement techniques Chapter. The measured values of the inductance and coil length were used for the calculations. Table 19.7 gives the specifications of our crystal radio.

Description	Value
Crystal radio name	CREX-1
Antenna designed for	Standard 25-metre antenna + standard trans line
Primary coil	PC-BUILD-1
Secondary coil	SC-BUILD-1
Diode	BAT85
Audio resistor	47 kΩ
Audio capacitor	none
Earphone	piezoelectric

Table 19.7
Crystal radio specifications.

The results of the measurements for both configurations are shown in Table 19.8. The difference between the calculated and the measured Q values range from 0.86 to 5.05%, with an average discrepancy of 2.76%. Note that the Q values when the switches were used were higher than the alligator clip use. This is most likely due to the alligator lead not making as good an electrical connection compared to the use of terminal blocks and solder. This possibility is one of the reasons I'm not too fond of slide arms. If there is any loss due to tap lead wires and switches, it is lost within the ability of accurate measurements in this instance.

Primary tap	**Q Calculated**	**No switches**		**Using switches**	
		Q measured	% difference from calculated Q	Q measured	% difference from calculated Q
1-5	116	111	4.31	117	0.86
2-5	99	94	5.05	97	2.02
3-5	90	87	3.33	88	2.22
4-5	93	96	3.22	94	1.08

Table 19.8
Crystal radio performance with and without tap leads and switches.

20

Building High Q Crystal Radios

20.1 MAKING A HIGH-PITCH COIL

Coil Analysis

From our coil analysis, we know that increasing the pitch length decreases the proximity effect, resulting in a coil with a higher Q. Using the same 90 mm PCV former and the same wire (0.63 mm), we will build a coil with 74 turns with a pitch of 1.4 mm. The resulting calculated inductance is 308 uH. Tap analysis was performed to determine the tap positions using this coil with the standard antenna, the previously created primary coil and the BAT85 as the diode. The graphs of the maximum earphone SPL as a function of coil separation are shown in Figure 20.1.

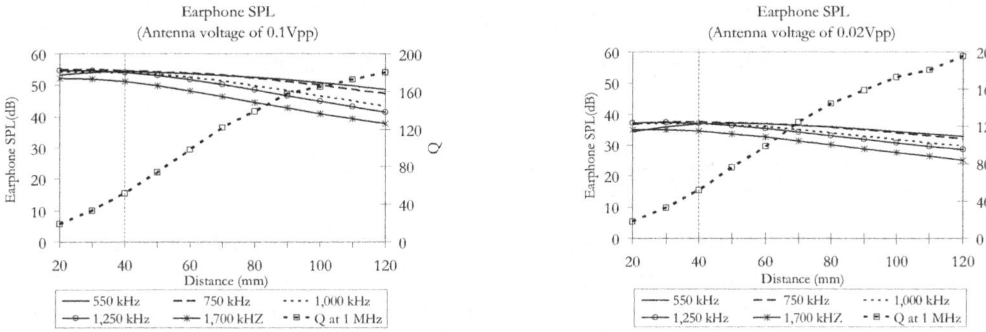

Figure 20.1
Maximum SPL as a function of separation distance.

From Figure 20.1, the distance where all the SPL values differ by less than 2 dB is about 40 mm. Figure 20.2 shows the SPL level as a function of loop number. Notice that the shapes of the curves are very similar to the tightly wound coil SPL curves. Based on Figure 20.2, loops 15, 20, 25, 35 and 50 were chosen as the taps. Using the rule of thumb, we would arrive at taps 15, 22, 30, 37 and 52, similar to the loop numbers we have chosen.

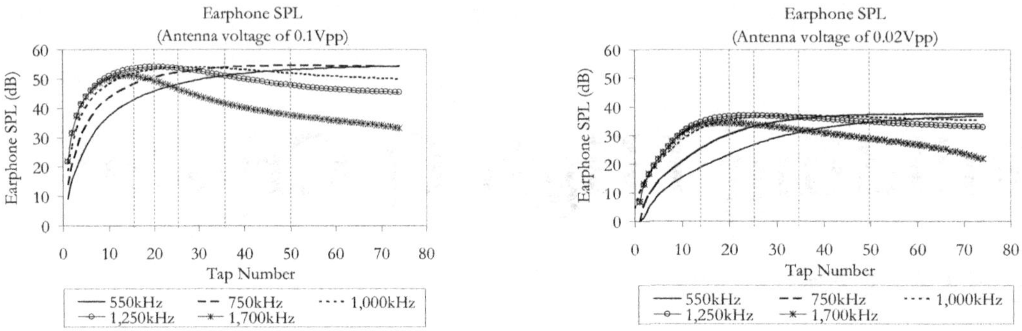

Figure 20.2
SPL as a function of winding number at 40 mm distance.

Description	Calculated values
Coil name	SC-BUILD-2
Coil inductance	308 uH
Coil diameter	90.68 mm
Coil length	103.6 mm
Number of windings	74
Taps at loop number	15, 20, 25, 35, 50
Pitch	1.4 mm
Wire	0.63mm enamelled copper wire (core D=0.63 mm, OD=0.68 mm)
Former material	PVC
Former diameter	90 mm
Former wall thickness	2 mm

Table 20.1
The high-pitch secondary coil specifications.

Constructing a Widely Spaced Coil

In tightly wound coils, each wire loop is butted up against the previous loop, thus keeping all the turns in place. Winding coils with a pitch greater than the wire diameter is a bit more difficult as we need something to keep the loops in place without relying on the wire itself. One way of doing this is to insert fishing line between the loops. Each turn of wire and line is butted up against the previous loop, keeping the wire firmly in place. The fishing line, however, does introduce a small amount of dielectric loss. Alternatively, one could wind a coil and glue each winding in place. Or one could cut grooves into the former for each loop. Either approach is impractical, and we will take this small dielectric loss for simplicity and ease of construction.

As the pitch increases, the length of the coil increases for a given inductance value, so we will need to cut a longer length of PVC piping. The beginning and ending lead holes need to be a bit bigger to accommodate both the wire and the fishing line. We will use the same enamelled copper wire as in our previous coil, with a total wire diameter (including the enamel) of 0.68 mm. And 0.7 mm diameter nylon fishing line as the spacer. The fishing line I bought is blue, so it contrasts nicely against the red copper wire.

Figure 20.3 shows the dimensions of the secondary coil former. Use the technique discussed in the previous Chapter to draw the registration lines, then cut and drill holes using Figure 20.3 as your guide.

To wind the coil, place the spool of wire and the spool of fishing line on the wire dispenser and feed both lines across the books to keep them taught. Start the winding by threading both the wire and fishing line through the left side lead holes on the former. Treating the wire and fishing line as a single strand, wind the coil. See Figure 20.4. When it comes to winding the taps, keep both the wire and line in place with the masking tape and create the taps using the wire only. Then continue winding with both the wire and line until you are done. The specifications for this coil are given in Table 20.1.

When the coil is completed, run some glue around both ends of the coil and around the lead holes to keep both the wire and line in place. Place some glue inside the former around the left and right lead holes. When the glue is dry, tie a knot in the fishing line, glue the knot and cut off any excess line.

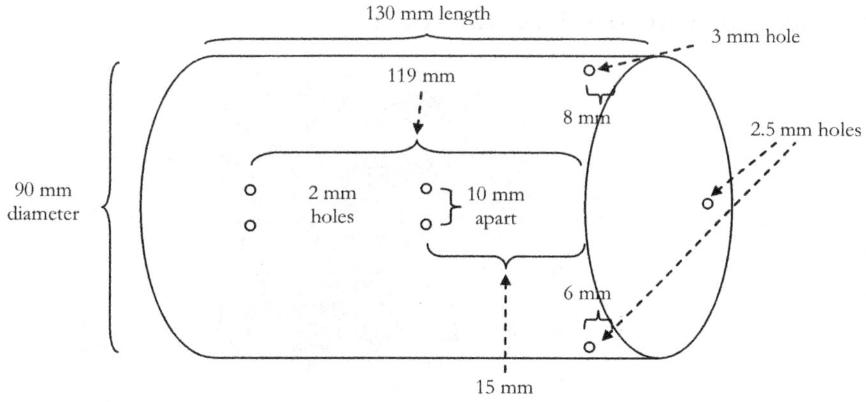

Figure 20.3
1.4 mm pitch secondary coil former.

Table 20.2 shows the difference between the calculated and measured values. There is a 10% difference between the calculated effective Q and the measured effective Q of the coil. This is most likely due to the dielectric loss in the fishing line. As the line is in between the wire loops, this effect is much greater than for the dielectric losses in the former. We shall examine this further when testing the performance of the crystal radio. However, even with this loss, this still represents a 77% increase in effective Q above the tightly wound coil.

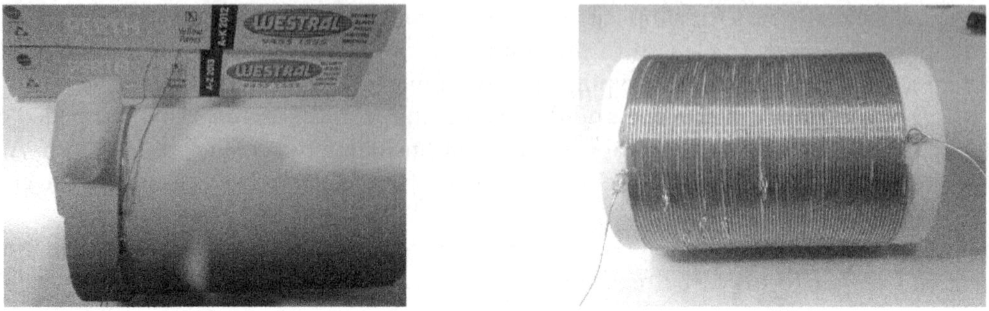

(a) (b)

Figure 20.4
Using fishing line to increase the pitch of a coil.

	Calculated	Measured	% difference
Inductance	307.56 uH	301.40	2.0
Effective Q (1 MHz)	381	342	10.2
Self-resonant frequency	5.21 MHz	5.2 MHz	0.2
Length	103.6 mm	104.3	0.7
Pitch	1.40 mm	1.41	0.7

Table 20.2
Difference between calculated and measured values.

Inserting the Coil into the Crystal Radio

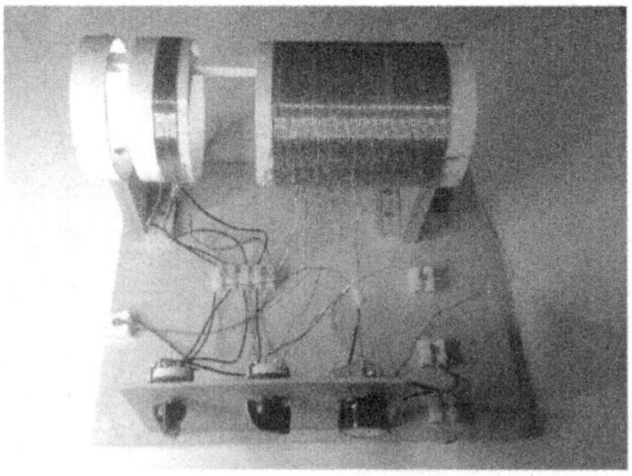

Figure 20.5
The widely spaced coil in the crystal radio.

To place this coil in the existing crystal radio,

1. Remove the tightly wound coil from the radio.

2. Disconnect the old coil from the tuning capacitor and unsolder the secondary tap leads from the rotary switch.

3. Unscrew the old coil from the coil stand and install the new coil.

4. Solder the tap leads to the rotary switch and connect the leads to the tuning capacitor. See Figure 20.5.

Performance

The frequency range covered by this crystal radio is from 478 kHz to 2,540 kHz using primary taps 1-5. The Q of the crystal radio was measured by injecting an unmodulated 0.1 Vpp one-MHz signal from a function generator with 50 Ω output impedance. A 2.2 nF capacitor was connected in series to simulate the antenna capacitance. Different primary tap configurations were used, and the distance between the primary and secondary coil was kept constant at 65 mm. The demodulator was connected to tap 3 on the secondary (winding number 15). The measured Q values were divided by 0.93 to account for the measuring equipment loss, as discussed in the measurement techniques Chapter. The measured values of the inductance and coil length were used for the calculations.

Primary tap	Q Measured	Q Calculated	% Difference
1-5	210	213	1.4
2-5	170	185	8.8
3-5	153	167	9.1
4-5	163	176	7.9

Table 20.3
Difference between calculated and measured values

Notice the model predicts a decrease in Q until primary taps 4-5 are used, at which time it increases, as is confirmed by the measurements. The average percent difference between the measured Q and the calculated Q is 6.8%. Therefore, the dielectric loss due to the fishing line is somewhere between 7% and 10%.

20.2 MAKING A LITZ WIRE COIL

Coil Analysis

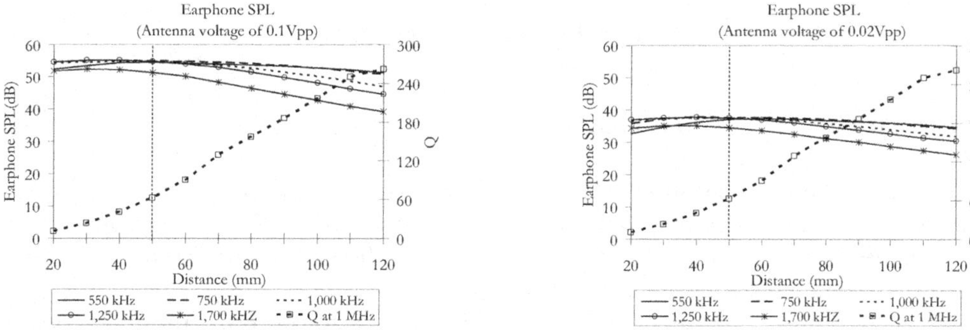

Figure 20.6
Maximum SPL as a function of separation distance.

We will now wind a coil using 175/46 Litz wire, using the same 90 mm former as before. This coil consists of 59 tightly wound turns (using a single equivalent wire diameter of 0.53 mm) with a pitch of 0.89 mm, giving us an inductance of 303 uH. The tap analysis was performed, and the results are shown in Figure 20.6. The distance of 50 mm was chosen as all the SPL levels at a closer distance are less than 2 dB. The SPL as a function of loop number is shown in Figure 20.7. The loops chosen for the taps are 10, 15, 20, 30 and 40.

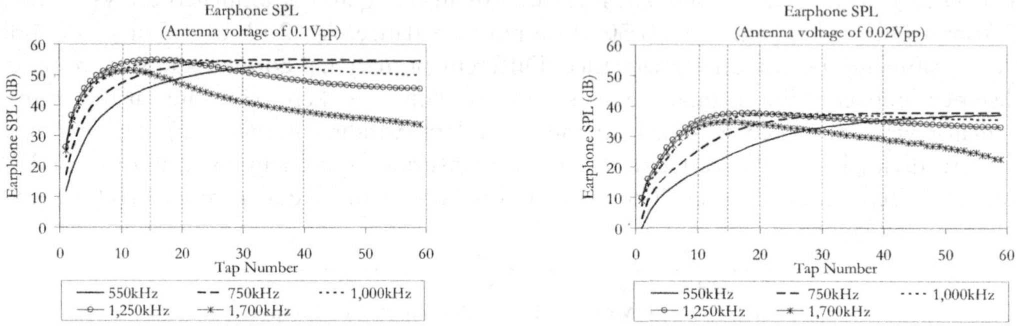

Figure 20.7
SPL as a function of winding number at 50 mm distance.

Description	Calculated values
Coil name	SC-BUILD-3
Coil inductance	302.9uH
Coil diameter	90.89 mm
Coil length	52.51 mm
Number of windings	59
Taps at loop number	10, 15, 20, 30, 40
Pitch	0.89 mm
Wire	175/46 Litz wire (175 strands of 46 gauge copper wire) equiv core D=0.53 mm, OD=0.89 mm)
Former material	PVC
Former diameter	90 mm
Former wall thickness	2 mm

Table 20.4
The secondary Litz wire coil specifications.

Constructing the Litz Wire Coil

Cut and drill the coil former, as shown in Figure 20.8. Place the spool of Litz wire on the wire feeder and feed the wire through the books. Prepare the former for winding as per our normal coil winding technique. Feed the Litz wire through the starting holes, then pull out the lead wire to about 130 mm in length and wind the coil.

To create the taps, place a 60 mm length of 5 mm diameter dowel across the coil and wind the loop over the rod. Do not twist; see Figure 20.9a. Keep the rod in place and continue winding by sliding the wire under the rod for normal loops and over the rod for tap loops. When completed, glue the lead wires and the edges as usual, but also place some glue where the taps meet the former to keep the tap loops in place. Wait for the glue to dry, and then carefully remove the rod. Just wiggle the rod slightly back and forth and pull it out from under the taps. Take your time and make sure you do not disturb the other loops.

Now it is time to remove the outer covering and the enamel from the individual strands for the lead wires and taps. Set your soldering iron to 300° C and run it smoothly over the lead ends. The outer insulation will melt and is easily removed. Removing the enamel from each of the strands

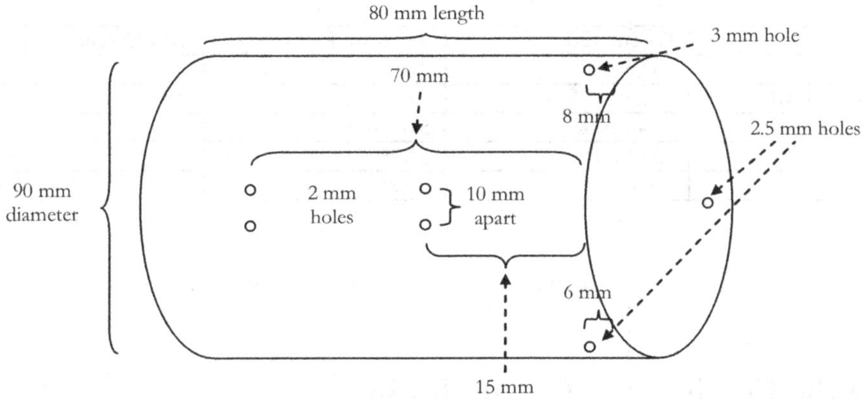

Figure 20.8
Secondary Litz wire coil former.

requires a bit more work. Place some solder on your iron and press the iron down on the wires. Hold it in place for five seconds or so. Place more solder on the iron and slowly run the solder up and down the exposed wires. Run the solder all around the wire and take off excess solder. Repeat this process three or four times. The heat from the soldering iron should melt the enamel leaving the exposed wire. Do this for both the lead wires and then test the DC resistance. For this coil, it should be around 1.4 ohms. Make sure you zero out the multimeter before testing.

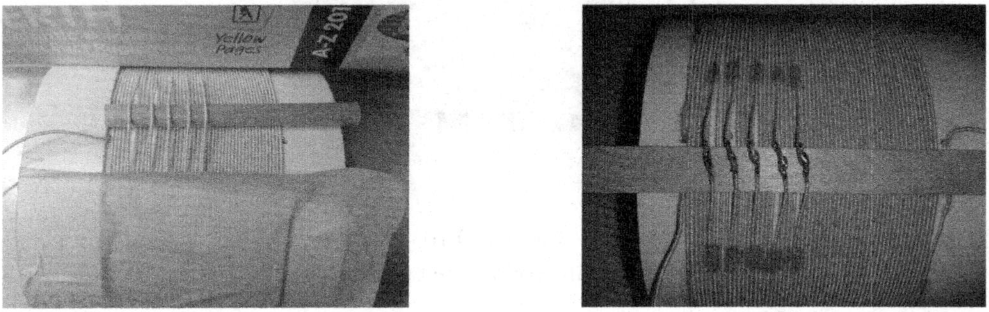

(a) (b)

Figure 20.9
Winding the Litz wire coil.

For the taps, place a Popsicle stick under the taps; see Figure 20.9b. Use the Popsicle stick to hold the underside of the tap loop and press down on the wire with the soldering iron to remove the outer coating. Run the iron all around the wire using the stick to keep the tap in place and to provide a backing. Then add some solder to the iron to remove the enamel in a similar fashion already described for the lead wires. It's a little tricky and fiddly to do, so take your time. When all taps have been soldered, check them with a multimeter. The resistance should follow a sequence with the first tap having most resistance and the last tap the least, and all less than the total DC resistance of the coil.

Table 20.5 shows the difference in calculated and measured values for this coil. The calculations were performed using the Litz wire modifications to the model given in Section 11.10.

	Calculated	Measured	% difference
Inductance	302.9 uH	303.1	0.07
Effective Q (1 MHz)	598.36	595	0.56
Self-resonant frequency	5.34 MHz	5.05 MHz	5.74
Length	52.51 mm	52.75	0.46
Pitch	0.890 mm	0.894	0.46

Table 20.5
Difference between calculated and measured values.

Inserting the Coil into the Crystal Radio

Remove the current secondary coil from your crystal radio. Now cut five 160 mm lengths of Litz wire to connect the taps to the rotary switch. Do not use the existing enamel copper wire, as it will reduce the Q. Remove the coating from the ends of the Litz wire lengths as described above and then solder them to the taps and rotary switch. Place the Litz wire coil in your crystal radio and attach the lead wires to the tuning capacitor; see Figure 20.10.

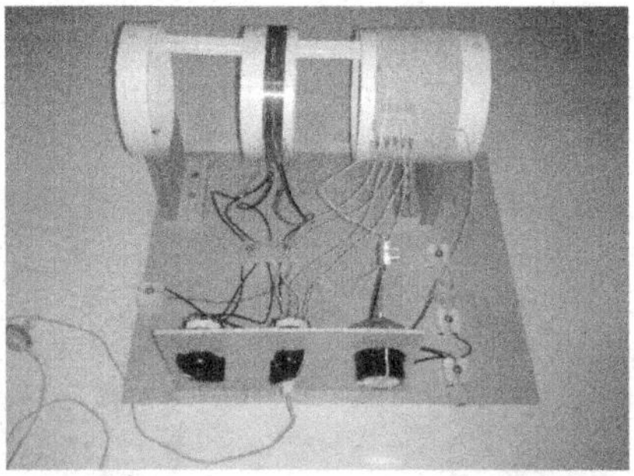

Figure 20.10
The Litz wire coil in the crystal radio.

Performance

The frequency range covered by the crystal radio is from 473 kHz to 2,600 kHz using primary taps 1-5. The Q of the crystal radio was tested using a separation distance of 70 mm. The demodulator was attached to winding 10. The calculations were made using the measured inductance and coil length. The results are given in Table 20.6.

Primary taps	Q Measured	Q Calculated	% Difference
1-5	262	267	1.9
2-5	209	232	11.0
3-5	186	207	11.2
4-5	224	224	0.0

Table 20.6
Difference between calculated and measured values.

20.3 THE BENEFIT OF HIGH Q COILS

In the three iterations of the crystal radio, the only change was the secondary coil. We are now in a good position to examine the effect that a high Q coil has on the radio's performance. A tap was chosen for each of the three coils that gave the maximum SPL at 1,000 kHz. The standard antenna (+transmission line) impedance at this frequency is 32 Ω with 1,612 pF of capacitance. The 19.7 uH tap was chosen for the primary coil, as it is close to uncoupled resonant at this frequency. The SPL and the Q were calculated and plotted as a function of distance, as given in Figure 20.11, for high and low power stations. The lines are SPL, and the symbolled lines are Q values. The maximum volumes are given in Table 20.7 for each of the coils, for the high and very low power stations.

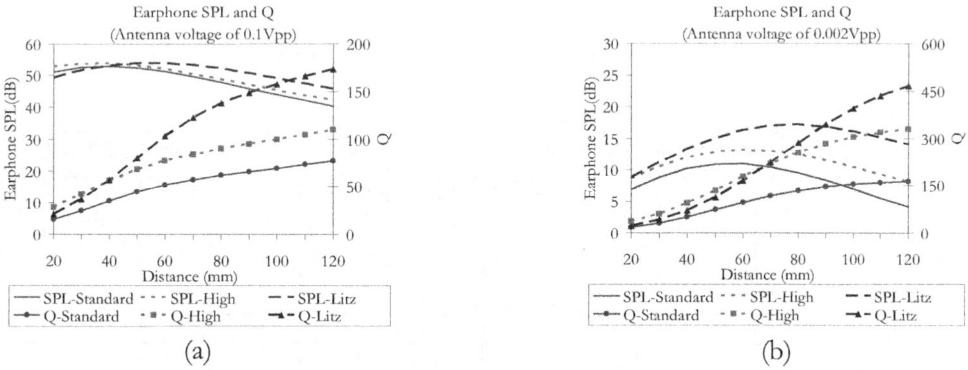

Figure 20.11
Crystal radio performance comparison for high and very low power stations.

For high-power stations, the difference in the maximum volume between the coils is not noticeable. If you have a single medium or high power station, or widely spaced medium or high power stations in your area, then the difference in volume between low Q coils and high Q coils will not be significant. However, for low-power stations, the difference is quite noticeable. Also, note that the Q for high power stations is lower than for low power stations. This is because a greater proportion of the energy contained in the secondary is converted into sound for high-power stations than it is for low-power stations.

Say you have multiple stations in your area, and you require a Q of at least 120 to separate a loud station without interference from nearby stations. To achieve a Q of around 120 for the tightly wound coil you will need to increase the separation distance to about 190 mm, see Table 20.8. For the higher Q coils this separation distance is a lot less. But there is now a significant difference in the volume of the low-Q coil as compared to the higher Q coils. In fact, the Litz wire coil will sound four times as loud compared to the low-Q coil, and twice as loud as the high-pitch coil.

For a station with low power, you will require an even higher Q value to prevent interference from nearby stations. Say we require a Q of 160 or greater to separate the low power station. From Table 20.8 we see that the separation distance for the low-Q coil is such that only a small amount of energy will be induced into the secondary. So small, that the SPL of 5.57 will not be audible. Both the high-Q coils can detect this signal, with the Litz wire coil having the loudest volume, and with Q to spare.

Coil	High power station					Very low power station				
	Loop #	D (mm)	Q	SPL (dB)	SPL diff from max	Loop #	D (mm)	Q	SPL (dB)	SPL diff from max
Tightly wound	20	40	35	52.89	-0.97	25	60	97	10.97	-6.22
High pitch	25	40	56	53.76	-0.10	30	60	178	13.17	-4.02
Litz	15	60	103	53.86	0.00	30	80	286	17.19	0.00

Table 20.7
Comparison of maximum SPL for each of the coils at 1,000 kHz.

Coil	High power station (Q > 120)					Very low power station (Q > 160)				
	Loop #	D (mm)	Q	SPL (dB)	SPL diff from max	Loop #	D (mm)	Q	SPL (dB)	SPL diff from max
Tightly wound	20	190	119	30.63	-22.62	25	110	160	5.57	-11.62
High pitch	25	140	122	39.32	-13.93	30	60	178	13.17	-4.02
Litz	15	70	123	53.25	0.00	30	80	286	17.19	0.00

Table 20.8
SPL for each of the coils with a required minimum Q of 120 and 160 at 1,000 kHz.

This clearly shows the benefit of a high Q coil. They provide the ability to separate close stations without a major sacrifice of volume, if any. Also, high Q secondary coils will make the difference between hearing and not hearing low-power or distant radio stations.

21

Double Resonant Crystal Radios

21.1 DETERMINE THE CAPACITANCE RANGE

For a circuit to be resonant, it must look resistive. If we place a tuning capacitor across the primary coil, as shown in Figure 21.1a, then at resonance only the tank circuit will look resistive. The impedance of a parallel tank circuit is very high. Therefore, very little current will flow across the primary circuit. We are also still left with the antenna capacitance, as shown in Figure 21.1b. Therefore we need to place the tuning capacitor in series with the coil and antenna capacitance, as shown in Figure 21.2.

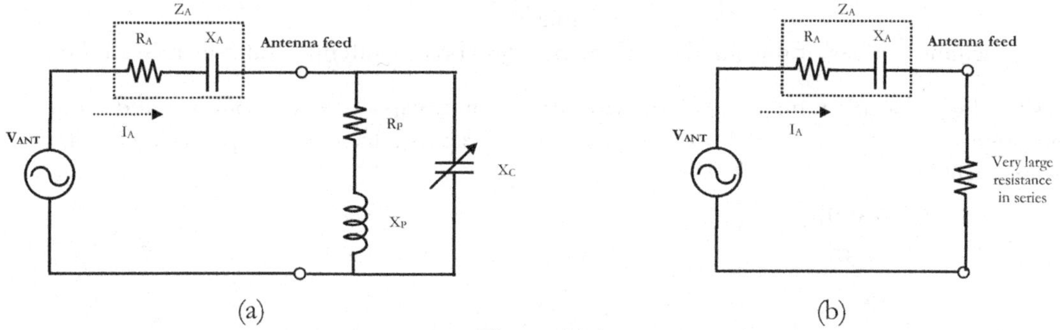

Figure 21.1
A parallel tank circuit will not make the primary circuit resonant.

Capacitors in series reduce the overall capacitance, and the total capacitance of the primary circuit for a given capacitance value will be

$$C_{TOT} = \frac{C_A C_C}{C_A + C_C} \tag{21.1}$$

where C_A is the antenna capacitance, and C_C is the capacitance value set by the tuning capacitor. By using a variable capacitor, we can adjust the total series capacitance of the circuit to equal the inductive reactance at a given frequency, thus making the circuit resonant.

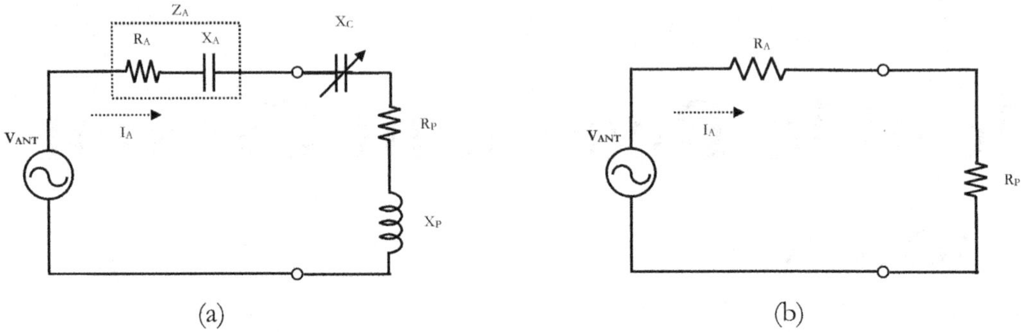

Figure 21.2
A tuning capacitor in series will make the circuit resonant.

To determine the tuning capacitor's range, first determine your antenna's minimum and maximum capacitance for the lower and upper frequencies using the equations given in Chapter 10 on Antennas. Consider our standard 25-metre antenna; the capacitance with and without the standard transmission line is shown in Table 21.1 for a low frequency of 530 kHz and a high frequency of 1,700 kHz (the AM band).

Frequency (kHz)	**Antenna only**		**+Transmission line**	
	Capacitance (pF)	Resistance	Capacitance (pF)	Resistance Ω
530	227	11.0	1,521	12.5
1,700	307	64.1	1,954	66.4

Table 21.1
Capacitance values of the standard 25-metre crystal radio antenna and transmission line.

Using the 7-360 pF tuning capacitor with a maximum capacitance of 360 pF and an antenna capacitance of 227 pF at 530 kHz. The maximum capacitance for the primary circuit at this frequency will be

$$C_{\max} = \frac{227(360)}{227 + 360}$$
$$= 139 pF$$

And the lowest capacitance value will be used to tune to 1,700 kHz with an antenna capacitance of 307 pF. So

$$C_{\min} = \frac{307(7)}{307 + 7}$$
$$= 6.8 pF$$

Using inequality 18.4, determine if we can tune across the entire frequency range.

$$\frac{f_{high}^{\ 2}}{f_{low}^{\ 2}} \le \frac{C_{\max}}{C_{\min}}$$

And

$$\left(\frac{1700}{530}\right)^2 = \frac{139}{6.8}$$
$$10.3 \le 20.4$$

The inequality is satisfied; now we can determine the inductor required for resonance.

21.2 DETERMINING THE PRIMARY COIL INDUCTANCE

The primary coil's inductance required for the series resonant circuit at a given frequency can be calculated using

$$L_{REQ} = \frac{1}{4\pi^2 f_{LOW}^2 C_{\max}} \qquad 21.2$$

where f_{LOW} is the lowest frequency in the range, and C_{max} is the maximum capacitance given by Equation 21.1. Using the lowest frequency and the value of C_{max} calculated above, we get

$$\begin{aligned} L_{REQ} &= \frac{1}{4\pi^2 f_{LOW}^2 C_{\max}} \\ &= \frac{1}{4\pi^2 (530x10^3)^2 (139x10^{-12})} \\ &= 648x10^{-6} \end{aligned}$$

We require a primary coil with 648 uH of inductance for the antenna-only situation. Performing the same analysis using the antenna with the transmission line, we get

$$\begin{aligned} \frac{C_{\max}}{C_{\min}} &= \frac{291}{6.9} \\ &= 42.2 \end{aligned}$$

Inequality 18.4 is still satisfied, and the calculated inductance required using Equation 21.2 is 309.9 uH. Note that the higher the antenna capacitance is, the greater the proportion of the tuning capacitor's range becomes usable.

Winding a high-value Q coil for the primary will only marginally increase the resonant voltage across the primary, as the Q of the coil will be greatly reduced by the total resistance of the antenna feed; which includes ground resistance, radiation resistance and the resistance of the wire. For instance, we could use the tightly wound coil we made in a previous Chapter. This coil has an AC resistance of 10 Ω at 1 MHz, and the antenna resistance at this frequency is 32 Ω. The in-circuit Q of the primary coil becomes

$$Q = \frac{1,948}{10+32} = 46$$

Using the high-pitch coil, having an AC resistance of 5 ohms at 1 MHz, we get an in-circuit Q of

$$Q = \frac{1,911}{5+32} = 52$$

Giving a 13% increase in Q at 1 MHz. The actual value will greatly depend on your ground resistance, and the Q may be better or worse.

If we chose a primary coil inductance of 310 uH, we can tune across the entire AM spectrum as long as we use the standard antenna and transmission line. If the capacitance of the antenna input feed is less than this, the highest tunable frequency will remain the same; however, the lowest tunable frequency will increase. For instance, if this inductor were used with an antenna capacitance of 227 pF, the lowest tunable frequency would be 778 kHz.

If you used an antenna with more capacitance, the highest tunable frequency would also remain the same, and the lowest tunable frequency would decrease (below 530 kHz). That is, our tunable range has increased, and we can still use this radio to tune across the entire AM spectrum.

21.3 MODIFYING THE RADIO FOR DOUBLE RESONANCE

I will use the tightly wound coil as the primary coil for this double-resonant crystal radio. Not having a spare 360 pF capacitor, I ran a bridge wire across a dual 340/25 pF tuning capacitor to make a 365 pF capacitor. The lowest capacitance was measured at 14.1 pF. Using this with the standard antenna and transmission line gives a capacitance range of 14.1 to 295 pF and a frequency range, using our 310 uH coil, of 526 kHz to 2,407 kHz.

To modify our crystal radio for double residence, first solder two 200 mm lead lines to the terminals of the capacitor as shown in Figure 21.3a. This tuning capacitor will be placed in series with the primary coil, so one lead will go to the antenna connector and one to the first lead wire of the primary coil, as shown in the circuit diagram Figure 21.2a. The other lead wire of the primary coil will go to the Earth connector.

Hot glue the variable capacitor to the three-ply panel. If the tuning capacitor shaft is too short to attach a knob, extend the shaft by screwing a 12 mm nylon standoff into the shaft of the capacitor as shown in Figure 21.3b.

Snip off the tap lead wires from the existing tightly wound coil and insert a cardboard sleeve for the slide rail to go through. Then add 80 mm of flexible hook-up wire to the primary coil's end leads so it can move along the rail, see Figure 21.4a.

There is room on the base to extend the separation distance between the coils by 60 mm by repositioning the right-hand coil stand to the right edge of the base. If you did not cut the 290 mm long slider when building the radio, you must do this now to prevent over-coupling. The rule of thumb is to keep the distance to one coil diameter or more; this distance is about 90 mm for our radio.

Cut a 290 mm length of foamboard for the extended slider rail. Remove the coils. Unscrew the secondary stand and move it to the edge of the base. Redrill the base holes and screw the stand back on the base. The tap wires from the secondary may not be long enough to reach the switch on the front panel. If so, remove the front panel from the base, move it up against the resistor terminal block, and re-screw it onto the base at this position. Slip the slide rail through the coils and re-attach the coil assembly to the coil stands. See Figure 21.5.

(a)

(b)

Figure 21.3
The primary tuning capacitor.

(a)

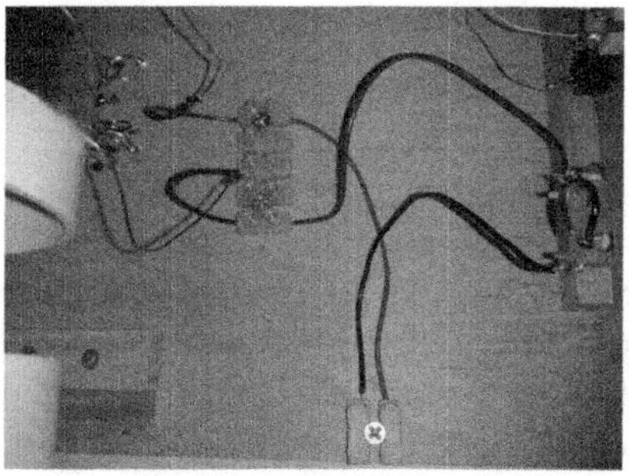

(b)

Figure 21.4
Using the tightly wound coil as a primary coil.

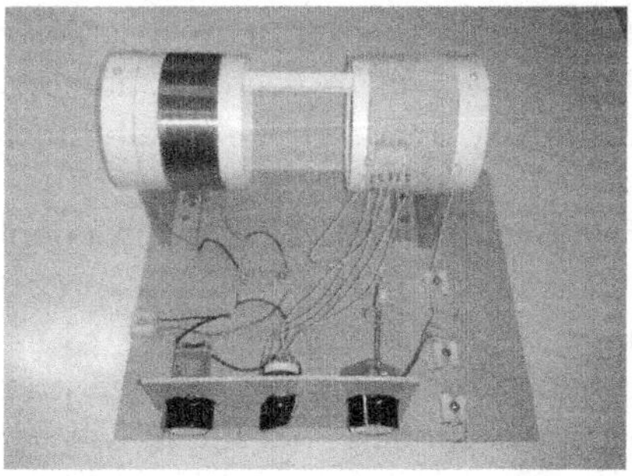

(a)

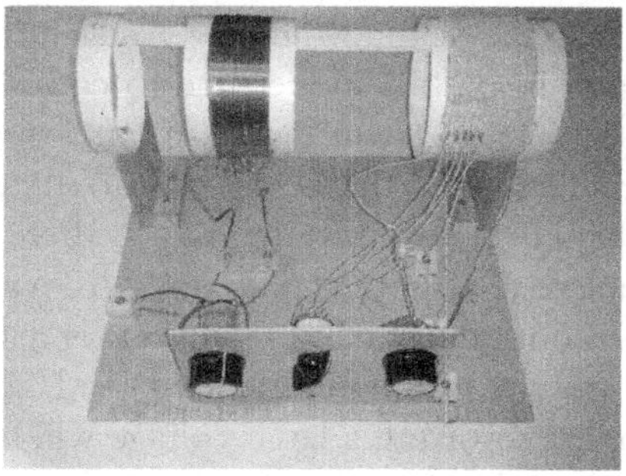

(b)

Figure 21.5
Double resonant crystal radio. a) unextended coil train. b) Extended coil train.

21.4 PERFORMANCE

The Q of the crystal radio was tested at various distances, with the demodulator attached to winding number ten. The calculations were made using the measured inductance and coil length. The primary coil was made uncoupled resonant at each separation distance by removing the secondary coil so no reflected impedance into the primary occurs. The secondary was then moved into position and tuned for maximum voltage, and the Q of the secondary coil was measured.

The Q measurements were taken again by making the primary coupled resonant by retuning the primary (with the secondary coil in place), for maximum voltage across the primary. The measurements and calculations are shown in Table 21.2.

Separation distance (mm)	Secondary Q Measured (primary uncoupled-resonant)	Secondary Q Calculated	% Difference	Secondary Q Measured (primary coupled-resonant)
90	31	30.21	2.5	224
110	58	57.20	1.4	194
130	91	95.45	3.8	161

Table 21.2
Difference between calculated and measured values of the double-resonant crystal radio.

When the primary is made uncoupled resonant, the initial induced voltage onto the secondary coil is at a maximum. However, the secondary will induce a large current across the primary coil resulting in large resistive losses (large mutual inductance loss) and a low Q value. As the coils are moved further apart, the mutual inductance losses decrease as expected, and the Q rises.

When the primary is tuned to coupled resonance, a smaller initial voltage is induced across the secondary. However, the impedance of the primary is now much greater than at uncoupled resonance, and the mutual inductance losses across the primary from the secondary are lower and the Q increases. Counter-intuitively, as the separation distance increases, the Q decreases. The opposite of what normally happens. As the separation distance increases, the reflected impedance from the secondary into the primary decreases and the impedance of the primary circuit decreases. That is, as the distance increases, the primary tends towards becoming more uncoupled resonant. The induced voltage from the secondary to the primary is much larger than for low-inductance primary coils of a few turns. And the separation distance does not result in a large enough reduction in mutual inductance to cause the mutual inductance losses to decrease, and so the Q decreases. The Q will decrease with distance until the primary becomes essentially uncoupled resonant, and then the Q will start to increase.

When using the crystal radio, there is no way of knowing if the primary is coupled or uncoupled resonant, and it doesn't matter. The primary and secondary tuning capacitors are adjusted until the maximum volume is reached. For instance, at a separation distance of 130 mm, a tuned station may sound just as loud at a separation distance of 90 mm. When the primary is uncoupled resonant, a large voltage will be induced across the secondary, but the Q will be lower. When the primary coil is made coupled resonant, less voltage is induced across the secondary, but the Q will be higher. As always, the station volume is given by the combination of both the induced voltage and Q of the secondary.

21.5 USING THE DOUBLE-RESONANT RADIO

Feed the antenna line to the tuning capacitor and the earth line to the coil, as shown in the circuit diagram Figure 21.2a. Having the lines swapped slightly reduces the volume of higher-frequency stations. In a lumped model where the antenna capacitance is treated as a discrete value, it should not matter. However, as the capacitance is distributed along the transmission line and antenna, there does seem to be some small effect, at least at resonance.

Set the separation distance to about one coil diameter, in this case, about 90 mm, to prevent over-coupling. To tune in a station, set both capacitors to their maximum capacitance values, fully anticlockwise. Slowly move the secondary capacitor clockwise for a small amount of travel. Then do the same for the primary until a station is heard. When a station is heard, move the secondary back and forth in both directions until the signal becomes loudest. I refer to this as ***rocking the capacitor***. Then do the same for the primary capacitor. You may need to perform this step several times to get the loudest signal. Achieving exact resonance for both coils is a little tricky for double resonant radios as the reflected impedance affects both the primary and secondary circuits, requiring them to be re-adjusted until both are resonant at precisely the same frequency.

The selectivity of this radio is insane. It is really high, and the higher the frequency, the greater the selectivity, as Q is directly proportional to frequency. Above 1 MHz, or with low power stations, the selectivity is so high that it is really easy to miss a station completely. You barely need to move both capacitors. Rocking the capacitors seems to be a good method for finding stations. This extreme selectivity is achieved because the primary is made resonant at a given frequency, causing the signal to be boosted by the primary Q. This allows us to have a large separation distance and still induce a strong signal into the secondary. The other frequencies are not boosted, and due to the separation distance, these induced voltages are very much smaller. Thus we can achieve very high levels of selectivity.

This high selectivity can be a benefit or a hindrance in terms of ease of operation, depending on your local station situation and if you are DXing or not. I listen primarily to local stations, and using a tapped primary coil gives me enough of a boost to achieve excellent selectivity. Using the double-resonant radio did not result in a noticeable difference in volume over the tapped radio, as my tapped primary is designed to be used with my antenna. And using the switch on the primary is a lot easier than constantly fiddling with tuning capacitors. Your needs and setup will likely differ, and you may prefer to use the tuned primary.

The tuned primary, however, is an excellent choice when a tapped primary is not well-matched for a given antenna. For instance, I used a metal shelf as an antenna and another shelf as the earth. I could tune in all but one very low power station using the tuned primary. Even though the capacitance of this antenna is very low, around 149 pF, and I could not tune the primary below 880 kHz, I could still pick up a station at 585 kHz as enough voltage was induced into the secondary. I could not achieve this using my tapped primary crystal radio as the primary coil inductance is so low that it does not even come close to being resonant. So if you want to listen to local radio stations and don't want to (or can't) attach your radio to an external antenna, a double-tuned crystal radio will work really well. Good if you live in a high-density housing area. For short, low-capacitance antennas, you will need to build a higher inductance primary coil. Calculate the value using the technique described above.

22

Ferrite Cores

22.1 FERRITE MATERIAL

Ferrite materials are ceramics (a mixture of materials fired at high temperatures like clay) composed of about 50% iron oxide (Fe_2O_3), with the balance made from other materials such as Nickel, Zinc and Manganese. Being ceramic, ferrites are quite brittle, and care should be taken not to drop them. Common compositions include Manganese and Zinc (MnZn), used for low-frequency work, and Nickel and Zinc (NiZn), used for high-frequency applications. The proportion of the compounds within a ferrite varies, and other trace elements can be added to change the electrical and magnetic characteristics.

Specific compositions of ferrites are given codes to designate the type of material. For example, Manganese and Zinc (MnZn) ferrite belong to code types 31, 33 and several other code groups. Nickel and Zinc (NiZn) compositions belong to code types 61 and S1 and others. As we deal with RF frequencies, we are more interested in type 61 and S1 ferrite materials.

Ferrite materials have high electrical resistivity and large relative permeabilities making them ideal to use as cores for inductors. Typically, these cores are shaped like rods or doughnuts. This high relative permeability allows coils of a given inductance to be wound using less wire and turns compared to air-core coils, resulting in physically much smaller inductors. Using ferrite rods, you can create small, compact crystal radios that are easier to construct than their larger air-core cousins. Because of the compact size of these coils, they were used in small pocket-sized crystal radios; which were quite popular in the 1960s.

Unfortunately, there are substantial energy losses that occur inside the ferrite material. Coils wound using ferrite cores will have lower Q values than an equivalent air-core coil. However, you can still build reasonably effective crystal radios using ferrite cores. Also, you can encase your radio in different enclosures to create fun and interesting sets. During the 1960s, you could get crystal radios in the shapes of rockets. How you encase your radio is only limited by your imagination.

22.2 EDDY CURRENTS

In materials containing free electrons, electrons will move in random paths due to thermal energy. Depending on the direction of travel of any free electron, when an external fluctuating magnetic field cuts the material, some of these electrons will be forced to travel in a circular motion called an eddy current (as given by the left-hand rule). The energy used to change the path of an electron comes from the magnetic field resulting in a loss of useful energy, a loss that does not occur in air coils.

The formation of eddy currents is inversely proportional to the material's resistivity. The magnitude of these eddy currents will be smaller for materials of high resistivity than for materials with low resistivity. Therefore, ferrite materials with high resistivity are more desirable for our purposes.

22.3 HYSTERESIS

Ferrite materials are ferromagnetic and can be magnetised to some degree, some much more than others. We will use the domain theory of magnetism in the following discussion. Consider a coil with a ferrite rod core connected to a power supply. Before current flows through the coil, we will consider the domains in the rod to be randomly oriented so no net magnetic field exists within the rod. When power is supplied to the coil, current flows through the wire, creating a magnetic field. The domains within the ferrite material will experience a magneto motive force and start to orient themselves along these lines of force. As more domains align themselves in the same direction, the magnetic field inside the core increases. From Equation 7.51, we know the magnetising force H is related to the magnetic field inside the core by

$$B = \mu_R \mu_0 H$$

As the current increases, the higher the magnetising force becomes, and the magnetic field flux density inside the core increases. Core materials with high permeability will result in higher flux densities for a given current. As the current increases, there will be a point where all of the domains will have oriented themselves in the same direction. At this point, magnetic saturation has been reached, and the magnetic field flux density will not increase any further. The magnetic field density has reached a maximum. This process is shown in graphical form in Figure 22.1a.

After reaching saturation, if we start to decrease H, some domains will randomly re-align themselves. However, when we have decreased H to zero, some domains will still be aligned, and some residual magnetism will exist inside the material, Figure 22.1b. If we reverse the current, the direction of the magnetising force reverses, and as we increase the negative current at some point, the magnetic flux density inside the core will go to zero. If we continue to increase the reverse current, the core will once again reach saturation, but the domains will be aligned in the opposite direction, Figure 22.1c. When we decrease the amplitude of the reverse current back to zero, there will still be some domains aligned, resulting in residual magnetism. And increasing the current in the positive direction will eventually result in flux density saturation again, Figure 22.1d, and we have gone full circle. However, the value of the magnetising force at saturation is now different from when we first hit saturation. We can think of the magnetic field in the core lagging behind the magnetising force. As such, the curve shown in 22.1d is called the ***hysteresis curve***, from the Greek ***hysterein***, meaning to lag.

To move the magnetic domains in the material requires work, and work requires energy. This energy is taken from the magnetising force from the coil and thus will result in a reduction of Q.

The area inside the hysteresis curve represents this loss of energy. So for crystal radios, we require ferrite material with the smallest possible area inside the hysteresis curve. This means we require ferrite material with the smallest possible residual magnetic field remaining after the magnetising force has been removed, as shown in Figure 22.2.

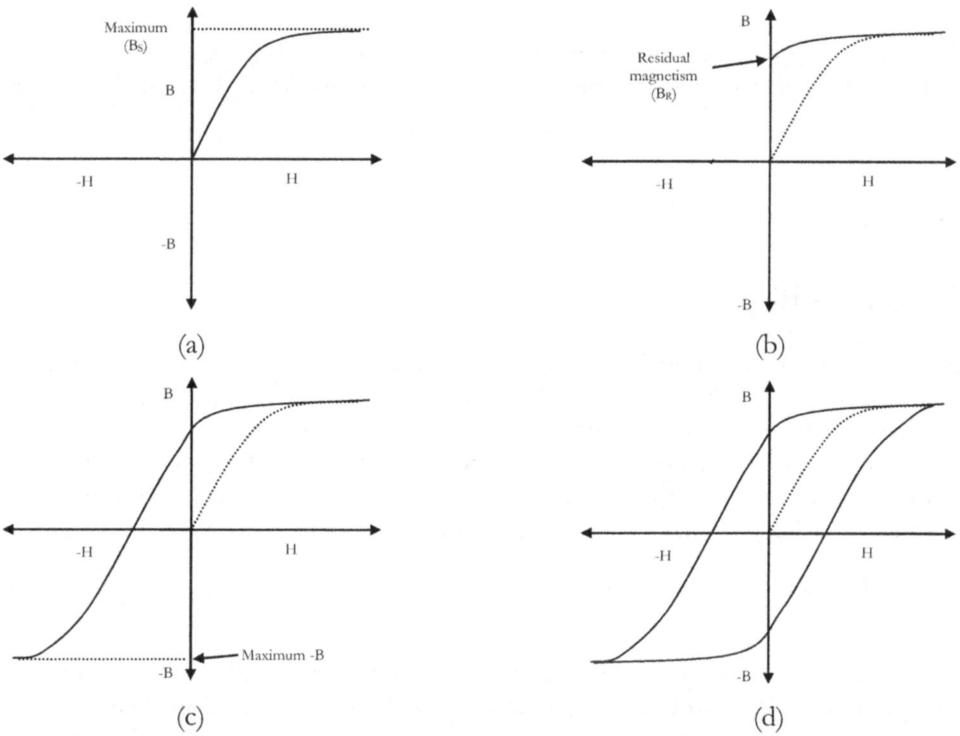

Figure 22.1
The response of ferrite material to the magnetising force.

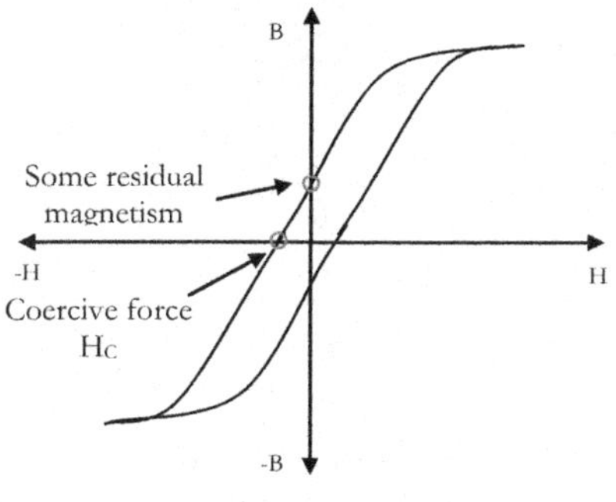

Figure 22.2
A small area enclosed by the hysteresis curve is required by crystal radios.

The value of the magnetising force (H_C) required to completely remove any residual magnetic field, after being driven into saturation, is called the ***coercive force***; and is a measure of the material's ***coercivity***. We want ferrite material with low coercivity.

22.4 FERRITE CORE PERMEABILITY

Given the relationship between the magnetising force H and the magnetic flux density B, we can calculate the relative permeability of the ferrite material. Rearranging Equation 7.51 for relative permeability μ_R, we get

$$\mu_R = \frac{B}{\mu_0 H} \qquad 22.1$$

Initial Permeability

The permeability of ferrite material is not constant but is a function of the magnetising force H, the magnetic flux density within the material and the frequency. However, for a small change in H (a small current change), the permeability can be treated as constant. The permeability given in specification sheets is the ***Initial (relative) permeability***, μ_i. This is actually the relative permeability measured at some small value of the magnetising force, typically between 0.1At/m to 0.4At/m on a toroidal core. This initial permeability is calculated as follows.

$$\mu_i = \frac{L}{\mu_0 N^2}\frac{\ell}{A} \qquad 22.2$$

where L is the self-inductance of the coil wound on the ferrite core in Henrys, N is the number of turns, ℓ is the average length of the magnetic path in the core material in metres, and A is the cross-sectional surface area of the core in metres squared. It is assumed that the magnetic flux is mostly contained within the core and that the coil is wound directly onto a toroid without a gap. The average magnetic length value for a toroid is the circumference of the toroid as measured at half its thickness, as shown in Figure 22.3.

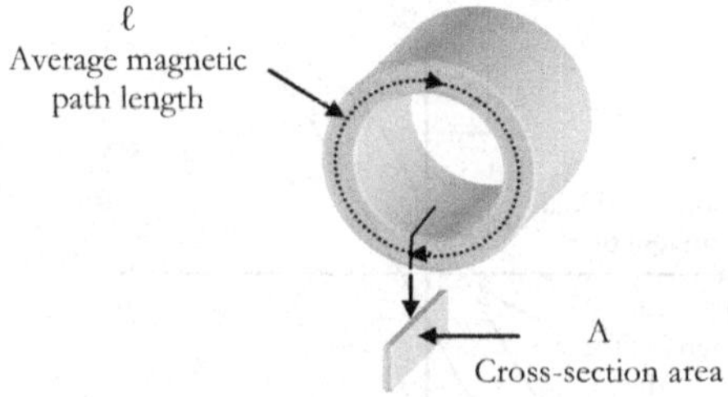

Figure 22.3
Toroidal ferrite core.

Initial permeability values for ferrite materials are measured using toroids. The permeability stated in any given specification sheet is only for the type of material for a given geometry. For instance, I will be using a rod core made from S1 (like type 61) material. The permeability for S1 material is given as 120 in specification sheets and on various websites selling the rod. If this value is used to calculate the inductance of a coil wound on this rod, the inductance value will be completely wrong.

Effective Permeability

When some of the magnetic flux is not contained within the ferrite material, the inductance of the coil will differ from that obtained using a closed magnetic circuit. The flux not passing through the core is called ***leakage flux***. Any measurement of the permeability of the ferrite material will be measuring the combination of the permeability of the ferrite material and any other material the magnetic flux is passing through. For instance, a toroid may have an air gap between the ends and not be closed. The permeability measurement will be a combination of the permeability of the ferrite material and the permeability of air. This permeability is called the ***effective (relative) permeability*** (μ_e).

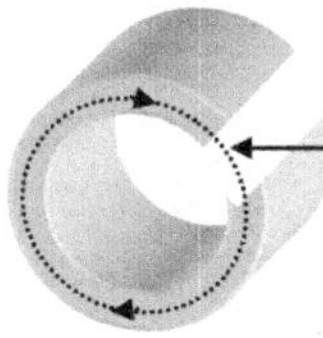

Figure 22.4
Opened toroidal ferrite core.

We need to modify Equation 22.2 to take this into account.

$$\mu_e = \frac{L}{\mu_0 N^2} \sum \frac{\ell}{A} \tag{22.3}$$

The last term sums the ratio of the magnetic path length to the cross-sectional area for each of the different materials the flux is passing through. This is generally just air, and this value is sometimes given in specification sheets and is called the core constant C_1. That is

$$C_1 = \sum \frac{\ell}{A} \tag{22.4}$$

Apparent Permeability

When a significant proportion of the magnetic flux is not passing through the ferrite material, such as in a rod, the measured permeability is called the ***apparent (relative) permeability***, μ_a. A significant proportion of the measured permeability will not be due to the ferrite material. Some of the flux lines will pass through the core material, some through any former material used, and some through the air for at least part of the magnetic path. This value will be significantly different from the initial permeability given in specification sheets that assumes all the flux lines are contained within the core material itself. For this reason, initial permeability values cannot be used for rods.

We can determine the apparent permeability by measuring the inductance of a coil with the ferrite rod inserted, and then by measuring the inductance value without the rod. The ratio of the two is the apparent permeability.

$$\mu_a = L_{CORE} / L_{AIR} \tag{22.5}$$

where L_{CORE} is the coil's inductance with the ferrite rod inserted, and L_{AIR} is the inductance without any core, just air. This measurement, however, is only suitable for the coil and rod combination you are measuring. If you change the coil's geometry or rod material, this measurement will not apply.

22.5 FERRITE ROD COIL PERMEABILITY

A formula for calculating the apparent permeability of a rod was given by Khomich[5]. This takes into account the reduction of the initial permeability due to the geometry of the rod, and is given by

$$\mu_a = \frac{\mu_i}{1 + 0.84\left(\dfrac{D_{rod}}{\ell_{rod}}\right)^{1.7}(\mu_i - 1)} \tag{22.6}$$

where μ_i is the initial permeability. D_{rod} is the diameter in metres of the rod, and ℓ_{rod} is the length of the rod. However, this assumes the coil is wound directly onto the rod, and does not take into account any flux leakage due to any gap between the coil and the rod.

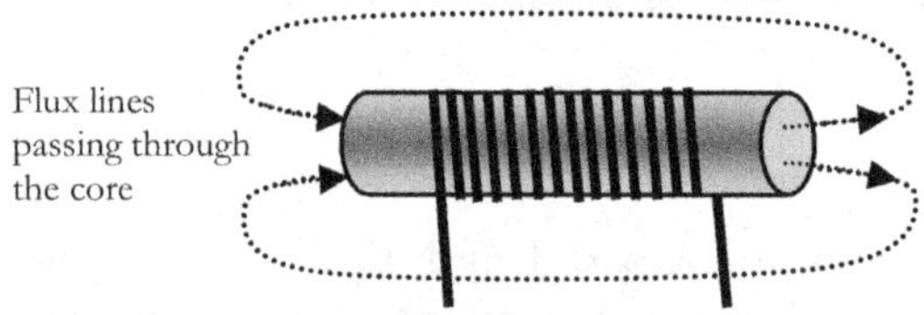

Figure 22.5
Ferrite rod coil magnetic flux path.

Apparent permeability is dependent not only on the material and geometry of the core but also on the geometry of the coil itself. Let some function γ (gamma) represent the reduction of apparent permeability due to the coil geometry. We can write

$$\mu_a = \left(\frac{\mu_i}{1 + 0.84\left(\dfrac{D_{rod}}{\ell_{rod}}\right)^{1.7}(\mu_i - 1)}\right)\gamma \tag{22.7}$$

Rod apparent permeability for a given coil

[5] Khomich, V. I. (1989) Ferrite Antennas, 3rd Edition, Energy. Moscow Russia.

Gamma must be greater than 0 and less than 1, as the coil geometry will always reduce the apparent permeability, but the apparent permeability will never go to zero.

Suppose the coil is wound on a former. In this case, there will be a gap between the coil windings and the rod allowing flux to pass through the air as well as the former material, thus reducing the flux density through the rod and reducing the apparent permeability. We know the flux density is related to the surface area, so we can assume the reduction in permeability is proportional to the ratio of the coil diameter to the core diameter. We will write this reduction as a power function. Power functions fit many physical phenomena such as diode I-V curves and the reduction in apparent permeability when a coil is offset from the core. We shall write this term as

$$\Delta\mu_g = \left(\frac{D_{rod}}{D_{coil}}\right)^{\alpha} \tag{22.8}$$

We will refer to the change in permeability due to the gap as the ***gap factor*** and the exponent α as the ***gap exponent***. Also, there is a relationship between the length of the coil and the length of the rod. The longer the coil is with respect to the rod, the greater the apparent permeability. And this relationship is in itself affected by the gap between the coil and rod. We will use the gap term to modify the exponent of the power equation. Writing this term as a power function we have

$$\Delta\mu_\ell = \left(\frac{\ell_{coil}}{\ell_{rod}}\right)^{\left(\frac{D_{coil}}{D_{rod}}\right)\beta} \tag{22.9}$$

We will refer to the change in permeability due to the coil length, the ***length factor*** and the exponent β as the ***length exponent***. Both these terms will be used to derive the function γ. If we sum or subtract these terms, then there is a possibility that the value of the function will be zero or greater than one. To ensure that the value is greater than zero and less than one, the terms need to be multiplied. The product of both the gap and length factors will change the apparent permeability of the coil. We will refer to this product as the ***rod permeability coefficient***, γ.

$$\gamma = \Delta\mu_g \Delta\mu_\ell \qquad \text{Rod permeability coefficient} \tag{22.10}$$

In deriving the rod permeability coefficient, we have taken educated guesses as to the form the function takes. We shall put this to the test to see how well it fits real-world data.

22.6 FERRITE CORE COIL WINDING PROBLEMS

Winding coils using ferrite rods has always been problematic. Equations to determine the inductance and the required number of loops cannot be performed with any accuracy as manufacturers supply relative permeability values for toroids and not for rods. The problem is amplified as the given initial permeability in speciation sheets typically has a range of ±20%. For example, I bought several ferrite rods of the same length-to-diameter ratio made from the same ferrite material. The rods were inserted into coils wound on a former, and the resulting inductance measurements differed by nearly 10%. And the use of formers makes a bad situation even worse.

A common technique for winding coils using ferrite rods was to guess at the number of turns required. Then by using an inductance meter to measure the inductance, loops would be added or removed until the coil was at the correct value. This is obviously a laborious and time-wasting technique. And you also need to strip some enamel off the copper wire to measure the inductance.

If you are going to work with ferrite materials, particularly ferrite rods, invest in a good inductance meter. These can be purchased online quite cheaply. Meters with an accuracy of 1 uH will be more than adequate for the inductances we are dealing with. Meters with an accuracy of 10 uH will do at a pinch and can be had for under $20.

If you don't have an inductance meter, use Equation 22.6 to get an estimate of the apparent permeability of your rod and then follow the steps outlined in Chapter 11 to determine the number of loops required for your coil. The apparent permeability given by Equation 22.6 will be greater than the real apparent permeability, so add about 20% more loops for your coil. The inductance of the coil can be reduced, if required, by moving the rod away from the centre position. By moving the coil toward the outer edge of the rod, you can drop the inductance. The further you move the rod out from the centre, the greater the reduction in inductance. You can use this technique to drop the inductance value if it is too high. By a combination of rod position and tuning the variable capacitor, you should be able to tune into all your stations. If you cannot tune into the higher frequency stations then the inductance of the coil is too large, therefore, remove a loop and try again. If you cannot tune into the lower frequencies, then the inductance is too small, so add a loop. By repeating this process you should get a coil of the correct inductance.

Either technique is pretty time-consuming. We shall look at using equation 22.7 to derive an equation (for a particular rod) that can be used to accurately determine the number of turns required for a given inductance. This, however, does require the use of an inductance meter.

An Improved Technique

From Equation 22.5, we can see that the apparent permeability of a rod is not a fixed value. Different apparent permeability values will be measured for different coil geometries. If we make two test coils with known geometries and measure the apparent permeability for each coil, we can use these values to derive an equation for the apparent permeability for any coil we wind on the rod.

Property	Test Coil 1	Test Coil 2
Coil number of turns (N)	60	89
Coil length	40.8 mm	60.8 mm
Coil diameter	17.18 mm(former + wire OD)	17.18 mm(former + wire OD)
Wire core diameter	0.63 mm	0.63 mm
Wire outside diameter	0.68 mm	0.68 mm
Pitch	0.68 mm	0.68 mm
Former outside diameter	16.5 mm	16.5 mm
Former inside diameter	11.0 mm	11.0 mm
Former material	PEX	PEX
Coil inductance (no core)	21.5 uH	33.9 uH

Table 22.1
Test coils specifications.

To determine the rod apparent permeability for any given coil (Equation 22.7), we need to calculate the rod permeability coefficient γ using Equation 22.10. Equation 22.10 has two unknowns; the gap and length exponents α and β. We will therefore require two equations to solve. If we wind two different test coils on a former, we can then use these coils for any rod we purchase by simply sliding the rod into the coils and measuring the inductance for each coil with the core inserted. I wound two test coils onto lengths of PEX (cross-linked polyethylene) piping; their inductances without any core inserted were measured. As this air-core inductance won't change, we only have to do this once.

Property	**Symbol**	**Unit**	**Value/Description**
Material			S1(61/64 material)
Initial permeability	μi		120
Relative loss factor	Tan δ/μ	10^{-6}	429.5 @1MHz
Gap exponent			0.81
Length exponent			0.3682
Inductance coefficient a		10^{-4}	0.1313
Inductance coefficient b		10^{-4}	1.8383
Coupling k_0			0.6953
Coupling exponent a		10^{-4}	4.5631
Rod diameter	Drod	mm	9.68
Rod length	ℓrod	mm	180

Table 22.2
S1 ferrite rod specifications.

Now to calculate α and β for a particular rod, insert the rod into each of the test coils and measure the inductance value of each. Then use Equation 22.5 to calculate the apparent permeability for each coil. The specifications for each test coil are given in Table 22.1, and the specifications for the ferrite rod material used are given in Table 22.2. The measured inductance for each test coil, with the core inserted, was 362.6 uH for test coil 1 and 742.0 uH for test coil 2.

The apparent permeability for test coil 1 is

$$\begin{aligned}\mu_{a1} &= L_{CORE} / L_{AIR} \\ &= 362.6 / 21.5 \\ &= 16.9\end{aligned}$$

And for test coil 2 is

$$\begin{aligned}\mu_{a2} &= 742.0 / 33.9 \\ &= 21.9\end{aligned}$$

Now to solve the two equations, we first set

$$\beta_0 = \left(\frac{D_{coil}}{D_{rod}}\right)\beta$$

Now substituting our values for the rod and coil into equation 22.7, for test coil 1 we get

$$\mu_{a1} = 120 \Bigg/ 1 + 0.84\left(\frac{9.68}{180.0}\right)^{1.7} (119)^{\left(\frac{9.68}{17.18}\right)^{\alpha}\left(\frac{40.8}{180.0}\right)^{\beta_0}}$$

$$16.9 = (70.805)(0.5634)^{\alpha}(0.2267)^{\beta} \qquad 22.11$$

And the equation for test coil 2 becomes

$$\mu_{a2} = \frac{120}{1+0.84\left(\frac{9.68}{180.0}\right)^{1.7}(119)}\left(\frac{9.68}{17.18}\right)^{\alpha}\left(\frac{60.8}{180.0}\right)^{\beta_0}$$

$$21.8 = (70.805)(0.5634)^{\alpha}(0.3378)^{\beta}$$

If we divide the apparent permeability of test coil 1 by the apparent permeability of test coil 2, we get the following ratio

$$\frac{\mu_{a1}}{\mu_{a2}} = \frac{16.8651}{21.8879} = \frac{(70.805)(0.5634)^{\alpha}(0.2267)^{\beta_0}}{(70.805)(0.5634)^{\alpha}(0.3378)^{\beta_0}}$$

Simplifying

$$\left(\frac{0.2267}{0.3378}\right)^{\beta_0} = 0.77052$$

And

$$\beta_0 \log(0.6711) = \log(0.77052)$$

$$\beta_0(-0.1732) = -0.1132$$

$$\beta_0 = 0.6535$$

Substitute β into the equation for test coil 1 (22.11), and solving for α, we get

$$(70.805)(0.5634)^{\alpha}(0.2267)^{0.6535} = 16.9$$

$$(0.5634)^{\alpha} = \frac{16.9}{(70.805)(0.2267)^{0.6535}}$$

$$\alpha \log(0.5634) = \log(0.6282)$$

$$\alpha = 0.8100$$

Now determine β

$$\left(\frac{17.18}{9.68}\right)\beta = 0.6535$$

$$\beta = 0.3682$$

Finally, the equation for determining the apparent permeability of this ferrite rod becomes

$$\mu_a = 70.805\left(\frac{9.68}{D_{coil}}\right)^{0.8100}\left(\frac{\ell_{coil}}{180}\right)^{\left(\frac{D_{coil}}{9.68}\right)0.3682} \qquad 22.12$$

Using Equation 22.12 to analyse test coil 2, we get an apparent permeability of 21.8 (measure

21.8) and a calculated inductance of 740.27 uH (measured 742.0 uH), a 0.25% difference. I now wound a coil of 20 turns directly onto the rod (with some plastic wrap around the rod as a thin former) using the same 0.63 mm wire. The diameter of the rod and wrapping was 10.5 mm giving a coil diameter of 11.18 mm and a coil length of 13.6 mm. The calculated apparent permeability was 22.07 giving a calculated inductance of 55.61 uH, and the measured inductance was 55.4 uH.

By entering the measured values into the CrystalRadioDesign program you can quickly determine the appropriate equation for a rod, allowing you to calculate the apparent permeability for any coil you wish to wind using the rod as a core.

22.7 INDUCTANCE CALCULATION

For ferrite rods, use the apparent permeability value to determine the inductance. Rewriting Equation 11.6 to include the apparent permeability, we have

$$L_{ROD} = \frac{\mu_a \mu_0 \pi D^2 N^2}{4\ell} K$$

Inductance of a ferrite rod coil 22.13

For coils wound directly onto a toroidal core, you can use the given initial or effective permeability. The inductance of a coil wound on a toroid is

$$L_{TOROID} = \frac{\mu_e \mu_0 N^2}{2\pi} h \ln\left(\frac{d_o}{d_i}\right)$$

Inductance of a ferrite toroid coil 22.14

The inductance is in Henrys with the height h, the outside diameter d_o, and the inside diameter d_{in} in metres. See Figure 22.6.

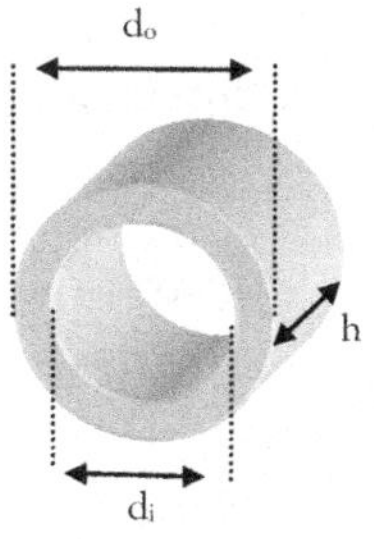

Figure 22.6
Ferrite toroid core dimensions.

Inductance Due to Coil's Relative Position

The inductance of a ferrite rod coil depends on the coil's relative position to the rod. The maximum inductance occurs when the coil is centred along the rod. When the coil is moved off-centre, to the left or right, the inductance decreases. The further the coil moves to the outer edge of the rod, the lower the inductance. This decrease in inductance is due to a reduction in the apparent permeability and is a function of offset distance (the distance from the centre of the rod to the centre of the coil).

If we treat this reduction in inductance as an apparent reduction in rod length, we can modify the apparent permeability equations 22.7 and 22.9 using a modified rod length. Taking several measurements of a coil's inductance at different offset distances, the length of the rod used in equation 22.7 was adjusted until the calculated inductance value equalled the measured inductance. The ratio of the apparent rod length reduction as a function of offset distance is shown in Figure 22.7.

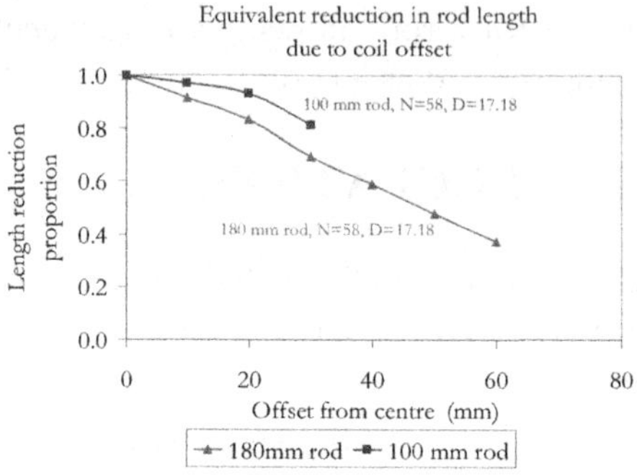

Figure 22.7
Ratio of apparent rod length reduction as a function of coil offset distance.

The reduction of apparent rod length is quite linear. The following is an empirical equation taking into account rod length and offset distance.

$$\ell'_{ROD} = (1 - 59.63\ell_{ROD}D)\ell_{ROD} \qquad \text{Apparent rod length} \qquad 22.15$$

Where ℓ_{ROD} is the rod length in metres, and D is the offset coil distance from the centre of the rod in metres (as measured from the centre of the rod to the centre of the coil). Figure 22.8 shows various graphs of inductance as a function of offset distance for a number of coils wound on 180 mm and 100 mm length rods. The squares are the measured inductance values and the curves are the calculated values.

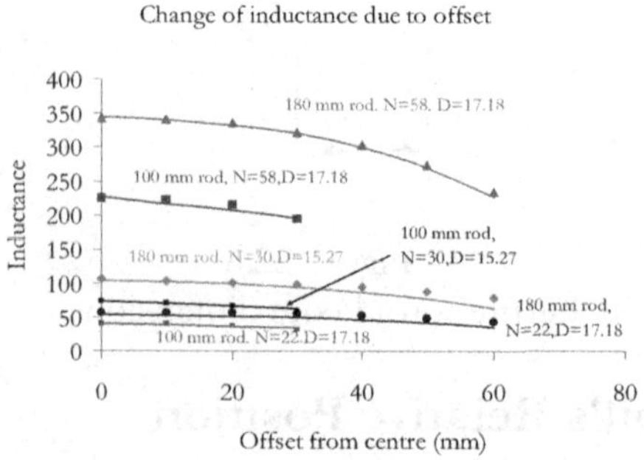

Figure 22.8
Inductance as a function of coil offset distance.

Errors in the calculated inductance increase as the offset distance increases. The calculated inductance values at zero offset are typically within 3%, or less, of the measured value. With the error increasing out to a worst case of 18% for a short coil of 30 turns at 60 mm distance.

Inductance Coefficient

Another property sometimes given in specification sheets is the ***inductance coefficient***, A_L. This specifies the inductance per loop squared, usually in nano-Henrys (10^{-9}). This value assumes you have wound the coil onto the core. It is defined as

$$A_L = \frac{L_{CORE}}{N^2} \qquad \text{Inductance coefficient} \qquad 22.16$$

You can determine this value by counting the number of loops (N) and measuring the inductance of the ferrite core coil. The value of A_L can be used to determine the inductance of a coil.

$$L_{CORE} = A_L N^2 \qquad \text{Inductance of a ferrite core coil} \qquad 22.17$$

where L_{CORE} is the inductance of the coil with the core inserted. Once again, this value is only suitable for a particular core material and geometry. If the A_L is given for a toroid and you use a rod, then the value will be incorrect.

22.8 LOSS FACTOR AND Q

The ***loss factor*** is defined as the ratio of the energy losses to the magnetic energy stored in the material. That is, it is the inverse of the Q.

$$\tan\delta = \frac{R_M}{2\pi f L} = \frac{1}{Q_{core}} \qquad \text{Loss factor} \qquad 22.18$$

It takes into account all the losses of the material at a specific frequency and includes the hysteresis and eddy current losses. If we divide the loss factor by the initial permeability, we have the loss per unit of relative permeability. This is **called the *relative loss factor*.**

$$\boxed{\frac{\tan\delta}{\mu_i} = \frac{R_M}{2\pi f L \mu_i} = \frac{\mu_S''}{(\mu_S')^2}} \qquad \text{Relative loss factor} \qquad 22.19$$

The value of μ'' is the imaginary part of the complex value of the permeability, and μ' is the real part. If a graph of the complex permeability is given as a function of frequency, you can calculate the relative loss factor for a given frequency.

The relative loss factor is sometimes given in specification sheets as a single value for a toroidal core, without a gap, at a specific frequency. To calculate the loss factor for a coil wound on a toroidal core, multiply the relative loss factor by the initial permeability or the effective permeability if a small air gap exists. Also, as the relative loss factor is given for a particular frequency, we must add a frequency modification factor. The actual increase depends on the ferrite material, but core losses tend to increase linearly with increasing frequency for ferrite materials designed for high-frequency use. This means the Q of the material decreases as the frequency increases. Accounting for the frequency, the Q of the material becomes

$$Q_{M(TOROID)} = \frac{1}{((\tan\delta)/\mu_i)\mu_e}\left(\frac{f_\delta}{f}\right) \qquad 22.20$$

where f_δ is the frequency at which the loss factor was measured, and f is the frequency we are interested in.

For rods, once again, the situation is a little more complicated. We need to take into account the gap between the coil and the rod as a source of flux leakage and also the length of the coil compared to the rod length. That is, the relative loss factor needs to be given in terms of apparent permeability. We can determine the relative loss factor by measuring the Q of a test coil using the ferrite rod. If you do not have the equipment to measure the Q of a coil, then modify the given relative loss factor in the specification sheet by multiplying the relative loss factor by the given initial permeability μ_i and then dividing through by the apparent permeability. That is

$$(\tan\delta)/\mu_{a0} = \frac{\tan\delta}{\mu_i}\frac{\mu_i}{\mu_{a0}} \qquad 22.21$$

If the Q of a ferrite rod coil is measured, we can determine the loss due to the core as follows. First, calculate the Q of the coil (due to Ohmic losses only) by dividing the measured Q by the self-resonant coefficient α.

$$Q_{coil} = \frac{Q_{measured}}{\alpha} \qquad 22.22$$

Then determine the Ohmic losses due to the core as follows,

$$R_{M0} = \frac{2\pi f L_{coil}}{Q_{coil}} - R_{coil} \qquad 22.23$$

where L_{coil} is the inductance of the coil with the rod inserted. R_{coil} is the total of all the Ohmic wire and former dielectric losses of the coil. Finally, determine the relative loss factor in terms of apparent permeability.

$$(\tan\delta)/\mu_{a0} = \frac{R_{M0}}{\mu_{a0} 2\pi f L_{coil}} \qquad 22.24$$

Where μ_{a0} is the measured apparent permeability for the coil used in the Q measurement. We can now use the relative loss factor to determine the core losses when calculating the Q of any coil wound on this particular rod.

$$Q_M = \frac{1}{((\tan\delta)/\mu_{a0})\mu_a}\left(\frac{f_\delta}{f}\right) \qquad 22.25$$

Where μ_a is the apparent permeability of the coil, f_δ is the frequency at which the loss factor was measured, and f is the frequency of interest. The core loss is then calculated using

$$R_M = \frac{2\pi f L_{coil}}{Q_M} \qquad \text{Equivalent Ohmic core loss value} \qquad 22.26$$

And the effective Q of the coil with a ferrite core becomes

$$Q_{effective} = \frac{2\pi f L_{coil}}{R_M + R_{coil}}\alpha \qquad \text{Effective Q of a ferrite core coil} \qquad 22.27$$

22.9 THE COUPLING COEFFICIENT

In Chapter 12, we developed an equation to determine the mutual inductance between two air-core cylindrical coils. Simply substituting the rod's permeability into this equation will result in a very low value for mutual inductance. Not only is the permeability much higher for ferrite rod coils, but the magnetic flux lines tend to be trapped in the ferrite material resulting in less flux leakage. As we have adopted an empirical method for determining the coil's apparent permeability, we will adopt an empirical method to determine the coupling coefficient.

Figure 22.9 graphs the measured coupling coefficient as a function of the separation distance between two coils ($N_P = 22$, and $N_S = 58$) on a 180 mm rod of type S1 material (with a diameter of 9.68 mm).

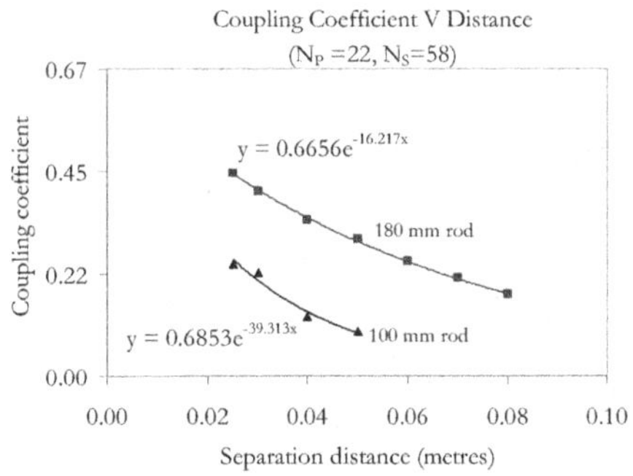

Figure 22.9
Coupling coefficient as a function of separation distance.

The coupling coefficient is a shallow exponential curve as a function of separation distance, taking the form

$$\boxed{k = k_0 e^{-aD}} \qquad \text{Coupling coefficient between ferrite rod coils} \qquad 22.28$$

where k_0 is the coupling coefficient between the coils when the separation distance is zero, and D is the separation distance in metres. The exponent "a" determines the rate at which the coupling coefficient reduces with distance. Note that Equation 12.28 is only valid when both the coils are situated on the rod, that is $\ell_P + \ell_S + D < \ell_{core}$.

Firstly let's examine the initial coupling coefficient k_0. The magnetic flux density at the centre of a single loop (D is zero) is given by Equation 12.11, and we get

$$B = \frac{\mu I A}{2\pi(r^2 + D^2)^{3/2}}$$

$$= \frac{\mu I \pi r^2}{2\pi(r^2)^{3/2}}$$

$$= \frac{\mu I}{2r}$$

Assuming the flux density is relatively constant throughout the core of the coil, then the flux through the coil is

$$\begin{aligned}\Phi_{coil} &= B_{coil}A_{coil}\\ &= \frac{\mu I}{2r_{coil}}(\pi r_{coil}^2)\\ &= \frac{\mu I \pi r_{coil}}{2}\end{aligned}$$

This is close to being true when the core is a ferrite material. The flux through the ferrite rod is

$$\begin{aligned}\Phi_{rod} &= B_{coil}A_{rod}\\ &= \frac{\mu I}{2r_{coil}}(\pi r_{rod}^2)\\ &= \frac{\mu I \pi r_{rod}^2}{2r_{coil}}\end{aligned}$$

The proportion of the total flux through the ferrite rod is thus

$$\frac{\Phi_{rod}}{\Phi_{coil}} = \frac{\mu I \pi r_{rod}^2}{2r_{coil}} \frac{2}{\mu I \pi r_{coil}}$$

And we can write the portion of the initial coupling coefficient due to the core as

$$\boxed{k_0^{core} = \left(\frac{r_{rod}}{r_{coil}}\right)^2 = \left(\frac{d_{rod}}{d_{coil}}\right)^2}$$

When the primary and secondary coils are touching, the proportion of the flux in the rod will be coupled between the two coils without much loss. The flux proportion through the air (the flux leakage) is

$$1 - k_0^{core}$$

The flux leakage reduction, from Equation 12.11, is proportional to the cube of the distance. The distance between the loops of the primary coil and the loops of the secondary coil means there is a reduction in the overall contribution of the flux leakage. So the contribution for the flux leakage term, on the initial coupling coefficient, will also be cubed. We can write the initial coupling coefficient between two ferrite rod coils as

$$\boxed{k_0 = k_0^{core} + \left(1 - k_0^{core}\right)^3} \qquad \text{Initial coupling coefficient} \qquad 22.29$$

For the coils used in Figure 22.9,

$$\begin{aligned}k_0^{core} &= \left(\frac{d_{rod}}{d_{coil}}\right)^2\\ &= \left(\frac{9.68}{17.18}\right)^2\\ &= 0.3175\end{aligned}$$

And k_0 is

$$k_0 = k_0^{core} + \left(1 - k_0^{core}\right)^3$$
$$= 0.3175 + (0.6825)^3$$
$$= 0.6354$$

This is within 4.5% of the value for the coefficient in the best-fit exponential given in Figure 22.9 for the 180 mm rod, and 7.3% for the 100 mm rod.

Now examining the exponent "a". From Figure 22.9 we can see that this value is quite different for the different rod lengths. This exponent determines the rate at which the coupling coefficient declines as distance increases. The smaller the gap between the coil and the rod, the greater the coupling coefficient, as more of the flux is now contained within the core. We would expect the rate at which the coupling coefficient declines with distance to decrease. Therefore, the exponent is related to k_0^{core}. Also, the same coils were used for both measurements, so the difference in the exponent is related to the rod geometry. Using these factors to determine the exponent, an empirical formula was arrived at, and

$$a = \frac{1}{k_0^{core\left[\frac{1}{\beta}\left(1-\frac{D_{rod}}{\ell_{rod}}\right)^2\right]}} \quad \text{Coupling exponent} \quad 22.30$$

Where β is the rod length exponent. For the 180 mm rod, we get

$$a = \frac{1}{0.3175^{\left[\frac{1}{0.3682}\left(1-\frac{9.68}{180}\right)^2\right]}}$$
$$= \frac{1}{0.3175^{2.4317}}$$
$$= 16.28$$

This is within 0.4% of the best-fit curve. For the 100 mm rod, we get the same k_0 and an exponent of 41.20, or a 4.8% difference between the calculated and best-fit exponent. The calculated coupling coefficient errors range from less than 1% up to about 20% of the measured coupling coefficient. If you require greater accuracy, then you will need to determine your own best-fit coupling coefficient equation from your data.

Rearranging Equation 12.18, the mutual inductance can be calculated using

$$M = k\sqrt{L_P L_S} \quad \text{Mutual inductance between ferrite rod coils} \quad 22.31$$

where L_P is the primary coil's inductance due to its position on the core, and L_S is the secondary coil's inductance due to its position on the core, as calculated using Equation 22.13.

This mutual inductance value is used when calculating the mutual inductance loss between the coils. And along with the core loss, it is used when calculating the Q of the radio, with the inductive reactance value given by the inductance value of the coil due to its position on the rod.

23

Building a Ferrite Rod Crystal Radio

23.1 MAKING FERRITE COILS

Apart from the shortcomings of a ferrite rod radio, they are easier to build and can be fun to experiment with. In this Chapter, we will build a ferrite core crystal radio with tapped primary and secondary coils, wound on separate former sections but sharing a single ferrite rod core. The coils will be wound on the PEX former using an S1 ferrite rod as the core.

Coil Analysis

The specifications for the primary coil are given in Table 23.1, and the specifications for the

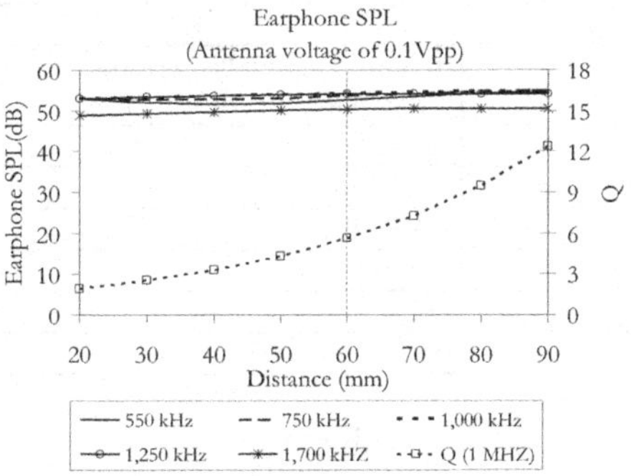

Figure 23.1
SPL as a function of separation distance.

secondary are given in Table 23.2. Tap analysis was performed using the standard antenna and transmission line, with the results shown in Figure 23.1. The values used for the analysis of the S1 ferrite core rod are given in Table 22.2.

For the tap analysis, a distance of 60 mm was chosen as all the SPL levels are less than 2 dB from the maximum SPL. The SPL as a function of loop number is shown in Figures 23.2 and 23.3. Note that the SPL level for most of the frequencies increases with tap number, with the maximum SPL achieved at the last loop of the coil. However, choosing lower loop numbers increases the Q with only a small reduction in SPL. As always with crystal radios, increasing the separation distance between the primary and secondary coils will increase the Q and aid in selectivity. However, due to the nature of the ferrite core, the boost in Q is not as great as it is for air-core coils.

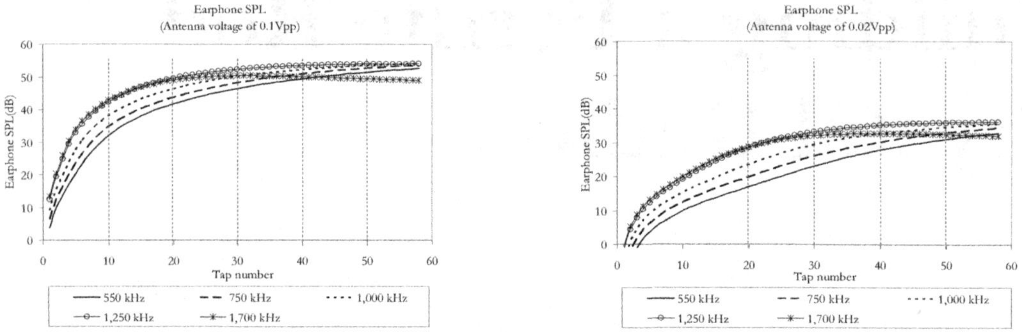

Figure 23.2
SPL as a function of winding number at 60 mm distance.

Description	Calculated values
Coil name	PC-BUILD-2
Coil inductance	55 uH
Coil diameter	17.18 mm
Coil length	14.96 mm
Number of windings	22
Taps at loop number	6, 9, 12
Pitch	0.68 mm
Wire	0.63mm enamelled copper wire (core D=0.63mm, OD=0.68mm)
Former material	PEX
Former outside diameter	16.5 mm
Former inside diameter	11.0 mm
Former length	30 mm
Rod material	S1
Rod dimensions	9.68 x 180.00 mm

Table 23.1
The primary ferrite coil specifications.

Description	Calculated values
Coil name	SC-BUILD-4
Coil inductance	344 uH
Coil diameter	17.18 mm
Coil length	39.44 mm
Number of windings	58
Taps at loop number	10, 20, 30, 40, 50
Pitch	0.68 mm
Wire	0.63mm enamelled copper wire (core D=0.63mm, OD=0.68mm)
Former material	PEX
Former outside diameter	16.5 mm
Former inside diameter	11.0 mm
Former length	60 mm
Rod material	S1
Rod dimensions	9.68 x 180.00 mm
Q effective	125 (measured 119)

Table 23.2
The secondary ferrite coil specifications.

Winding the Primary Coil

The specifications for the primary coil are given in Table 23.1. We will add taps at loops 6, 9 and 12. Cut a 30 mm length of former and drill two 1 mm holes 7 mm from both ends. Feed the wire through the starting hole and move the wire to one side. As the former is so small, wind the coil without any backstop. Use a small object like a 10 mm length of thin wooden skewer or toothpick to wind the taps around. Try and stagger the tap positions. When the coil is wound, feed the end lead wire through the end hole and run glue around both ends of the coil and the former to keep the loops in place. When the glue is dry, pull the lead wires out from their holes and add more glue to the ends. See Figure 23.3.

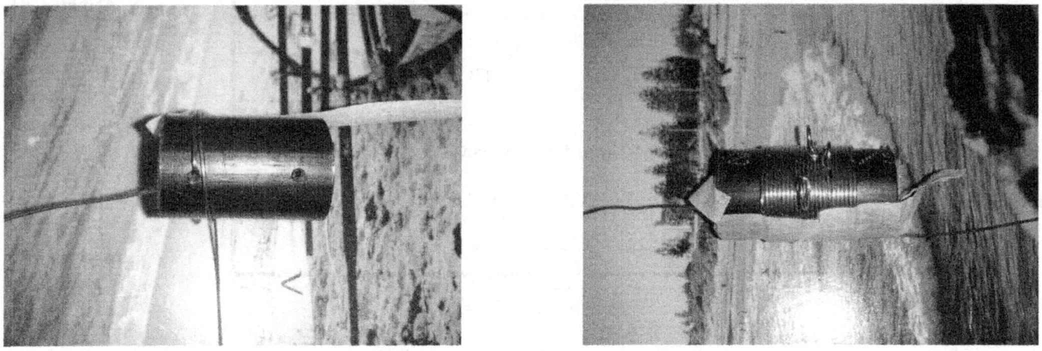

Figure 23.3
Winding the primary coil.

Winding the Secondary Coil

Cut a 60 mm length of PEX former and drill two 1 mm holes 10 mm from the ends. Use the same technique to wind the coil as for the primary coil with taps at windings 10, 20, 30, 40 and 50. Note the primary coil inductance is a bit higher than for the air-core oils. When the secondary is moved to the side of the ferrite rod, the inductance will drop.

23.2 MAKING THE CHASSIS

The chassis for this crystal radio is somewhat smaller than our previous chassis build. The base is made from 7 mm plywood. Drill and cut the base using Figure 23.4 as your guide. The coil stands are made from 30 x 12 mm pine (Figure 23.5a), and the faceplate is three-ply (Figure 23.5b).

Drill the holes using a 2 mm drill bit. Only the coil stands, faceplate connector and faceplate are screwed together. Attach the coil stands to the base using 3 x 12 mm self-tapping screws from underneath the base and into the stands. Attach the faceplate connector using 3 x 20 mm screws to the base. Screw the faceplate onto the faceplate connector using 3 x 12 mm screws. The terminal blocks are glued to the base. I use hot glue as it is quick and easy. It makes a good bond but not so strong that you can't remove them if necessary. Finally, cut two 10 mm lengths of the former to be inserted into the coil stand former holes. Drill these holes to the outside diameter of the former you are using. See Figure 23.6.

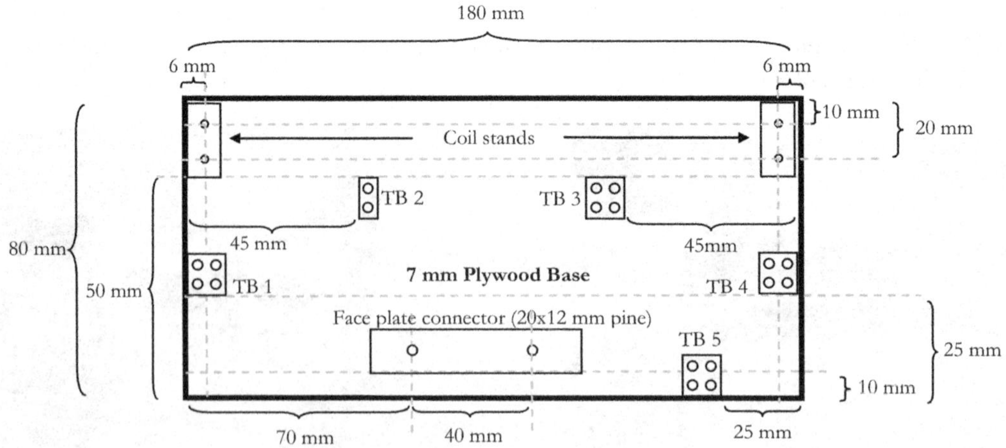

Figure 23.4
The base.

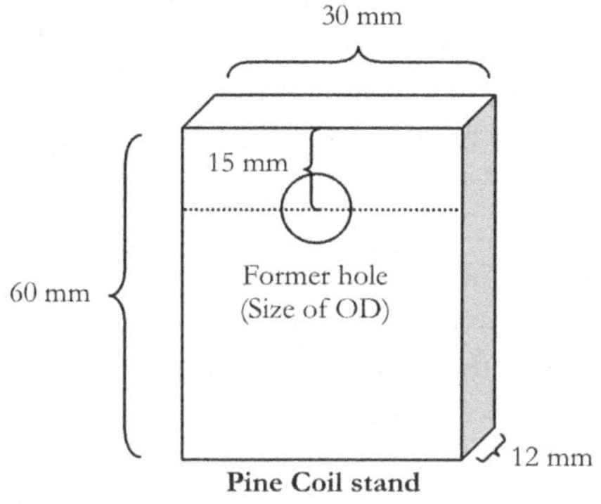

(a)

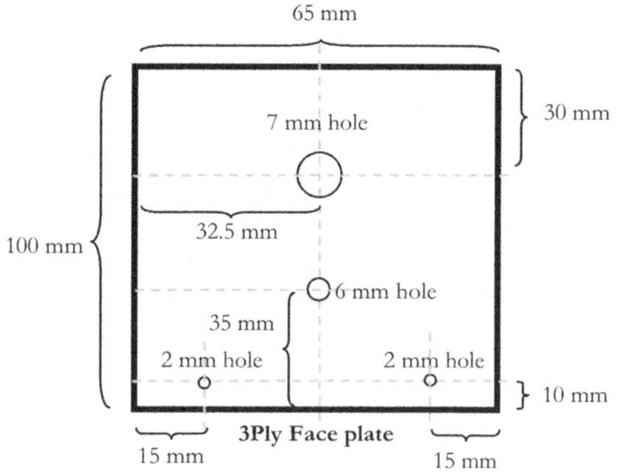

(b)

Figure 23.5
Coil stands and faceplate.

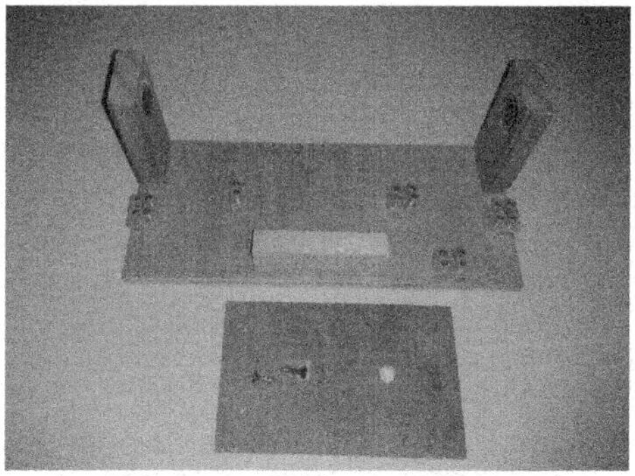

(a)

(b)

Figure 23.6
The chassis.

23.3 WIRING THE CRYSTAL RADIO

Follow the wiring setup shown in Figure 23.7.

Wiring steps:

1. Cut the end leads of the primary and secondary coil down to 30 mm and file away the enamel coating. Tin the taps on the primary and secondary coils with solder to ensure a good connection for the alligator clips. Also, tin all the wires going into the terminal blocks to give the terminal block screw something to bite against.

2. The primary and secondary taps are selected using an alligator clip. Cut an alligator clip hook-up lead in half and tin the exposed ends. Cut three 80 mm lengths of flexible hook-up wire and solder them onto the coil leads. Screw the primary coil's right side flexible lead wire into terminal block 2 and the flexible wires from the secondary into terminal block 3. Cut an 80 mm length of wire and screw one end into terminal block 1 and the other side into terminal block 2.

3. Cut two 110 mm lengths of hook-up wire and solder them onto the tuning capacitor. Cut a 50 mm length of wire and twist it together with the earth wire (centre tap) of the tuning capacitor. Screw the tuning capacitor earth wire into the left side of terminal block 3. Screw the other tuning capacitor wire into the right side of terminal block 3.

4. Cut an 80 mm length of hook-up wire and solder it onto the earth terminal of the earphone socket. Cut a 110 mm length of wire for the positive side of the earphone socket and solder it in place. Connect the earphone earth wire and the earth wire coming from terminal block 3 and screw them into the left side of terminal block 5. Screw the positive side earphone wire into the right side of terminal block 5. Cut a 50 mm length of wire and screw one side into terminal block 4 and the other into the right side of terminal block 5.

Now that the wiring is complete, screw the earphone socket into the faceplate and hot glue the tuning capacitor onto the faceplate. Slide the ferrite rod through the coil stand, through the secondary and primary coils and finally through the other coil stand. Now screw the faceplate onto the faceplate-to-base connector. Attach a diode to terminal block 4 and a resistor across terminal block 5. Plug your earphone into the earphone socket. Screw the earth and antenna wires to terminal block 1, and you are ready to go. The tuning capacitor I used ranges from 15 to 270 pF. A flat knob was screwed onto the tuning capacitor; see Figure 23.8.

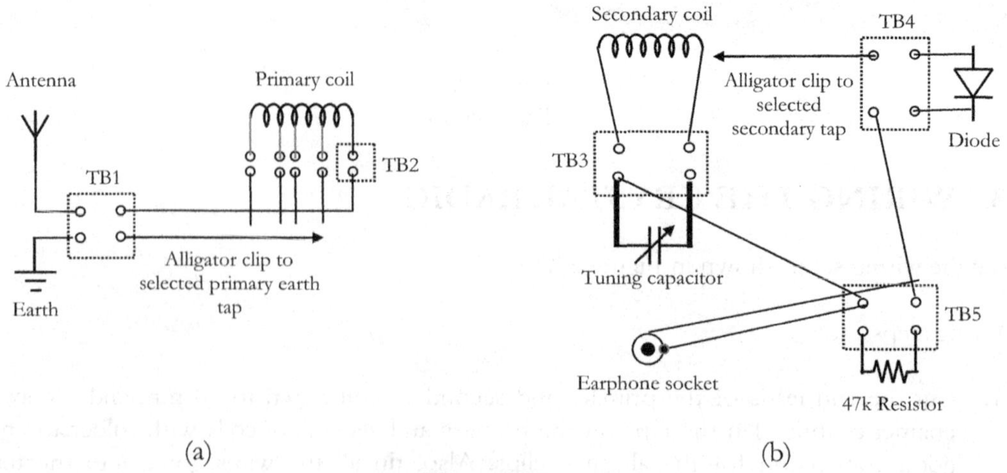

Figure 23.7
Wiring layout.

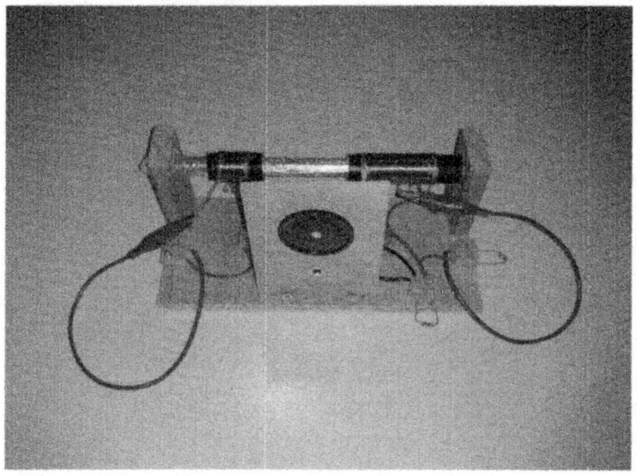

Figure 23.8
The completed ferrite rod crystal radio.

23.4 THE CRYSTAL RADIO PERFORMANCE

Start by separating both coils to the maximum separation distance and select tap four on the secondary for your demodulator feed. Experiment with the separation distance and demodulator tap selection to determine the settings that give you the best volume with the least amount of interference.

The performance of this radio, when attached to the standard antenna, is quite good. The station volumes are strong and quite audible. For medium to high power stations, the maximum volume levels are similar to high Q radios. The major drawback of this crystal radio is its lack of Q, so the ability to separate close stations is not as good as air-core radios. You may have to sacrifice a lot more volume to separate stations than you would for air-core radios. Also, the volume of distant or low-power stations for the radio is not as good as air-core radios. However, if your local stations are not that close in frequency, and depending on their individual output power levels, this radio can work very well and may be quite adequate for your needs.

24

Accessories and Modifications

24.1 CAT'S WHISKER DETECTOR

Before vacuum tubes or solid-state diodes became widely available, early radio enthusiasts used naturally occurring materials such as galena (Figure 24.1a) to detect signals. Galena is a naturally occurring ore of lead (lead sulphide). Due to random formations of impurities in the ore, when a thin wire is brought into contact with the surface, the junction between the wire and the galena may result in the formation of a potential barrier, in the same way as a barrier potential is formed in a Schottky diode. The galena crystal and wire act like a diode, allowing an AC signal to pass easily in one direction while presenting a high resistance in the other direction. This thin wire resembles the appearance of a cat's whisker, and detectors of this type are called ***cat's whisker*** detectors or diodes.

Common Name	Composition	Chemical formula
Galena	Lead sulphide	PbS
Iron pyrite (fool's gold)	Iron sulphide	F_eS_2
Zincite	Zinc oxide	ZnO
Chalcopyrite	Copper iron sulphide	$C_uF_eS_2$
Moissanite (Carborundum)	Silicon carbide	S_iC
Rusty razor blades	Rust and steel	

Table 24.1
Materials to try as a cat's whisker diode.

There are a number of materials that can be used to form cat's whisker diodes. Table 24.1 lists some of these materials. Iron pyrite (iron sulphide – see Figure 24.1b), sometimes called fool's gold, is a commonly available ore that makes for a good cat's whisker detector. The cat's whisker does not have to be wire; graphite from pencils or rusty nails can be used. Soldiers used rusty razor blades and nails during WWII to make crystal radios called foxhole radios. Also, not every surface location on the material will form the necessary junction to act as a suitable semiconductor. It is random in nature, and some locations are better than others.

Typically a cat's whisker detector consisted of a crystal of galena potted in a brass pot using Woods metal (an alloy of bismuth, lead, tin and cadmium). Woods metal has a low melting point of around 70 °C and acts as a form of solder. Rose metal, another alloy of bismuth, lead and tin, with a higher melting point of about 98 °C, can also be used to pot crystals. An arm held in a ball socket that could pivot in three dimensions (up, down, sideways, back and forth) was used to move the cat's whisker over the surface of the galena crystal to find the best contact point.

(a) Galena.

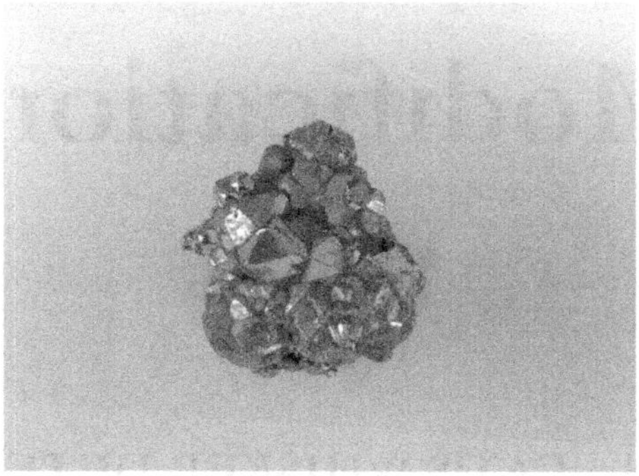

(b) Iron Pyrite.

Figure 24.1

Adding a Cat's Whisker Detector

It's a lot of fun to experiment with different objects to see if they can be used to detect signals. Therefore we will make our own cat's whisker detector. As the objects to test will vary in size, a metal plate will be used to rest the object on and to make electrical contact. You will want to experiment with different materials to use as the cat's whisker, so an alligator clip will serve to hold the whisker while allowing for easy replacement.

The base electrode is made by removing three sections of a flat "make-a-bracket' strip that can be readily purchased from any hardware store. These bracket strips are made from a series of sections that can easily be removed by bending the section back and forth until it snaps off. Select a length of three sections and snap them off. File the edges to remove any burrs.

Position the plate to the right edge of the crystal radio base and place a 1/8" washer over the top hole of the bracket and screw in a 4Gx9 mm sheet metal self-tapping screw. Do not drive the screw all the way in just yet. Cut a 60 mm length of 0.8 mm diameter enamel copper wire and strip the enamel from both ends. Bend one end of the wire so it wraps around the screw between the metal base and the washer. Now drive the screw all the way in so the wire forms a good electrical connection.

An alligator clip will be used to hold the cat's whisker. Buy a cheap alligator test lead and cut it in two, with one length being 50 mm long. Strip the insulation from the end to expose the bare wires. Cut a 160 mm length of 0.8 mm copper wire. Remove the enamel about 50 mm from one end of the copper wire and solder the end of the alligator test lead to this point, see Figure 24.2b. Twist some of the enamelled wire around the test lead all the way up to the clip to provide some support. Now remove the enamel coating from the other end of the copper wire.

Place two washers over a metal screw and drive it partly into the base about 10 mm from the end of the metal base. Make sure it is not touching the metal base. Wrap the copper wire coming from the clip around the screw and in between the two washers and drive the screw all the way in.

Prepare some cat's whiskers to test. Cut an 80 mm length of 0.5 mm copper wire and strip the enamel from both ends. Make a small spring in the middle of the wire by looping the wire around a nail three or four times. Cat's whiskers can be made from old steel guitar strings as well. You can also try steel pins and pencil graphite.

To use the cat's whisker detector, first tune into a loud station with the normal diode in place. Then remove the diode from the terminal blocks and insert the wires from the cat's whisker detector. Insert the cat's whisker wire into the alligator clip and place an object, such as a galena crystal, onto the base plate. Move the cat's whisker slowly around the object's surface to see if a signal is detected. Use the stiffness of the wire to keep the whisker in place. Bend the alligator clip wire as needed to aid in positioning. Sometimes it's best to rest the cat's whisker on top of the object and then move the object itself. Use both techniques and experiment. Also, flip the object around so that a different side of the surface is touching the plate. Try using different cat's whiskers on the same object. I have found that graphite seems to work best on metal objects like razor blades.

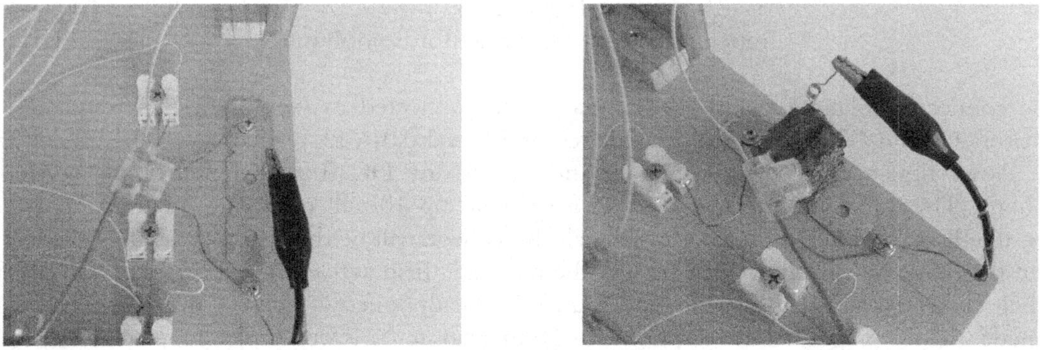

(a) (b)

Figure 24.2
The cat's whisker detector.

24.2 USING LOW-IMPEDANCE HEADPHONES

Typically, crystal radios use piezoelectric earphones due to their high impedance and sensitivity. However, it is possible to use low-impedance headphones in a crystal set by impedance matching. An autotransformer can be used to present a high-impedance load to the secondary coil while presenting a matching low impedance to the earphone.

The circuit shown in Figure 24.3 is an impedance-matching circuit for headphones with various audio impedances. The parallel RC circuit in series with the diode is called a ***benny***; named after Ben Tongue, a retired electrical engineer credited with first adding this to the transformer impedance matching circuit. The benny reduces audio distortion that may be encountered with strong stations.

This circuit uses a KPB-02 audio transformer specifically designed for crystal radios and can be purchased online. The inductance values of each tap of the KPB-02 transformer are given in Table 24.2. The impedance values shown for the taps in Figure 24.4 are for low audio frequencies of around 40 Hz. Other audio transformers can be used, provided they present a high impedance to the secondary coil. A popular audio transformer often used is the Bogen T-725.

First, let us consider the circuit at an RF frequency of 1 MHz. Say we have the output from the

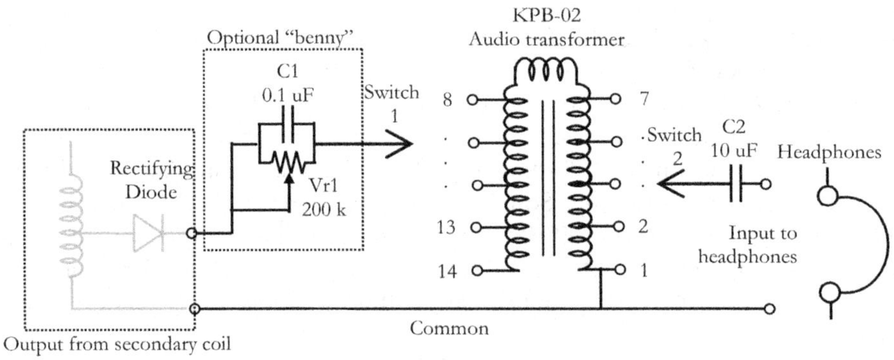

Figure 24.3
Impedance matching circuit for headphones.

diode connected to tap 10, and a 16 Ω earphone connected to tap Bl. At this frequency, the capacitors C1 and C2 will have an impedance of 1.6 and 0.016 Ω, respectively. The capacitor C2 only allows time-varying signals to pass and blocks any DC from being dropped across the earphone. The impedance across the transformer at tap 10 will be about 240 MΩ. Therefore, when the diode is forward biased, virtually all the voltage will be dropped across the transformer. When the diode becomes reversed biased, the magnetic field across the transformer will collapse, sending current through the capacitor C2 and the headphones (it tries to maintain the existing magnetic field strength). The earphone is connected to a lower tap value, and due to the transformer action, the voltage across the output tap is less than the primary tap, so the current through the output tap will be larger.

For a given voltage, five time constants are required for the voltage to drop to zero. The time constant for an inductor is given by

$$\tau = \frac{L}{R} \qquad 24.1$$

And using the transformer values given in Table 24.2, one time-constant at the output tap through the earphone (treating the C2 as a short) for the given tap configuration will be about 3.5 milliseconds. This is much slower than the RF period of 1 microsecond, and the voltage will build up across the transformer in a similar way it builds up across a piezoelectric earphone. The capacitor C2 only allows time-varying signals to pass and blocks any DC from being dropped across the earphone. As the voltage across the transformer rises and falls due to the RF-modulated signal, the discharge signal will follow the edges of the RF signal due to the slow time-constant. That is, it demodulates the signal, and this signal appears as an audio signal across the earphones.

Any charge placed across the Benny (C1 capacitor) will discharge across the parallel resistor and through the transformer, earphone and the reverse bias impedance of the diode. The current passing through the earphones will depend on the impedances of the discharge paths and may have some effect at audio frequencies. However, if very little charge is placed across the C1, the Benny will have little or no effect on the audio output. Given the very high impedance of this transformer, almost no charge will be placed across C1, and the Benny will have no noticeable effect.

Tap	Inductance (mH)	DC resistance Ω
1	0.00	0.00
R	14.20	0.63
Y	27.70	1.20
Bl	56.40	2.04
W	117.90	3.87
2	243.00	6.49
3	655.00	11.31
4	1,173.00	23.35
5	2,100.00	36.11
6	2,990.00	51.20
7	5,760.00	78.44
8	9,100.00	105.40
9	19,400.00	166.30
10	37,800.00	250.20
11	76,300.00	474.00
12	155,000.00	808.40
13	389,900.00	2,033.00
14	779,000.00	3,537.00

Table 24.2
KPB-02 Autotransformer measured tap inductance values.

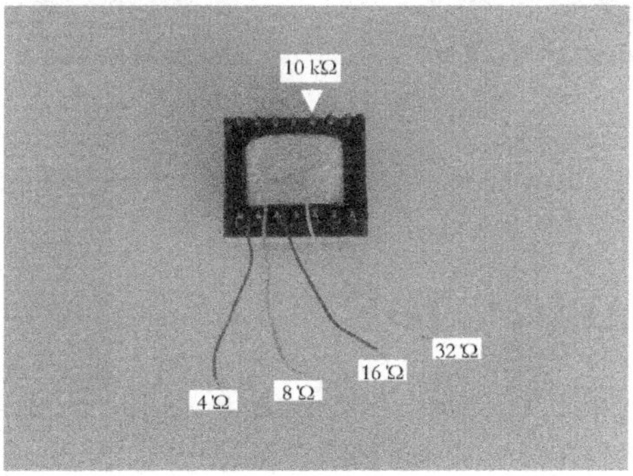

Figure 24.4
The KPB-02 audio-matching transformer.

Adding Audio Impedance Matching

To connect the audio transformer, remove the hook-up wire going from terminal block 4 to terminal block 5. This is the output from the diode. Connect a hook-up wire from the diode (terminal block 4) to the input tap of the transformer (or the Benny if you are using one). In my case, the input was to tap 10 (10 kΩ audio impedance) of the transformer. Remove any resistor you have across terminal block 5. Connect the earth (the left side of terminal block 5) to the earth of the transformer (tap 1), and connect the output tap from the transformer (the white wire, in my case) to the right side of terminal block 5. See Figure 24.5. Terminal block 5 is connected across the earphone socket on the front panel, so we can use this socket to plug in the earphone.

If your headphone plug is the same diameter as the earphone socket (3.5 mm), then plug your headphones into the earphone socket. Generally, stereo sockets will work in a mono socket. If the headphone plug is too large or does not work in the mono-socket, use a stereo-to-mono plug converter. Make sure the stereo-to-mono converter has the proper input and output sizes. I bought a cheap set of stereo earphones that I plugged into a stereo-to-mono converter plug. I then plugged this into the crystal radio's mono socket.

For my setup, the loudest volume was achieved by selecting tap 10 (10 kΩ audio impedance) as the input from the crystal radio, with the headphones connected to the white tap (32 Ω audio). Neither the Benny nor the C2 capacitor was required. Initially, I tried the benny, but there was a slight increase in volume when the benny was removed. Overall, however, the volume was noticeably lower than the standard piezoelectric earphone setup. Using a transformer introduces extra wire and ferrite core losses. Also, in-ear headphones will sound louder than over-the-ear headphones.

Figure 24.5
Wiring the low-impedance earphone matching circuit.

24.3 POWER AMPLIFICATION

The usefulness of a crystal radio can be increased by adding audio amplification. An audio amplifier allows you to listen to the radio while you go about your business, just like a regular radio. Also, if your hearing is not as good as it used to be, or you wish to use low-impedance headphones instead of the piezoelectric earphone, you can use an audio amplifier.

A simple, cheap, but effective audio amplifier can be made using an LM386 op-amp chip and a few extra external parts. The parts list for this amplifier is given in Table 24.3. The LM386 has been around for a long time and is still readily available from any electronics parts store. The descriptions of the pin-outs are shown in Figure 24.6. There is a notch on the case of the chip. Use this notch to orient the chip to determine the pin numbers.

Material	Qty	Material	Qty
LM386N-1 op-amp	1	Mono headphone socket	1
Veroboard	1	8 Ω 1-watt speaker (OD 66 mm)	1
100 uF electrolytic capacitors	2	Terminal blocks	4
10 uF electrolytic capacitors	1	Battery holder (6 x AA)	1
250 uF electrolytic capacitor	1	2.5 mm DC panel mount connector	1
0.1 uF polyester capacitors	2	8-pin IC socket	1
50 kΩ pot (log type)	1	3.5 mm mono plug and cable	1
1 kΩ resistor	1	6 x 12 mm circuit board standoffs	4
Light emitting diode (LED)	1	Hook-up wire	
Single pole single throw switch	1		

Table 24.3
Audio amplifier parts list.

Wiring the Amplifier

The circuit diagram of the amplifier is shown in Figure 24.7. The amplifier can be assembled on stripboard (also known as a Veroboard). Veroboards come in different shapes and sizes. The Veroboard used for this project was 60 x 40 mm. The on/off power switch, the LED indicator, the volume potentiometer, the speaker and the headphone socket will be attached to the front panel of an enclosure, see Figure 24.13. The amplifier can be powered by a 9-volt battery (made from 6 x 1.5 volt AA batteries) or an external 9-volt DC power supply capable of drawing 150 mA or more. Using an external power supply with a maximum draw of less than 150 mA may cause the power supply transformer to overheat when the speaker volume is high.

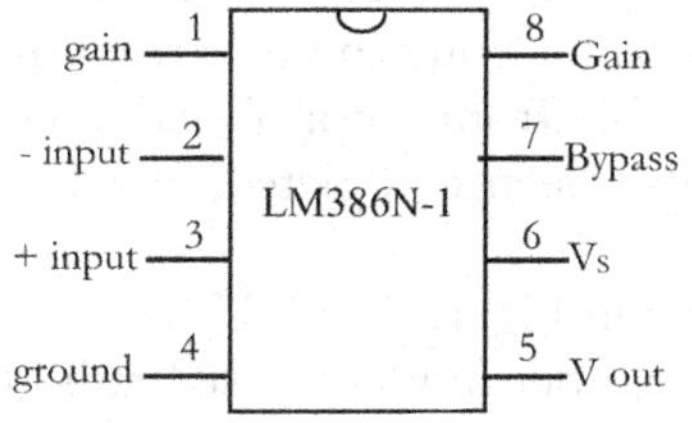

Figure 24.6
LM386 chip.

The DC power socket is centre pin positive. To determine if a power supply is centre pin positive or negative, there should be a diagram printed on the power supply case, as shown in Figure 24.8a.

Before soldering the components onto the Veroboard, drill four 3 mm holes at the corners of the Veroboard to be used for standoffs. Standoffs are used to secure the circuit board to the base of the enclosure. Do not solder the LM386 directly onto the Veroboard. Solder an 8-pin IC socket onto the board near the middle of the board. This prevents possible damage during soldering and allows easy chip removal if you need to replace it. Add components, starting from the smallest to the largest, and lastly, add the hook-up wires to the external components. Use the circuit diagram shown in Figure 24.7 and Figure 24.8b as your guide.

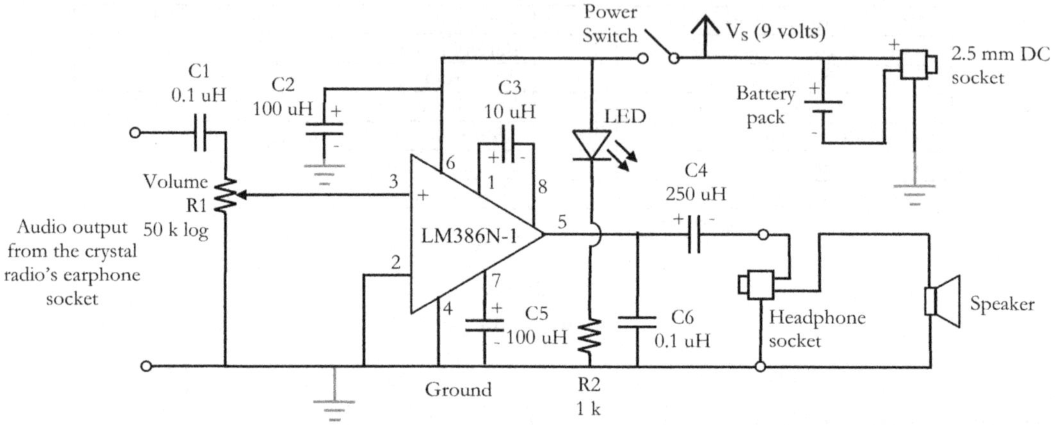

Figure 24.7
LM386 audio amplifier circuit diagram.

The volume control (R1), the LED and R2, the DC socket, the on/off switch, the headphone socket and the speaker are external to the Veroboard, so cut hook-up wire of around 150 mm in length to attach these components to the board. Also, the input from the crystal radio and the power supply is connected to the circuit via terminal blocks. Note the DC socket and the headphone sockets have internal switches in them. The battery is disconnected from the circuit when an external power supply plug is inserted into the DC socket. When a headphone is plugged into the headphone socket, the speaker is disconnected from the circuit.

The amplifier is connected to the audio socket on the crystal radio, so the amplifier requires a 3.5 mm mono audio input plug and cable. You can buy a ready-made cable and cut one end off, or you can make your own. I cut the cable from a non-functioning piezoelectric earphone and used that. When the amplifier circuitry has been completed, attach the standoffs to the circuit board using four 3 x 6 mm metal screws.

The completed amplifier is shown in Figure 24.13. Use this as a guide for attaching the external components. You should test the amplifier in your crystal radio before you install it into a case to make sure it is working.

Building the Enclosure

The case is made from three-ply. Cut and drill the three-ply panels as shown in figures 24.9 to 24.12. Cut four 115 mm lengths of 20 x 12 mm pine to use as braces to connect the panels. Screw them to the top and bottom panels, as shown in Figure 24.10, using 4G x 12 mm self-tapping screws. If you wish to stain or varnish the enclosure, do this now before installing the amplifier circuit.

To enclose the amplifier, attach the variable pot, the earphone socket, the power switch and

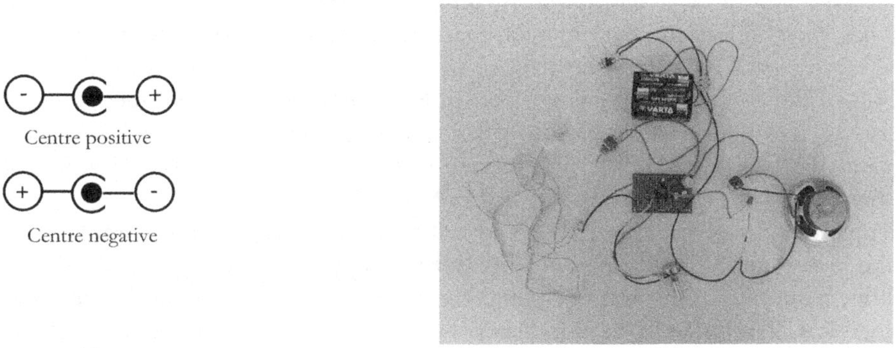

(a) (b)

Figure 24.8
The completed amplifier circuit.

the LED to the front panel. Use hot glue to attach the speaker to the faceplate so that it covers the speaker holes.

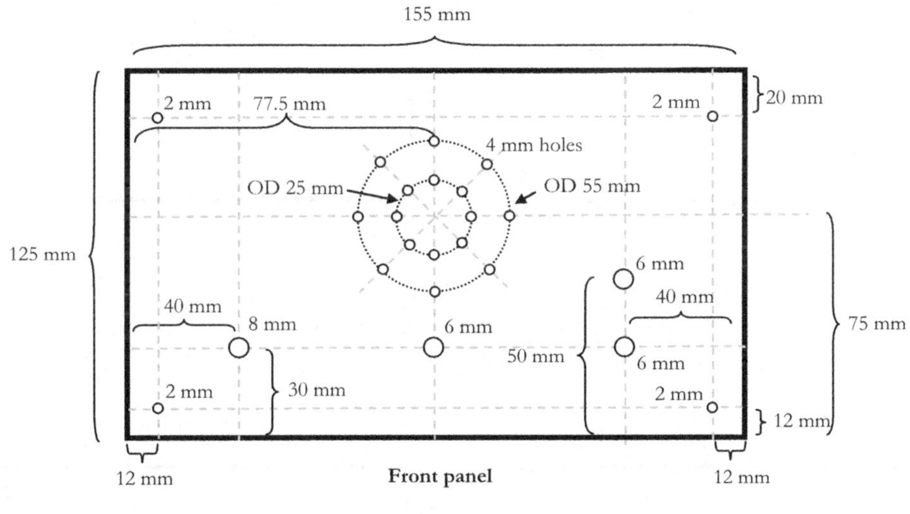

Figure 24.9
Audio amplifier enclosure front panel.

Attach the circuit board to the base. Place the circuit board with the standoffs attached to the base about 20 mm from the front and to the right; see Figures 24.13a and 24.13b. The exact location does not matter just as long as it is clear from the components sticking out from the front panel. Mark the position of the standoffs and drill four 3 mm holes through the base. Attach the circuit board to the base from underneath using four 3 x 6 mm metal screws.

Pass the 3.5 mm audio jack wire through the hole on the left panel and screw it into the audio input terminal block. Screw the terminal blocks to the base using self-tapping 4G x 12 mm screws. Screw the DC power supply terminal block to the base towards the back of the base using self-tapping 4G x 12 mm screws.

To keep the battery pack in place, create a battery compartment by gluing a 70 x 50 mm piece of three-ply onto a 70 x 10 mm strip of styrene. Align the battery compartment up against the right side panel. Place the battery pack (narrow end up) against the compartment so that the end of the battery pack is flush with the back of the enclosure. Remove the battery pack and mark the location of the compartment using a pencil. Run some glue (or hot glue) along the bottom of the battery compartment and glue it into place along the pencil mark just made. Now attach the DC socket to the back panel. Place the battery in the battery compartment and screw the enclosure together using self-tapping 4G x 12 mm screws.

When completed, plug the 3.5 audio jack into your crystal radio's earphone socket. Turn the volume down and switch on the amplifier. Turn the volume up to the level you want. This amplifier is more than capable of filling a room with sound and can be heard quite clearly from other rooms. If you wish to use headphones, first turn down the volume and then plug the headphones into the amplifier's audio socket and then turn up the volume.

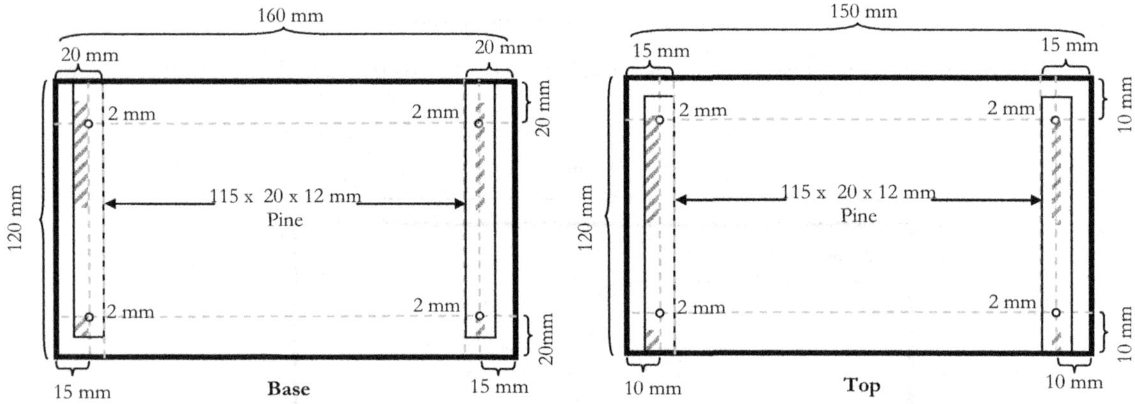

Figure 24.10
Audio amplifier enclosure base and top.

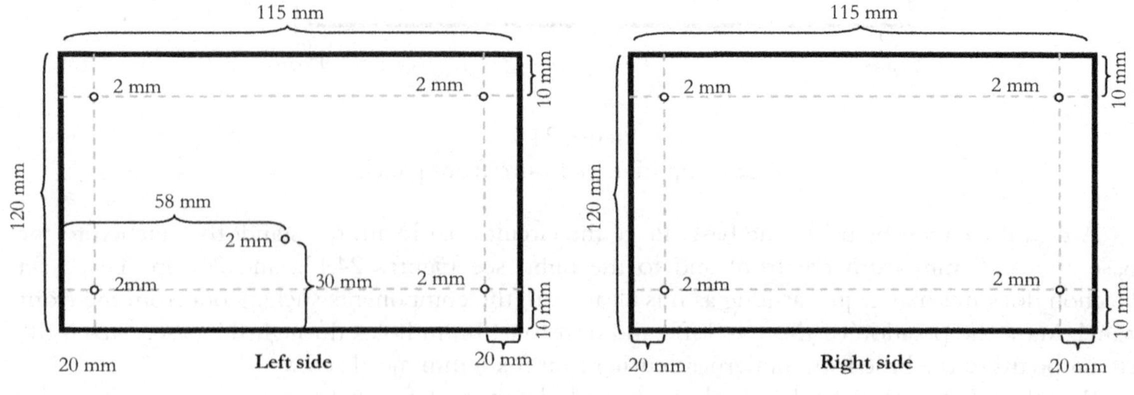

Figure 24.11
Audio amplifier enclosure left and right sides.

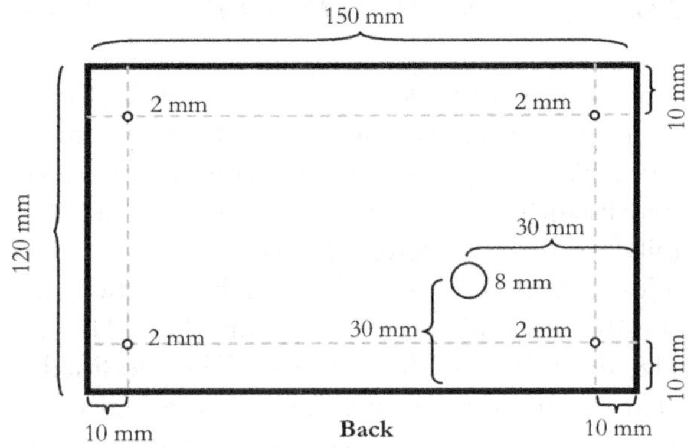

Figure 24.12
Audio amplifier enclosure back.

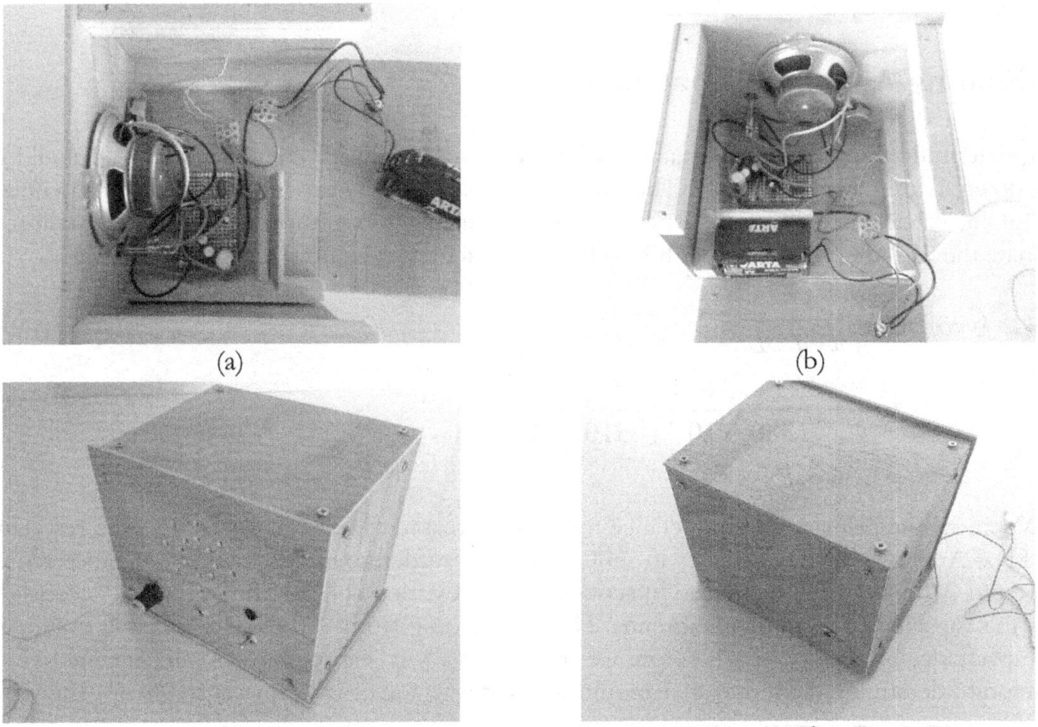

(a) (b)

(c) (d)

Figure 24.13
Completed audio amplifier.

24.4 ADDING INDUCTANCE FOR SHORT ANTENNAS

If you do not have enough room to run a long external antenna, a crystal radio can be made to work with short antennas at a much reduced antenna voltage and audio signal. Although crystal radios using tapped primary coils can be made to work, the double-resonant crystal radio is more suited as it is easier, in this situation, to achieve primary resonance. You could then connect the audio amplifier and use it like a normal powered radio.

I connected the double-resonant crystal radio to a metal shelf as an antenna. The lowest frequency the primary could be made resonant was around 880 kHz. To resonate the primary all the way down to 530 kHz, the primary circuit needs to be modified. You may be tempted to simply add a capacitor across the series tuning capacitor to raise the maximum capacitance. However, this will still be in series with the antenna capacitance, and the maximum capacitance that can be achieved will simply approach the lowest series capacitance value, which is the antenna capacitance in this instance. So no matter how much capacitance you add across the tuning capacitor, the maximum series capacitance will simply be the antenna capacitance.

Adding a large capacitor in parallel across the antenna and earth will result in a lot of current flowing directly to earth, bypassing the coil. If you now attempt to tune the primary coil using the tuning capacitor, it will be in parallel with the added capacitor resulting in a reduced or non-existent ability to tune across the frequency range. We must now look at changing the inductance instead.

Estimating Antenna Capacitance

First, determine the maximum capacitance as seen by the primary coil. In this case, I could only tune down to a station at 880 kHz. The total capacitance seen by the primary coil (the antenna and the tuning capacitance value in series) can be calculated using the resonant LC equation. To resonate the 310 uH primary coil at 880 kHz, the capacitance required would be

$$
\begin{aligned}
C_{TOT} &= \frac{1}{4\pi^2 f^2 L} \\
&= \frac{1}{4\pi^2 (880\ x10^{3})^2\, 310\ x10^{-6}} \\
&= 106\ pF
\end{aligned}
$$

We know the primary coil inductance must be increased for the circuit to become resonant at 530 kHz. And as inductance values in series add, we can determine the extra inductance required above our 310 uH coil to be placed in series with our existing coil. As the frequency decreases, so does the capacitance for a given antenna. However, going from 880 to 530 kHz will only reduce the capacitance by about 1%. Therefore we shall use the total capacitance as calculated above. The equation to determine the inductance required in series to tune to the lowest frequency is given by

$$
L_{REQ} = \frac{1}{4\pi^2 f^2 C_{TOT}} - L_P \qquad 24.2
$$

And for our setup, this will be

$$L_{REQ} = \frac{1}{4\pi^2(530x10^3)^2(106x10^{-12})} - 310uH$$
$$= 851uH - 310uH$$
$$= 541uH$$

We need to add a 541 uH inductor in series with the existing 310 uH primary coil to tune down to 530 kHz. Compact high-value inductors can easily be made using ferrite material. Ferrite coils have much lower Q values than air coils. However, as this will be used for the primary circuit with a low Q anyway, this may not be a problem. The 541 uH inductor is added in series, with the antenna input connected to one side of the coil and the existing primary coil connected to the other.

The current primary tuning capacitor may not be able to tune across the entire AM spectrum with the addition of this new coil. See Chapter 18 to determine whether the tuning capacitor can span the entire frequency range. However, this is not a problem, as the inductance of the added coil can be modified by changing the rod position. Move the core further inside the coil when tuning to low-frequency stations, and retract or totally remove the core when tuning to higher-frequency stations.

If interested, once the total capacitance has been calculated, the antenna capacitance can be calculated using

$$C_A = \frac{C_C}{\left(\frac{C_C}{C_{TOT}} - 1\right)} \qquad 24.3$$

where C_C is the maximum tuning capacitance value. For this example, the maximum capacitance C_C of the tuning capacitor is 365 pF, and the total capacitance seen by the primary coil was calculated to be 106 pF. The antenna capacitance is then

$$C_A = \frac{365}{\left(\frac{365}{106} - 1\right)}$$
$$= 149pF$$

at 880 kHz.

Building the Coil

When using the metal shelf antenna, we calculated an inductor of 541 uH was required in series with the primary coil to tune down to 530 kHz. Using the same former and rod used in our previous ferrite core coils would require a coil of 75 loops. However, a benefit of using ferrite rod cores is the ability to change the inductance value. This property gives us some flexibility. We could instead wind a coil of 90 turns, giving us a 755 uH inductor. This gives us a wider range of short antennas to use. This coil's inductance drops to 34 uH when the rod is removed, giving us a variable inductor ranging from 34 – 752 uH; quite some range.

Using the shelving as an antenna, I can pull out the rod until the inductance matches the required inductance for tuning. Using a TV rabbit ear antenna with even lower capacitance, I can push the rod in to increase the inductance for tuning. Using test-coil-2 in this fashion, I used a rabbit ear antenna and could tune and hear all the stations except a 1-kilowatt station, which is reasonably faint, even using the standard antenna. If you cannot set up the standard antenna, this is a very good option; especially for people with limited space, such as those living in flats or duplexes. Adding this coil to your toolbox will greatly enhance the range of conditions where the crystal radio can be used. Ferrite core rods excel in this situation, and if you wind your coil on a former, you can remove the ferrite rod when needed for other projects.

To build the 742 uH coil (see Table 24.4 for specifications), cut a 100 mm length of your chosen former material. Drill two 1 mm holes 19 mm from each end of the former. Cut two thin strips of masking tape slightly longer than the former to keep the loops in place as you wind the coil. Pull a length of about 100 mm of copper wire as a lead and thread it through the starting hole. Start winding the coil using the same coil winding technique already described. Wind 90 turns and then pull out 100mm of wire as the end lead. Place some masking tape along the top and bottom of the coil to keep the loops in place. Run beads of epoxy glue around the exposed ends of the coil and wait half an hour for the glue to dry. Then pull the lead wire out from the hole and place more glue around the ends. When dry, remove about a centimetre length of enamel from both lead wires.

Description	Calculated values
Coil name	SC-BUILD-5
Coil inductance	752 uH (34 uH rod removed)
Coil diameter	17.18 mm
Coil length	61.20 mm
Number of windings	90
Taps at loop number	None
Pitch	0.68 mm
Wire	0.63mm enamelled copper wire (core D=0.63mm, OD=0.68mm)
Former material	PEX
Former outside diameter	16.5 mm
Former inside diameter	11.0 mm
Former length	100 mm
Rod material	S1
Rod dimensions	9.68 x 180.00 mm

Table 24.4
Ferrite rod short antenna coil specifications.

The inside diameter of the former is about 1 mm wider than the rod, so I wound some plastic wrap around the rod to build up its outside diameter, see Figure 24.14. Ferrite rods are quite brittle, and adding some padding offers some protection against cracking.

Making a Coil Stand

A stand needs to be built to hold the coil; see Figure 24.14. To place this coil in series, connect one end of the coil to the antenna and the other to the first lead of the primary coil. Slide the rod out a bit until you have achieved the inductance required to tune at a given frequency for your antenna.

To make the stand, cut two 60 mm lengths of 30 x 12 mm pine. Drill a hole with the same outside diameter as your former about 10 mm down from the top of each stand. The hole, in this case, was 16 mm. Cut a 100 mm length of 65 x 12 mm pine as the base. Fix the stands to the base by drilling two 3 mm holes through the base and into the stands. Screw the left stand onto the base using 3 x 15 mm screws. Push the former into the hole on the stand and then do the same for the other stand. Position the free stand onto the base and screw it in place. Hot glue two terminal blocks onto the base and attach the coil lead wires.

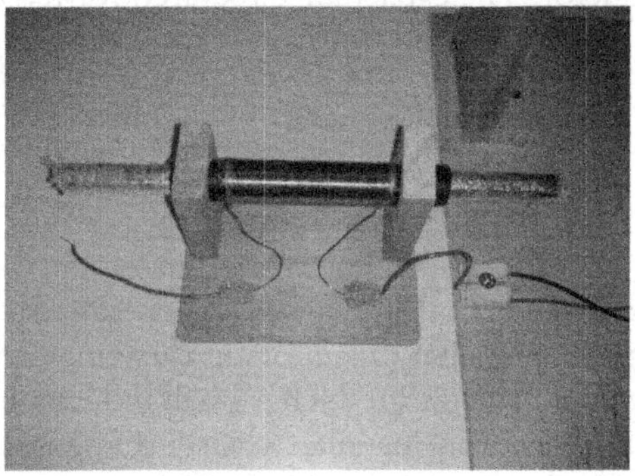

Figure 24.14
The 752 uH ferrite core coil and stand.

24.5 BLOCKING RADIO STATIONS

Close or high-powered radio stations may induce such a large signal across your antenna that even high Q radios have trouble tuning into other stations without bleed-through. These stations are called ***boomers***. We can use a resonant circuit to greatly attenuate or remove a signal at a given frequency before it reaches our primary coil.

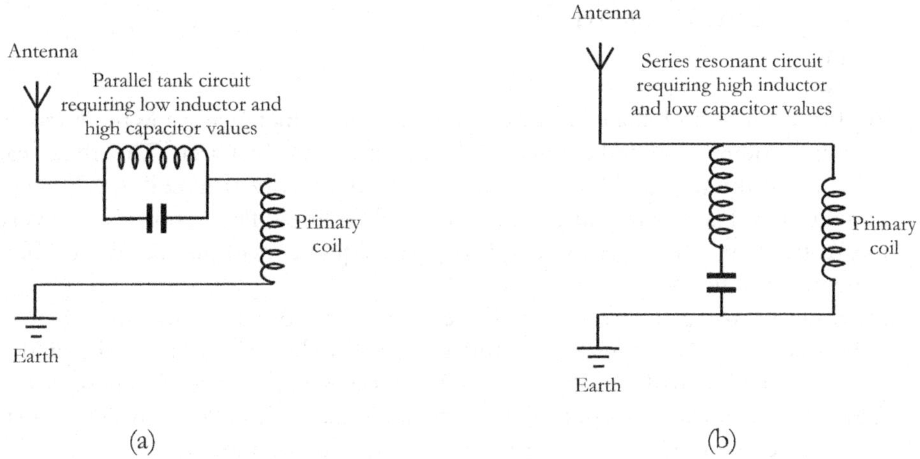

Figure 24.15
Blocking stations using parallel or series resonant circuits.

If we place a parallel tank circuit on the antenna line, when resonant, it will appear as a very high resistor at the resonant frequency, and very little current will pass through to the primary coil, Figure 24.15a. At non-resonance, the capacitive and inductive reactance should be as low as possible; if not, there will be a large voltage drop across the tank circuit relative to the primary coil reducing the received signal. For example, I have a very loud station at 880 kHz, although not a boomer. Say I wish to remove this station. I need to choose an inductance value very much smaller than the inductance of the primary coil. The inductance of my tapped primary coil is 55 uH, so I will select a 5 uH inductor. The capacitance required for resonance at this frequency is

$$\begin{aligned} C &= 1 \Big/ 4\pi^2 L f^2 \\ &= 1 \Big/ 4\pi^2 (5.0x10^{-6})(880x10^3)^2 \\ &= 6{,}542\,pF \end{aligned}$$

The problem with this approach is both the inductor and capacitor must be exactly matched if we wish to use fixed components. Variable capacitors in this range are not generally available, and using a ferrite rod inductor will be difficult for such a small inductance value. Alternatively, we can use a series-resonant circuit placed in parallel across the antenna, Figure 24.15b. In this configuration, we require high-value inductors and small-value capacitors. We wish to make the impedance at non-resonance as high as possible so most of the current will pass through the primary coil. And at resonance, the impedance of the resonant circuit will be so small that most of the signal will be shorted to ground without passing through the primary coil, thus removing the unwanted signal. With a high-value inductor, we can use a ferrite rod to adjust the inductance until resonance is achieved.

I have the high-value inductor of 742 uH, wound for the short antenna, so I will use that. For resonance at 880 kHz, the capacitor required is

$$\begin{aligned} C &= \frac{1}{4\pi^2 L f^2} \\ &= \frac{1}{4\pi^2 (752.0x10^{-6})(880x10^3)^2} \\ &= 43\,pF \end{aligned}$$

I have a 56 pF capacitor, so I shall use that and adjust the inductance using the ferrite rod. Using a 56 pF capacitor requires the inductance to be dropped to 584 uH. The fixed capacitor was connected to one of the terminal blocks in series with the ferrite core coil, see Figure 24.16. With the series circuit placed across the antenna and earth feed lines. This setup worked very well, and I was able to greatly reduce the signal strength at that frequency without any noticeable drop in the volume of the other stations.

The downside of using a ferrite rod is the large core loss reducing the Q at the resonant frequency. This core loss at resonance (in this case, 880 kHz) will be in parallel with the primary coil and may be a significant fraction of the primary coil's impedance. As these two components are in parallel, the current will be split by the current divider rule. The equivalent resistive losses at 880 kHz for this ferrite core coil is around 36 ohms, and the inductive reactance of the 55 uH primary coil is 305 ohms at this frequency. About 10% of the signal will pass across the primary coil. This may result in some signal bleed-though. If the signal is not reduced enough for your needs, then you may need to build a high Q air-core coil and use a variable capacitor in place of the ferrite core coil and fixed capacitor.

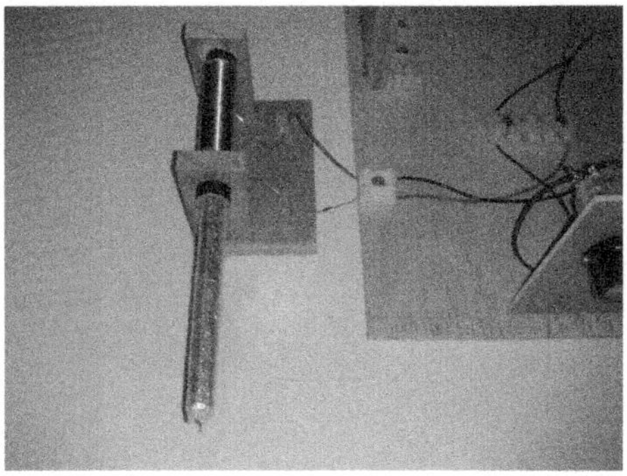

Figure 24.16
Blocking stations using a series-resonant circuit.

25

Measurement Techniques

25.1 COIL MEASUREMENTS

Measuring a coil's self-resonant frequency and Q once required dedicated and expensive test equipment designed specifically for this purpose. Luckily the price of electronic test equipment has fallen dramatically over recent years, and these measurements can now be taken using relatively inexpensive test equipment that many electronics enthusiasts will have on their workbench, namely a function generator and an oscilloscope.

When performing voltage or current measurements on a circuit, one must consider the effect the measuring equipment has on the circuit. For instance, a typical digital volt meter has an input impedance of 10 MΩ or greater. If we were to take a voltage measurement across a 1 kΩ ohm resistor, the volt meter's impedance would be parallel to the resistor, and the total impedance across the resistor would still be very close to the original 1 kΩ, so that the volt meter will have very little effect on the circuit. If we were to measure the voltage across a 10 MΩ resistor, then the total resistance would be halved, and we would measure 5 MΩ, and the volt meter has a large impact on the circuit.

Oscilloscopes typically have a large resistance in parallel with some small capacitance at the input terminal. The cable plugged into the oscilloscope has a switch that uses or bypasses a 9 MΩ resistor in parallel with a small trim capacitor. And the cable itself has some resistance and capacitance to ground. Figure 25.1 represents a circuit diagram of a typical passive probe connected to an oscilloscope. The actual values will differ between makes and models of probes and oscilloscopes, but they will typically be around this order of magnitude.

The oscilloscope input and cable capacitance make the voltage displayed on the oscilloscope frequency dependent. The capacitance and resistance of the entire circuit form a voltage divider with the device under test.

When the probe switch is closed, the probe impedance is bypassed, and the cable is said to be in 1x mode. The impedance of the oscilloscope is very much higher than the cable impedance, and most of the voltage (up to a certain frequency) is dropped across the oscilloscope input, and the oscilloscope displays the voltage it measures.

When the probe switch is open, the current passes across the parallel probe resistor and capacitor. Ninety percent of the voltage is dropped across the probe, and 10% of the voltage is dropped across the input impedance of the oscilloscope. The probe is said to be in 10x mode. The oscilloscope measures one-tenth of the actual voltage across the device under test. The oscilloscope multiplies the measured voltage value by ten and displays this value on the screen. The oscilloscope will either auto-detect or have a manual setting to take this into account when displaying voltages.

In 1x mode, the relatively large cable capacitance is in parallel with the input capacitance and sets an upper limit to the frequency at which the oscilloscope can accurately measure. In 10x mode, the upper-frequency limit will be increased due to the lowering of the total capacitance value by the inclusion of the smaller probe capacitance in series with the cable and input capacitance. These frequency limits will be given in the oscilloscope documentation.

The cable capacitance and resistance are distributed throughout the length of the cable but are represented as a lumped component capacitor and resistor. For the values given in Figure 25.1, the cable capacitance is in parallel to the oscilloscope input capacitance. So at 1x, the input capacitance to the oscilloscope (from the tip of the probe) will be 104 pF (88 + 16 pF). At 1 MHz, the input impedance to the oscilloscope will be 1,567 Ω (340 – j1,530). Say we wish to measure the voltage across a 300 uH coil at 1 MHz. At this frequency, the coil impedance will be 1,885 Ω. The oscilloscope input (from the probe) and coil are in parallel, and almost half the current will flow through the cable and oscilloscope. The cable resistance is around 340 Ω, and as this is in series, all the current will flow through this resistance. Very little current will flow across the 1 MΩ resistance at the input of the oscilloscope as this is in parallel with the 16 pF capacitance. The current through the cable resistance represents extra Ohmic losses above that of the coil. As such, any Q measurement of the coil will be significantly lower than its actual value.

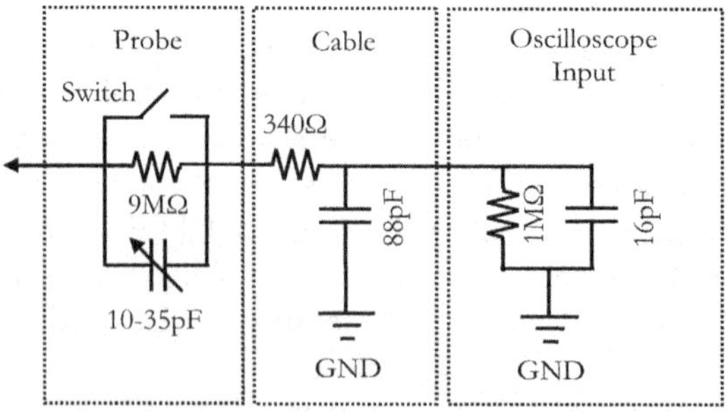

Figure 25.1
Circuit input impedance diagram of an oscilloscope.

Analysis of Measurement Loss

Consider the circuit shown in Figure 25.2. The coil under test forms a tank circuit with the variable capacitor C_{RES}. We supply a 1 MHz sine wave signal to the coil and measure the voltage across the tuning capacitor using the oscilloscope. We bring the tank circuit to resonance by adjusting the tuning capacitor for maximum voltage.

Say the coil under test has an inductance of 300 uH, an AC resistance of 9 Ω at 1 MHz, and a self-resonant frequency of 4 MHz. We will examine the effect the measuring equipment has on the Q of the coil. First, calculate the effective Q of the coil at this frequency. Using the coil specifications, the effective Q of this coil should be

$$Q_{effl} = \frac{X_L}{R_{coil(ac)}}\left(\frac{f_0^2 - f_Q^2}{f_0^2}\right)$$

$$= \frac{2\pi(1x10^6)(300x10^{-6})}{9}\frac{(4x10^6)^2 - (1x10^6)^2}{(4x10^6)^2}$$

$$= \frac{1885}{9}\frac{15}{16}$$

$$= 196$$

If we attach the oscilloscope probe across the tuning capacitor with the probe in 1x mode, we add 1,567 Ω of impedance in parallel with the tuning capacitor. The resonant circulating current flows through the inductor and its internal resistance and across the tuning capacitor. As the probe is in parallel across the tuning capacitor, the current splits, with some of the current flowing into the oscilloscope.

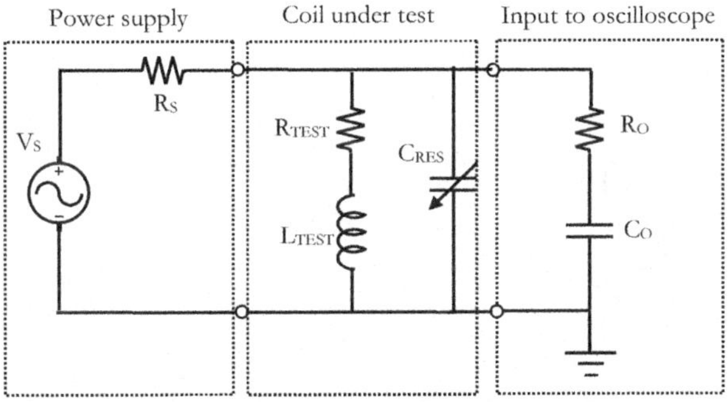

Figure 25.2
Circuit diagram of Q measurement.

We can use the current divider rule to get the proportion of current flowing into the oscilloscope.

$$\mathbf{I_{OSC}} = \frac{\mathbf{Z_{test}I}}{\mathbf{Z_{test}} + \mathbf{Z_{OSC}}}$$

$$= \frac{-j1{,}885\mathbf{I}}{-j1{,}885 + 340 - j1{,}530}$$

$$= \frac{1{,}885\angle -90^\circ\mathbf{I}}{340 - j3{,}415}$$

$$= \frac{1{,}885\angle -90^\circ\mathbf{I}}{3432\angle -84^\circ}$$

$$= 0.55\angle -6^\circ\mathbf{I}$$

All the circulating current will pass through the coil, and 55% will pass through to the oscilloscope. Using the definition of Q, we can write

$$Q_{measured} = \frac{X_L}{R_{coil(ac)} + (0.55)^2 Rosc}\alpha$$

$$= \frac{1885}{9 + (0.3025x340)}(0.9375)$$

$$= 16$$

This is a massive 92% drop in Q. What will be the effect on Q if we switch the probe to 10x? Setting the probe to 10x mode will result in an input capacitance of around 9.5 pF (the trim capacitor value of the probe in series with 104 pF). And at 1MHz, the input impedance to the oscilloscope will be about 16,748 Ω (620 – j16,737).

$$\mathbf{I}_{\mathbf{OSC}} = \frac{1{,}885\angle -90^\circ I}{-j1{,}885 + 620 - j16{,}748}$$

$$= \frac{1{,}885\angle -90^\circ I}{18{,}643\angle -88^\circ}$$

$$= 0.10\angle -2^\circ I$$

And the measured Q will be

$$Q_{measured} = \frac{X_L}{R_{coil(ac)} + (0.10)^2 Rosc}\alpha$$

$$= \frac{1885}{9 + (0.01x340)}(0.9375)$$

$$= 142$$

This results in a 28% drop in Q; better than in 1x mode but still not very good.

To reduce the measurement loss even further, add a 1-pF capacitor in series with the probe. The 1-pF capacitor will have a high impedance at 1 MHz and will reduce the amount of current flowing into the oscilloscope even further. As the stated value of a capacitor can vary, and to take into account stray capacitance due to wiring, the input capacitance from the probe across the oscilloscope was measured (the input capacitance). The capacitance value measured was 3 pF. At 1 MHz, this gives a reactance of 53,052 Ω. Recalculating the proportion of current flowing into the oscilloscope, we have

$$\mathbf{I}_{\mathbf{OSC}} = \frac{1{,}885\angle -90^\circ I}{-j1{,}885 + 340 - j53{,}052}$$

$$= \frac{1{,}885\angle -90^\circ I}{54{,}938\angle -90^\circ}$$

$$= 0.034\angle 0^\circ I$$

And the Q becomes

$$Q_{measured} = \frac{X_L}{R_{coil(ac)} + R_{wiring} + (0.034)^2 Rosc}\alpha$$

$$= \frac{1885}{9 + (0.001156x340)}(0.9375)$$

$$= 183$$

This is only a 7% drop in Q due to the measuring device and extra wiring. If your oscilloscope values differ significantly, you may want to recalculate using your values. Note these values are dependent on the inductance of the coil and frequency. However, most of the coils we deal with will be around the 300 uH value, and our Qs are always given at 1 MHz.

In this analysis, we have assumed an ideal power supply with no losses. A typical function generator will have an output impedance of 50 Ohms. Connecting this power supply directly to a tank circuit will result in a substantial drop in Q. If, instead, we connect the power supply to a primary coil and use mutual inductance to drive a current in the secondary circuit, we can eliminate this loss. However, we will now have losses due to the mutual inductance. Figure 25.3 shows typical Q and coil voltage curves as a function of primary and secondary coil distance.

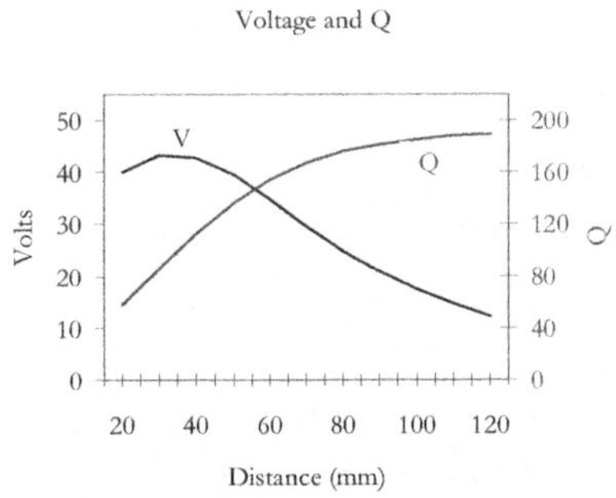

Figure 25.3
Q and voltage as a function of distance.

By using a primary coil with a small self-inductance of around 7 uH and a diameter of 103mm, and if we keep the separation between coils to about 150 mm, then the loss due to mutual inductance can be kept to around 5% or less. We can now use these rules of thumb to give a reasonable approximation for a coil's Q from a measurement. That is a 5% drop due to mutual inductance and a 7% drop due to the measuring device.

Q Measurement Correction Factors

For the effective Q of a coil (only resistance and self-resonance losses), we can get the effective Q from measurement by applying correction factors for losses due to the oscilloscope and the mutual inductance losses. That is

$$Q_{eff} = \frac{Q_{measured}}{(0.95)(0.93)} = \frac{Q_{measured}}{0.88}$$ Effective Q correction factor 25.1

And the circuit Q of a coil is reduced by the oscilloscope losses only, that is

$$Q_{circuit} = \frac{Q_{measured}}{0.93}$$ Circuit Q correction factor 25.2

Voltage Measurement Correction Factors

Using a 1-pF capacitor in series with the oscilloscope probe will result in the voltage measurements being attenuated. We can calculate the actual voltage across the variable capacitor by using the voltage divider rule. The voltage measured across the input of the oscilloscope is given by

$$V_{measured} = \frac{Z_{OSC}V}{Z_s + Z_{OSC}}$$

And

$$V_{measured} = \frac{\frac{1}{\omega C_{OSC}}}{\frac{1}{\omega C_s} + \frac{1}{\omega C_{OSC}}} V$$

$$= \frac{1}{C_{OSC}(\frac{C_s + C_{OSC}}{C_s C_{OSC}})} V$$

Simplifying gives

$$V = \frac{C_S + C_{OSC}}{C_S} V_{measured}$$ 25.3

where C_S is the capacitance value placed in series with the probe, and C_{OSC} is the capacitance value across the input of the oscilloscope. We have ignored resistances as the series resistance is small compared to the capacitive reactance at this frequency, and the parallel resistance is large.

For the above example, the series input capacitor was measured to be 1.3 pF, and the input to the oscilloscope was measured to be 15.6 pF at 10x. Say we measured a voltage across the variable capacitor to be 1.2 volts. The actual voltage would be

$$V = \frac{1.3 + 15.6}{1.3} 1.2$$

$$= 13(1.2)$$

$$= 15.6$$

The voltage correction factor is 13. However, this voltage is reduced to the reduction in Q. Apply both the Q correction factor and the voltage factor we get

$$V = \frac{13}{0.93} V_{measured} \qquad 25.4$$

Measuring Coil Q

To reduce interference from external objects, we need a test jig that provides some space between the coil and other things, including the bench. Figure 25.4 is a diagram of the setup I use to measure coil Q. The test stand is made by cutting a sheet of Styrofoam board into the appropriate sections. Styrofoam boards are cheap and easy to cut using a utility knife, but any plastic should do. The height of the test stand is 205 mm, and the width is 510 mm. The actual dimensions aren't particularly important. The function of the stand is to keep the coils away from the workbench and any materials that may interfere with the measurement.

The suspension bar is attached to the test legs by a couple of plastic screws, and the legs are connected to each foot by two countersunk screws underneath the foot. Coils are placed in the stand by first removing the suspension bar and sliding the coils over the bar. The bar is then replaced on the stand, and using a ruler, the distance between the primary coil and the coil under test is set to 150 mm.

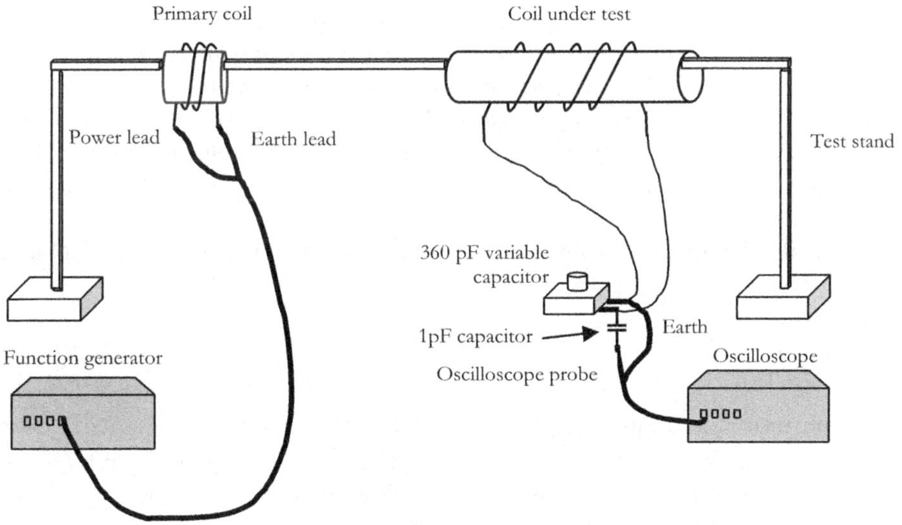

Figure 25.4
Coil Q measurement stand diagram.

Figure 25.5 shows a photo of the test stand. The primary coil was made by winding six loops of 1.25 mm enamel-coated copper wire on a 103 mm polyethylene coupling pipe. Hot glue was applied to the loops to keep them in place. The coil was then slid off the former. The self-inductance of the coil was measured to be 7.4 uH with a self-resonant frequency of 27 MHz.

The exact number of loops for the primary does not matter that much, just so long as there is enough mutual inductance between the coils to induce a measurable voltage in the coil-under-test (the secondary coil). The variable capacitor was mounted onto an empty wire spool using a bit of hot glue. Two alligator clips were soldered onto the capacitor terminals, red for the active and black for the earth (common) terminal. A 1 pF capacitor was also soldered onto the active terminal with a small socket to accept the oscilloscope probe.

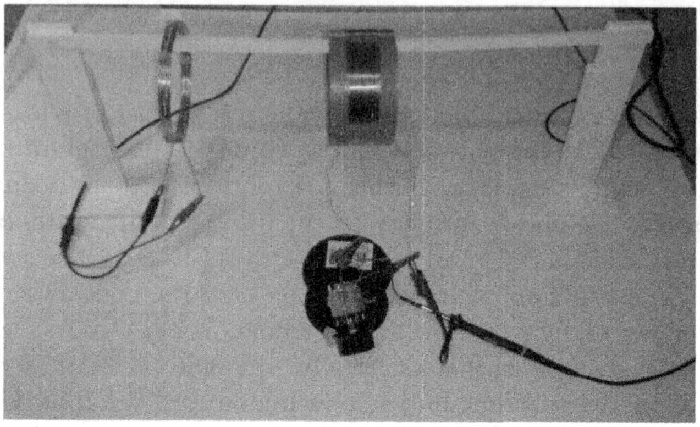

Figure 25.5
Coil Q measurement stand setup.

To make a Q measurement, set the function generator to output a 1 MHz sine wave at about 5 volts peak-to-peak. Adjust the variable capacitor until the maximum voltage is shown on the oscilloscope. It is sometimes a little challenging to get the maximum voltage by adjusting the variable capacitor. Get it as close as you can and then twig the frequency output of the function generator up and down a bit; the voltage should drop on either side of the 1 MHz signal if you have made the circuit resonant. If it goes up, you are not precisely resonant at 1 MHz. Set the function generator back to 1 MHz and make very small adjustments to the tuning capacitor until the peak voltage is reached.

We will use the bandwidth definition of Q as discussed in section 8.1, which is given by Equation 8.12. That is

$$Q_{coil} = \frac{f_Q}{f_h - f_l}$$

where f_Q is the resonant frequency (in this case 1 MHZ). f_h is the frequency where the peak voltage drops by a factor of 0.707 (-3 dB) above the resonant frequency, and f_l is the frequency where the peak voltage drops by a factor of 0.707 below the resonant frequency.

To take the measurements, note the peak-to-peak voltage on the oscilloscope and multiply it by 0.707. Adjust the frequency downwards until the peak-to-peak voltage drops to this value. This frequency is the lower bandwidth frequency f_l. Now increase the frequency above the resonant frequency until the same voltage level is reached. This is your upper bandwidth frequency f_h. Now calculate the Q using Equation 8.12.

For example, say the peak voltage across the primary coil was 2.15 volts at 1,000 kHz. The 3 dB down voltage would be

$$\begin{aligned} V_{3dB} &= 2.15(0.707) \\ &= 1.52 \end{aligned}$$

And say the frequency below the resonant frequency where the peak-to-peak voltage is 3 dB down from the maximum was 997.5 kHz (f_l). And the frequency above the resonant frequency where the peak-to-peak voltage is also 3 dB down was 1003.2 kHz (f_h), then the measured Q would be

$$\begin{aligned} Q_{measured} &= \frac{f_Q}{f_h - f_l} \\ &= \frac{1000.0}{1003.2 - 997.5} \\ &= 175 \end{aligned}$$

Using the rule of thumb and Equation 25.1, we could get a good estimate of the coil's effective Q as

$$\begin{aligned} Q_{eff} &= Q_{measured} / 0.88 \\ &= 175 / 0.88 \\ &= 199 \end{aligned}$$

Measuring Coil Self-resonance

To take self-resonant frequency measurements will require three coils; the primary coil attached to the function generator, the coil-under-test and another coil attached to the oscilloscope to view the voltage. The three coils are placed on the coil jig, with each coil separated by about one coil diameter (in this case, about 100 mm). When a signal is supplied to the primary coil, an emf (voltage) will be induced across the coil-under-test. Note the coil-under-test is not connected to anything and is open-circuited. The induced voltage across the coil-under-test will, in turn, induce an emf in the third coil attached to the oscilloscope. For the primary coil, you can wind another coil similar to the primary used in the Q measurements. I opted to wind three loops of 1.25 mm enamelled copper wire with a 103 mm diameter. See Figure 25.6.

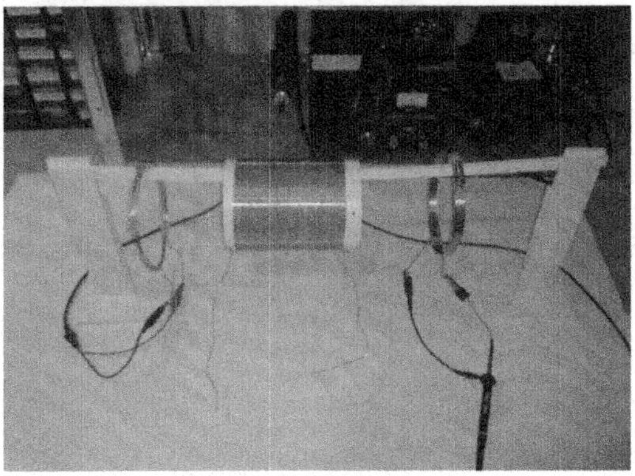

Figure 25.6
Self-resonant frequency measurement setup.

Set the function generator to produce a 5-volt peak-to-peak signal starting at 1 MHz. Ramp up the frequency of the function generator until a peak voltage is seen on the oscilloscope. The frequency at which this peak voltage occurs is the first self-resonant frequency of the coil-under-test. Note, that due to the transmission line nature of self-resonance, there will be higher frequencies at which the coil will become self-resonant again. The first self-resonance is when the coil stops acting like an inductor and is the frequency we are interested in.

25.2 DIODE MEASUREMENTS

Turn-on Voltage

Digital multimeters measure the turn-on voltage across a diode by applying a fixed voltage across a known resistor in series with the diode under test. The voltage across the resistor is measured and subtracted from the supplied voltage to display the diode voltage drop as the turn-on voltage. Before the widespread use of Light Emitting Diodes (LED), this voltage was around 2.7 volts with a 4.7 kΩ resistor in series. As LEDs became common, this source voltage was increased to accommodate the larger turn-on voltages for these devices of around 1.7 volts or higher. Digital multimeter manufacturers increased the supplied voltage to measure these diodes, resulting in higher currents passing through small signal diodes with higher displayed turn-on voltage.

To measure the turn-on voltage of a diode, select the diode symbol on the multimeter dial and clip the common lead (black) to the cathode of the diode and the red lead to the anode. The cathode will generally be marked on the diode's case by a black mark.

The I-V Curve

Producing an I-V curve requires taking a series of voltage and current measurements and plotting these measurements. Figure 25.7 shows the circuit diagram for taking voltage and current measurements. Two multimeters are required; one to measure the voltage and another to measure the current. Most power supplies have an internal resistance of 50 ohms which will limit the current for a given voltage. So we must adjust the power supply until we get the required voltage drop across the diode. To get decent results at low voltages and in reverse bias, you will need a multimeter to read currents down to at least the micro amp level.

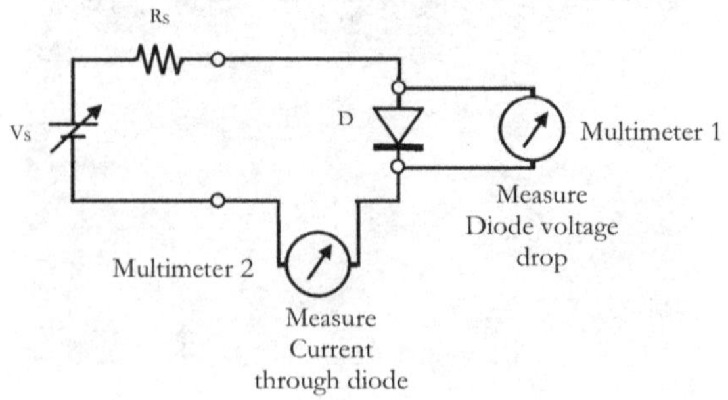

Figure 25.7
Circuit for measuring diode I-V curve.

For the forward bias measurements, start with a diode voltage drop of 0.05 volts and note the current passing through the circuit. Increase the voltage across the diode by 0.05 volts for each measurement and record the current. Take as many sets of measurements as you require, but don't exceed 30 mA of current. Most small signal diodes will handle this amount of current without damage. Enter these measurements into a spreadsheet to graph the forward bias section of the I-V curve.

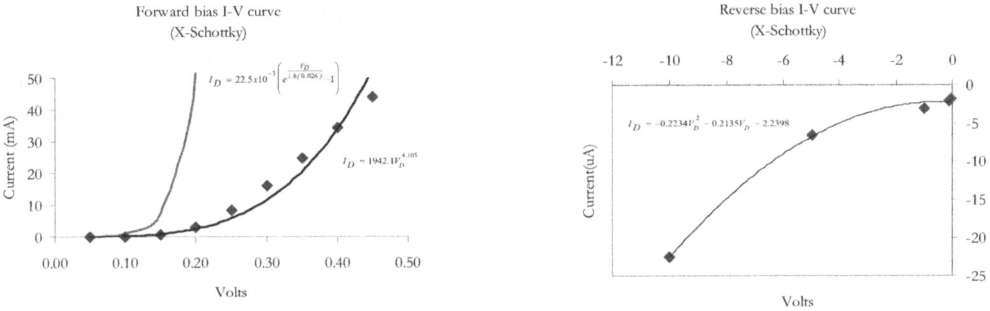

Figure 25.8
X-Schottky I-V curve.

For the reverse bias measurements, swap the polarity of the power supply and start with a diode voltage drop of -0.05 volts, then -0.1, then -1.0, -5.0 and finally -10 volts. Record both voltage and current. Enter these measurements into a spreadsheet to graph the reverse bias section of the I-V curve.

Measurements taken for the unknown Schottky are given in Table 25.1. The I-V curves for forward and reverse bias are shown in Figure 25.8. The measured values are shown as square markers. Two equations are plotted for the forward bias graph; one uses the ideal diode Equation 14.6, and the other a best-fit power curve. Notice how much the idealised diode equation differs from the measured curve. For this diode, the best-fit equation would be used to calculate the forward current for a given diode voltage drop. For the reverse bias voltage graph, a best-fit polynomial was used. Note the forward bias current is in milliamps, and the reverse current is in micro amps.

V_D	I_D(mA)
0.05	0.0122
0.10	0.0931
0.15	0.7010
0.20	2.8000
0.25	7.7800
0.30	15.8300
0.35	24.7000

(a) Forward bias.

V_D	I_D(μA)
-0.05	-1.8
-0.10	-2.1
-1.00	-3.0
-5.00	-6.5
-10.0	-22.5

(b) Reverse bias.

Table 25.1
X-Schottky diode voltage and current.

Determining the best equation to use is a trial-and-error process. This, however, is not too bad if you use a spreadsheet with the ability to add trend lines. Try various types of trend lines to see which fits best, and use the best-fit forward and reverse equations to calculate the current for a given voltage. Note that the I-V curves will differ slightly for individual diodes, even within a given family of diodes.

Capacitance

A diode's junction and diffusion capacitance are very important characteristics when dealing with small AC signals in crystal radios. We need to determine the total impedance of the diode to calculate both the forward and reverse AC currents and voltages, not just the DC resistance. Getting precise capacitance values is very difficult unless you have expensive and specialised equipment. The measuring equipment will have some capacitance, and as we are dealing with small capacitance values of the diode, the equipment capacitance cannot be ignored and will form part of the circuit analysis.

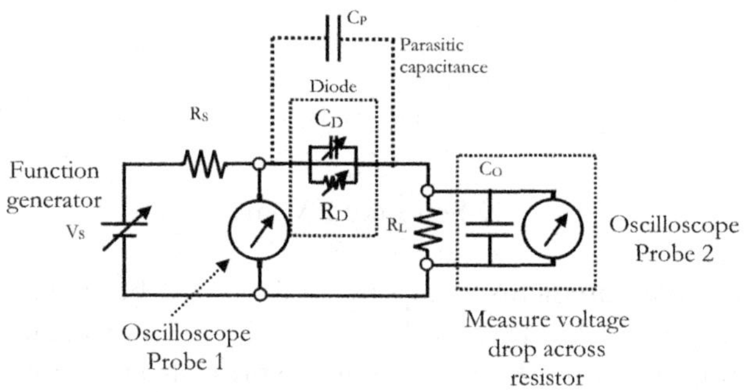

Figure 25.9
Circuit for measuring diode junction capacitance.

Figure 25.9 shows the circuit diagram for taking diode capacitance measurements. The diode is biased to a DC voltage to set a static condition. A small AC signal is then fed across the diode to determine the capacitance value at this DC offset. The AC signal needs to be small enough that it does not change the operating condition too much. From the circuit diagram, the diode-under-test is placed in series with a load resistor. The voltage across the load resistor is given by

$$\mathbf{V_L} = \mathbf{V_S}\frac{\mathbf{Z_L}}{\mathbf{Z_L} + \mathbf{Z_D}}$$

Rearranging, we can write the impedance of the diode as

$$\mathbf{Z_D} = \frac{\mathbf{V_S}}{\mathbf{V_L}}\mathbf{Z_L} - \mathbf{Z_L} \qquad 25.5$$

Given the source AC voltage and the load impedance, if we measure the voltage across the load, we can determine the diode's impedance. The diode capacitance is in parallel with the diode's DC resistance, so we can write

$$\mathbf{Z_D} = \frac{-jR_D X_C}{R_D - jX_C}$$

And

$$\frac{-jR_D X_C}{R_D - jX_C} = \frac{\mathbf{V_S}}{\mathbf{V_L}}\mathbf{Z_L} - \mathbf{Z_L}$$

Rearranging, we can write the capacitive reactance of the diode as

$$X_C = \left| \frac{R_D \mathbf{Z_D}}{R_D - \mathbf{Z_D}} \right| \qquad 25.6$$

To determine the diode's reactance, we need to know the DC resistance of the diode for the given voltage to solve the equation. We can get the DC résistance for a given voltage from the I-V curve, that is

$$R_D = \frac{I_D}{V_D}$$

Finally, knowing the capacitive reactance, we can calculate the diode capacitance using

$$\boxed{C_D = \frac{1}{2\pi f X_C} - C_P} \qquad 25.7$$

Measuring Junction Capacitance

To take measurements, set the function generator to supply a 1MHz sine wave of 0.05 volts peak-to-peak with a DC offset sufficient to reverse bias the diode for a given value. A 0.05 volts peak-to-peak sine wave amplitude was chosen so that the amplitude of the sine wave does not change the diode junction capacitance to any great extent. Place a multimeter across the diode and adjust the DC offset until the required reverse voltage across the diode is reached, then remove the multimeter as there will be substantial capacitance in the cable. Use a load resistor R_L of around 470 k Ohms to provide some measurable drop across the load resistor. Measure the peak-to-peak voltage across the load resistor with the oscilloscope. Now attach the second oscilloscope probe across the circuit to measure the phase difference between the input sine wave voltage and the load sine wave voltage. Also, note the actual peak-to-peak value of the sine wave and use this as the source sine wave voltage. Take two load voltage and phase shift measurements at a reverse DC diode bias voltage of -0.1 and -1.0 volts.

As we are dealing with small capacitance values, we must keep the parasitic capacitance across the diode to a minimum. So keep the leads of the diode straight. This means a breadboard is not suitable for this measurement, as you will need to bend the leads into the holes. Use two terminal blocks to connect the diode to the circuit. If you have a good capacitance meter, before connecting the diode, measure the capacitance across the terminals where the diode will be inserted, but without the oscilloscope or power supply cables connected. I measured this capacitance as 0.4 pico farads. Also, there is some small case capacitance across the diode; this is around 0.1 pF. So the total parasitic capacitance across the diode is 0.5 pF. You can use this Figure if you don't have a good capacitance meter.

If there is significant noise on the waveform, take a single shot and use the cursors to measure voltage by placing the cursors in the middle of the interference at the top and bottom values. If the output signal is too small to read, replace the load resistor with a higher value, reset the voltage across the diode and repeat the measurement. This technique was used to measure the junction capacitance of the unknown Schottky diode. The measured values are given in Table 25.2.

Diode voltage drop V_{DC} (Measured)	Diode DC resistance R_D (calculated)	Load resistance R_L (chosen)	Oscilloscope probe capacitance C_0 (measured)	Parasitic capacitance C_P (measured)	Phase angle	Measured peak-to-peak voltage across the load resistor V_L
-0.1	47.62k	470k	15pF	0.5pF	12.6°	0.0160
-1.00	333.33k	470k	15pF	0.5pF	10.8°	0.0132

Table 25.2
X-Schottky results.

The load impedance consists of the load resistor in parallel with the capacitance of the oscilloscope probe. The load impedance at 1 MHz (R_L in parallel with C_o) is

$$\begin{aligned}\mathbf{Z_L} &= R_L \,//\, C_o \\ &= 239 - j10{,}605 \\ &= 10{,}607\angle -89^\circ\end{aligned}$$

Using Equation 25.5 to determine the load impedance at -0.1 volt, we get

$$\begin{aligned}\mathbf{Z_D} &= \frac{\mathbf{V_S}}{\mathbf{V_L}}\mathbf{Z_L} - \mathbf{Z_L} \\ &= \left(\frac{0.05}{0.016\angle 12.6^\circ}10{,}607\angle -89^\circ\right) - (10{,}607\angle -89^\circ) \\ &= 22{,}913\angle -107^\circ\end{aligned}$$

Now using Equation 25.6 to calculate the capacitive reactance, we get

$$\begin{aligned}X_C &= \left|\frac{R_D\mathbf{Z_D}}{R_D - \mathbf{Z_D}}\right| \\ &= \left|\frac{47{,}620(22{,}913\angle -107^\circ)}{47{,}620 - 22{,}913\angle -107^\circ}\right| \\ &= 18{,}618\end{aligned}$$

Finally, using Equation 25.7 to calculate the junction capacitance, we get

$$\begin{aligned}C_j &= \frac{1}{2\pi f X_C} - C_P \\ &= \frac{1}{2\pi 10^6 18{,}618} - 0.5x10^{-12} \\ &= 8.0x10^{-12}\end{aligned}$$

The junction capacitance for this diode at -0.1 volts is 8 pF. Performing the same analysis for the reverse bias voltage of -1.0 volts, we get a junction capacitance of 4.7 pF. Typical junction capacitance values are given in the specification measured at -1.0 volts.

Measuring Built-in Voltage

By rearranging Equation 14.10, we can express the junction capacitance as a linear function of applied voltage.

$$\frac{1}{C_j^2} = \frac{(V_{bi} - V_D)}{\beta_D}$$

And

$$\boxed{\frac{1}{C_j^2} = \frac{-V_D}{\beta_D} + \frac{V_{bi}}{\beta_D}} \qquad 25.8$$

where $1/\beta_D$ is the slope of a straight line. If we can measure the junction capacitance at two reverse bias voltages, we can plot the $1/C^2$ values and calculate the straight line passing between these two values. From this line, we can calculate the slope ($1/\beta_D$) and the built-in voltage V_{bi}.

Expressing $1/C^2$ as C´, we can write 25.8 in the form of a straight line given by

$$C' = mV_D + b$$

where m is the slope of the line. Figure 25.10 shows the shape of the plot.

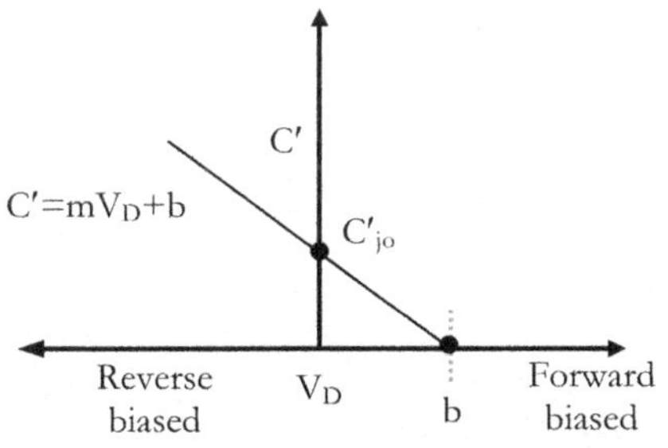

Figure 25.10
C′ as a function of diode voltage.

We can calculate the slope using

$$m = \frac{1}{\beta_D} = \frac{C_2' - C_1'}{V_{D2} - V_{D1}} = \frac{\frac{1}{C_2^2} - \frac{1}{C_1^2}}{V_{D2} - V_{D1}} \qquad 25.9$$

where C_2 and C_1 are the junction capacitance values at voltages V_{D2} and V_{D1} respectively. We can then calculate b using

$$b = C_1' - mV_{D1} \qquad 25.10$$

And finally, the built-in voltage V_{bi} is

$$V_{bi} = \frac{-b}{m} \qquad 25.11$$

Using the previously determined junction capacitance values at -0.1 and -1.0 volts for the unknown Schottky, we can derive the straight-line equation and extrapolate to get the built-in voltage. Using Equation 25.9 to calculate the slope

$$m = \frac{\frac{1}{C_2^2} - \frac{1}{C_1^2}}{V_{D2} - V_{D1}}$$

$$= \frac{\frac{1}{(8.0x10^{-12})^2} - \frac{1}{(4.7x10^{-12})^2}}{-1.0 - (-0.1)}$$

$$= -3.29x10^{22}$$

And b is

$$b = C_1' - mV_{D1}$$

$$= 1.56x10^{22} - (-3.29x10^{22})(-0.1)$$

$$= 1.231x10^{22}$$

The built-in voltage V_{bi} is

$$V_{bi} = \frac{-b}{m}$$

$$= \frac{-1.231x10^{22}}{-3.29x10^{22}}$$

$$= 0.37$$

Measuring Diffusion Capacitance

To determine the diffusion capacitances use the same procedure used in calculating the junction capacitance, except now the voltages are positive, and the diode is forward biased. Set the voltage bias across the diode to 0.1 volts DC. Select a lower load resistance of around 1 kΩ as the forward impedance will be very much lower than the reverse impedance. Using the unknown Schottky as an example, at a DC forward bias voltage of 0.1 volts with a 1 MHz sine wave of 0.05 volts peak-to-peak, 11.0 mV peak-to-peak AC voltage was measured across the load resistor, with a 0° phase shift with respect to the input voltage sine wave. Using Equations 25.5, 25.6 and 25.7, the total diode capacitance was calculated to be 102.9 pF. The total forward capacitance is the addition of the junction and diffusion capacitance. So if we calculate the junction capacitance when the diode is forward biased at 0.1 volts, and subtract the calculated junction capacitance from the total capacitance, we will have the diffusion capacitance.

$$C_D = C_{TOTAL} - C_J \qquad 25.12$$

Using Equation 14.10 to calculate the junction capacitance at 0.1 volts of the unknown Schottky, we get a value of 10.6 pF. Using Equation 25.12, the diffusion capacitance is

$$\begin{aligned} C_D &= C_{TOTAL} - C_J \\ &= 102.9pF - 10.6pF \\ &= 92.3pF \end{aligned}$$

To use Equation 14.15 to determine the diffusion capacitance, we need to know the average transit time. From Equation 14.6, we can determine the average transit time using the diode current when the diode voltage is 0.1 volts. From Table 25.1, the current at 0.1 volts DC is 0.0931 mA. Plugging these values into Equation 14.16, the average transit time becomes

$$\begin{aligned} \tau_t &= C_D \frac{\Delta V_D}{\Delta I_D} \\ &= \frac{92.3x10^{-12}(0.1)}{0.0931x10^{-3}} \\ &= 99.1x10^{-9} \end{aligned}$$

25.3 TRANSMISSION LINE CHARACTERISTIC IMPEDANCE

Measuring the characteristic impedance of a transmission line requires an LCR meter. An LCR meter measures inductance L, capacitance C, and resistance R. LCR meters typically measure L and C values at low frequencies (1 kHz or less), so the transmission line will appear as a lumped component to the meter. The characteristic impedance of a transmission line is given by Equation 9.23.

$$\mathbf{Z_0} = \sqrt{\frac{R' + j\omega L'}{G' + j\omega C'}}$$

To measure the capacitance of a transmission line, connect the LCR meter to one end of the line while the other end is open-circuited and measure the capacitance value. To measure the resistance and inductance of a line, short one end of the line and use the LCR meter to measure the total resistance and inductance. The conductance G is typically so small it can be ignored.

To get the values per meter, divide the measured values by the length of the transmission line.

$$R' = \frac{R}{\ell}$$

$$L' = \frac{L}{\ell}$$

$$C' = \frac{C}{\ell}$$

And use these values in Equation 9.23.

25.4 MUTUAL INDUCTANCE

From our discussion on mutual inductance, we know that the total inductance of inductors in series with the same winding direction is given by Equation 12.5

$$L_{T+} = L_1 + L_2 + 2M$$

For two inductors with opposite winding directions (or the current is flowing in opposite directions) is given by Equation 12.6

$$L_{T-} = L_1 + L_2 - 2M$$

Subtracting the two equations, we get

$$L_{T+} - L_{T-} = (L_1 + L_2 + 2M) - (L_1 + L_2 - 2M)$$
$$= 4M$$

Rearranging for mutual inductance, we get

$$M = \frac{1}{4}(L_{T+} - L_{T-}) \qquad 25.13$$

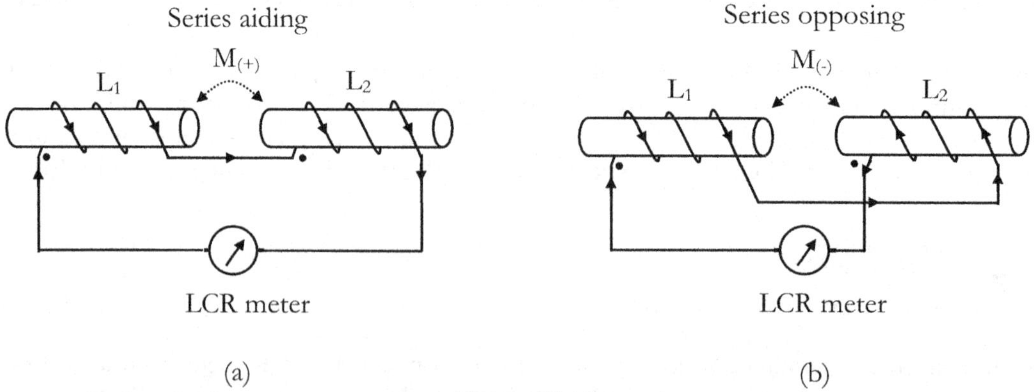

Figure 25.11
Measuring mutual inductance.

To measure the mutual inductance between two coils at a given distance, first, connect the LCR meter and the two coils in series aiding as shown in Figure 25.11a, and measure the total inductance L_{T+}. Next, connect the two coils in series opposing, as shown in Figure 25.11b and measure the total inductance L_{T-}. Now use Equation 25.13 to calculate the mutual inductance.

Appendix 1

Deriving Ampere's Law from Biot-Savart

Given the Biot-Savart law,

$$dB = \frac{\mu I \sin(\theta)}{4\pi r^2} dl$$

Summing up the contributions to the magnetic field density along the length of the wire, we can write

$$\int dB = \int \frac{\mu I \sin(\theta)}{4\pi r^2} dl$$

$$B = \frac{\mu I}{4\pi} \int \frac{\sin(\theta)}{r^2} dl \qquad \text{A1}$$

From Figure 5.9 and using trigonometric functions, we can write the distance r as

$$r = \frac{d}{\sin(\theta)}$$

Subsisting r into Equation A1, we can write

$$B = \frac{\mu I}{4\pi} \int \frac{\sin^3(\theta)}{d^2} dl$$

The length of the wire ℓ is

$$\ell = d\cot(\theta)$$

and

$$d\ell = -d\csc^2(\theta)d\theta$$

and

$$B = \frac{\mu I}{4\pi} \int \frac{\sin^3(\theta)}{d^2} (-d\csc^2(\theta))d\theta$$

$$= -\frac{\mu I}{4\pi} \int \frac{\sin(\theta)}{d} d\theta$$

Integrating for $\theta_1 = \pi$ and $\theta_2 = 0$, we can write

$$B = -\frac{\mu I}{4\pi d}\int_{\pi}^{0} \sin(\theta) d\theta$$
$$= \frac{\mu I}{4\pi d}[\cos(0) - \cos(\pi)]$$
$$= \frac{\mu I}{4\pi d}[2]$$
$$= \frac{\mu I}{2\pi d}$$

This is Ampere's law by Equation 5.6.

Appendix 2

Specifications of Coils Used in this Book

Description	Value
Coil name	PC1
Coil inductance	32.5 uH
Coil diameter	90.5 mm
Coil length	8 mm
Number of windings	13
Pitch	0.61
Wire	0.5 mm enamelled copper wire (core D=0.5 mm, OD=0.56 mm)
Former material	PVC
Former diameter	90 mm
Former wall thickness	2 mm
Q_{eff} (at 1MHz)	-

PC1
90 mm copper wire primary coil.

Description	Value
Coil name	PC2
Coil inductance	27.26 uH
Coil diameter	90.5 mm
Coil length	8 mm
Number of windings	12
Pitch	0.67
Wire	0.5 mm enamelled copper wire (core D=0.5 mm, OD=0.56 mm)
Former material	PVC
Former diameter	90 mm
Former wall thickness	2 mm
Q_{eff} (at 1MHz)	-

PC2
90 mm primary coil.

Description	Value
Coil name	PC3
Coil inductance	33.3 uH
Coil diameter	103.5 mm
Coil length	8 mm
Number of windings	12
Pitch	0.67
Wire	0.5 mm enamelled copper wire (core D=0.5 mm, OD=0.56 mm)
Former material	PVC
Former diameter	103 mm
Former wall thickness	2 mm
Q_{eff} (at 1MHz)	-

PC3
103 mm primary coil.

Description	Value
Coil name	SC1
Coil inductance	332 uH
Coil diameter	103.56 mm
Coil length	28 mm
Number of windings	48
Pitch	0.58
Wire	0.5 mm enamelled copper wire (core D=0.5 mm, OD=0.56 mm)
Former material	PVC
Former diameter	103 mm
Former wall thickness	2 mm
Q_{eff} measured (at 1MHz)	189

SC1
103 mm tightly wound copper wire secondary coil.

Description	Value
Coil name	SC2
Coil inductance	228 uH
Coil diameter	90.56 mm
Coil length	67 mm
Number of windings	56
Pitch	1.2 mm
Wire	0.5 mm enamelled copper wire (core D=0.5 mm, OD=0.56 mm)
Former material	PVC
Former diameter	90 mm
Former wall thickness	2 mm
Q_{eff} measured (at 1MHz)	301

SC2
90 mm widely spaced copper wire secondary coil.

Description	Value
Coil name	SC3
Coil inductance	174 uH
Coil diameter	75.56 mm
Coil length	65 mm
Number of windings	56
Pitch	1.16 mm
Wire	0.5 mm enamelled copper wire (core D=0.5 mm, OD=0.56 mm)
Former material	PVC
Former diameter	75 mm
Former wall thickness	2 mm
Q_{eff} measured (at 1MHz)	-

SC3
75 mm widely spaced copper wire secondary coil.

Description	Value
Coil name	SC4
Coil inductance	441.5 uH
Coil diameter	50.48 mm
Coil length	58 mm
Number of windings	120
Pitch	0.48 mm
Wire	0.4 mm enamelled copper wire (core D=0.4 mm, OD=0.48 mm)
Former material	Cardboard
Former diameter	50 mm
Former wall thickness	4 mm
Q_{eff} measured (at 1MHz)	-

SC4
50 mm tightly wound copper wire secondary coil.

Description	Value
Coil name	SC5
Coil inductance	229 uH
Coil diameter	75.9 mm
Coil length	55 mm
Number of windings	60
Pitch	0.92
Wire	175/46 Litz wire (core D=0.53 mm, OD=0.92 mm)
Former material	PVC
Former diameter	75 mm
Former wall thickness	2 mm
Q_{eff} measured (at 1MHz)	547

SC5
Litz wire 75 mm secondary coil specification.

Description	Value
Coil name	SC6
Coil inductance	302.5 uH
Coil diameter	75.5 mm
Coil length	38 mm
Number of windings	62
Pitch	0.61 mm
Wire	0.5 mm enamelled copper wire (core D=0.5 mm, OD=0.56 mm)
Former material	PVC
Former diameter	75 mm
Former wall thickness	2 mm
Q_{eff} measured (at 1MHz)	215

SC6
75 mm copper wire secondary coil.

Description	Value
Coil name	SC7
Coil inductance	315.5 uH
Coil diameter	75.5 mm
Coil length	34.72 mm
Number of windings	62
Pitch	0.56 mm
Wire	0.5 mm enamelled copper wire (core D=0.5 mm, OD=0.56 mm)
Former material	PVC
Former diameter	75 mm
Former wall thickness	2 mm
Q_{eff} measured (at 1MHz)	167

SC7
75 mm copper wire secondary coil.

Appendix 3

Calculating the Forward Bias Diode Voltage Drop

When the diode is forward biased, the tap voltage is dropped across the diode impedance Z_{DF} and the load impedance Z_L as shown in Figure 16.7, reproduced below.

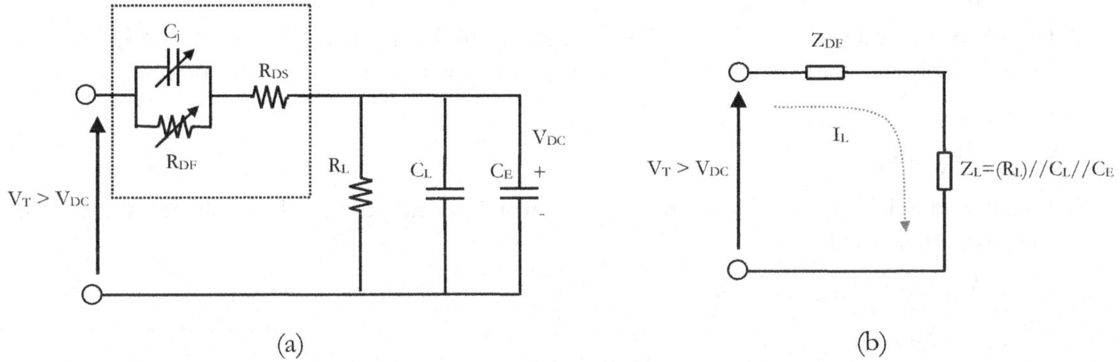

Diode forward biased – earphone charging.

The load impedance is given by

$$\mathbf{Z_L} = R_L \parallel C_L \parallel C_E$$

As the diode impedance is voltage dependent we need to determine the series current flowing through the circuit that will satisfy

$$\mathbf{V_{F(peak)}} = \mathbf{V_D} + \mathbf{V_L}$$
$$= \mathbf{I_L Z_{DF}} + \mathbf{I_L Z_L}$$

Applying an iterative approach, we set the voltage across the diode to be half the peak applied voltage and calculate the diode impedance for this given voltage. The impedance across a diode for a given voltage and frequency is the combination of the dynamic resistance in parallel with the capacitive reactance plus some small series resistance. The steps to calculate the diode voltage, current, and impedance are as follows.

1.0 Set the diode voltage to half the peak applied voltage.

$$\mathbf{V_D} = \frac{\mathbf{V_{A(peak)}}}{2}$$

2.0 Set the working variables V_{bottom} to zero and V_{top} to the magnitude of the peak applied voltage.

$$V_{bottom} = 0.0$$

$$V_{top} = V_{A(peak)}$$

3.0 Treating V_{DF} as a DC voltage, for the given diode voltage V_{DF}, calculate the diode's DC current I_{DF} and the total forward bias diode capacitance. Calculate the diode's average dynamic resistance R_D (V_{DF}/ I_{DF}) and the capacitive reactance. Now calculate the diode impedance for the given voltage as

$$\mathbf{Z_D} = (R_D \,//\, X_{C_j}) + R_{DS}$$

4.0 Calculate the current through the load. As the diode is in series with the load, we can write

$$\mathbf{I_L} = \frac{\mathbf{V_D}}{\mathbf{Z_D}}$$

5.0 Calculate the voltage drop across the load. Note the load voltage V_L is the voltage placed on the earphone. And set V_{DROP} as the voltage dropped across the diode and load.

$$\mathbf{V_L} = \mathbf{I_L Z_L}$$

$$\mathbf{V_{Drop}} = \mathbf{V_D} + \mathbf{V_L}$$

6.0 Calculate the difference between the voltage drop and the peak applied voltage as a ratio of the peak applied voltage.

$$V_{diff} = \frac{\mathbf{V}_{A(peak)} - V_{Drop}}{\mathbf{V}_{A(peak)}}$$

7.0 If the difference is some small value say 0.001, we have determined the current and voltages, so stop. Else if the difference is greater than zero, then set the bottom voltage to the current diode voltage

$$V_{bottom} = V_D$$

8.0 Else set the top voltage to the current diode voltage

$$V_{top} = V_D$$

9.0 Now set the diode voltage drop to

$$V_D = V_{bottom} + (V_{top} - V_{bottom})/2$$

Go back to step 3 and redo the calculations.

Index

Made in the USA
Monee, IL
08 November 2026

68113586R00273